CELL BIOLOGY: NEW INSIGHTS

Cell Biology: New Insights

Editor:

Dr. Prerna Pandey, Ph.D.

Delve Publishing

www.delvepublishing.com

Cell Biology: New Insights

Editor: **Dr. Prerna Pandey, Ph.D.**

708 3rd Avenue, 6th Floor
New York, NY 10017
United States of America
www.delvepublishing.com

ISBN: 978-1-68095-754-9
Library of Congress Control Number: 2016951510

About the Editor

Dr. Prerna Pandey, Ph.D.

Dr Prerna Pandey, a biotechnologist with several years of wet lab research experience, worked as a senior research fellow at International Center for Genetic Engineering and Biotechnology, New Delhi. Her field of research in PhD included isolation and molecular characterization of Geminiviruses, genome sequencing and gene annotation, and gene silencing using the RNA interference technology. She has published papers in peer-reviewed journals in the field, and has a submitted a number of annotated Geminiviral genome sequences in the GenBank, including two novel ones. Prerna has also completed her editing and proof-reading courses from Society for the promotion of Editors and Proof-readers, UK and now works as a freelance scientific writer and editor.

Contents

List of Contributors

Philipp E. Geyer
Department of Proteomics and Signal Transduction, Max Planck Institute of Biochemistry, 82152 Martinsried, Germany
NNF Center for Protein Research, Faculty of Health Sciences, University of Copenhagen, 2200 Copenhagen, Denmark

Nils A. Kulak
Department of Proteomics and Signal Transduction, Max Planck Institute of Biochemistry, 82152 Martinsried, Germany

Garwin Pichler
Department of Proteomics and Signal Transduction, Max Planck Institute of Biochemistry, 82152 Martinsried, Germany

Lesca M. Holdt
Institute of Laboratory Medicine, Ludwig-Maximilians University Munich, 80539 Munich, Germany

Daniel Teupser
Institute of Laboratory Medicine, Ludwig-Maximilians University Munich, 80539 Munich, Germany

Matthias Mann
Department of Proteomics and Signal Transduction, Max Planck Institute of Biochemistry, 82152 Martinsried, Germany
NNF Center for Protein Research, Faculty of Health Sciences, University of Copenhagen, 2200 Copenhagen, Denmark

Helen Watson
University of Exeter Medical School, St. Luke's Campus, Magdalen Road, Exeter EX1 2LU, U.K.

Shaowei Yang
Wound Healing and Cell Biology Laboratory, Institute of Basic Medical Science, Trauma Center of Postgraduate Medical School, Chinese PLA General Hospital, 28 Fu Xing Road, Beijing 100853, P. R. China

Kui Ma
Key Research Laboratory of Tissue Repair and Regeneration of PLA, and Beijing Key Research Laboratory of Skin Injury, Repair and Regeneration, First Affiliated Hospital to the Chinese PLA General Hospital, 51 Fu Cheng Road, Beijing 100048, P.R. China

Zhijun Geng
Key Research Laboratory of Tissue Repair and Regeneration of PLA, and Beijing Key Research Laboratory of Skin Injury, Repair and Regeneration, First Affiliated Hospital to the Chinese PLA General Hospital, 51 Fu Cheng Road, Beijing 100048, P.R. China

Xiaoyan Sun
Wound Healing and Cell Biology Laboratory, Institute of Basic Medical Science, Trauma Center of Postgraduate Medical School, Chinese PLA General Hospital, 28 Fu Xing Road, Beijing 100853, P. R. China

Xiaobing Fu
Wound Healing and Cell Biology Laboratory, Institute of Basic Medical Science, Trauma Center of Postgraduate Medical School, Chinese PLA General Hospital, 28 Fu Xing Road, Beijing 100853, P. R. China
Key Research Laboratory of Tissue Repair and Regeneration of PLA, and Beijing Key Research Laboratory of Skin Injury, Repair and Regeneration, First Affiliated Hospital to the Chinese PLA General Hospital, 51 Fu Cheng Road, Beijing 100048, P.R. China

Dapeng Bi
Department of Physics and Soft Matter Program, Syracuse University, Syracuse, New York 13244, USA
Center for Studies in Physics and Biology, Rockefeller University, New York 10065, USA

Xingbo Yang
Department of Physics and Soft Matter Program, Syracuse University, Syracuse, New York 13244, USA
Department of Physics, Northwestern University, Evanston, Illinois 60208, USA

M. Cristina Marchetti
Department of Physics and Soft Matter Program, Syracuse University, Syracuse, New York 13244, USA
Syracuse Biomaterials Institute, Syracuse University, Syracuse, New York 13244, USA

M. Lisa Manning
Department of Physics and Soft Matter Program, Syracuse University, Syracuse, New York 13244, USA
Syracuse Biomaterials Institute, Syracuse University, Syracuse, New York 13244, USA

Mohan S. Maddur
Institut National de la Santé et de la Recherche Médicale, Unité 872, 15 rue de l'Ecole de Médicine, Paris, F-75006, France
Centre de Recherche des Cordeliers, Equipe 16- Immunopathology and Therapeutic Immunointervention, Université Pierre et Marie Curie – Paris 6, UMR S 872, 15 rue de l'Ecole de médicine, Paris, F-75006, France
Université Paris Descartes, UMR S 872, 15 rue de l'Ecole de médicine, Paris, F-75006, France

Janakiraman Vani
Institut National de la Santé et de la Recherche Médicale, Unité 872, 15 rue de l'Ecole de Médicine, Paris, F-75006, France
Centre de Recherche des Cordeliers, Equipe 16- Immunopathology and Therapeutic Immunointervention, Université Pierre et Marie Curie – Paris 6, UMR S 872, 15 rue de l'Ecole de médicine, Paris, F-75006, France
Université Paris Descartes, UMR S 872, 15 rue de l'Ecole de médicine, Paris, F-75006, France

Jordan D. Dimitrov
Institut National de la Santé et de la Recherche Médicale, Unité 872, 15 rue de l'Ecole de Médicine, Paris, F-75006, France
Centre de Recherche des Cordeliers, Equipe 16- Immunopathology and Therapeutic Immunointervention, Université Pierre et Marie Curie – Paris 6, UMR S 872, 15 rue de l'Ecole de médicine, Paris, F-75006, France
Université Paris Descartes, UMR S 872, 15 rue de l'Ecole de médicine, Paris, F-75006, France

Kithiganahalli N. Balaji
Department of Microbiology and Cell Biology, Indian Institute of Science, Bangalore 560012, India

Sébastien Lacroix-Desmazes
Institut National de la Santé et de la Recherche Médicale, Unité 872, 15 rue de l'Ecole de Médicine, Paris, F-75006, France
Centre de Recherche des Cordeliers, Equipe 16- Immunopathology and Therapeutic Immunointervention, Université Pierre et Marie Curie – Paris 6, UMR S 872, 15 rue de l'Ecole de médicine, Paris, F-75006, France
Université Paris Descartes, UMR S 872, 15 rue de l'Ecole de médicine, Paris, F-75006, France

Srini V. Kaveri
Institut National de la Santé et de la Recherche Médicale, Unité 872, 15 rue de l'Ecole de Médicine, Paris, F-75006, France
Centre de Recherche des Cordeliers, Equipe 16- Immunopathology and Therapeutic Immunointervention, Université Pierre et Marie Curie – Paris 6, UMR S 872, 15 rue de l'Ecole de médicine, Paris, F-75006, France
Université Paris Descartes, UMR S 872, 15 rue de l'Ecole de médicine, Paris, F-75006, France

Jagadeesh Bayry
Institut National de la Santé et de la Recherche Médicale, Unité 872, 15 rue de l'Ecole de Médicine, Paris, F-75006, France
Centre de Recherche des Cordeliers, Equipe 16- Immunopathology and Therapeutic Immunointervention, Université Pierre et Marie Curie – Paris 6, UMR S 872, 15 rue de l'Ecole de médicine, Paris, F-75006, France
Université Paris Descartes, UMR S 872, 15 rue de l'Ecole de médicine, Paris, F-75006, France

Pei Xiong Liew
Snyder Institute for Chronic Diseases, University of Calgary, Alberta, Canada

Pierre Marquet
Centre Hospitalier Universitaire Vaudois (CHUV), Centre de Neurosciences Psychiatriques, Département de Psychiatrie, Site de Cery, Prilly/ Lausanne CH-1008, Switzerland
Brain Mind Institute, École Polytechnique Fédérale de Lausanne, Lausanne CH-1015, Switzerland

Christian Depeursinge
Brain Mind Institute, École Polytechnique Fédérale de Lausanne, Lausanne CH-1015, Switzerland
King Abdullah University of Science and Technology (KAUST), Division of Biological and Environmental Sciences and Engineering, Thuwal 23955-6900, Kingdom of Saudi Arabia

Pierre J. Magistretti
Centre Hospitalier Universitaire Vaudois (CHUV), Centre de Neurosciences Psychiatriques, Département de Psychiatrie, Site de Cery, Prilly/ Lausanne CH-1008, Switzerland
Brain Mind Institute, École Polytechnique Fédérale de Lausanne, Lausanne CH-1015, Switzerland
King Abdullah University of Science and Technology (KAUST), Division of Biological and Environmental Sciences and Engineering, Thuwal 23955-6900, Kingdom of Saudi Arabia

Mathias Uhlén
Science for Life Laboratory, KTH - Royal Institute of Technology, Stockholm, Sweden
Department of Proteomics, KTH - Royal Institute of Technology, Stockholm, Sweden
Novo Nordisk Foundation Center for Biosustainability, Technical University of Denmark, Hørsholm, Denmark

Björn M Hallström
Science for Life Laboratory, KTH - Royal Institute of Technology, Stockholm, Sweden
Department of Proteomics, KTH - Royal Institute of Technology, Stockholm, Sweden

Cecilia Lindskog
Department of Immunology, Genetics and Pathology, Science for Life Laboratory, Uppsala University, Uppsala, Sweden

Adil Mardinoglu
Science for Life Laboratory, KTH - Royal Institute of Technology, Stockholm, Sweden
Department of Biology and Biological Engineering, Chalmers University of Technology, Gothenburg, Sweden

Fredrik Pontén
Department of Immunology, Genetics and Pathology, Science for Life Laboratory, Uppsala University, Uppsala, Sweden

Jens Nielsen
Science for Life Laboratory, KTH - Royal Institute of Technology, Stockholm, Sweden
Novo Nordisk Foundation Center for Biosustainability, Technical University of Denmark, Hørsholm, Denmark
Department of Biology and Biological Engineering, Chalmers University of Technology, Gothenburg, Sweden

Simon A Joosse
Department of Tumor Biology, Center of Experimental Medicine, University Medical Center Hamburg-Eppendorf, Hamburg, Germany

Tobias M Gorges
Department of Tumor Biology, Center of Experimental Medicine, University Medical Center Hamburg-Eppendorf, Hamburg, Germany

Klaus Pantel
Department of Tumor Biology, Center of Experimental Medicine, University Medical Center Hamburg-Eppendorf, Hamburg, Germany

Simon P. Jones
St. Vincent's Centre for Applied Medical Research, The University of New South Wales, Sydney, Australia.

Gilles J. Guillemin
Australian School of Advanced Medicine, Macquarie University, Sydney, Australia.

Bruce J. Brew
St. Vincent's Centre for Applied Medical Research, The University of New South Wales, Sydney, Australia.
Department of Neurology, St. Vincent's Hospital, Sydney, Australia.

James I. McDonald
Center for Pharmacogenomics, Department of Medicine, Washington University in St. Louis School of Medicine, St. Louis, MO, USA

Hamza Celik
Section of Stem Cell Biology, Division of Oncology, Department of Medicine, Washington University in St. Louis School of Medicine, St. Louis, MO 63110, USA

Lisa E. Rois
Center for Pharmacogenomics, Department of Medicine, Washington University in St. Louis School of Medicine, St. Louis, MO, USA

Gregory Fishberger
College of Arts and Science, Washington University in St. Louis, St. Louis, MO 63130, USA

Tolison Fowler
Center for Pharmacogenomics, Department of Medicine, Washington University in St. Louis School of Medicine, St. Louis, MO, USA

Ryan Rees
College of Arts and Science, Washington University in St. Louis, St. Louis, MO 63130, USA

Ashley Kramer
Section of Stem Cell Biology, Division of Oncology, Department of Medicine, Washington University in St. Louis School of Medicine, St. Louis, MO 63110, USA

Andrew Martens
Section of Stem Cell Biology, Division of Oncology, Department of Medicine, Washington University in St. Louis School of Medicine, St. Louis, MO 63110, USA

John R. Edwards
Center for Pharmacogenomics, Department of Medicine, Washington University in St. Louis School of Medicine, St. Louis, MO, USA

Grant A. Challen
Section of Stem Cell Biology, Division of Oncology, Department of Medicine, Washington University in St. Louis School of Medicine, St. Louis, MO 63110, USA
Developmental, Regenerative and Stem Cell Biology Program, Division of Biology and Biomedical Sciences, Washington University in St. Louis School of Medicine, St. Louis, MO 63110, USA

Anagha Joshi
The Roslin Institute and Royal (Dick) School of Veterinary Studies, The University of Edinburgh, Scotland, United Kingdom;

Christopher Pooley
The Roslin Institute and Royal (Dick) School of Veterinary Studies, The University of Edinburgh, Scotland, United Kingdom;

Tom C. Freeman
The Roslin Institute and Royal (Dick) School of Veterinary Studies, The University of Edinburgh, Scotland, United Kingdom;

Andreas Lennartsson
Department of Biosciences and Nutrition, Karolinska Institute, Huddinge, Sweden;

Magda Babina
Department of Dermatology and Allergy, Charit´e Universit¨atsmedizin Berlin, Germany;

Christian Schmidl
Department of Internal Medicine III, University Hospital, University of Regensburg, Germany;

Teunis Geijtenbeek
Department of Experimental Immunology, Academic Medical Center, Amsterdam, the Netherlands; and

FANTOM Consortium
RIKEN Preventive Medicine and Diagnosis Innovation Program, Tsurumi-ku, Yokohama, Japan;

Tom Michoe
The Roslin Institute and Royal (Dick) School of Veterinary Studies, The University of Edinburgh, Scotland, United Kingdom;

Jessica Severin
RIKEN Center for Life Science Technologies, Division of Genomic Technologies, Tsurumi-ku, Yokohama, Japan;
RIKEN Omics Science Center, Tsurumi-ku, Yokohama, Japan

Masayoshi Itoh
RIKEN Preventive Medicine and Diagnosis Innovation Program, Tsurumi-ku, Yokohama, Japan;

RIKEN Center for Life Science Technologies, Division of Genomic Technologies, Tsurumi-ku, Yokohama, Japan;
RIKEN Omics Science Center, Tsurumi-ku, Yokohama, Japan

Timo Lassmann
RIKEN Center for Life Science Technologies, Division of Genomic Technologies, Tsurumi-ku, Yokohama, Japan;
RIKEN Omics Science Center, Tsurumi-ku, Yokohama, Japan

Hideya Kawaji
RIKEN Preventive Medicine and Diagnosis Innovation Program, Tsurumi-ku, Yokohama, Japan;
RIKEN Center for Life Science Technologies, Division of Genomic Technologies, Tsurumi-ku, Yokohama, Japan;
RIKEN Omics Science Center, Tsurumi-ku, Yokohama, Japan

Yoshihide Hayashizaki
RIKEN Preventive Medicine and Diagnosis Innovation Program, Tsurumi-ku, Yokohama, Japan;
RIKEN Omics Science Center, Tsurumi-ku, Yokohama, Japan

Piero Carninci
RIKEN Center for Life Science Technologies, Division of Genomic Technologies, Tsurumi-ku, Yokohama, Japan;
RIKEN Omics Science Center, Tsurumi-ku, Yokohama, Japan

Alistair R. R. Forrest
RIKEN Center for Life Science Technologies, Division of Genomic Technologies, Tsurumi-ku, Yokohama, Japan;

Michael Rehli
Department of Internal Medicine III, University Hospital, University of Regensburg, Germany;

David A. Hume
The Roslin Institute and Royal (Dick) School of Veterinary Studies, The University of Edinburgh, Scotland, United Kingdom;

Ekaterina Kldiashvil
Georgian Telemedicine Unio (Association), Tbilisi, Georgia

Sophio Bojgua
New Vision University, Tbilisi, Georgia

Elena F. Brachtel
Department of Pathology, Massachusetts General Hospital, Boston, Massachusetts, USA

Theresa N. Operaña
Biotheranostics, Inc., San Diego, California, USA

Peggy S. Sullivan
Department of Pathology and Laboratory Medicine, David Geffen School of Medicine, University of California Los Angeles, Los Angeles, California, USA

Sarah E. Kerr
Department of Laboratory Medicine and Pathology, Mayo Clinic, Rochester, Minnesota, USA

Karen A. Cherkis
Biotheranostics, Inc., San Diego, California, USA

Brock E. Schroeder
Biotheranostics, Inc., San Diego, California, USA

Sarah M. Dry
Department of Pathology and Laboratory Medicine, David Geffen School of Medicine, University of California Los Angeles, Los Angeles, California, USA

Catherine A. Schnabel
Biotheranostics, Inc., San Diego, California, USA

Mauro Corrado
Dulbecco-Telethon Institute, Venetian Institute of Molecular Medicine, Padova, Italy
IRCCS Fondazione Santa Lucia, Roma, Italy

Luca Scorrano
Dulbecco-Telethon Institute, Venetian Institute of Molecular Medicine, Padova, Italy
Department of Biology, University of Padova, Padova, Italy

Silvia Campello
IRCCS Fondazione Santa Lucia, Roma, Italy
Department of Biology, University of Roma Tor Vergata, Roma, Italy

Ben Loos
Department of Physiological Sciences, Stellenbosch University, South Africa

Zhaleh Shariati Sarabi (MD)
Rheumatic Diseases Research Center, School of Medicine, Mashhad University of Medical Sciences, Mashhad, Iran

Jalil Tavakol Afshari (PhD)
Immunology Research Center, Avicenna Research Institute,Mashhad University of Medical Sciences, Mashhad, Iran

Ali Ghassemi (MD)
Department of Bone marrow stem cell transplantation, School of Medicine, Mashhad University of Medical Sciences, Mashhad, Iran

Mehdi Yaghobi (MD)
Department of Internal medicine, School of Medicine, Mashhad University of Medical Sciences, Mashhad, Iran

Angione
Computer Laboratory, University of Cambridge, Cambridge, United Kingdom

Giovanni Carapezza
Department of Mathematics and Computer Science, University of Catania, Catania, Ital

Jole Costanza
Department of Mathematics and Computer Science, University of Catania, Catania, Ital

Pietro Lió
Computer Laboratory, University of Cambridge, Cambridge, United Kingdom

Giuseppe Nicosia
Department of Mathematics and Computer Science, University of Catania, Catania, Ital

Fan An,
School of Pharmaceutical Science and Technology, Dalian University of Technology, Dalian, China.
State Key Laboratory of Fine Chemicals, Department of Chemical Engineering, Dalian University of Technology,
Yueyang Qu

Xianming Liu
Dalian Institute of Chemical Physics, Chinese Academy of Sciences, Dalian, China.

Runtao Zhong
Dalian Institute of Chemical Physics, Chinese Academy of Sciences, Dalian, China.

Yong **Luo**
School of Pharmaceutical Science and Technology, Dalian University of Technology, Dalian, China.
State Key Laboratory of Fine Chemicals, Department of Chemical Engineering, Dalian University of Technology,

Claudia Altamirano
Pontificia Universidad Católica de Valparaíso, Escuela de Ingeniería Bioquímica, Valparaíso, Chile
Creas Conicyt Regional GORE Valparaíso, Chile

Julio Berrios
Pontificia Universidad Católica de Valparaíso, Escuela de Ingeniería Bioquímica, Valparaíso, Chile

Mauricio Vergara
Pontificia Universidad Católica de Valparaíso, Escuela de Ingeniería Bioquímica, Valparaíso, Chile

Silvana Becerra
Creas Conicyt Regional GORE Valparaíso, Chile

Preface

Cell biology, an important area of biology is the study of the cell and its components. The chapters in this book are related to cell division, biomarkers, membrane structures and functions, a few molecular analytical techniques, and cytological techniques that are useful in modern biomedical research. The chapters are original research articles and reviews dealing with advance concepts, and principles of cell biology, cell division guiding wound repair, fundamental concepts of membrane transport, dendritic cells, transcriptomics, clinical implications of metastatic cells, cellular biomarkers of psychiatric disorders, and Keynurine pathway in stem cells. A few chapters focus on analytical techniques like gene editing or CRISPR, application of Chromogenic In Situ Hybridization (CISH) for HPV Genotyping, molecular classification of cancer with the 92-gene assay panel, a model that simultaneously captures cell motility and multibody cell-cell interactions in a confluent tissue, and phase-digital holographic microscopy as a promising imaging technique to identify cellular biomarkers of psychiatric disorders. In one of the reviews, the use of alternative substrates and genetic modifications (host cell engineering) are analyzed as tools to overcome factors limiting productivity. The main objective is to provide easier access and sufficient information to the readers to enrich their existing knowledge base.

Editor
Dr. Prerna Pandey, Ph.D.

CHAPTER 1

PLASMA PROTEOME PROFILING TO ASSESS HUMAN HEALTH AND DISEASE

Philipp E. Geyer[1,2] Nils A. Kulak[1], Garwin Pichler[1], Lesca M. Holdt[3], Daniel Teupser[3] and Matthias Mann[1,2]

[1]Department of Proteomics and Signal Transduction, Max Planck Institute of Biochemistry, 82152 Martinsried, Germany

[2]NNF Center for Protein Research, Faculty of Health Sciences, University of Copenhagen, 2200 Copenhagen, Denmark

[3]Institute of Laboratory Medicine, Ludwig-Maximilians University Munich, 80539 Munich, Germany

INTRODUCTION

Blood, plasma, and serum are the predominant samples used for diagnostic analyses in clinical practice and are available in biobanks from thousands of clinical studies (Végvári et al., 2011). The quantitative analysis of individual plasma proteins by immunoassays is used in daily clinical diagnostics. However, immunoassays have inherent limitations with regard to multiplexing, their specificity for protein isoforms, and their incompatibility with hypothesis-free investigations. Mass spectrometry (MS)-based proteomics is a technology that could address all of these limitations and that should be capable of discovering biomarkers in this easily accessible body fluid (Anderson, 2014). However, MS-based plasma proteomics is extremely challenging for a number of reasons, most prominently the extremely large dynamic range of protein abundances (Anderson and Anderson, 2002, Omenn, 2005). There is also a lack of very reproducible, robust, and high-throughput proteomic workflows to identify and verify potential biomarker in large cohorts. As a result, only few novel biomarkers have been established—fewer than 1.5 per year in the 15 years before 2010 (Anderson, 2010)—and this has generally been done by immunoassay-based technologies, such as prostate-specific antigen, one of the best known biomarkers in medicine (Vihko et al., 1978).

Dramatic improvements in the technology of MS-based proteomics over the last few years (Cox and Mann, 2011, Geiger et al., 2010, Muñoz and Heck, 2014) have rekindled an interest in plasma proteomics. Using such technology and combining it with immunodepletion of high- and medium-abundance proteins as well as very extensive peptide fractionation methods, it has now become possible to identify more than 1,000 (Addona et al., 2011, Cao et al., 2012,Paczesny et al., 2010) or even more than 5,000 proteins (Keshishian et al., 2015) in plasma. However, immunodepletion may lead to biases because of cross-reactions of the antibodies used or by proteins bound to carrier proteins such as albumin (Bellei et al., 2011, Tu et al., 2010). Furthermore, extensive pre-fractionation decreases throughput, which is undesirable in clinical practice. Accordingly, the paradigm in biomarker discovery by MS has been to analyze a small number of samples in as much depth as possible, whereas the verification phase was to be done on larger cohorts but with targeted methods and a small number of candidate markers. The final clinical test for a biomarker identified by MS was to be performed with classical

Citation: Philipp E. Geyer, Nils A. Kulak, Garwin Pichler, Lesca M. Holdt, Daniel Teupser, Matthias Mann, Plasma Proteome Profiling to Assess Human Health and Disease, DOI: http://dx.doi.org/10.1016/j.cels.2016.02.015

immunoassays (Anderson et al., 2009,Surinova et al., 2011). Although this scheme is practical with current technology, it is very laborious and loses much of the promise of systemwide and unbiased investigation of the plasma proteome. Using another approach, Liu et al. (2015))constructed a list of plasma peptide transitions, which they used to interpret the signals in sequential window acquisition of all theoretical MS (SWATH) runs of plasma samples of twins. In this way, the contribution of heritable and environmental changes to the plasma proteome could be distinguished.

In contrast to previous approaches, we here focused on developing a robust and highly streamlined shotgun plasma proteomics workflow. For the MS readout, we used very short liquid chromatography (LC)-MS/MS gradients and recent advances in label-free quantification (Cox et al., 2014). We hypothesized that the resulting "plasma proteome profile" would have a high yield of information about the health state of an individual and that it can be obtained for a large number of clinical samples.

RESULTS

Rapid, Robust, and Highly Reproducible Plasma Proteomic Workflow

Past efforts in shotgun plasma proteomics endeavored to maximize protein identifications, whereas generally less emphasis was placed on quantitative accuracy or throughput. Here we wished to develop a convenient workflow, from sample preparation to data analysis, that can potentially be used in a clinical context. We reasoned that such a workflow should be rapid, optimal for high-throughput, robust, and highly reproducible. Therefore, it should minimize all preparation and analysis steps, while still quantifying clinically interesting proteins accurately. With this in mind, we decided to omit any depletion steps of high-abundance plasma proteins.

Building on the recently described in-StageTip (iST) method (Kulak et al., 2014), we further streamlined the procedure for plasma (Experimental Procedures;Figure 1A). Starting with 1 μl of plasma from a single finger prick, all preparation steps were performed in a single reaction vial. Using ordinary amounts of digestion enzymes, we found that adequate protein digestion had already occurred after 1 hr (protein coefficients of variation [CVs] and tryptic missed cleavage rates were similar to overnight digestion). Peptides were then eluted and ready for LC-MS/MS analysis. The entire up-front procedure took less than 2 hr and can readily be performed in a 96-well format and automated in a liquid handling platform, if desired.

Starting with single-run gradient times typical of proteomics experiments, we successively reduced them to determine the maximum information content per unit time. We found that the number of identified proteins decreased very slowly with). Below this time, loading and equilibration times become dominant, and therefore we chose 20 min gradients as our standard (33 min between injections, about 50 samples/day). The combination of optimized sample preparation and LC setup allowed for hundreds of plasma proteome analyses, whereas previously clogging of columns was a common occurrence with plasma samples.

We used MaxQuant for quantitative label-free analysis of the LC-MS/MS data (Cox et al., 2014, Cox and Mann, 2008) and for transferring peptide identifications from one LC run to other LC runs in which the peptide was not sequenced ("match between runs"). In combination with a matching library consisting of undepleted plasma of ten different individuals as well as plasma depleted of the 20 highest abundant proteins, this boosted protein identification by 39% (Experimental Procedures). Of the 347 protein groups identified in total in the 20 min gradients, 285 were detected in all ten individuals. The entire workflow, including the finger-prick procedure and the data analysis, takes less than 3 hr (Figure 1A).

Accuracy of Label-free Quantification of the Plasma Proteome

To investigate the quantitative reproducibility of our workflow (intra-assay variability), we sampled blood by venous puncture from one individual and harvested plasma after centrifugation. We performed the entire workflow 15 separate times on 1 μl aliquots of this stock and correlated protein abundances across the whole measuring range of each of the replicates. The mean R^2correlation value of the quantified protein signals between individual replicates was excellent at 0.980, with a range of 0.966–0.994 (Figure 1B, excluding keratins. We performed 96 blood plasma analyses using multiplexed preparation on a liquid handling robot and short measurement times (5 hr and 51 hr in total, respectively) and achieved a mean R^2 value of 0.97.

On average, 284 ± 5 different proteins were quantified (total 313): the large majority in all 15 samples and only 3% uniquely in single LC runs (Figure 1C). We picked six well-characterized plasma proteins across a million-fold abundance range and found that quantification was highly reproducible (Figure 1D). We compared different conditions by the proportion of proteins with CVs less than 20%, because this is a commonly used cutoff for in vitro diagnostic assays (U.S. Department of Health and Human Services, Food and Drug Administration, Center for Drug Evaluation and Research, Center for Veterinary Medicine, 2001). Notably, 67% of quantified proteins were within the 20% cutoff range, and 30% had CVs below 10% (Figure 1E).

To determine the variability caused by LC-MS/MS analysis alone (analytical variability), peptides from one sample preparation were injected and measured 15 times. This resulted in only slightly better reproducibility (71% with a CV less than 20% and 37% with a CV less than 10%), indicating that up-front sample preparation contributed little to overall quantitative variability (Figure 1F). A notable exception to this trend were certain keratin proteins, which had very small analytical variability but sometimes had a large quantitative difference between repeated analysis of the same sample. This is readily explained by contamination with exogenous keratins during sample preparation. Nevertheless, it is clinically relevant, because we found that plasma proteomes of the same person clustered together much better after excluding keratins and other proteins introduced by sample processing such as hemoglobins (see below).

Intra- and Inter-individual Variability of the Plasma Proteome

The high-throughput of our workflow allowed us to extensively characterize the quantitative variation within and between individuals. To determine inter-individual variability, we performed finger pricks on one person four times a day over 8 consecutive days and analyzed all 32 blood proteomes with less than 24 hr of measuring time. This revealed stability of the plasma proteome over time (55% of proteins below 20% CV; Figure 2A). The proteins with large CVs were the aforementioned keratins, as well as high-abundance erythrocyte-specific proteins. The latter are caused by a slightly different extent of erythrocyte lysis during plasma preparation or by contamination of plasma with erythrocytes during plasma harvesting.

To determine inter-individual variability, we harvested plasma from five female and five male donors in triplicate by finger pricks. The average R^2 value within the technical workflow triplicates was 0.976, excluding keratins and erythrocyte-specific protein groups. For CVs of the technical replicates of all individuals. Of 345 proteins quantified, only a minority was under the CV cutoff. This indicates that overall, the plasma proteome has much higher inter- than intra-individual variability (19% and 55% of proteins within a CV of 20%, respectively; Figure 2B). These general trends have been observed previously (for a recent example, see Liu et al., 2015). Here they suggest that our label-free workflow is well suited to capture the natural or pathological variation of protein levels between individuals.

To directly test this notion, we asked if we could discern systematic differences between the plasma proteomes of women and men, a question that to our knowledge has not been addressed by shotgun proteomics before. Indeed, one-dimensional principal-component analysis (PCA) was already sufficient for complete separation (Figure 2C). Inspection of the drivers of the PCA separation revealed that several of them are known to be regulated by estrogen (Figure 2D) (Christensen et al., 1989, Ottosson et al., 1981, Sand et al., 1985). Direct comparison of the plasma proteome profiles of a woman and a man shows that pregnancy zone protein (PZP) and sex hormone-binding globulin (SHBG) are of high absolute abundance in the plasma proteome of women and can be as high as 1% of human serum albumin (Figure 2E). This suggests a functional role in plasma, and indeed, SHBG binds estrogen, whereas PZP traps proteases (Figure 2F). On average, PZP levels were 33-fold higher in women compared with men. Furthermore, two women had 10- to 100-fold higher levels than the other three, likely because of highly elevated levels of estrogen (Figure 2G).

Rapid Assessment of Sample Quality by Plasma Proteomes

A frequently discussed issue in plasma proteomics as well as in clinical laboratory medicine is the potentially deleterious effects of inconsistent sample handling, such as variable time between blood taking and workup. We reasoned that our rapid and highly reproducible workflow might also allow the determination of protein markers of sample quality. In clinical practice, a certain degree of hemolysis is not uncommon. Starting from our observation that high-abundance erythrocyte-specific proteins often showed high variability (Figures 2A, 2B, and 3A), we deliberately spiked in increasing amounts of erythrocyte lysates to pure plasma. We obtained a proportionate increase of erythrocyte-specific proteins, specifically, hemoglobin subunits alpha, beta, delta, and carbonic anhydrase 1. Notably, these proteins increased linearly (R^2 = 0.99), and even an admixture of 1 in 10,000 could easily be spotted (Figure 3B). This demonstrates that plasma proteome profiling readily indicates even small amounts of cellular contamination, in which case the values of pertinent proteins could be disregarded or corrected. The importance of this analysis step is illustrated by triplicate plasma proteome analysis, in which the samples from individual donors clustered together much more tightly in PCA when keratins and prominent red blood cell proteins were removed.

The blood coagulation system is primed for clotting in case of injury and wound repair. Although serum is harvested by inducing coagulation, harvesting of plasma requires addition of appropriate amounts of anticoagulants. Our plasma proteome profile contained many proteins with a function in the coagulation cascade, and we next evaluated the coordinate behavior of these proteins as a quality control for appropriate plasma preparation. Plasma from each of the fingers of one individual was processed (Experimental Procedures). The levels of fibrinogen alpha (FGA), fibrinogen beta (FBA), and fibrinogen gamma (FGG) were lower in two of the samples. In addition, platelet basic protein (PPBP) and platelet factor 4 variant (PF4V1), which are released from activated blood platelets, were increased only in these same samples (Figure 3C), suggesting that partial coagulation had occurred. To test this hypothesis, we collected plasma and serum from two individuals and carried out sample preparation in triplicates. Indeed, levels of FGA, FGB, and FGG were much lower and levels of PPBP and PF4V1 much higher in serum compared with plasma (Figure 3D).

These observations prompted us to investigate coagulation and erythrocyte status in optimally prepared plasma. For this purpose, we obtained reference samples from a blood bank, which had gone through an extremely rigorous sample collection procedure (Experimental Procedures). They had very low and constant levels of red blood cell-specific proteins, and none had evidence of partial clotting. Although our plasma samples were also virtually coagulation free, this is in our experience not always the case with samples obtained from clinical studies.

Quantification of Clinically Interesting Markers in Short Gradients

Apolipoproteins are functional blood proteins involved in lipid homeostasis. They therefore reflect an individual's metabolic status, and some of them are classical markers of cardiovascular risk and metabolic disorders such as diabetes (Jenkins et al., 2014, Jensen et al., 2014). We quantified 15 apolipoproteins at each of 32 different time points in one individual. Apolipoprotein-a (LPA) had the strongest variation (CV = 20%), whereas APOB had the lowest (CV = 6%). The distribution of LPA levels in the population is skewed toward zero, with most individuals having low LPA levels but some (~20%) having higher levels. The successful quantification of the apolipoproteins in 32 plasma proteomes demonstrates the feasibility of a longitudinal measurement of risk factors known to be associated with an individual's propensity for certain diseases (Figure 3E).

Some of the apolipoproteins have allelic variants occurring with high frequency in populations that can easily be detected by MS (Krastins et al., 2013, Martínez-Morillo et al., 2014). The apolipoprotein allele APOE4 in the homozygous form is the largest known risk factor for late-onset Alzheimer's disease with a 10-fold higher risk compared with the homozygous APOE3 form (Tanzi, 2012). APOE4 has an arginine at position 112 instead of a cysteine residue in APOE2 and APOE3. In the 20 min LC-MS/MS data, we were able to clearly distinguish between the peptides LGADMEDVR (APOE4) and LGADMEDVCGR (APOE2, APOE3). In our group of ten individuals, two had one APOE4 allele (Figures 3F and 3G). The second allele was either an APOE3 or the APOE2 allele.

Serum amyloid A-1 protein (SAA1) and C-reactive protein (CRP) are acute phase proteins that are routinely measured in the clinic. Both are correlated with inflammatory states, and chronic elevation is strongly associated with increased risk for future cardiovascular events (Hua et al., 2009, Wilson et al., 2008). We found that their expression levels varied up to 1,000-fold among the ten individuals, and in a correlated manner ($R^2 = 0.6$; Figures 3H and 3I). In the plasma proteome with the highest levels of SAA1 and CRP, these are by far the largest differences to the plasma proteomes of the other healthy individuals, and this is presumably caused by recovery from a common cold (Figure 3J).

Next, we asked if our rapid proteome profiles contained information on any further known biomarkers. We scanned the raw data of the 15 technical workflow replicates to calculate the CVs for Food and Drug Administration (FDA)-cleared or FDA-approved biomarkers, as listed Anderson (2010)). In total, 49 FDA-approved biomarkers were present in this data set (46 quantified in all 15 workflow replicates); 41 of them had CVs of less than 20%, and 28 had CVs even less than 10% (Figure 3K). When dividing these FDA-approved biomarkers into different classes (Anderson, 2010), 45 fell into "act in plasma," 2 into "tissue leakage," and 1 into "receptor ligand," and 1 was lysozyme, which had not been assigned to any category. The 20 min gradients already covered 45 of a total of 54 proteins among the "act in plasma" biomarkers. Interestingly, 42 of them were among the 180 highest abundance proteins, whereas the next 133 proteins contained only 7 known biomarkers.

Plasma Protein Epitope Signature Tags as Internal Standards for Protein Quantification

In clinical applications, quantification is almost always performed with internal standards. To add this capability to our fast workflow, we investigated the use of stable isotope labeling of amino acids in cell culture (SILAC)-protein epitope signature tags (PrESTs), which are recombinant expressed stable isotope-labeled protein fragments. This approach has the advantage that it controls for digestion efficiency, alkylation rate, and other workflow aspects and that a "master mix" of dozens of proteins of interest can be readily prepared and quantified (Edfors et al., 2014, Zeiler et al., 2012). We used APOA1, APOA4, APOB, APOE, and SHBG to construct a master mix for quantification of multiple plasma proteins in short gradients. Samples from ten individuals were prepared in triplicate and measured. This resulted in low CVs for these proteins (APOA1 = 2.3%, APOA4 = 3.8%, APOB = 5.3%, APOE = 3.8%, and SHBG = 14.7%). Optimized targeted methods applied to peptides resulting from the PrESTs could improve these CVs even further.

A Quantitative Proteome of 1,000 Plasma Proteins

The above experiments highlight the value of quantifying hundreds of proteins in a very short analysis time. To obtain estimates of abundances for a deeper plasma proteome, we used a combination of peptide pre-fractionation, a matching library consisting of depleted plasma, and 100 min high-performance LC (HPLC) gradients. With 16 hr of measurement time, we identified 1,040 proteins in non-depleted plasma, of which 965 had label-free protein quantification (LFQ) values. Although MS signals for these proteins span more than six orders of magnitude, the majority of them were confined to a 100-fold abundance range (Figure 4B). The deep proteome data can be assessed and in the MaxQB database (Schaab et al., 2012), which also displays the mass spectrometric evidence and MS/MS transitions for all identified peptides.

Unexpectedly, the deep plasma proteome contained only 14 additional FDA-approved biomarkers compared with the 49 already found in the 20 min gradients. Nine of them were classified as "tissue leakage," only 4 as "act in plasma," and 1 was not assigned to any category. A depth of 450 plasma proteins would be sufficient to cover 87% of the FDA-approved biomarkers present in our deep data set according to Anderson (2010)).

The fact that we did not use protein depletion allowed us to investigate the quantitative nature of the plasma proteome. Bioinformatics analysis revealed that 457 of all quantified proteins had an extracellular annotation and 651 an intracellular one, with 221 overlapping proteins. Interestingly, the 430 proteins with exclusive intracellular annotation, which also have an independent abundance estimate in the Plasma Proteome Database (Nanjappa et al., 2014), are almost completely excluded from the top three order of magnitudes of protein abundance (Figure 4C). These proteins are likely of tissue origin and have been released by normal tissue damage, without necessarily having a function in blood. In contrast to the deep proteome, these intracellular proteins were largely absent in the 20 min measurements. Reassuringly, 91% of proteins identified in plasma by us had also been identified in at least one of the studies collected in the PeptideAtlas repository (Farrah et al., 2014). In the absence of MS-derived quantitation of a deep, non-depleted proteome, we turned to the Plasma Proteome Database, which lists absolute concentrations of 597 of the proteins that are also quantified in our data set. Although these concentrations derive from the literature from a wide variety of individuals, health states, and quantification methods, we found a reasonable correlation, with an R^2value of 0.53. This analysis also confirmed that we had quantified many proteins of clinical interest in the lower abundance range, such as the plasma protein ferritin (FTL) (12 ng/ml), which is widely used to diagnose dysregulation of iron homeostasis, or the cytokine macrophage migration inhibitory factor (MIF)

(10 ng/ml). A total of 183 proteins in our data set have reported concentrations below 10 ng/ml.

To bioinformatically analyze the functional nature of the plasma proteome, we used the "1D annotation" algorithm in the MaxQuant software, to assign UniProtKB keywords to distinct abundance ranges. This resulted in 58 statistically significant features. Classical characteristics of the plasma proteome were typically located in the high-abundance range. These include "glycation," "immunoglobulin," and "chylomicrons," which are expected because functional plasma proteins are typically glycated, a large proportion of functional plasma proteins are antibodies, and apolipoproteins are the structural components of chylomicrons. In the low-abundance tail of the distribution, we found highly abundant intracellular complexes such as the proteasome as well as RNA-binding and processing proteins. "Phosphoprotein" was situated close to the middle of the distribution, and above "membrane," "cytoplasm," and "nucleus," presumably because most intracellular proteins have by now been shown to be phosphorylatable, in addition to some of the extracellular ones. The mean of functionally important Gene Ontology annotated biological processes, such as "protein-lipid complex assembly," "sterol and cholesterol transport," "acute phase response," and "regulation of coagulation" processes all scored in the upper third of the distribution, highlighting that these functions are overwhelmingly carried out by high-abundance plasma proteins. The above analysis can also be used to infer the likely function or lack thereof of a protein found at a certain concentration in normal plasma. For instance, the hormone-binding protein SHBG is in the upper range of the plasma proteome, which correlates well with its carrier function for an abundant circulating hormone (Ottosson et al., 1981) (Figure 4D).

DISCUSSION

Using state-of-the-art shotgun proteomics technology, in particular the recently described iST preparation (Kulak et al., 2014), Orbitrap instrumentation with very high sequencing speed (Kelstrup et al., 2014, Scheltema et al., 2014), and advances in label-free quantification (Cox et al., 2014), we here developed a streamlined and robust workflow for shotgun plasma proteomics. Sample preparation steps are minimized without loss of performance, and the procedure can be performed in 96-well format by a liquid handling platform. In this way, hundreds of plasma proteomes can be processed and sample preparation is not a limiting step for plasma proteomics in our workflow. We found that even extremely short measurements of 20 min still allowed the identification of more than 300 proteins, which was aided by a reference data set and the "match between runs" functionality. Accuracy and precision of the label-free workflow were excellent with intra-assay correlation of about $R^2 = 0.98$ and CVs smaller than 20% for the majority of quantified proteins.

Starting from only a finger prick of blood, the entire workflow, including database search and label-free quantification, can be performed in less than 3 hr. Previous plasma proteome studies typically started from milliliter amounts of blood, used depletion, and extensive pre-fractionation and therefore required days for completion (Cao et al., 2012, Keshishian et al., 2015, Liu et al., 2015, Such-Sanmartín et al., 2014). The ability to use small sample amounts makes blood testing much less invasive, improves cost-efficiency, and is clinically attractive in many situations, including the testing of infants as well as elderly patients (Bai et al., 2013). Likewise, fast response time is frequently important such as in the case of myocardial infarction. Our procedure uses a very short digestion time (60 min), which could be reduced by further optimization, so that the entire procedure could conceivably be performed in less than 1 hr.

Our very short LC-MS/MS runs contain nearly 50 proteins that are already subject to FDA-approved diagnostic tests, whereas the deep proteome only added few additional ones. Furthermore, the proportion of functional plasma proteins was very high, in contrast to the lower abundance range, which was dominated by tissue-derived "leakage proteins." Nevertheless, the deeper proteome still contained many proteins of known clinical significance, and it is interesting to speculate whether the relative paucity of approved biomarkers in this range is due to the greater difficulties associated with studying these proteins. Even in the short analysis runs, the lower half of the distribution has not yet been associated with specific patient states. We suggest that mixtures of recombinant isotope-labeled protein fragments, so-called SILAC-PrESTs (Edfors et al., 2014, Zeiler et al., 2012), could routinely be added to plasma samples. This would enable very high accuracy in absolute quantification for the discovery and validation of such biomarkers at high-throughput.

Further throughput improvements can be achieved with chemical labeling strategies, for instance with isobaric chemical tags such as TMT (Thompson et al., 2003). For 10-plex encoding, this could increase throughput to hundreds of patients per day. This compares favorably with metabolomics studies, which are already performed at large scale in plasma cohorts (Suhre et al., 2011), while providing equally useful and complementary information. Alternatively, TMT could be used to label patient samples before peptide pre-fractionation. This should result in deep proteome coverage, while keeping effective MS measurement time reasonably short at 1–2 hr per patient and compatible with large-scale studies.

The proteins characterized in our short workflow already contain a plethora of useful information. For example, it was easy to distinguish the gender of the donor and to obtain some risk-associated genotype information. The spectrum of apolipoproteins, as well as inflammatory markers, was excellently quantified, reflecting the cardiovascular and metabolic health state. Unexpectedly, the global nature of shotgun proteomics supplies us with valuable information about sample quality, which is not tested in routine clinical practice but can influence test results and medical decisions.

As mentioned above, the current strategy in plasma biomarker discovery by MS-based proteomics involves a narrowing down and widening strategy: a small number of patients and controls are analyzed in great depth with unbiased and relatively low-throughput methods. Resulting potential biomarkers are then envisioned to be validated with targeted MS-based methods or classical immunoassays in much larger cohorts.

Here we suggest an additional strategy, which we term "plasma proteome profiling." It consists of the measurement of large numbers of plasma proteomes at the greatest possible depth with streamlined and high-throughput technologies as described in this paper. This allows us to retain one of the basic attractions of unbiased, systemwide methodologies, namely, that associations do not have to be predefined but emerge naturally from "big data mining." Although our current work is only a first step in this direction, we believe that rapid development in the underlying technology will make this strategy more and more attractive. Given the low resource requirements, large cohorts could be investigated in the future, and one can even envision individuals routinely and repeatedly have their plasma proteome profile recorded. These high-dimensional profiles could indicate current disease risk as well as efficacy of lifestyle changes or pharmacological interventions and thereby contribute to individual and public health.

ACKNOWLEDGMENTS

We thank all members of the Proteomics and Signal Transduction Group for help and discussions and in particular Igor Paron, Korbinian Mayr, Gaby Sowa for MS technical assistance, Jürgen Cox for bioinformatic tools, and Niklas Grassl and Sean Humphrey for fruitful discussions. Nils Kulak and Garwin Pichler received an m^4 award from the Bio^M Munich Biotech Cluster funded by the Bavarian government. The work carried out in this study was partially supported by the Max Planck Society for the Advancement of Science and by the Novo Nordisk Foundation (grant NNF15CC0001).

REFERENCES

1. Addona, T.A., Shi, X., Keshishian, H., Mani, D.R., Burgess, M., Gillette, M.A., Clauser, K.R., Shen, D., Lewis, G.D., Farrell, L.A. et al.A pipeline that integrates the discovery and verification of plasma protein biomarkers reveals candidate markers for cardiovascular disease.*Nat. Biotechnol.* 2011; 29:635–643
2. Anderson, N.L.The clinical plasma proteome: a survey of clinical assays for proteins in plasma and serum.*Clin. Chem.* 2010; 56: 177–185
3. Anderson, L.Six decades searching for meaning in the proteome.*J. Proteomics*. 2014; 107: 24–30
4. Anderson, N.L. and Anderson, N.G.The human plasma proteome: history, character, and diagnostic prospects. *Mol. Cell. Proteomics*. 2002;1: 845–867
5. Anderson, N.L., Anderson, N.G., Pearson, T.W., Borchers, C.H., Paulovich, A.G., Patterson, S.D., Gillette, M., Aebersold, R., and Carr, S.A.A human proteome detection and quantitation project.*Mol. Cell. Proteomics*. 2009;8: 883–886
6. Bai, J.P., Barrett, J.S., Burckart, G.J., Meibohm, B., Sachs, H.C., and Yao, L.Strategic biomarkers for drug development in treating rare diseases and diseases in neonates and infants.*AAPS J.* 2013; 15: 447–454
7. Bellei, E., Bergamini, S., Monari, E., Fantoni, L.I., Cuoghi, A., Ozben, T., and Tomasi, A.High-abundance proteins depletion for serum proteomic analysis: concomitant removal of non-targeted proteins.*Amino Acids*. 2011; 40: 145–156
8. Cao, Z., Tang, H.Y., Wang, H., Liu, Q., and Speicher, D.W.Systematic comparison of fractionation methods for in-depth analysis of plasma proteomes.*J. Proteome Res.* 2012; 11:3090–3100
9. Christensen, U., Simonsen, M., Harrit, N., and Sottrup-Jensen, L.Pregnancy zone protein, a proteinase-binding macroglobulin. Interactions with proteinases and methylamine.*Biochemistry*. 1989; 28: 9324–9331
10. Cox, J. and Mann, M.MaxQuant enables high peptide identification rates, individualized p.p.b.-range mass accuracies and proteome-wide protein quantification. *Nat. Biotechnol.* 2008; 26:1367–1372
11. Cox, J. and Mann, M.Quantitative, high-resolution proteomics for data-driven systems biology.*Annu. Rev. Biochem.* 2011;80: 273–299
12. Cox, J. and Mann, M.1D and 2D annotation enrichment: a statistical method integrating quantitative proteomics with complementary high-throughput data.*BMC Bioinformatics*. 2012; 13:S12
13. Cox, J., Neuhauser, N., Michalski, A., Scheltema, R.A., Olsen, J.V., and Mann, M.Andromeda: a peptide search engine integrated into the MaxQuant environment.*J. Proteome Res.* 2011; 10:1794–1805
14. Cox, J., Hein, M.Y., Luber, C.A., Paron, I., Nagaraj, N., and Mann, M.Accurate proteome-wide label-free quantification by delayed normalization and maximal peptide ratio extraction, termed MaxLFQ.*Mol. Cell. Proteomics*. 2014;13: 2513–2526
15. Edfors, F., Boström, T., Forsström, B., Zeiler, M., Johansson, H., Lundberg, E., Hober, S., Lehtiö, J., Mann, M., and Uhlen, M.Immunoproteomics using polyclonal antibodies and stable isotope-labeled affinity-purified recombinant proteins.*Mol. Cell. Proteomics*. 2014;13: 1611–1624
16. Farrah, T., Deutsch, E.W., Omenn, G.S., Sun, Z., Watts, J.D., Yamamoto, T., Shteynberg, D., Harris, M.M., and Moritz, R.L.State of the human proteome in 2013 as viewed through PeptideAtlas: comparing the kidney, urine, and plasma proteomes for the biology- and disease-driven Human Proteome Project.*J. Proteome Res.* 2014; 13:60–75
17. Geiger, T., Cox, J., and Mann, M.Proteomics on an Orbitrap benchtop mass spectrometer using all-ion fragmentation.*Mol. Cell. Proteomics*. 2010;9: 2252–2261
18. Hua, S., Song, C., Geczy, C.L., Freedman, S.B., and Witting, P.K.A role for acute-phase serum amyloid A and high-density lipoprotein in oxidative stress, endothelial dysfunction and atherosclerosis.*Redox Rep.* 2009; 14: 187–196
19. in: A.J. Jenkins, P.P. Toth, T.J. Lyons (Eds.) Lipoproteins in Diabetes Mellitus. Humana Press, ; 2014
20. Jensen, M.K., Bertoia, M.L., Cahill, L.E., Agarwal, I., Rimm, E.B., and Mukamal, K.J.Novel metabolic biomarkers of cardiovascular disease.*Nat. Rev. Endocrinol.* 2014;10: 659–672
21. Kelstrup, C.D., Jersie-Christensen, R.R., Batth, T.S.,

Arrey, T.N., Kuehn, A., Kellmann, M., and Olsen, J.V.Rapid and deep proteomes by faster sequencing on a benchtop quadrupole ultra-high-field Orbitrap mass spectrometer.*J. Proteome Res.* 2014; 13:6187–6195

22. Keshishian, H., Burgess, M.W., Gillette, M.A., Mertins, P., Clauser, K.R., Mani, D.R., Kuhn, E.W., Farrell, L.A., Gerszten, R.E., and Carr, S.A.Multiplexed, quantitative workflow for sensitive biomarker discovery in plasma yields novel candidates for early myocardial injury.*Mol. Cell. Proteomics*. 2015;14: 2375–2393
23. Krastins, B., Prakash, A., Sarracino, D.A., Nedelkov, D., Niederkofler, E.E., Kiernan, U.A., Nelson, R., Vogelsang, M.S., Vadali, G., Garces, A. et al.Rapid development of sensitive, high-throughput, quantitative and highly selective mass spectrometric targeted immunoassays for clinically important proteins in human plasma and serum.*Clin. Biochem.* 2013; 46: 399–410
24. Kulak, N.A., Pichler, G., Paron, I., Nagaraj, N., and Mann, M.Minimal, encapsulated proteomic-sample processing applied to copy-number estimation in eukaryotic cells. *Nat. Methods*. 2014; 11: 319–324
25. Liu, Y., Buil, A., Collins, B.C., Gillet, L.C., Blum, L.C., Cheng, L.Y., Vitek, O., Mouritsen, J., Lachance, G., Spector, T.D. et al.Quantitative variability of 342 plasma proteins in a human twin population.*Mol. Syst. Biol.* 2015; 11: 786
26. Martínez-Morillo, E., Nielsen, H.M., Batruch, I., Drabovich, A.P., Begcevic, I., Lopez, M.F., Minthon, L., Bu, G., Mattsson, N., Portelius, E. et al.Assessment of peptide chemical modifications on the development of an accurate and precise multiplex selected reaction monitoring assay for apolipoprotein e isoforms.*J. Proteome Res.* 2014; 13:1077–1087
27. Muñoz, J. and Heck, A.J.From the human genome to the human proteome.*Angew. Chem. Int. Ed. Engl.*2014; 53: 10864–10866
28. Nagaraj, N., Kulak, N.A., Cox, J., Neuhauser, N., Mayr, K., Hoerning, O., Vorm, O., and Mann, M.System-wide perturbation analysis with nearly complete coverage of the yeast proteome by single-shot ultra HPLC runs on a bench top Orbitrap.*Mol. Cell. Proteomics*. 2012;11
29. Nanjappa, V., Thomas, J.K., Marimuthu, A., Muthusamy, B., Radhakrishnan, A., Sharma, R., Ahmad Khan, A., Balakrishnan, L., Sahasrabuddhe, N.A., Kumar, S. et al.Plasma Proteome Database as a resource for proteomics research: 2014 update.*Nucleic Acids Res.* 2014; 42:D959–D965
30. Omenn, G.S.Exploring the Human Plasma Proteome. Volume 5. John Wiley, ; 2005
31. Ottosson, U.B., Damber, J.E., Damber, M.G., Selstam, G., Solheim, F., Stigbrand, T., Södergård, R., and von Schoultz, B.Effects of sex hormone binding globulin capacity and pregnancy zone protein of treatment with combinations of ethinyl-oestradiol and norethisterone. *Maturitas*. 1981; 3: 295–300
32. Paczesny, S., Braun, T.M., Levine, J.E., Hogan, J., Crawford, J., Coffing, B., Olsen, S., Choi, S.W., Wang, H., Faca, V. et al.Elafin is a biomarker of graft-versus-host disease of the skin.*Sci. Transl. Med.* 2010; 2:13ra2
33. Sand, O., Folkersen, J., Westergaard, J.G., and Sottrup-Jensen, L.Characterization of human pregnancy zone protein. Comparison with human alpha 2-macroglobulin.*J. Biol. Chem.* 1985; 260:15723–15735
34. Schaab, C., Geiger, T., Stoehr, G., Cox, J., and Mann, M.Analysis of high accuracy, quantitative proteomics data in the MaxQB database.*Mol. Cell. Proteomics*. 2012;11
35. Scheltema, R.A. and Mann, M.SprayQc: a real-time LC-MS/MS quality monitoring system to maximize uptime using off the shelf components.*J. Proteome Res.* 2012; 11:3458–3466
36. Scheltema, R.A., Hauschild, J.P., Lange, O., Hornburg, D., Denisov, E., Damoc, E., Kuehn, A., Makarov, A., and Mann, M.The Q Exactive HF, a Benchtop mass spectrometer with a pre-filter, high-performance quadrupole and an ultra-high-field Orbitrap analyzer. *Mol. Cell. Proteomics*. 2014;13: 3698–3708
37. Such-Sanmartín, G., Ventura-Espejo, E., and Jensen, O.N.Depletion of abundant plasma proteins by poly(N-isopropylacrylamide-acrylic acid) hydrogel particles. *Anal. Chem.* 2014; 86: 1543–1550
38. Suhre, K., Shin, S.Y., Petersen, A.K., Mohney, R.P., Meredith, D., Wägele, B., Altmaier, E., Deloukas, P., Erdmann, J., Grundberg, E...., and CARDIoGRAM. Human metabolic individuality in biomedical and pharmaceutical research.*Nature*. 2011; 477: 54–60
39. Surinova, S., Schiess, R., Hüttenhain, R., Cerciello, F., Wollscheid, B., and Aebersold, R.On the development of plasma protein biomarkers.*J. Proteome Res.* 2011; 10: 5–16
40. Tanzi, R.E.The genetics of Alzheimer disease.
41. *Cold Spring Harb. Perspect. Med.* 2012; 2: 2
42. Thompson, A., Schäfer, J., Kuhn, K., Kienle, S., Schwarz, J., Schmidt, G., Neumann, T., Johnstone, R., Mohammed, A.K., and Hamon, C.Tandem mass tags: a novel quantification strategy for comparative analysis of complex protein mixtures by MS/MS.*Anal. Chem.* 2003; 75: 1895–1904
43. Tu, C., Rudnick, P.A., Martinez, M.Y., Cheek, K.L., Stein, S.E., Slebos, R.J., and Liebler, D.C.Depletion of abundant plasma proteins and limitations of plasma proteomics.*J. Proteome Res.* 2010; 9:4982–4991
44. U.S. Department of Health and Human Services, Food and Drug Administration, Center for Drug Evaluation and Research, Center for Veterinary Medicine.Guidance for Industry, Bioanalytical Method Validation.
45. U.S. Department of Health and Human Services, ; 2001
46. Végvári, A., Welinder, C., Lindberg, H., Fehniger, T.E., and Marko-Varga, G.Biobank resources for future patient care: developments, principles and concepts.*J. Clin. Bioinforma.* 2011; 1:24
47. Vihko, P., Sajanti, E., Jänne, O., Peltonen, L., and Vihko, R.Serum prostate-specific acid phosphatase: development and validation of a specific radioimmunoassay.*Clin. Chem.* 1978; 24: 1915–1919
48. Wilson, P.W., Pencina, M., Jacques, P., Selhub, J., D'Agostino, R. Sr., and O'Donnell, C.J.C-reactive protein and reclassification of cardiovascular risk in

the Framingham Heart Study.*Circ Cardiovasc Qual Outcomes*. 2008; 1: 92–97

49. Zeiler, M., Straube, W.L., Lundberg, E., Uhlen, M., and Mann, M.A protein epitope signature tag (PrEST) library allows SILAC-based absolute quantification and multiplexed determination of protein copy numbers in cell lines.

CHAPTER 2

BIOLOGICAL MEMBRANES

Helen Watson[1]

[1]University of Exeter Medical School, St. Luke's Campus, Magdalen Road, Exeter EX1 2LU, U.K.

ABSTRACT

Biological membranes allow life as we know it to exist. They form cells and enable separation between the inside and outside of an organism, controlling by means of their selective permeability which substances enter and leave. By allowing gradients of ions to be created across them, membranes also enable living organisms to generate energy. In addition, they control the flow of messages between cells by sending, receiving and processing information in the form of chemical and electrical signals. This essay summarizes the structure and function of membranes and the proteins within them, and describes their role in trafficking and transport, and their involvement in health and disease. Techniques for studying membranes are also discussed.

STRUCTURE AND ORGANIZATION OF MEMBRANES

Membranes are Composed of Lipids, Proteins and Sugars

Biological membranes consist of a double sheet (known as a bilayer) of lipid molecules. This structure is generally referred to as the phospholipid bilayer. In addition to the various types of lipids that occur in biological membranes, membrane proteins and sugars are also key components of the structure. Membrane proteins play a vital role in biological membranes, as they help to maintain the structural integrity, organization and flow of material through membranes. Sugars are found on one side of the bilayer only, and are attached by covalent bonds to some lipids and proteins.

Three types of lipid are found in biological membranes, namely phospholipids, glycolipids and sterols. Phospholipids consist of two fatty acid chains linked to glycerol and a phosphate group.

Citation: Helen Watson, Biological membranes, doi: 10.1042/BSE0590043

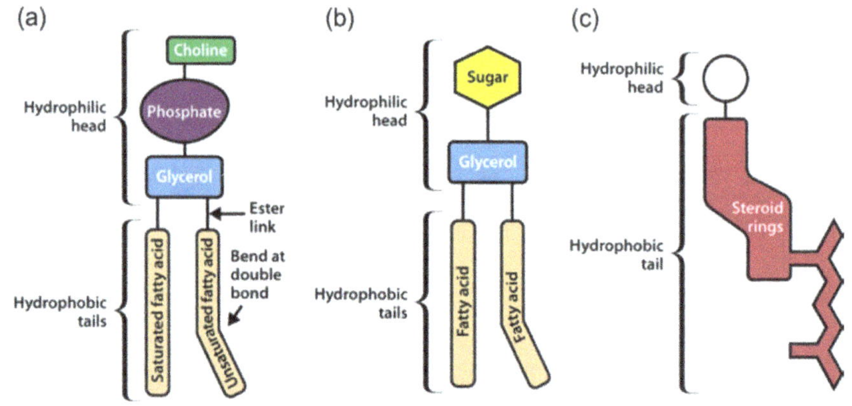

Figure 1: Schematic representations of three types of membrane lipid. (a) Phosphatidylcholine, a glycerophospholipid. (b) Glycolipid. (c) A sterol.

Phospholipids containing glycerol are referred to as glycerophospholipids. An example of a glycerophospholipid that is commonly found in biological membranes is phosphatidylcholine (PC) (Figure 1a), which has a choline molecule attached to the phosphate group. Serine and ethanolamine can replace the choline in this position, and these lipids are called phosphatidylserine (PS) and phosphatidylethanolamine (PE), respectively. Phospholipids can also be sphingophospholipids (based on sphingosine), such as sphingomyelin. Glycolipids can contain either glycerol or sphingosine, and always have a sugar such as glucose in place of the phosphate head found in phospholipids (Figure 1b). Sterols are absent from most bacterial membranes, but are an important component of animal (typically cholesterol) and plant (mainly stigmasterol) membranes. Cholesterol has a quite different structure to that of the phospholipids and glycolipids. It consists of a hydroxyl group (which is the hydrophilic 'head' region), a four-ring steroid structure and a short hydrocarbon side chain (Figure 1c).

The sugars attached to lipids and proteins can act as markers due to the structural diversity of sugar chains. For example, antigens composed of sugar chains on the surface of red blood cells determine an individual's blood group. These antigens are recognized by antibodies to cause an immune response, which is why matching blood groups must be used in blood transfusions. Other carbohydrate markers are present in disease (e.g. specific carbohydrates on the surface of cancer cells), and can be used by doctors and researchers to diagnose and treat various conditions.

Amphipathic Lipids form Bilayers

All membrane lipids are amphipathic—that is, they contain both a hydrophilic (water-loving) region and a hydrophobic (water-hating) region. Thus the most favourable environment for the hydrophilic head is an aqueous one, whereas the hydrophobic tail is more stable in a lipid environment. The amphipathic nature of membrane lipids means that they naturally form bilayers in which the hydrophilic heads point outward towards the aqueous environment and the hydrophobic tails point inward towards each other (Figure 2a). When placed in water, membrane lipids will spontaneously form liposomes, which are spheres formed of a bilayer with water inside and outside, resembling a tiny cell (Figure 2b). This is the most favourable

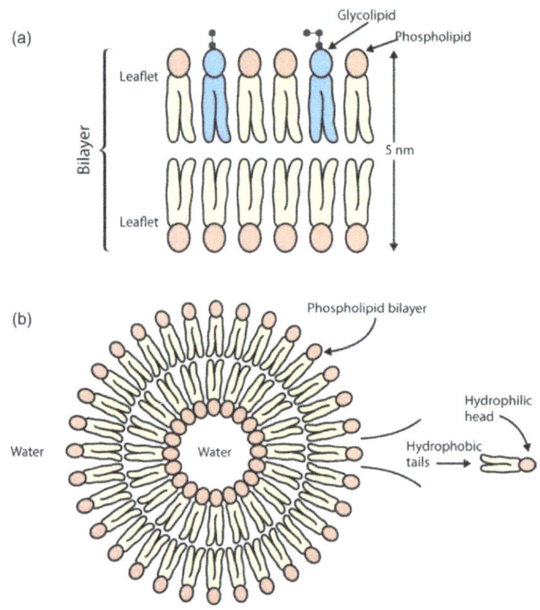

Figure 2: A membrane bilayer and liposome. Spontaneous formation of bilayers by membrane lipids. The hydrophilic heads (pink circles) will always face the aqueous environment in bilayers (a) and liposomes (b). The hydrophobic tails will face inward away from the water.

Configuration for these lipids, as it means that all of the hydrophilic heads are in contact with water and all of the hydrophobic tails are in a lipid environment. Early experiments by E. Gorter and F. Grendel in 1925 were the first to demonstrate that biological membranes are bilayers. These researchers extracted the lipids from red blood cells and found that they occupied a space that was twice the surface area of the cell. Red blood cells contain no internal membranes, so they deduced that the plasma membrane must be composed of two layers of lipids.

Biological Membranes and the Fluid Mosaic Model

The fluid mosaic model proposed by Jonathan Singer and Garth Nicolson in 1972 describes the dynamic and fluid nature of biological membranes. Lipids and proteins can diffuse laterally through the membrane. Phospholipids can diffuse relatively quickly in the leaflet of the bilayer in which they are located. A phospholipid can travel around the perimeter of a red blood cell in around 12 s, or move the length of a bacterial cell within 1 s. Phospholipids can also spin around on their head-to-tail axis, and their lipid tails are very flexible. These different types of movements create a dynamic, fluid membrane which surrounds cells and organelles. Membrane proteins can also move laterally in the bilayer, but their rates of movement vary and are generally slower than those of lipids. In some cases, membrane proteins are held in particular areas of the membrane in order to polarize the cell and enable different ends of the cell to have different functions. One example of this is the attachment of a glycosyl-phosphatidylinositol (GPI) anchor to proteins to target them to the apical membrane of epithelial cells and exclude them from the basolateral membrane.

Fluorescence photobleaching is one experimental method that is used by scientists to demonstrate visually the motility of proteins and lipids in a bilayer (Figure 3). A lipid or membrane protein located on the surface of a cell is tagged with a fluorescent marker such as green fluorescent protein (GFP). A beam of laser light is then focused on to a small area of the cell surface using a fluorescence microscope in order to bleach the fluorescent tags in this area so that they no longer emit a fluorescence signal. This small area of membrane is observed over time and gradually the fluorescence increases again, indicating that other tagged proteins or lipids are diffusing into this region from elsewhere in the membrane. This demonstrates that the lipid bilayer surrounding cells is fluid in nature and allows lateral diffusion of both lipids and membrane proteins.

Despite all this movement of lipids and proteins in the bilayer, vertical movement, or 'flipflop', of lipids and proteins from one leaflet to another occurs at an extremely low rate. This is due to the energetic barrier encountered when forcing the hydrophilic head (in the case of lipids) or hydrophilic regions (in the case of proteins) through the hydrophobic environment of the inside of the membrane. This near absence of vertical movement allows the inner and outer leaflets of the bilayer to maintain different lipid compositions, and enables membrane proteins to be inserted in the correct orientation for them to function.

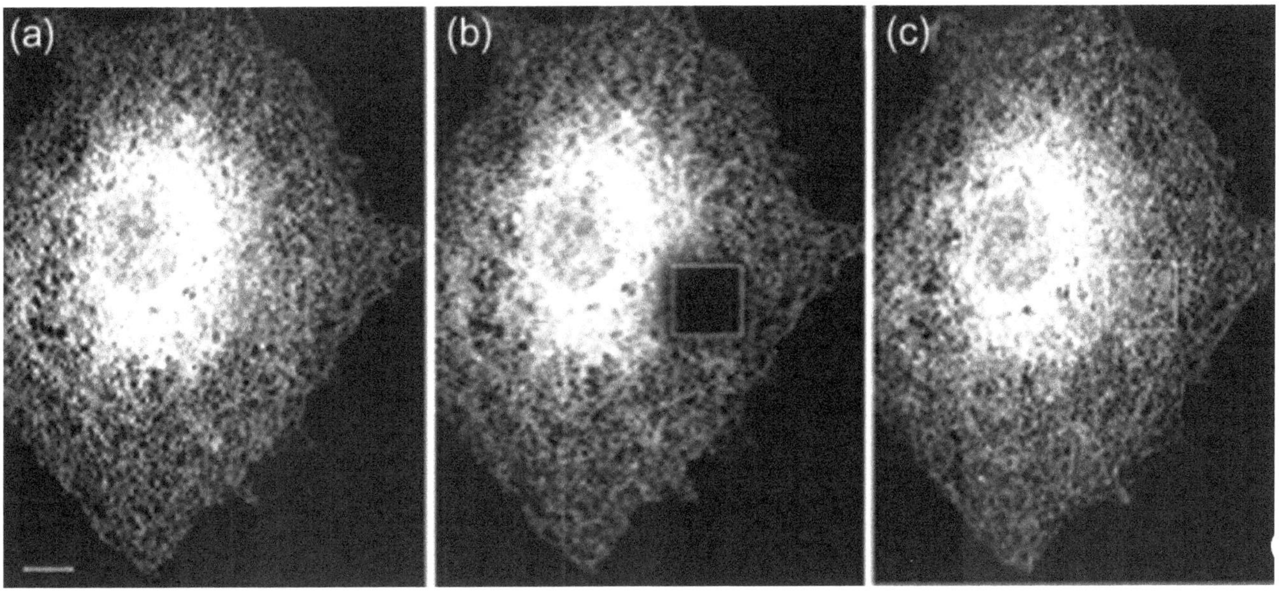

Figure 3: Photobleaching. Cells expressing a GFP-labelled protein in the endoplasmic reticulum were subjected to photobleaching. (a) A cell before bleaching. (b) The same cell immediately after bleaching of the square section shown. (c) The same cell 5 min after photobleaching. Adapted from Figure 1b from Lippincott-Schwartz, J., Snapp, E. and Kenworthy, A. (2001) Studying protein dynamics in living cells. Nat. Rev. Mol. Cell. Biol. 2, 444–456.

However, some enzymes facilitate the process of lipid flip-flop from one leaflet to another. These flippases, or phospholipid translocators, use ATP to move lipids across the bilayer to the other leaflet. In eukaryotic cells, flippases are located in various organelles, including the endoplasmic reticulum (ER), where they flip-flop newly synthesized lipids.

How Membranes are Made

Biological membranes are formed by adding to a pre-existing membrane. In prokaryotes this occurs on the inner leaflet of the plasma membrane, facing the cytoplasm. In eukaryotes, membrane synthesis takes place at the ER on the cytoplasmic leaflet of the ER membrane (termed the 'inside' of the cell). Lipids then leave the ER and travel through the secretory pathway for distribution to various subcellular compartments or the plasma membrane. In eukaryotic cells, enzymes that span the ER catalyse the formation of membrane lipids. In the cytoplasmic leaflet of the ER membrane, two fatty acids are bound, one by one, to glycerol phosphate from the cytoplasm. This newly formed diacylglycerol phosphate is anchored in the ER membrane by its fatty acid chains. The phosphate is then replaced by the head group (e.g. phosphate and choline). Flippases in the ER membrane can then move some of these newly formed lipids to the luminal side of the ER membrane. Similarly, flippases in prokaryotes can transfer new lipids from the inner leaflet of the plasma membrane to the outer leaflet. These flippases are responsible for adjusting the lipid composition of each layer of the membrane. In eukaryotes, lipids must then be distributed to the various intracellular membranes. The traffic of vesicles between organelles in combination with signals that direct particular lipids to specific locations is required to create the correct lipid composition in all of the cellular membranes (Figure 4). Vesicles bud from the ER and travel via the ER–Golgi intermediate compartment (ERGIC) to join with the Golgi, where sorting of lipids takes place. The Golgi then sends lipids in vesicles to various destinations, including the plasma membrane and lysosomes. Lipids and proteins are internalized from the plasma membrane into endosomes. Organelles, such as mitochondria, acquire lipids from the ER by a different mechanism. Watersoluble proteins called phospholipid-exchange proteins remove phospholipids from the ER membrane and deposit them in the membranes of the appropriate organelles.

Distribution of Lipids

The inner and outer leaflets of bilayers differ in their lipid composition. In mammalian cells, the outer leaflet of the plasma membrane contains predominantly PC and sphingomyelin, whereas PS and PE are found on the inner leaflet. During programmed cell death (apoptosis), PS is no longer restricted to the inner leaflet of the plasma membrane. It is exposed on the outer leaflet by the action of an enzyme called scramblase which is a type of flippase enzyme. PS is negatively charged, unlike PC, which has no net charge. The movement of PS into the outer leaflet therefore changes the charge of the plasma membrane as viewed from the outside of the cell. This change in surface charge labels the apoptotic cell for phagocytosis by phagocytic cells such as macrophages.

Lipid composition also varies between the organelles within eukaryotic cells. Cholesterol is synthesized in the ER, but the ER membrane has a relatively low cholesterol content, as much of the cholesterol is transported to other cellular membranes. The prevalence of cholesterol in membranes increases through the secretory pathway, with more in the Golgi than in the ER (the trans-Golgi network is richer in cholesterol than the cis-Golgi), and most in the plasma membrane. This increase in cholesterol through the secretory pathway results in slightly thicker membranes in the late Golgi and plasma membrane compared with the ER, and is thought to be a contributing factor to protein sorting through the pathway, as membrane proteins in the plasma membrane generally have longer hydrophobic transmembrane domains than membrane proteins that reside in the ER.

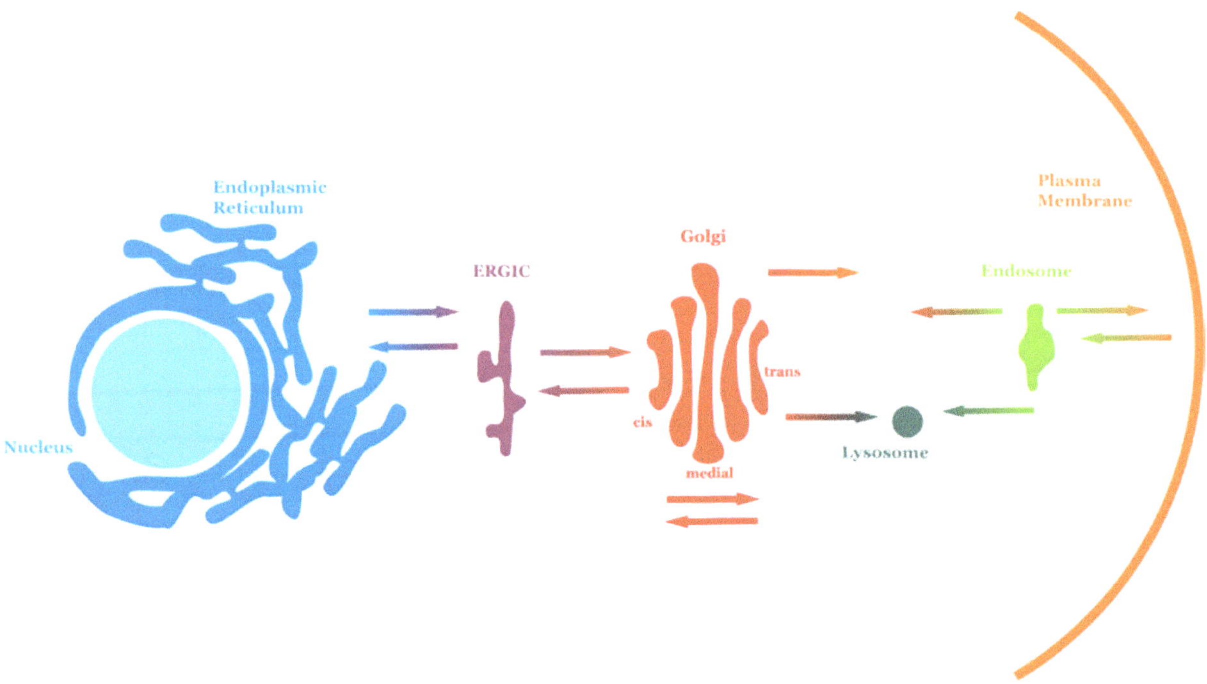

Figure 4: Membrane traffic in eukaryotic cells. The main compartments of eukaryotic cells are shown. Arrows indicate movement of lipid vesicles between them, with colours at the tail end indicating origin and those at the head end indicating destination.

Membrane Proteins

Membrane proteins are the nanomachines that enable membranes to send and receive messages and to transport molecules into and out of cells and compartments. Without membrane proteins the phospholipid membrane would present an impenetrable barrier and cells would be unable to communicate with their neighbours, transport nutrients into the cell or waste products out of it, or respond to external stimuli. Both unicellular and multicellular organisms need membrane proteins in order to live. The membrane proteins that are present in a particular membrane determine the substances to which it will be permeable and what signal molecules it can recognize.

Synthesis of Membrane Proteins

In eukaryotic cells, the synthesis of membrane proteins destined for the plasma membrane, ER or any other membrane-bound compartment begins on cytosolic ribosomes. After a short segment of protein has been synthesized, the ribosome, mRNA and nascent protein chain associate with the ER, where the rest of the protein is made and simultaneously inserted into the membrane. This phenomenon was first explained by Günter Blobel, David Sabatini and Bernhard Dobberstein in the

1970s. These scientists proposed that there is a 'binding factor' which recognizes the emerging protein chain and can dock the ribosome at the ER membrane. We now know that there is an N-terminal signal sequence within membrane proteins. These signal sequences are not identical but share a common motif, namely a hydrophobic stretch of 20–30 amino acids, a basic region at the N-terminus and a polar domain at the C-terminus of the signal. These N-terminal signal sequences are recognized by the signal recognition particle (SRP), which has binding sites for the signal sequence, ribosome and the SRP receptor which is embedded in the ER membrane. Upon binding the SRP, the ribosome pauses protein synthesis. The SRP binds to the SRP receptor, adjacent to a translocon pore in the ER membrane. The translocon is a protein pore through which membrane protein chains can be threaded into the membrane. It has a laterally opening gate to allow newly synthesized proteins into the ER membrane. Once the ribosome is at the translocon, the SRP dissociates and protein synthesis resumes. This process is referred to as co-translational targeting, and the main events are summarized in Figure 5. Co-translational targeting is the dominant mechanism for protein delivery to the ER in higher eukaryotes, whereas yeast and prokaryotes favour post-translational targeting, whereby proteins are delivered to the ER after completion of synthesis. Post-translational targeting also occurs in higher eukaryotes, often when a membrane protein is so small that the signal sequence does not emerge until the whole protein has been synthesized. Post-translational targeting can be carried out both by SRP-dependent and by SRP-independent mechanisms.

Structure and Function of Membrane Proteins

Membrane-spanning proteins are diverse in structure and function. They can be constructed of α-helices or from β-barrels. The β-barrel membrane proteins often function as pores, with hydrophobic amino acids facing out into the bilayer. In addition, there are other non-spanning proteins which associate with the bilayer, often using a hydrophobic anchor. Here we shall focus on the α-helical membrane proteins. These proteins have at least one α-helical hydrophobic stretch of amino acids, around 20 residues in length, which corresponds to around 30 Å (the thickness of an average phospholipid bilayer). If an α-helical membrane protein spans the membrane more than once, it will have more than one of these hydrophobic sections. For example, the Ca2+-ATPase of the ER and sarcoplasmic reticulum (SR) spans the membrane 10 times, so it has 10 hydrophobic stretches of around 20 amino acids each.

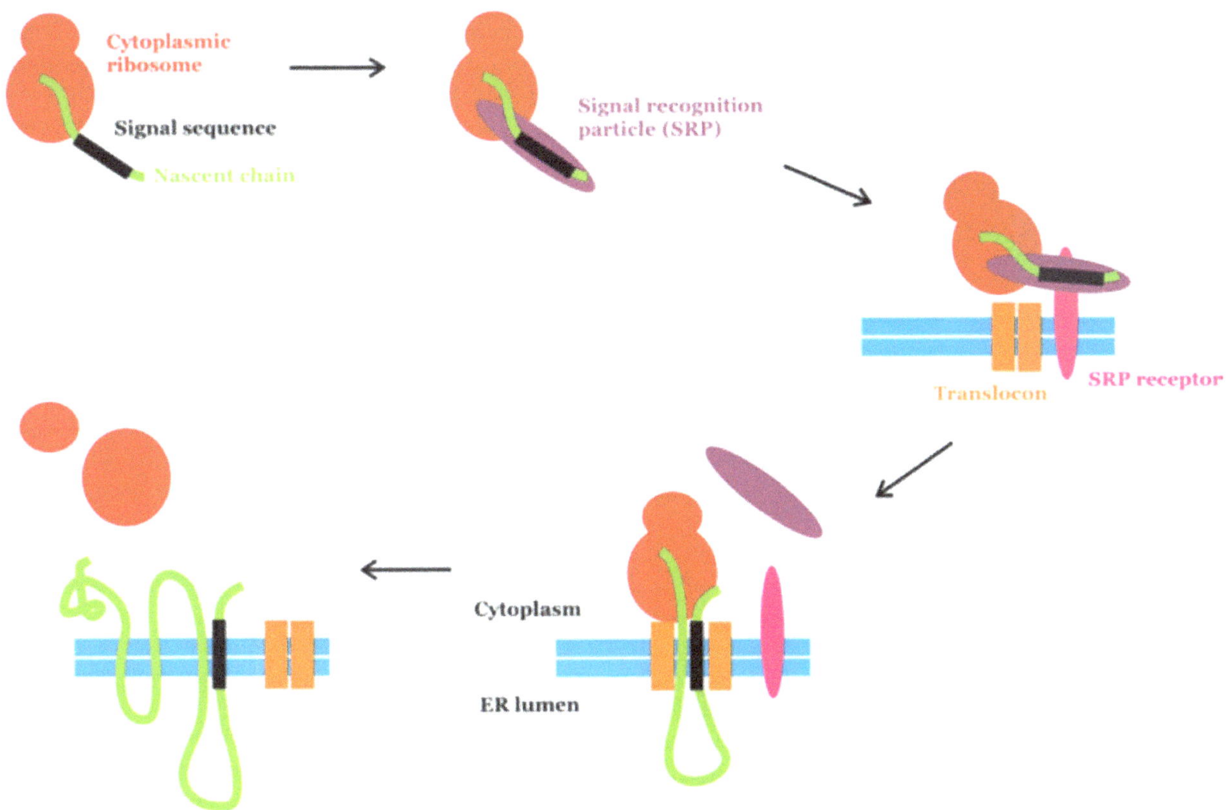

Figure 5: Co-translational ER protein targeting. The key steps of ER targeting are summarized. Each component is labelled and the ER membrane is represented by double blue lines. The signal sequence (shown in black) becomes the first transmembrane domain of the protein in this example.

MEMBRANE PROTEINS CONTROL WHAT ENTERS AND LEAVES THE CELL

A vital class of membrane proteins are those involved in active or passive transport of materials across the cell membrane or other subcellular membranes surrounding organelles. For a cell or an organism to survive, it is crucial that the right substances enter cells (e.g. nutrients) and the right substances are transported out of them (e.g. toxins).

Passive and Active Transport

Molecules can cross biological membranes in several different ways depending on their concentration on either side of the membrane, their size and their charge. Some molecules, including water, can simply diffuse through the membrane without assistance. However, large molecules or charged molecules cannot cross membranes by simple diffusion. Charged molecules such as ions can move through channels passively, down electrochemical gradients. This movement is described as 'downhill', as the ions or molecules travel from an area of high concentration to an area of low concentration. This requires channel proteins but no energy input.

Passive transport can also be mediated by carrier proteins that carry specific molecules such as amino acids down concentration gradients, again without any requirement for energy. Active transport moves species against concentration gradients and requires energy, which is obtained from ATP, from light, or from the downhill movement of a second type of molecule or ion within the same transporter (Figure 6).

Passive Transport

Passive transport is the movement of molecules across biological membranes down concentration gradients. This type of transport does not require energy. Channels form water-filled pores and thus create a hydrophilic path that enables ions to travel through the hydrophobic membrane. These channels allow downhill movement of ions, down an electrochemical gradient. Both the size and charge of the channel pore determine its selectivity. Different channels have pores of different diameters to allow the selection of ions on the basis of size. The amino acids that line the pore will be hydrophilic, and their charge will determine whether positive or negative ions travel through it. For example, Ca^{2+} is positively charged, so the amino acids lining the pores of Ca^{2+} channels are generally basic (i.e. they carry a negative charge). Channels are not always open. They can be gated by ligands which bind to some part of the protein, either by a change in membrane potential (voltage gated) or by mechanical stress (mechanosensitive). The nicotinic acetylcholine receptor is an example of a ligand-gated ion channel which opens upon binding the neurotransmitter acetylcholine (Figure 7). The nicotinic acetylcholine receptor is a pentameric membrane protein composed of five subunits arranged in a ring, with a pore through the centre. In the closed state, the pore is blocked by large hydrophobic amino acid side chains which rotate out of the way upon acetylcholine binding to make way for smaller hydrophilic side chains, allowing the passage of ions through the pore. Opening of the nicotinic acetylcholine receptor allows rapid movement of Na^+ ions into the cell and slower movement of K^+ ions out of the cell, in both cases down the electrochemical gradient of the ion. The difference in gradients between Na^+ and K^+ across the membrane means that more Na^+ enters the cell than K^+ leaves it.

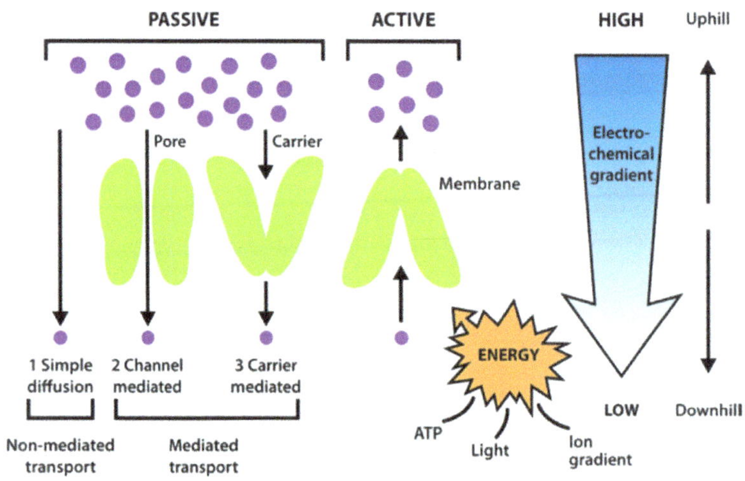

Figure 6: Passive and active transport. The different types of membrane proteins involved in passive and active transport are shown.

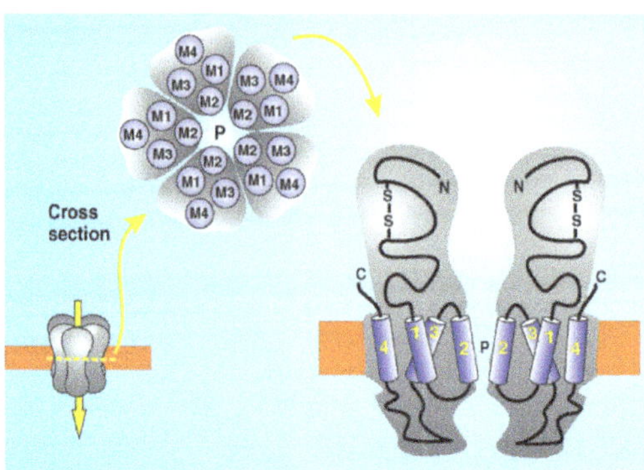

Figure 7: The nicotinic acetylcholine receptor. The pentameric structure of the receptor is shown, with the pore region (P) indicated. Transmembrane helices (M1–M4) are labelled in each subunit. The bilayer is shown in orange. Reproduced from Berridge, M.J. (2012) Cell Signalling Biology; doi:10.1042/csb0001003, with permission.

This creates a net movement of positive charges into the cell, resulting in a change in membrane potential. Acetylcholine released by motor neurons at the neuromuscular junction travels across the synapse and binds to nicotinic acetylcholine receptors in the plasma membrane of the muscle cells, causing membrane depolarization. This depolarization of the muscle cells triggers Ca^{2+} release and muscle contraction. Carrier proteins are the other class of membrane proteins, apart from channels, which can facilitate passive transport of substances down concentration gradients. Carrier proteins transport molecules much more slowly than channels, as a number of conformational changes in the carrier are required for the transport of the solute across the membrane. A molecule such as a sugar binds to the carrier protein on one side of the membrane where it is present at a high concentration. Upon binding, the carrier changes conformation so that the sugar molecule then faces towards the opposite side of the membrane. The concentration of sugar on this side is lower, so dissociation occurs and the sugar is released. This is downhill movement and, although slower than movement through channels, it requires no energy.

The cystic fibrosis transmembrane conductance regulator (CFTR) is an ATP-dependent chloride ion (Cl−) channel that has an important role in regulating the viscosity of mucus on the outside of epithelial cells. ATP is used to gate the channel, but the movement of Cl^- occurs down its electrochemical gradient, so does not require energy. A heritable change in the CFTR gene which results in a single amino acid deletion in the protein causes cystic fibrosis. This is a serious illness in which thick mucus accumulates in the lungs, causing a significantly lower than average life expectancy in patients who have the disease. Unimpaired ion transport is vital for our survival and health, and conditions such as cystic fibrosis highlight the need for research into these types of proteins.

Active Transport

The transport of molecules across a membrane against a concentration gradient requires energy, and is referred to as active transport. This energy can be obtained from ATP hydrolysis (primary active transport), from light (as, for example, in the case of the bacterial proton pump bacteriorhodopsin), or from an electrochemical gradient of an ion such as Na^+ or H^+ (secondary active transport). Calcium ions signal many events, including muscle contraction, neurotransmitter release and cellular motility. However, high cytoplasmic concentrations of Ca^{2+} are toxic to the cell. Therefore Ca2+ must be tightly regulated and removed from the cytoplasm either into internal stores (the ER, and the SR in muscle cells) or into the extracellular space. This Ca^{2+} removal is carried out by a family of Ca^{2+}-ATPases, including the sarco/endoplasmic reticulum Ca^{2+}-ATPase (SERCA), which hydrolyse ATP to move Ca^{2+} against its electrochemical gradient into the ER and SR (Figure 8). There are Ca2+-ATPases in the ER, Golgi and plasma membrane, and despite their sequence similarity, these proteins are differentially targeted to the appropriate membrane. These Ca^{2+} pumps are primary active transporters. SERCA moves two Ca^{2+} ions into the ER or SR for every ATP molecule that is hydrolysed. The pump undergoes a cycle of binding ATP and phosphorylation, and undergoes large conformational changes every time it transports a pair of Ca^{2+} ions. SERCA is a P-type ATPase (so called because it is phosphorylated during ion transport). There are many P-type ATPases, and they are conserved in evolution across many species. The Na^+/K^+-ATPase is one of these P-type ATPases, and it works in a similar way to SERCA to pump Na+ out of the cell and K^+ into the cell using energy derived from the hydrolysis of ATP. We have now obtained three-dimensional structures of SERCA in a number of conformational states, which allow scientists to visualize the transport process.

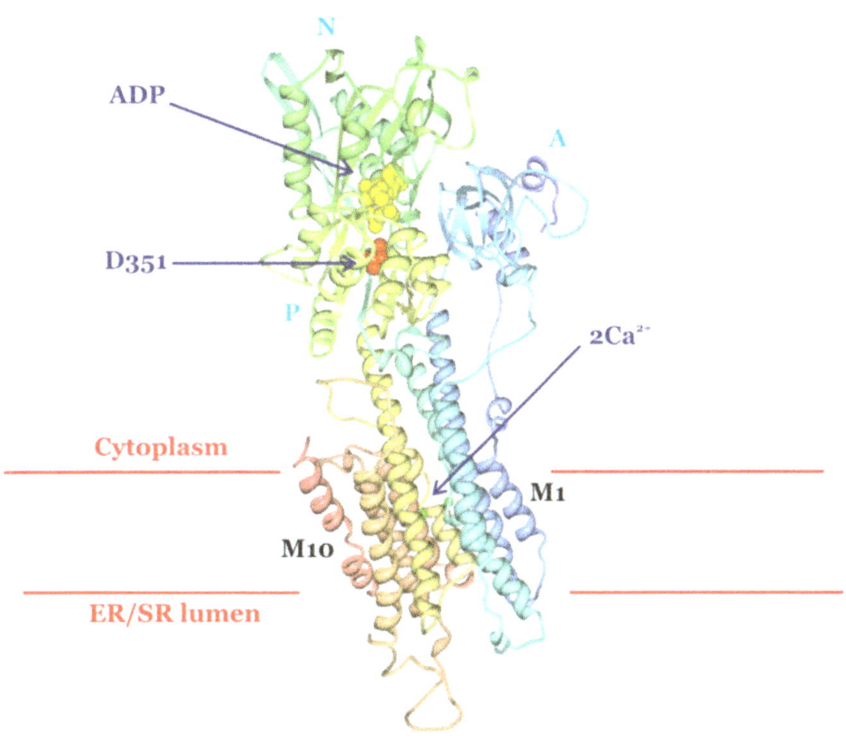

Figure 8: The sarco/endoplasmic reticulum Ca^{2+}-ATPase (SERCA).

The crystal structure of SERCA in the ADP- and Ca2+-bound state is shown. D351 (in red) is the residue phosphorylated during the movement of Ca^{2+} ions into the ER or SR. The three cytoplasmic domains, phosphorylation (P), nucleotide binding (N) and actuator (A) are labelled. ADP is shown in yellow and Ca^{2+} ions in green. Protein Data Bank (PDB) code 2ZBD, rendered using PDB Protein Workshop.

Secondary active transport requires an ion electrochemical gradient to drive the uphill transport of another solute. The downhill movement of one species drives the uphill movement of the other. This can be symport (in which both types of molecule or ion travel across the membrane in the same direction) or antiport (in which the two species travel in opposite directions), as shown in Figure 9. In order to transport glucose into cells, the Na+–glucose symporter uses the electrochemical gradient of Na+ across the plasma membrane. The concentration of Na^+ is much higher outside the cell, and the inside of the cell is negatively charged relative to the outside, so by allowing Na+ to travel down its electrochemical gradient, these transporters can move glucose uphill, into the cell and against its concentration gradient. This is referred to as symport, as both Na^+ and glucose travel in the same direction—in this case into the cell. In order for this symport to be sustainable, the Na+ gradient must be maintained. This is done by the Na^+/K^+-ATPase, which uses ATP to pump the Na+ back into the extracellular space, thus maintaining a low intracellular Na+ concentration. Both Na^+ and Ca^{2+} are present at much higher concentrations outside the cell than inside it. Like the Na+–glucose symporter, the Na^+–Ca^{2+} exchanger uses the electrochemical gradient of Na+ across the plasma membrane to move a second species (Ca^{2+}) against its electrochemical gradient. However, in this case the transporter is an antiporter, as it uses the concentration gradient of one substance moving in (Na+) to move another (Ca^{2+}) out of the cell. This antiporter has an exchange rate of three Na^+ ions in to two Ca^{2+} ions out. It moves Ca^{2+} out of the cell faster than the plasma membrane equivalents of SERCA, but has a lower affinity for Ca^{2+} than these P-type ATPases. Again this transporter relies on the Na^+/K^+-ATPase to maintain the low intracellular Na+ concentration.

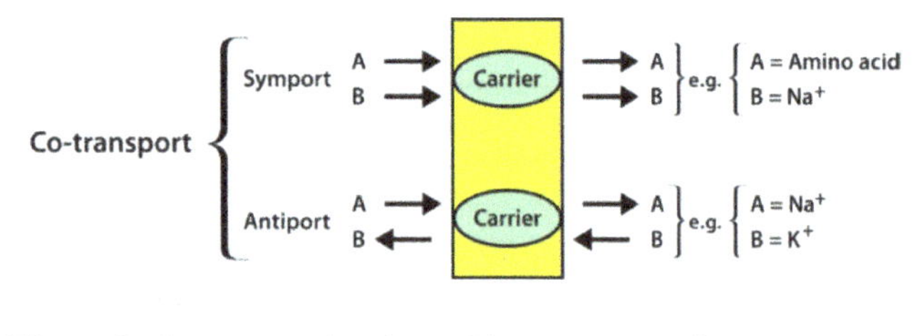

Figure 9: Symport and antiport. The two types of co-transport are shown, with examples.

SOLVING THE STRUCTURE OF MEMBRANE PROTEINS

In order to understand more fully the mechanisms of action of membrane proteins such as the transporters described here, we can determine their three-dimensional protein structures. As a result of huge advances in structural biology in the last 50 years, we now have access to many thousands of protein structures in online databases. This enables researchers to visualize the structure of their protein of interest, and thus gain insight into its mechanism.

X-ray Crystallography

The structure of whale myoglobin was solved in 1958 using X-ray crystallography, earning John C. Kendrew and Max Perutz the Nobel Prize in Chemistry. This was the first protein structure to be solved using this technique, and since then thousands of proteins have been solved using this method. X-ray crystallography works by firing a beam of X-rays at a crystalline structure and measuring the diffraction of the X-rays after they have passed through the structure of interest. This generates an electron density map, showing where different atoms in the structure are located. For regular crystalline solids such as salts this is relatively straightforward, but for large irregular molecules such as proteins it can present many technical challenges. Before a protein is subjected to X-ray beams, it must first be purified and crystallized. In nature, proteins exist in the busy milieu of a cell, surrounded by thousands of other types of proteins, as well as lipids and other molecules. A common method of obtaining enough of the protein of interest involves expressing the relevant gene in a system such as bacteria. The gene is tagged with a small protein tag which can be used to isolate the protein of interest. Bacterial systems allow large amounts of protein to be produced cheaply and quickly. However, if the protein of interest is from a species that is only distantly related to that in which it is normally expressed (e.g. a human protein produced in Escherichia coli (E. coli)), the lack of correct glycosylation enzymes and the differences in protein folding and assembly may prevent the production of a biologically active protein. In addition, the expression of membrane proteins that make pores or channels can kill the host organism.

The number of solved crystal structures of proteins is constantly growing as technology improves and expertise is shared among scientists to help to optimize conditions for crystal production. The Protein Data Bank (PDB) is an online archive of protein structures which can be freely accessed by scientists worldwide. At the time of writing, 88% of the structures in the PDB have been solved by X-ray crystallography, and there are currently just under 70 000 X-ray crystal structures in the database. The number of membrane protein structures in the PDB is increasing rapidly with the refinement of crystallization techniques (Figure 10).

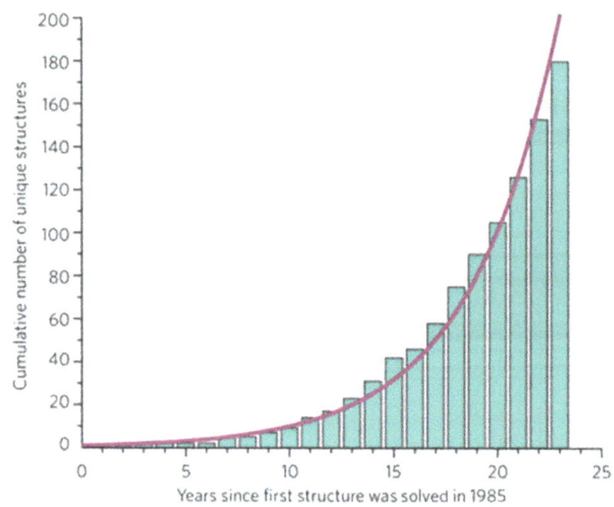

Figure 10: X-ray crystal structures of membrane proteins. The increase in the number of solved crystal structures of membrane pro-

teins is shown from 1985, when the first such structure was solved. Adapted from White, S.H. (2009) Biophysical dissection of membrane proteins. Nature 459, 344–346.

Other Structural Technique

Nuclear magnetic resonance (NMR) spectroscopy is another valuable technique for elucidating membrane protein structure. Molecules are placed in a magnetic field and the resonance properties of different atomic nuclei are measured, which gives an indication of where in a particular molecule different atoms are located. Generally, NMR is limited to smaller proteins, below around 35 kDa in size. It also offers the potential to visualize proteins in a more physiologically relevant environment (e.g. in lipid bilayers or micelles). Another advantage of NMR is that it does not require the protein to be locked in a crystal lattice—a structure which can distort the natural shape of the protein. Electron microscopy can also be used to study membrane protein structure. By freezing membrane proteins in their natural lipid environments, it is possible to investigate their structure using high-resolution electron microscopy. This provides a snapshot of the naturally occurring conformation of individual proteins in the bilayer.

INTERACTIONS BETWEEN LIPIDS AND PROTEINS IN BIOLOGICAL MEMBRANES

The lipids that surround membrane proteins in biological membranes play an important role in the activity of these proteins. As was mentioned earlier, some membrane protein crystal structures include lipids bound to the outside surface of the transmembrane domains of the proteins. It is thought that these lipids bind tightly to the protein, and have a long-lived interaction with the transmembrane region. In other cases, lipids are thought to interact briefly with membrane proteins, rapidly moving away and being replaced by other membrane lipids.

The activity of membrane proteins is considered to be dependent to some extent on the lipids that surround them in the membrane. Certain types of K^+ channel are thought to bind to negatively charged membrane lipids, as the activity of these channels increases at higher anionic lipid concentrations. These types of interaction can be studied by placing a purified form of the protein of interest in an artificial bilayer and measuring its activity. By altering the types of lipid present in the artificial bilayer, deductions can be made about the lipids that the protein requires in order to be active. Fluorescence spectroscopy and electron spin resonance are two techniques that are used to measure how strongly membrane proteins interact with specific lipids around them.

Molecular dynamics simulations use computer algorithms to work through theoretical problems. These simulated experiments are useful for investigating interactions between membrane proteins and lipids, as in real membranes these interactions are often so fleeting that they are very difficult to measure. Molecular dynamics simulations have predicted that in the case of the nicotinic acetylcholine receptor, the negatively charged lipid, phosphatidic acid, is required for activity. These simulations have also shown that cholesterol stabilizes the receptor and that the phosphatidic acid forms a shell around the protein which is more long-lasting than the interactions with other membrane lipids. Although molecular dynamics simulations are extremely useful, they are limited by the assumptions and approximations on which they are based. As in many areas of biology, a combination of experimental and computational research is required if real progress is to be made in understanding the complexity of biological membranes.

INTERNAL MEMBRANES IN EUKARYOTIC CELLS FORM ORGANELLES

Inside the plasma membrane that surrounds eukaryotic cells lie many other membranes which define the intracellular compartments, or organelles. Each of these organelles has distinct functions and contains specific complements of proteins adapted for these roles. With the exception of a few proteins that are coded for by the mitochondrial genome, synthesis of all of the proteins that are required in these organelles begins on ribosomes in the cytoplasm, and therefore the proteins must be directed to the correct destination. We have seen earlier how this is achieved with membrane proteins, and most organelles have some kind of signal sequence that can be recognized by various receptors and which ensures that the protein arrives at the correct organelle.

Organelles Have Distinct Lipid Compositions

Besides the specific protein complement of each organelle, the lipid make-up of the bilayers surrounding organelles varies. Lipids are synthesized in the ER, and flippases move lipid molecules between leaflets of the bilayer. For organelles in the secretory pathway and the plasma membrane, lipid transport into these compartments is mediated by vesicular membrane traffic through the pathway. The cholesterol concentration in membranes increases from the ER through the Golgi to the plasma membrane. Cholesterol makes membranes thicker and more rigid, so the low levels of cholesterol in the ER membrane render it thin and facilitate the insertion of newly synthesized membrane and secretory proteins. PC becomes relatively less abundant through this pathway, with more found in the ER than at the plasma membrane. PS and PE are found throughout the secretory pathway in the cytosolic leaflet of the membranes. This differential lipid composition through the secretory pathway is achieved by targeting specific lipids into transport vesicles. Proteins included in these vesicles act as labels and direct the lipids to the right compartment. Forward-moving (anterograde) vesicles destined for the plasma membrane are rich in cholesterol. Lipids also move backwards through the secretory pathway, from the plasma membrane towards the ER. This is known as retrograde traffic. Retrograde vesicles from the Golgi are enriched in lipids such as PC, which are concentrated in the ER.

The lipid composition of the mitochondria is very different from that of the secretory pathway compartments. Mitochondrial membranes are much richer in PE and cardiolipin than is the ER. Cardiolipin is synthesized in the mitochondria and is predominantly confined to this organelle. As membrane proteins have evolved along with their organelles and surrounding lipids, it follows that different lipid compositions are required in different organelles for the optimum activity of the proteins within their membranes. The structure of the ADP/ATP carrier in mitochondria has been solved and was found to include cardiolipin and PC molecules bound to the protein. The activity of this carrier protein is dependent on the presence of cardiolipin, which is relatively abundant in mitochondrial membranes.

Proteins Must be Targeted to the Correct Organelle for Cells to Function

The targeting of newly synthesized membrane and secretory proteins to the ER has already been briefly discussed. However, there are many different destinations within the cell to which a protein can be sent, and sometimes proteins are located in more than one of these. The signals and protein machinery that are required to target proteins to the correct compartment are many and various, and much of the detail of the exact mechanisms involved has yet to be clarified.

Vesicular Transport

Traffic through the secretory pathway is by vesicular transport in both anterograde and retrograde directions. Proteins and lipids can be included and excluded from vesicles by various means in order to selectively determine which molecules move forward or backward through the pathway. Vesicles are coated with proteins that determine their destination. Generally these coat proteins (COPs) are directional—COPII coats anterograde vesicles, and COPI coats retrograde vesicles. Proteins that travel in vesicles (referred to as cargo) are selected either by interacting with receptors in the vesicles or by directly interacting with the coat proteins. The selection of cargo occurs at the budding stage, when the coat proteins begin to distort the donor membrane (e.g. the ER) into a vesicle. Once the cargo has been selected and the coat proteins have been assembled, the vesicle buds off and travels to the acceptor membrane (e.g. the Golgi in the case of COPII vesicles from the ER) either by diffusion or with the help of motor proteins that 'walk' the vesicle along the cytoskeleton. The vesicle then fuses with the acceptor membrane, depositing its cargo and constituent lipids (Figure 11). The formation and fusing of vesicles are energetically demanding, as these processes require the stable bilayer to be broken in order to pinch off a new vesicle, and then fused with a different membrane. Both energy and specialized protein machinery are required to overcome this energy barrier.

Once the vesicle has budded, travelled through the cytosol and reached its destination, it must fuse with its receptor membrane. Again this is an energetically unfavorable process, and protein machinery has to be utilized in order to allow the fusion of two bilayers.

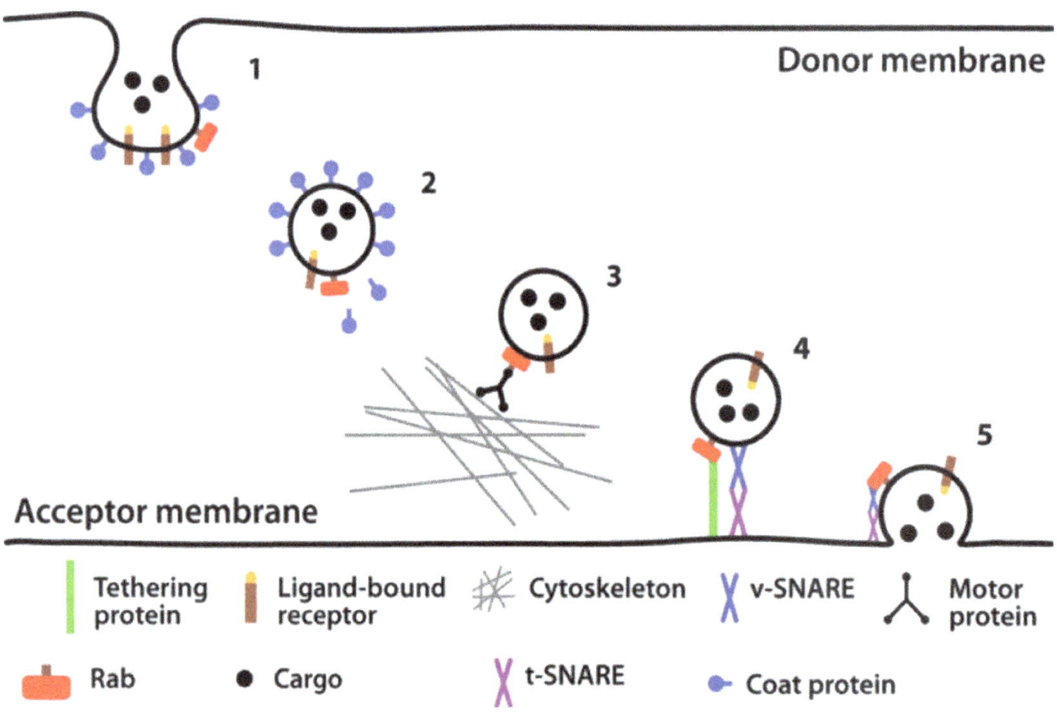

Figure 11: Vesicle production, transport and fusion.

The main events in vesicular transport of cargo are shown. Cargo is selected and packed into vesicles which are formed by coat proteins (1 and 2). The GTPase Rab is also incorporated on the outside of the vesicle and facilitates the steps illustrated. The vesicle then travels along proteins of the cytoskeleton towards its destination following dissociation of coat proteins (3). The vesicle is tethered to the donor membrane with the help of tethering proteins and the SNARE complexes (4), allowing membrane fusion and release of the cargo (5).

SNARE proteins are central to the process of vesicle fusion. Vesicles carry v-SNAREs that bind to specific t-SNAREs on the target membranes. Not only does this confer specificity in the targeting of vesicles, but also the SNAREs facilitate membrane fusion on arrival of the vesicle. The interacting v-SNAREs and t-SNAREs form a four-helix bundle at the interface between the two membranes, consisting of three helices from the t-SNARE and one helix from the v-SNARE. This stable interaction is thought to provide the free energy necessary to enable the two membranes to become very close and fuse. As the two bilayers become closer, the lipids in the two outer leaflets can come into contact with one another, thereby increasing the hydrophobic nature of the site and enabling the membranes to join, and overcoming the energy barrier. The transmembrane domains of the SNAREs are also believed to be involved in membrane fusion, as when they are replaced by lipids experimentally, fusion does not occur. Upon fusion, the cargo enters the target compartment, and the lipids and membrane proteins that formed the vesicle diffuse into the target membrane.

Determining the mechanisms of membrane budding is important for understanding how viruses such as the human immunodeficiency virus (HIV) produce new viral particles. Unlike the budding in the secretory pathway described earlier, HIV particles bud away from the cytoplasm, into the extracellular space. This viral budding occurs in the same orientation as the budding that occurs within endosomes. The proteins which enable this budding are referred to as endosomal sorting complexes required for transport (ESCRTs). HIV 'hijacks' the ESCRT machinery to enable it to bud from the plasma membrane, out of the cytoplasm and into the extracellular space. Interactions between HIV proteins and ESCRT proteins recruit the host cell ESCRT machinery to the budding vesicle, allowing membrane scission and vesicle release. Other viruses can bud without assistance from the ESCRTs, and it is thought that HIV may also be able to bud in an ESCRT-independent manner. Understanding more about these membrane budding and scission events is crucial to elucidating how viruses proliferate and how we can inhibit processes by means of drug interventions.

Protein Trafficking in the Secretory Pathway

As described earlier, a hydrophobic stretch of 20–30 amino acids with a basic N-terminus and a polar region at the C-terminus emerging from the ribosome causes the protein to be targeted to the ER, where synthesis is completed. This hydrophobic stretch can be cleaved in the case of soluble proteins, or it can remain attached. An uncleaved signal sequence is referred to as a signal anchor sequence, as it both signals ER targeting and then goes on to anchor a protein in the membrane, and becomes a transmembrane domain in the fully folded protein. The SRPdependent targeting step is common to ER proteins as well as proteins destined for the Golgi or the plasma membrane, or to be secreted from the cell. ER exit is thought to allow some selection of which proteins remain in the ER and which proteins leave and move in vesicles towards the Golgi. ER exit sites are located in areas of the ER close to the Golgi, and are rich in COPII coat proteins. It is not understood exactly which properties of a protein determine whether it will leave the ER in COPII vesicles, but it is currently thought that the transmembrane domain length is an important factor. Longer transmembrane domains appear to predispose proteins to exit the ER and travel towards the Golgi. This is consistent with the fact that membrane thickness increases through the secretory pathway due to increased cholesterol content, as described earlier.

Upon arrival at the cis-Golgi, proteins can then be retrieved to the ER, remain in the Golgi, or travel onward to the plasma membrane. Retrieval to the ER is not fully understood, but some proteins contain retrieval motifs, such as the KDEL four-amino-acid motif which is recognized by a receptor and enables packaging of the protein into retrograde COPI vesicles. Other proteins appear to cycle between the ER and the Golgi without known retrieval motifs. Proteins move in both anterograde and retrograde directions through the Golgi stack. They can then leave the trans-Golgi and move to the plasma membrane in vesicles. Proteins at the plasma membrane can move into the cell in vesicles by endocytosis (e.g. when surface receptors are internalized for degradation in lysosomes). Endocytic vesicles are often clathrin coated. Clathrin, like the COPs, distorts the membrane into curved structures, allowing vesicle formation. Clathrin forms a cage-like shape that promotes vesicle formation and scission by virtue of the rigid shape of the protein complexes which form at the membrane. Not all endocytosis is clathrin dependent, and there are other proteins, such as caveolin, which can facilitate the formation of endocytic vesicles

Mitochondrial and Nuclear Protein Targeting

Newly synthesized proteins destined for the mitochondria or the nucleus are targeted in a different way, independently of the secretory pathway. Some mitochondrial proteins are encoded by the mitochondrial genome, while others are encoded by the nuclear genome. Mitochondria have a double-layered membrane. Therefore targeting signals for mitochondrial proteins need to contain information not only to direct the protein to the organelle, but also to determine in which membrane it will be located (in the case of membrane proteins), or whether it will be located inside the mitochondria (the matrix) or in the intermembrane space between the inner and outer membranes (in the case of soluble proteins). Mitochondrial targeting motifs vary enormously, but generally are located at the N-terminus of the protein and are rich in positively charged and hydrophobic amino acids. Proteins destined for the nucleus are targeted by nuclear localization sequences that direct proteins which have been synthesized in the cytoplasm through nuclear pore complexes. Again these sequences are not very highly conserved, but generally contain clusters of positively charged amino acids.

SENDING MESSAGES ACROSS MEMBRANES

We have already seen how ion channels and other transport proteins can allow substances to cross the lipid bilayer. Knowledge of how fat-soluble and water-soluble substances cross membranes is important for gaining an understanding of how messages cross membranes and thus how one cell can communicate with another. Cells receive and send messages constantly (e.g. in order to respond to hormone signals, conduct action potentials, and sense external stimuli such as taste and smell).

Messengers: Lipid Soluble or Water Soluble?

Substances that send messages are known as messengers, and they vary enormously in their chemical composition, size and hydrophobicity. In order to understand how a cell receives a message, it is important to ascertain first whether the messenger is lipid or water soluble. Hormones are one example of messengers that are released by cells. The human body contains both lipid-soluble and water-soluble hormones. Lipid-soluble hormones are generally transported through the blood, bound to carrier proteins. Steroid hormones such as the androgens and oestrogens are lipid soluble by virtue of their ringed molecular structures, which are derived from cholesterol. This allows these hormones to diffuse freely through the plasma membrane of cells and bind to their receptors, which are located inside cells. In the case of oestrogen, the receptor is located in the cytoplasm and upon ligand binding relocates to the nucleus, where it binds DNA and acts as a transcription factor, altering gene expression. The receptor contains a nuclear localization sequence which is hidden until oestrogen binds, allowing it to be targeted to the nucleus. Other hormones, such as insulin and adrenaline, are water soluble and therefore cannot pass freely through the membranes of cells. Their receptors are located on the outside of the plasma membrane in order for them to be able to convey a message without entering cells. Insulin binds to the membrane-spanning insulin receptor on the surface of target cells, and initiates a signal cascade that results in an increase in the number of glucose transporters at the cell membrane, and a subsequent increase in glucose uptake.

G Proteins and Second Messengers

Many cell-surface receptors share structural features, including seven membrane-spanning helices. These 7TM receptors bind their ligand (the messenger molecule) on the extracellular side of the membrane, and bind a GTP-binding protein (G protein) on the intracellular side. Due to this interaction with G proteins, these receptors are called G-protein-coupled receptors(GPCRs). When the ligand binds the GPCR, the receptor undergoes conformational changes that are transferred through the membrane-spanning region to the bound G protein. This change in structure allows the G protein to exchange a bound GDP molecule for a GTP molecule, and thereby switch from an inactive state to an active state. G proteins consist of three subunits—α, β and γ. An inactive, GDP-bound G protein consists of all three subunits, with the nucleotide bound in the α subunit. When the GPCR binds the ligand, the G protein is activated and the α subunit, now with GTP bound to it, dissociates from the complex (Figure 12). This activated α subunit now has an exposed face (where the β and γ subunits were bound) and can bind proteins to propagate the signal. An example of this downstream signalling from GPCRs is the activation of adenylate cyclase by the GTP-bound α subunit in the case of the β-adrenergic receptor when it binds its ligand, adrenaline (epinephrine). The effect of this adenylate cyclase activation is an increase in cAMP production from ATP, leading to downstream effects. The dissociated βγ dimer also has downstream effects. The α subunit has GTPase activity so that it can convert the bound GTP back to GDP. The GDP-bound subunit then returns to and binds the β and γ subunits ready for another cycle of signalling.

After the initial ligand interaction with the GPCR and the G-protein dissociation, the message is then carried by second messengers activated by the signal cascade. In the example

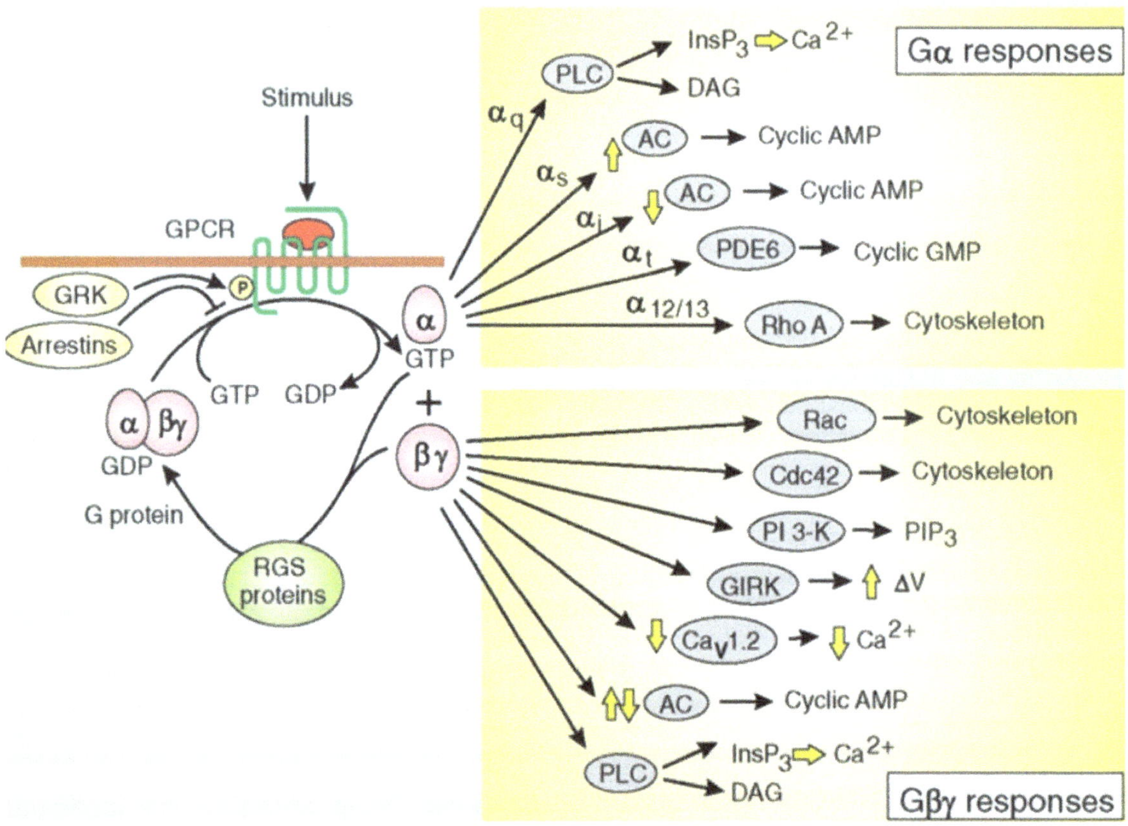

Figure 12: GPCR and heterotrimeric G-protein signalling. The ligand bound to the GPCR is shown in red. Binding allows the exchange of GDP for GTP by the associated G protein, and dissociation of the protein into Gα and Gβγ subunits. These then have downstream effects on a range of proteins, thereby propagating the signal from the bound ligand. Yellow arrows indicate either activation (up arrow) or inhibition (down arrow) of the targets. Regulators of G-protein signalling (RGS) proteins aid the GTPase activity of the G protein to turn off the signal. Arrestin can bind the receptor following GPCR phosphorylation by G-protein receptor kinase (GRK), desensitizing the receptor to further signalling. Reproduced from Berridge, M.J. (2012) Cell Signalling Biology; doi:10.1042/csb0001002, with permission.

that has just been given, the G protein associated with the β-adrenergic receptor activates adenylate cyclase, increasing the production of cAMP, which is a widely used second messenger. Most of the effects of cAMP are due to the activation of protein kinase A (PKA). PKA phosphorylates target enzymes to modify their activities. In the case of adrenaline, PKA activates enzymes involved in the production of glucose from glycogen stores, and inhibits enzymes involved in the production of more glycogen.

Around 25% of drugs are targeted at GPCRs, so an understanding of their structures and functions is crucial in the fight against disease. As explained earlier, membrane proteins are notoriously difficult to crystallize due to their hydrophobic nature, and GPCRs have a very small hydrophilic area. Some techniques, such as the production of an antibody–receptor complex to increase hydrophilicity, have been successful in aiding crystallization. Rhodopsin (Figure 13) was crystallized in 2000, followed by the related β2-adrenergic receptor in 2007. Since then, several more GPCR structures have been solved, providing valuable information that can help computational biologists to work out the detailed mechanisms of GPCR signalling.

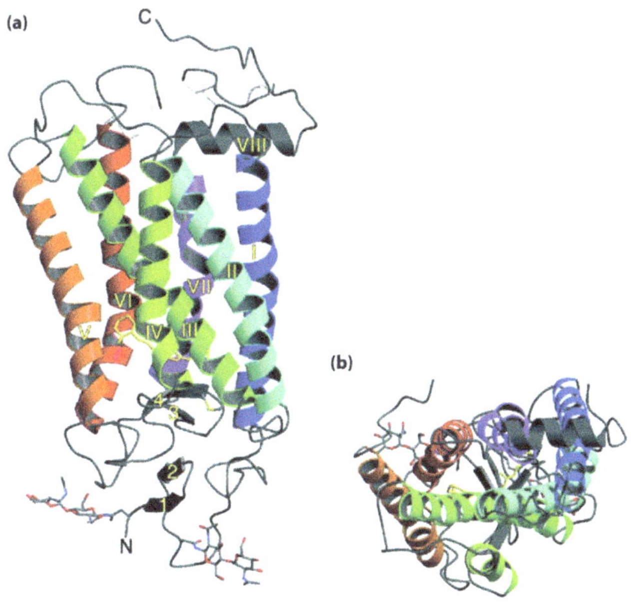

Figure 13: The crystal structure of rhodopsin. A ribbon representation of the first crystal structure of rhodopsin is shown in the plane of the membrane (a) and from the cytoplasmic side (b). The N- and C-termini are labelled, as are the seven transmembrane helices (I-VII). Adapted from Figure 2 from Palczewski, K., Kumasaka, T., Hori, T., Behnke, C.A., Motoshima, H., Fox, B.A., Le Trong, I., Teller, D.C., Okada, T., Stenkamp, R.E. et al. (2000) Crystal structure of rhodopsin: a G-protein-coupled receptor. Science 289, 739–745.

several more GPCR structures have been solved, providing valuable information that can help computational biologists to work out the detailed mechanisms of GPCR signalling. Molecular dynamics simulations have been performed on the interactions between GPCRs and their partner G proteins using the crystal structures available to inform the modelling process. These studies will play an important part in helping us to understand how the helices in the GPCRs move and twist in order to convey the extracellular signal to the intracellular G protein.

Nerve Impulses

Nerve impulses are able to occur because biological membranes are impermeable to ions, so a membrane potential can be generated across them, with more of one charge on one side than on the other. These membrane potentials are generated and altered by ion channels. A nerve impulse, or action potential, is generated when a membrane is depolarized upon influx of positively charged ions into the cell. The resting potential in a neuron is around –70 mV, maintained by K+ channels and the Na+/K+-ATPase. When an action potential is generated, voltage-dependent Na+ channels open once the cell membrane has crossed the threshold potential of around –60 mV. This allows a fast influx of Na+ down its electrochemical gradient, increasing the membrane potential (i.e. reducing its negative value). This influx of positive charges enables the inside of the cell to become positively charged compared with the extracellular environment, as the membrane potential exceeds 0 mV. The depolarization itself inhibits the Na+ channels, so no more ions enter the cell. To restore the negative resting potential, voltage-dependent K+ channels open, allowing K+ ions to move out of the cell, thus making the inside of the cell more negative. An after-potential (hyperpolarization) can then occur, whereby the membrane potential decreases below –70 mV before being restored by the action of ion channels and ATPases.

MEMBRANES IN HEALTH AND DISEASE

We have seen how membranes, and the membrane proteins within them, function in healthy cells and organisms. We shall now consider what happens in disease, and how we can use our knowledge of membrane proteins to make new drugs to treat disease.

Serious Disease Results from non-Functional ion Channels

Cystic fibrosis is an autosomal recessive disease that results from mutations in the CFTR gene. This gene encodes a Cl– channel that has a vital role in regulating the viscosity of mucus on membranes such as those in the lungs. In healthy individuals, transport of Cl– ions out of the cells through CFTR is followed by water, and mucus of the right viscosity is produced. However, lack of Cl– channels results in thick, dehydrated mucus, and consequently cystic fibrosis patients have difficulty in breathing and a predisposition to chest infections. Most cases of cystic fibrosis are caused by the ΔF508 mutation, which is a deletion of a phenylalanine residue at position 508 in the protein. Like nearly all membrane proteins, CFTR is translated on ribosomes at the ER and then moves through the secretory pathway to the plasma membrane, where it carries out its transport role. The single amino acid deletion of F508 causes the protein to misfold, and instead of moving out to the plasma membrane, it is held in the ER by the protein quality control machinery. Therefore very few CFTR molecules reach the plasma membrane in people with the ΔF508 mutation, and this results in serious disease.

Diseases such as this are not easily treated. Blocking the protein quality control machinery is not an option, because it would lead to the release of other misfolded proteins, with potentially disastrous consequences. Although some current drug treatments can ameliorate the symptoms of the disease, it is hoped that gene therapy might become routine as it addresses the cause of the problem. Treatment of patients with an artificial, functional version of the gene enables them to produce a working CFTR protein that can be expressed at the plasma membrane. Although this is not a complete cure, it is a potentially effective way to greatly reduce the symptoms of cystic fibrosis in the lungs. As DNA is a large hydrophilic molecule, it cannot be simply administered like many other drugs. Delivery of gene therapy is a challenge, and this is one reason why it is difficult to treat patients in this way, but methods of delivering new genetic material into cells have been developed. Viruses can be used to deliver the CFTR gene to cells, by harnessing their ability to inject cells with foreign DNA or RNA. Patients may also be able to be given liposomes containing the functional gene, which fuse with cell membranes and deliver the therapeutic gene. Gene therapy is a growing and important area of research, and it is hoped that many diseases, including some cancers, will eventually be able to be treated using DNA.

Membrane Proteins Provide an Entry Point for Viruses

Viruses that attack the human body can use the body's own membrane proteins to recognize their target cells. HIV attacks cells of the immune system. A protein on the surface of HIV called gp120 binds to CD4 protein molecules on the surface of T-cells that are involved in immunoregulation, and allows fusion of the virus with the host cell. Once the contents of the virus have entered the CD4-positive cell, the HIV genome is integrated with the host genome and uses the host machinery to make new copies of the virus. Over time, the numbers of CD4 T-cells are reduced by the virus, and the patient's immune system eventually becomes so compromised that they are unable to fight invading pathogens. Many therapeutic agents have been created to help to fight HIV, and the interaction between CD4 and gp120 is just one of the points at which drugs can be used to stop the progression of the virus.

Toxins use Endocytosis to Gain entry to Cells and Block Neurotransmission

Various toxins interfere with the transmission of messages across biological membranes. Tetanus neurotoxin (TeNT) and botulinum neurotoxin (BoNT) are both protein toxins that affect nerve impulse transmission between nerves and muscles. TeNT is produced by a soil bacterium and causes the skeletal muscle spasms that characterize tetanus infection. TeNTproducing bacteria generally enter the body through wounds, and TeNT binds glycolipids enriched at presynaptic membranes of motor neurons (Figure 14). TeNT then undergoes endocytosis and moves up the axon to the dendrites that connect the motor neuron to an inhibitory interneuron. TeNT is released into the synapse between these two cells and is endocytosed into the inhibitory interneuron. Acidification of vesicles containing TeNT causes the protein toxin to break apart into two domains. One of these, the L domain, is translocated into the cytoplasm of the interneuron, where it uses its proteolytic activity to cleave vesicle-associated membrane protein (VAMP). Under normal circumstances, VAMP is part of the protein complex that allows synaptic vesicles to fuse with the presynaptic membrane and release inhibitory neurotransmitters. The action of the L-domain protease of TeNT means that VAMP can no longer function, inhibiting neurotransmitter release across the synapse. As this occurs in inhibitory interneurons, the resulting effect is prolonged skeletal muscle contraction, as no inhibition is conveyed to the motor neuron to allow relaxation.

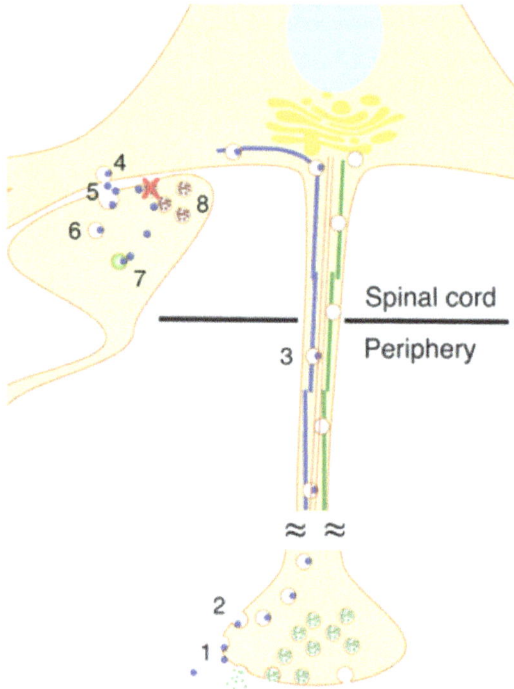

Figure 14: Mechanism of action of tetanus toxin. Tetanus toxin (blue circles) enters the presynaptic membranes of motor neurons by endocytosis, and moves up the axon to the dendrites that connect the motor neuron to an inhibitory interneuron. Microtubules (blue and green lines) and actin filaments (red lines) allow retrograde transport of the toxin. TeNT acts on the inhibitory interneuron, where it prevents the release of glycine (red dots), shown by a red cross. Small green dots represent neurotransmitter both inside and being released from vesicles. Adapted from Figure 2 from Rossetto, O., Scorzeto, M., Megighian, A. and Montecucco, C. (2013) Tetanus neurotoxin. Toxicon 66, 59–63.

Botulinum neurotoxin acts in a similar way to TeNT, but has the opposite effect. Like TeNT, it is released by bacteria. BoNT binds to and is internalized by the presynaptic membrane of motor neurons at the neuromuscular junction. It is released from endocytic vesicles into the cytoplasm of the motor neuron, where it acts on the SNARE complex to inhibit the fusion of synaptic vesicles and release of excitatory neurotransmitters at the neuromuscular junction. This has the effect of blocking muscle contraction and causing paralysis in people infected with botulism. Despite its sometimes lethally toxic nature, BoNT is increasingly used by people who wish to look younger. The toxin is injected into the muscles of the face to cause paralysis, thereby reducing wrinkles and lines in the skin. When used in this way it is referred to as Botox.

Membrane Proteins are the Target of Many Drugs

Membrane proteins are important drug targets. As our structural and functional knowledge of membrane proteins expands, it is becoming possible to design more effective medicines. Computational tools are becoming an increasingly important part of the process. One important class of drug targets are pore-forming membrane proteins encoded by viruses. HIV, influenza and polio, among other viruses, encode membrane proteins that form pores in the host cell membranes in order to cause leakage and promote infection. One of these pore-forming proteins was formerly used as a drug target in the treatment of influenza. NMR studies have provided structural information about the pore-forming Vpu protein from HIV-1. Using these data together with structural information about pores with similar sequences, computational models of the structure of the channel in the host membrane can be produced. These models, combined with advanced biophysical techniques, are invaluable for predicting sites for potential drug molecule binding which can then be tested both computationally and experimentally

More drugs target GPCRs than any other single group of proteins. As explained earlier, conformational changes in the GPCRs permit signals to cross the membrane. However, these large changes in conformation give the proteins flexibility, which makes it difficult for researchers to pinpoint the structures of the GPCRs in any one conformation. Solving the structures of different conformations using X-ray crystallography is a challenging task. As GPCRs are such important drug targets, much research has been focused on solving their structures in order to inform the discovery of new drugs. Computational methods have again proved crucial to understanding the detailed molecular structure of these proteins. Molecular dynamics simulations have been used to aid our understanding of the molecular changes that occur during GPCR activation, and also which lipids are required for the GPCR to function. Although molecular dynamics simulations are a key technique in this area of research, they have some limitations. The biological membranes that surround cells are extremely complex and contain different types of lipids and proteins, both within and associated with the membrane. At present, the time and financial resources required to provide the computational power to simulate such a complex environment are often prohibitive. Simpler models are therefore produced which, although they are able to predict conformational changes in receptors such as GPCRs, may omit other interactions that are important in the activity of membrane proteins. As is often the case, a combination of techniques will be required to gain greater insight into the conformational flexibility of the GPCRs. It is apparent from research conducted to date that these receptors can adopt many different conformations, rather than having just a simple 'on' and 'off' mechanism. Using different drug molecules to stabilize different conformations in different signalling pathways may be the best approach to finding more effective medicines in the future. There is now much pressure on researchers to replace, refine and reduce the use of animals in drug discovery (an approach referred to as the 'three Rs'). By using computers in the early stages of the process to model drug–target interactions, researchers can produce much more promising compounds to test in experiments and drug trials.

Closing Remarks

Biological membranes allow life to exist. From simple unicellular prokaryotes to complex multicellular eukaryotes such as humans, the properties of the membranes that surround cells are remarkably similar. Our understanding of the structure of these lipid bilayers is now expanding rapidly as a result of significant advances in biophysical techniques and the huge computational power now available to researchers. The proteins that inhabit these membranes allow messages to be sent and received so that the cell can communicate with the external environment. Many messages are relayed by hydrophilic molecules that require receptors to transmit information across the bilayer. It is this step that is targeted by the majority of drugs which are on the market today, as it enables us to modify the message before it enters the cell. An understanding of how membrane proteins work, how they reach the correct destinations and how we can alter their functions is key to the fight against human disease.

REFERENCES

1. Berman, H.M., Coimbatore Narayanan, B., Di Costanzo, L., Dutta, S., Ghosh, S., Hudson, B.P., Lawson, C.L., Peisach, E., Prlic´ , A, Rose, P.W. et al. (2013) Trendspotting in the Protein Data Bank. FEBS Lett. 587, 1036–1045.
2. Blobel, G. and Dobberstein, B. (1975) Transfer of proteins across membranes. 1. Presence of proteolytically processed and unprocessed nascent immunoglobulin light chains on membrane-bound ribosomes of murine myeloma. J. Cell Biol. 67, 835–851.
3. Blobel, G. and Sabatini, D.D. (1971) Ribosome–membrane interaction in eukaryotic cells. Biomembranes 2, 193–195.
4. Bonifacino, J.S. and Glick, B.S. (2004) The mechanisms of vesicle budding and fusion. Cell 116, 153–166.
5. Butler, J., Watson, H.R., Lee, A.G., Schuppe, H.J. and East, J.M. (2011) Retrieval from the ER-Golgi intermediate compartment is key to the targeting of C-terminally anchored ER-resident proteins. J. Cell Biochem. 112, 3543–3548.
6. Caffrey, M. (2003) Membrane protein crystallization. J. Struct. Biol. 142, 108–132. Devaux, P.F., López-Montero, I. and Bryde, S. (2006) Proteins involved in lipid translocation in eukaryotic cells. Chem. Phys. Lipids 141, 119–132.
7. Fischer, W.B. and Hsu, H.-J. (2011) Viral channel forming proteins — modeling the target. Biochim. Biophys. Acta 1808, 561–571.
8. Gorter, E. and Grendel, F. (1925) On bimolecular layers of lipoids on the chromocytes of the blood. J. Exp. Med. 41, 439–443.
9. Grossfield, A. (2011) Recent progress in the study of G protein-coupled receptors with molecular dynamics computer simulations. Biochim. Biophys. Acta 1808, 1868–1878.
10. Hurley, J.H., Boura, E., Carlson, L.-A. and Róz˙ycki, B. (2010) Membrane budding. Cell 143, 875–887.
11. Hurst, R., Rollema, H. and Bertrand, D. (2013) Nicotinic acetylcholine receptors: from basic science to therapeutics. Pharmacol. Ther. 137, 22–54.
12. Inesi, G., Prasad, A.M. and Pilankatta, R. (2008) The Ca2+ ATPase of cardiac sarcoplasmic reticulum: physiological role and relevance to diseases. Biochem.

Biophys. Res. Commun. 369, 182–187.

13. Kendrew, J.C., Bodo, G., Dintzis, H.M., Parrish, R.G., Wyckoff, H. and Phillips, D.C. (1958) A threedimensional model of the myoglobin molecule obtained by X-ray analysis. Nature 181, 662–666.
14. Lange, Y. (1991) Disposition of intracellular cholesterol in human fibroblasts. J. Lipid Res. 32, 329–339.
15. Lee, A.G. (2004) How lipids affect the activities of integral membrane proteins. Biochim. Biophys. Acta 1666, 62–87.
16. Lippincott-Schwartz, J., Snapp, E. and Kenworthy, A. (2001) Studying protein dynamics in living cells. Nat. Rev. Mol. Cell. Biol. 2, 444–456.
17. Lundmark, R. and Carlsson, S.R. (2010) Driving membrane curvature in clathrin-dependent and clathrin-independent endocytosis. Semin. Cell Dev. Biol. 21, 363–370.
18. Lynes, E.M. and Simmen, T. (2011) Urban planning of the endoplasmic reticulum (ER): how diverse mechanisms segregate the many functions of the ER. Biochim. Biophys. Acta 1813, 1893–1905.
19. Palczewski, K., Kumasaka, T., Hori, T., Behnke, C.A., Motoshima, H., Fox, B.A., Le Trong, I., Teller, D.C., Okada, T., Stenkamp, R.E. et al. (2000) Crystal structure of rhodopsin: a G proteincoupled receptor. Science 289, 739–745.
20. Risselada, H.J. and Grubmüller, H. (2012) How SNARE molecules mediate membrane fusion: recent insights from molecular simulations. Curr. Opin. Struct. Biol. 22, 187–196.
21. Rossetto, O., Scorzeto, M., Megighian, A. and Montecucco, C. (2013) Tetanus neurotoxin. Toxicon 66, 59–63.
22. Singer, S.J. and Nicolson, G.L. (1972) The fluid mosaic model of the structure of cell membranes. Science 175, 720–731.
23. Smith, A.W. (2012) Lipid–protein interactions in biological membranes: a dynamic perspective. Biochim. Biophys. Acta 1818, 172–177.
24. Tate, C.G. (2012) A crystal clear solution for determining G-protein-coupled receptor structures. Trends Biochem. Sci. 37, 343–352.
25. Tighe, A.P. and Schiavo, G. (2013) Botulinum neurotoxins: mechanism of action. Toxicon 67, 87–93.
26. Vaidehi, N. (2010) Dynamics and flexibility of G-protein-coupled receptor conformations and their relevance to drug design. Drug Discov. Today 15, 951–957.
27. van Meer, G. and de Kroon, A.I. (2011) Lipid map of the mammalian cell. J. Cell Sci. 124, 5–8.
28. Watson, H.R., Butler, J., Schuppe, H.J., Lee, A.G. and East, J.M. (2011) The localization of the ER retrieval sequence for the calcium pump SERCA1. Mol. Membr. Biol. 28, 216–226.
29. White, S.H. (2009) Biophysical dissection of membrane proteins. Nature 459, 344–346.
30. Wimley, W.C. (2003) The versatile β-barrel membrane protein. Curr. Opin. Struct. Biol. 13, 404–411.

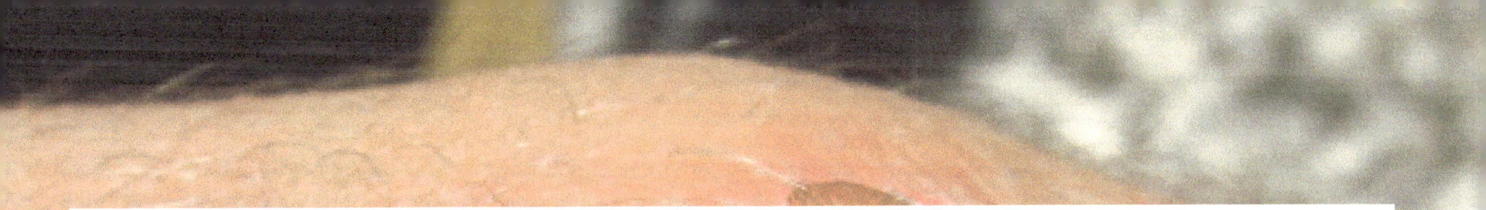

CHAPTER 3

ORIENTED CELL DIVISION: NEW ROLES IN GUIDING SKIN WOUND REPAIR AND REGENERATION

Shaowei Yang[1,2], Kui Ma[2], Zhijun Geng[2], Xiaoyan Sun[1],and Xiaobing Fu[1,2]

[1]Wound Healing and Cell Biology Laboratory, Institute of Basic Medical Science, Trauma Center of Postgraduate Medical School, Chinese PLA General Hospital, 28 Fu Xing Road, Beijing 100853, P. R. China

[2]Key Research Laboratory of Tissue Repair and Regeneration of PLA, and Beijing Key Research Laboratory of Skin Injury, Repair and Regeneration, First Affiliated Hospital to the Chinese PLA General Hospital, 51 Fu Cheng Road, Beijing 100048, P.R. China

ABSTRACT

Tissue morphogenesis depends on precise regulation and timely co-ordination of cell division and also on the control of the direction of cell division. Establishment of polarity division axis, correct alignment of the mitotic spindle, segregation of fate determinants equally or unequally between daughter cells, are essential for the realization of oriented cell division. Furthermore, oriented cell division is regulated by intrinsic cues, extrinsic cues and other cues, such as cell geometry and polarity. However, dysregulation of cell division orientation could lead to abnormal tissue development and function. In the present study, we review recent studies on the molecular mechanism of cell division orientation and explain their new roles in skin repair and regeneration.

INTRODUCTION

Tissues develop in a precisely controlled course to produce adult structures with the correct size and shape. Much progress has been made in understanding the molecular pathways that control tissue size [1–3]. However, much less is known about the mechanisms that control tissue shape. Recent evidence indicates that the orientation of the cell division plane plays an important role in co-ordinating division rate with cell fate choices and cell position, which further specifies the repertoire of cell types, the organizational structure and shape of a tissue or organ.

Oriented cell division mainly depends on the location of the mitotic spindle. Through most of the mitosis cycle, the spindle apparatus has no fixed structure and is constantly changing. However, during metaphase, the mitotic spindle forms two spindle poles constructed with three types of microtubules: kinetochore microtubules which attach to chromosomes; astral microtubules that radiate out from spindle poles and anchor the mitotic spindle to the cytoplasm and cell cortex; and interpolar microtubules that form an antiparallel array between the spindle poles [4,5]. Proper location and direction of the mitotic spindle apparatus requires pulling forces acting on astral microtubules [6,7]. Although the regulation of microtubule dynamic instability is necessary for correct spindle positioning, the alignment of the spindle relies on the interactions of astral microtubules with the cell cortex and cytoplasmic anchor sites

Citation: Yang, S., Ma, K., Geng, Z., Sun, X., & Fu, X. (2015). Oriented cell division: new roles in guiding skin wound repair and regeneration. Bioscience Reports, 35(6), e00280. http://doi.org/10.1042/BSR20150225

[8]. Cell–cell contacts and cell adhesion to their extracellular matrix (ECM) is also a key determinant of mitotic spindle orientation. Meanwhile, tissue morphogenesis is determined by the precise balance between asymmetric and symmetric cell divisions. Asymmetric cell divisions generate cellular diversity, whereas symmetric cell divisions expand the pool of similar cells [9]. Spindle orientation couples intrinsic cues of polarity, but also locates relative to cell-extrinsic cues such as niche-derived signals, which affects the choice of symmetric or asymmetric cell fates. Thus, dysregulation of oriented cell division could result in abnormal development and function of the tissue and cell fate mis-specification.

In the present review, we first present an overview of oriented cell division process in different developmental contexts. Then, we summarize a panel of proteins and molecular signals involved in orienting cell division. We speculate that facultative regulation of symmetric or asymmetric divisions by targeting the molecules responsible for oriented cell division, focusing specifically on epithelial polarization, may provide a new alternative to improve adult regenerative capacity in the skin during wound healing.

ORIENTED CELL DIVISION IN THE DEVELOPMENT OF ORGANISMS

All metazoans are faced with two basic developmental challenges: generating cellular diversity and shaping tissue morphology. Oriented cell division regulates embryogenesis, organogenesis and cellular differentiation and functions from a single cell zygote to the adult tissues and organs.

Shaping Tissues and Organs

Oriented cell divisions could drive morphogenesis by shaping tissues. These include the elongation of tissues, branching morphogenesis and maintenance of single lumens within epithelial tubes. Oriented cell division during growth and development has been studied in great detail and to great effect in *Drosophila*.

During early *Drosophila melanogaster* embryogenesis, the germ band extends and elongates along the anterior–posterior axis of the embryo after gastrulation. This process requires both cell intercalations and oriented cell divisions. In relation to the long axis of the extending tissue, cell division is executed preferentially along the anterior–posterior axis. Tissue elongation is not completely prevented by blocking cell divisions, but the amount of extension is reduced. Moreover, randomized spindle orientation causes an isotropic increase in tissue size in mutant embryos on account of lacking segmentation pattern [10]. Oriented cell division is involved in germ band extension. However, it is difficult to specify its individual role in the process because of the overlap between cell division and cell intercalations. Later in development, measurement of spindle orientation in the wing blade and in the eye disc further verifies that cell division plays a greater role than cell relocation in defining clonal shape [11,12].

Another example of oriented cell divisions in morphogenesis occurs in zebrafish. During gastrulation, the majority of cell divisions are oriented along the animal–vegetal axis in the dorsal region of the midline and later in the ventral region of the epiblast surface layer [13–15]. Similarly, during neurulation, most of neural plate cells undergo midline-crossing divisions [16]. The mitotic spindle rotates 90 degrees in the oriented cell division across the midline [17]. Cell division is essential for these midline crossing events, as blocking cell division prevents most cells from crossing the midline [18,19]. But proper orientation of these divisions is required, as perturbation of oriented cell divisions in the neuroepithelium results in severe defects in the neural rod midline [13,20]. Even unicellular organisms, such as yeast, exploit spindle orientation to grow as hyphae [21,22].

Generating Cellular Diversity

In addition to shaping tissues and organs, oriented cell division can generate cellular diversity, which mainly involves asymmetric cell division [23]. This is achieved by unequal partitioning of cell fate determinants and into asymmetrical daughter cells, which is required for the proper alignment of the mitotic spindle corresponding to an internal or external polarity axis. Studies performed in model organisms such as the early*Caenorhabditis elegans* embryo [24–26], *Drosophila* neuroblasts [23] and *Drosophila* sensory organ precursor (SOP) cells [27], have made great contributions in understanding asymmetric cell division.

Known as neural stem cell-like cells, *Drosophila* neuroblasts orient their mitotic spindle along an established axis of internal polarity. Segregation of the cell fate determinants asymmetrically into two daughter cells, depends upon the stereotypical spindle orientation. One is a small differentiating ganglion mother cell (GMC), the other is an apical neuroblast retaining self-renewed [10,28]. Cell dissociation experiments in *C. elegans* show that embryonic neuroblasts related neuroepithelial cells divide along a stabilized division axis over successive rounds of divisions, whereas unrelated neuroblasts divide along random division axes, which reveals that unknown extrinsic factors are essential for maintaining correct neuroblasts division orientation in the fly embryo, other than neuroblasts intrinsic polarity cues [29].

Likewise, soon after fertilization, the proper orientation of the mitotic spindle is required to the polarized early*C. elegans* embryo firstly. Spindle orientation and displacement then proceeds in two stages: first, the nuclear–centrosome complex moves to the centre of the cell and rotates 90° during prophase; second, the spindle is pulled to the posterior of cell during metaphase and anaphase. Finishing the process requires not only the interactions between the mitotic spindle and the cortex, but also the intrinsic polarity cues [30,31]. The proper segregation of cell fate determinants is also essential, which needs the appropriate spindle orientation [28]. These studies illustrate that oriented cell division could control the organ development as a ubiquitous morphogenetic manner in a variety of species.

MECHANISMS OF SPINDLE ORIENTATION IN CELL DIVISION

Since oriented cell division is critical to the development and growth of the organism and the division axis is a basic regulator in cell division, what are the cellular and molecular determinants modulating correct cell division orientation and how do these determinants differ among the different cell types and species? In metazoan development, oriented cell division depends upon the mitotic spindle orientation. However, spindle orientation can be controlled by intrinsic and external cues. Intrinsic cues rely on the partitioning of cell components that determine cell fate. External cues involve the placement of daughter cells associated with external cues. Furthermore, physical constraints, such as cellular environment and cell geometry, may influence cell division orientation [32,33].

Extrinsic Cues

Drosophila SOP cells use extrinsic signals to establish their anterior–posterior axis of division (Figures 1A and and1B).1B). The core planar cell polarity (PCP) pathway is used to define this axis within the epithelium plane. The seven-pass transmembrane protein receptor frizzled (Fz) and its cortical effector dishevelled (Dsh) localize to the posterior cortex of the SOP cell. This localization restricts Fz signalling to the posterior of the cell, which ultimately translates into planar polarization. At the same time, another PCP protein, Strabismus (Stbm), localizes to the anterior cortex and recruits partner of Insc (inscuteable; Pins), where they both work to restrict partitioning defective protein (Par) 3/6 to the posterior cortex [34,35]. Like the *C. elegans* zygote, because of lacking Insc expression in SOP cells, Pins and the Par proteins localize to opposite sides of the cell. Interestingly, spindle orientation along the anterior–posterior axis does not require Pins or the Par complex. Instead, both Fz and Dsh have been shown to be required [34,36]. This is possibly due to the binding of Dsh to mushroom body defect (Mud), linking extrinsic signalling through the Fz pathway to regulation of the mitotic spindle [37]. Oriented cell divisions during zebrafish gastrulation have been shown to use this pathway as well [14].

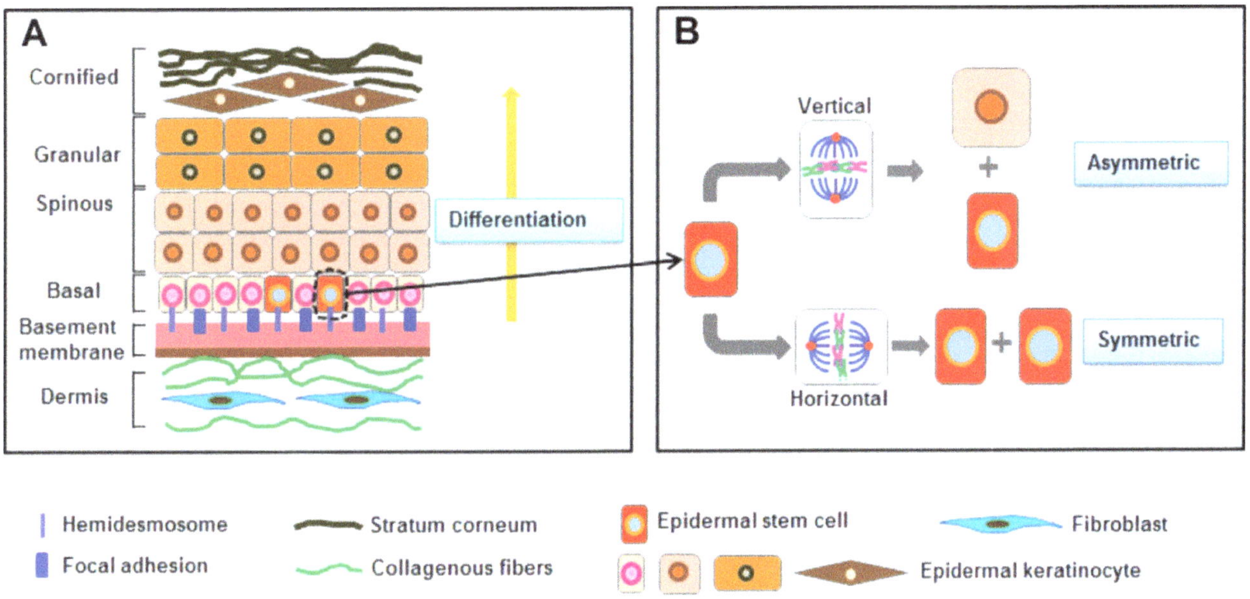

Figure 1: Extrinsic and intrinsic polarity cues in oriented cell division.

During zebrafish gastrulation, epiblast cells generate the neural ectoderm on the dorsal side of the epidermis [14,38,39]. Inhibition of Dsh could lead to defective polarity of elongation and randomization of cell division orientation in all epiblast layers [17]. Dsh participates in the canonical Wnt/β-catenin pathway and PCP pathway. Inhibiting classical Wnt signalling did not affect oriented cell division. Oriented cell division is not influenced by inhibiting classical Wnt signalling, but by blocking the two PCP ligands Wnt11 or Wnt5 [17]. Furthermore, lacking Fz7 (receptor for Wnt11) [40] and Stbm show randomized spindle orientation in zebrafish [13,17].

Addition to relying on Fz and Dsh, Fat and Daschsous can also regulate the PCP pathway. For example, in the *Drosophila* wing, proper oriented cell divisions require Fat and Daschsous [11]. A gradient of Daschsous is expressed along the proximal-distal axis of the wing [41]. Daschsous can bind to Fat and transduce the signal to atypical myosin Dachs. In response to the signal, Dachs becomes localized distally [42]. Mutant clones of Fat, Daschsous or Dachs show defects in spindle orientation [11]. Additionally, Daschsous is sufficient to induce oriented cell divisions in the wing, because the direction of cell division alters as the direction of Daschsous› gradient alters [12]. These studies demonstrate that PCP and Ft/Dsh signalling system conducting extrinsic cues plays part in oriented cell division.

Intrinsic Cues

Cell intrinsic polarity mainly functions in asymmetric cell division. Par/Insc pathway plays a crucial role in intrinsic polarity of cell division orientation.

Drosophila neuroblasts have the ability to orient their mitotic spindle even in absence of extrinsic cues. When cultured and removed from their normal environment, mitotic neuroblasts were still able to localize the key components required for spindle orientation. Furthermore, the resulting asymmetric division produces one neuroblast and one GMC [43]. Other than extrinsic cues, the intrinsic polarization of the neuroblasts is required to refine the process. Only when in contact with epithelial cells are the neuroblasts able to maintain proper positioning of the centrosome and polarize the Par complex at prophase [29]. The initial polarity of these cells comes from the polarized neuroectoderm from which the neuroblasts are derived. In the neuroectoderm, the polarity proteins Par3/6 are localized apically just above the adhesion junction. Neuroblasts then delaminate from the epithelium and inherit their polarity cues in this process. Par-3/6 in the stalk can extend into the epithelium as delamination occurs [44]. As a result, a specified neuroblast cell is formed with a clear apical–basal axis along which the spindle can align.

How does a mitotic spindle align along a predetermined polarity axis? A core component that connects the polarity cues to the mitotic spindle in the neuroblasts is the protein Insc. Insc is not expressed in the neuroectoderm, but during neuroblast delamination, Insc co-localizes with Par3/6 in the stalk. When the stalk is retracted and delamination is complete, Insc forms an apical crescent at the cortex [45]. This localization is achieved through Insc directly binding to Par3. In neuroblasts lacking Par3, Insc remains cytoplasmic and is not recruited to the apical cortex [46]. Once at the cortex, Insc recruits Pins, which is in an inactive during interphase, to the membrane during mitosis [47–49]. At the apical cortex, Pins is able to efficiently recruit Mud [50]. Interestingly, after their initial recruitment, both Insc and Pins are required to maintain each other localization at the apical cortex [47]. This range of interactions and the crystal structures of some of the interactions are clear; however, neither the molecular players nor their dynamic interactions throughout the spindle orientation process have been solved.

Moreover, the recruitment of Mud by Pins provides a link between the polarity cues and the mitotic spindle (Figure 1C). This interaction resulted in a Gα–Pins–Mud complex (homology of a Gα–LGN–NuMA complex in mammals) to recruit Mud to the apical cortex [51–53]. However, previous structural and biochemical data have clearly demonstrated that isolated LGN-binding domains of Insc and nuclear mitotic apparatus protein (NuMA) compete with one another for binding to LGN [54]. This suggests that they exist in two distinct complexes required for recruitment to the apical cortex. However, it also shows that either in the context of the full-length proteins or in the presence of additional factors, could exist in a single complex. Thus, the regulation of binding partners would not affect the integrity/localization of the complex, but could affect its activity.

Evidence indicates that the complex is able to interact with members of the dynein complex in the *C. elegans*zygote (Figure 1C). In accord with this, pulling forces can be reduced by various mutations resulting in an inactive dynein complex [55]. In addition to molecular motors, microtubule dynamics themselves can generate forces [56]. Thus a combination of forces from dynein motors and microtubule depolymerization may control spindle positioning. Therefore, it will be important to determine how these work together and/or are co-ordinated with each other.

Other cues on Spindle Orientation

Over 120 years ago, Oscar Hertwig recognized that cells divide along their long cell axis, which known as the 'long axis rule' [57]. According to the rule, through determining the long axis and the centre of the cell, cell geometry maybe one of cues for spindle orientation. Next studies verify the observation in a variety of systems [32,33]. In normal rat kidney cells, spindle positioning and orientation are preferentially along the long axis of the cell. Once cell shapes are altered, spindle will re-orient along the new long axis of the cell, which rely on in part microtubule. Experiment shows that inhibition of dynein enriched in astral microtubules could cause the spindle to deviate from the long axis of cell, which suggests that the spindle positioning and orientation depend on in part the interactions between astral microtubules and cortical factors [58].

Many studies indicate that cell geometry and ECM is essential for the spindle orientation [18,59,60]. Temporal and spatial coherence of cell adhesion and division appeared vital characteristics of tissue homoeostasis [61]. Cell geometry reflects the spatial structure of internal forces controlled by focal adhesions and actin fibres, which correspond to external cues from the ECM or from the neighbour cells. For embryo development and tissue integrity, cell division provides steadily a continuity of the traction forces [62]. Localization of adhesion related cues on the mitotic cell cortex ensures correct the alignment of mitotic spindle, which guarantees reset the traction forces in daughter cells [63]. The intimate relationship between cell adhesion and division make orientation of the division axis correctly with respect to other cues and then make the two daughter cells inherit the adhesion pattern equally from the mother cell and help to restore mechanical integrity of the tissue controlling the shapes of cells and tissues, which regulate the cell division orientation.

In conclusion, spindle orientation requires intrinsic cues, extrinsic cues and other cues, such as cell polarity, cell geometry, cell–cell adhesion and cell–ECM adhesion. But how these cues translate into correct spindle orientation is not well understood. Particularly, the cues link between the spindle and adhesive molecular determinants should be elucidated in future studies.

ORIENTED CELL DIVISION IN SKIN

As the outermost layer of the body, the stratified mammalian skin is consist of epidermis and dermis, which distinguished by the basement membrane [64]. The epidermis is mainly composed of basal layer, spinous layer, granular layer and cornified envelope (Figure 2A) and acts as a barrier function to prevent the physical and chemical damage and retain essential fluids. Throughout life, the epidermis must constantly keep self-renewal for homoeostasis, in which cells from a single inner basal layer periodically withdraw from the cell cycle, undergo differentiating and move upward.

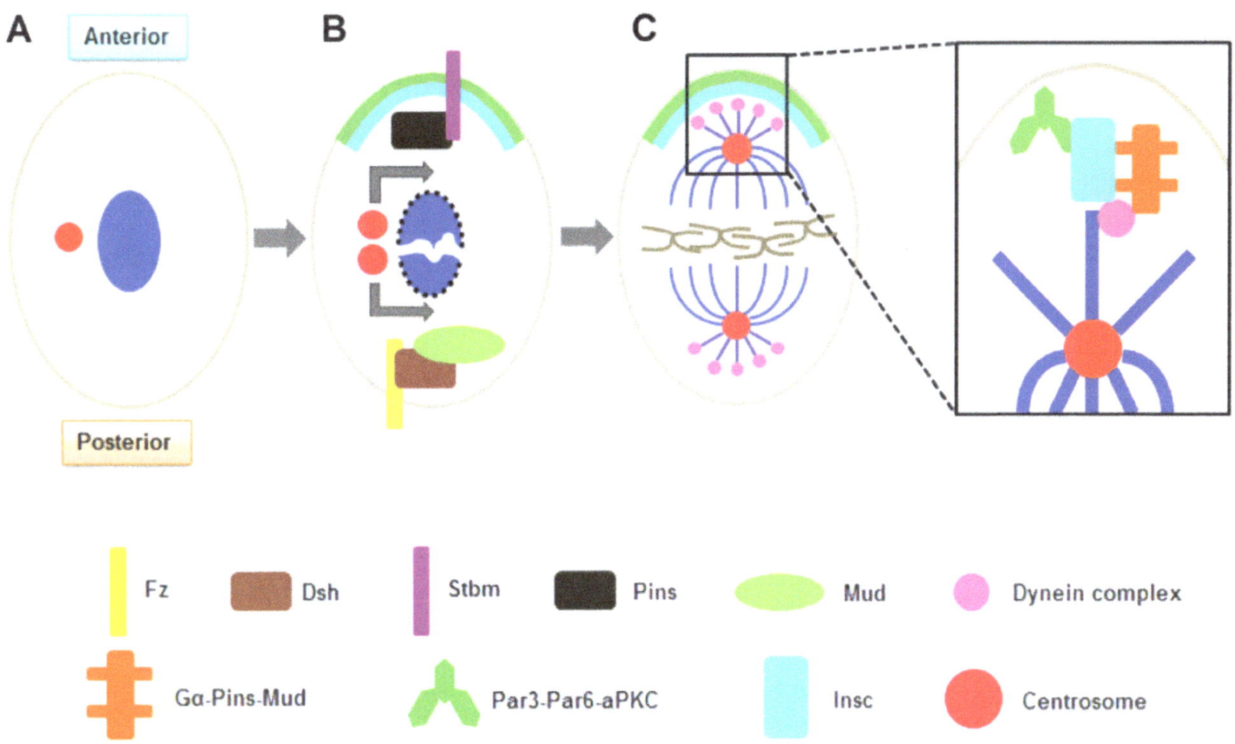

Figure 2: Stratification of the epidermis and oriented division of the ESCs.

Stratification of the Epidermis

Morphological studies show that early basal epidermal progenitors reorient their mitotic spindles, as they begin to develop tissue. From early embryo period to birth, most of mitotic spindles orient perpendicular rather than parallel relative to the underlying basement membrane in the epidermis [64–66]. But, perpendicular spindle orientation in the back skin of adult mice has not been observed after birth [67]. A few divisions in suprabasal located to the first suprabasal layer, which is on the account of transient window of early embryonic epidermal stratification. Both daughter cells of such divisions seem be prepare for differentiation [64]. The basal progenitors are the source of basal and suprabasal cells. On one hand, some basal cells divide symmetrically to generate daughter cells and few of which result in eventual delamination by reason of surrounding microenvironment. On the other hand, some basal cells undergo asymmetric division and one of daughter cells inherits more progenitor factors, whereas the other receives more preferential differentiating factors for delamination.

However, the choice between an asymmetric and symmetric division of basal cells is somewhat random in nature, which adjusts by the needs of the dynamic skin epithelium [66–68]. Thus, understanding how a single cell type choose can elucidate the molecular mechanisms that orient divisions in a particular tissue and how to change artificially the cell division orientation during skin wound healing to achieve the perfect tissue repair.

Apical–basal polarity and Related Molecules

We have discussed the possibilities of the mechanism of oriented cell division in *Drosophila* SOP cells and neuroblasts, zebrafish gastrulation and *C. elegans* zygote. The main signalling pathways involved are Par/Insc and PCP pathway. Which of these pathways involve in orienting the spindle apical–basally in epidermis?

Several lines of genetic evidence and localization assays have suggested that Par3, Par6, atypical protein kinase C (aPKC), LGN (a genuine Pins homologue) and adaptor protein mInscuteable (mInsc) play a key role in recruiting Par3/aPKC, the apical crescent of LGN and mInsc during epidermal stratification and differentiation [64–66]. Perpendicular spindle orientation depends upon Gα, LGN, NuMA and dynactin 1 (Dctn1) by localization assays. However, mammalian LGN isoform is required for asymmetric cell divisions in neural progenitors, is not necessary for that in epidermis [65,69].

Whereas the role of LGN and NuMA in asymmetric divisions is clear, we still have only a few minor hints of the role of mInsc. At least, mInsc could establish a direct relationship with the cortical polarity by interacting with Par3, although how it functions remains unknown [70–72]. Experiment showed that alignment of NuMA in the cell cortex position for locating the spindle apical–basally requires uncoupling of mInsc [50]. However, NuMA does not bind to mInsc directly and its interaction with LGN is mutually exclusive from the binding of mInsc to LGN. Thus, it is need to be answered how the Gα–LGN–NuMA pathway interacts with cortical Par3–mInsc–LGN during spindle orientation. In addition, discs large (Dlg) may function in regulating apical polarity complexes in mammalian epithelial tissues, whereas the exact molecular mechanism needs to be further clarified [65,73–76].

Asymmetric compared with symmetric divisions: how to choose?

Significant progress has been made in understanding the issue of the positioning of orientation in asymmetric divisions; however, how spindles in lateral divisions become oriented and switch to perpendicular spindle orientation is not fully understood. Then, what are the factors influencing oriented cell division in the skin.

As stated above, centrosome migration and spindle rotation play the key role in oriented cell division. Unlike other tissues, both symmetric and asymmetric cell divisions occur in the stratifying epidermis. Many studies have verified that spindle orientations could be either in perpendicular or in parallel to the basement membrane, which respectively corresponded to asymmetric and symmetric cell division, in the interfollicular epidermis (IFE) [77], hair follicles (HFs) [78,79] and in sebaceous gland development [80], where the epidermal stem cells (ESCs) reside. The balance between asymmetric and symmetric cell division is required for increasing the surface area of the epidermis for covering the body and generating the required differentiated layers for a right thickness of epidermis. ESCs in the skin divide symmetrically and a transit amplifying cell divides asymmetrically. The different proliferation rates of each cell type could be regulated as various cell types were needed. Every epidermal progenitor is equivalent and has the ability to divide in either orientation [66], which indicates that the microenvironment plays an important role in the overall outcome of cell divisions. For example, changes of the adhesive cues could result in a random spindle orientation that leads to altered differentiation patterns in mouse epidermis [64]. Therefore, the ratio of asymmetric/symmetric cell division would be the sum of individual cell choices. Besides, the switch from predominant symmetric divisions to asymmetric divisions coincides with the onset of stratification, indicating that asymmetric cell division promotes epidermal differentiation.

Study has shown that depletion of NuMA, LGN and Dctn1 causes a bias towards lateral spindle orientation in planar analyses, which indicates that 2D spindle orientation may be an abnormal pathway with respect to the original cues [65]. During a perpendicularly oriented division, NuMA presents at the apical pole of dividing cells, but enriched at both spindle poles in most lateral divisions, suggesting that it is also insufficient in expression and stability of asymmetric division machinery itself [64]. The difference between NuMA and LGN position is unclear, whereas all signs imply their crucial role in the molecular mechanism underlie in the division orientation.

Another possible factor is the positioning of the centrosomes, because the centrosomes exist at each spindle pole as the organelles and microtubule organizing centre. After duplication, centrosomes separation could be achieved either by moving from one pole to the opposite one or by the simultaneous separation followed by the spindle rotation of into two poles of cell and then the bipolar spindle mitotic spindle are formed. As observed in *Drosophila* neuroblasts, one centrosome anchor at the cell cortex, the other migrates and then spindle orientation is accomplished, which implies that positioning of the spindle poles takes place before the mitosis [81–83]. However, experiments of immunostaining and by tracking fluorescently-labelled centrosomes have verified that centrosomes of basal cells in the epidermis positioning apically and pull the spindle into a perpendicular orientation during the early stages of mitosis, even as the mitotic spindle is forming [65,66,84]. Because positioning of centrosomes is not pre-determined prior to mitotic entry, the spindle of basal epidermal cell could align either laterally or perpendicularly.

Oriented cell division and skin wound healing

As mentioned above, the skin is the protective, defensive barrier from the outside world. Once injured, the skin should restore the normal organizational morphology and physiological function efficiently. Wound healing biology refers to initiating relatively inactive cell lineages to the wound margin for the proliferation, invasion and reconstruction of new matrix in the wound gap [85]. Studies have illustrated that many people suffered from severe burns, occurrence of venous leg ulcer and chronic foot wounds from diabetes and the United States government cost very much for these [86–88]. Similarly, wounds in China are also severe problems and China has a huge demand for tissue repair and regeneration [89,90].

Wound healing is a complex event and consists of an inflammatory phase, re-epithelialization and remodelling phase [85,91–93]. At first, tissue injury disrupts vascular vessels and the release of growth factors, cytokines and components of ECM initiate inflammatory response. Then, tissue re-epithelialization is achieved through the migration and proliferation of keratinocytes to resurface the wound area with a layer of new epithelium. Finally, the collagen fibres reorganize and mature to gain tensile strength in the end. Generally speaking, skin injuries could be healed more quickly, but defective re-epithelialization may leave a connective tissue scar. During the re-epithelialization, keratinocytes at the wound edge conduct migration, proliferation and differentiation and reconstitute the epidermal integrity over the granulation tissue finally [94].

It is reported that skin-derived precursor cells (SKPs) and ESCs are an important component in the process of skin wound healing [95–98], which are located in different regions in the skin and express specific molecular markers (Figure 3). SKPs are a novel population of neural crest-related precursor cells that mainly reside in dermal papilla and can be isolated from embryonic and adult skin. SKPs express proteins of nestin and fibronectin and also express a variety of neural crest (NC) associated transcription factors, including Slug, Snail, Twist, Pax3, Sox9, Sox10, SHOX2, Dermo1 [97,98]. The multiple populations of ESCs have been identified to reside in different locations with the skin. As a major reservoir for epidermal keratinocytes, the bulge region of the HFs expresses the following marker molecules: CD34/α6-integrin, CK15, leucine rich repeat containing G protein coupled receptor 5 (Lgr5) [99,100]. In conditions such as wounding, bulge stem cells rapidly migrate toward the upper bulge to help with the rapid regeneration of wounded skin and these progeny of slow cycling bulge cells express CK15 and β1-integrin in the IFE [101–103], placenta-expressed transcript 1 (MTS24/Plet-1) and α6-integrin isthmus region of the HFs [104], Blimp1 in the sebaceous gland [105], Lrig1 in the junctional zone between upper isthmus and IFE adjacent to the sebaceous glands of the HFs [106]. Additionally, resent study is reported that Lgr6 marked a group of stem cells directly above the follicle bulge can produce cells of both the HFs and the sebaceous gland during embryonic development and can be activated by wounding and migrated toward the epidermis for wound repair [107]. Studies have illustrated that ESCs play an important role in the re-epithelialization and remodelling stage during wound healing [108–110]. In general, ESCs undergo symmetric division to produce either two stem cells (symmetric, non-differentiative division) or two transient amplifying cells (symmetric, differentiative division). Although symmetric division could replace cells rapidly for healing wounds, inappropriate symmetric (differentiative) division may deplete stem cells and cause fibrosis, neurodegeneration, failure to repair damaged tissues, osteoporosis and degenerative changes associated with aging. Rather than proliferate symmetrically into two identical cells, asymmetric division can preserve the stabilization of the pool of the ESCs (Figure 2B). One of the daughter cells remains in a niche as a stem cell for future use, whereas the other daughter cell is transient amplifying cell for current use [111]. The transient amplifying cells undergo subsequent rounds of symmetric proliferative mitotic divisions and differentiate further into a specific lineage such as keratinocytes [110].

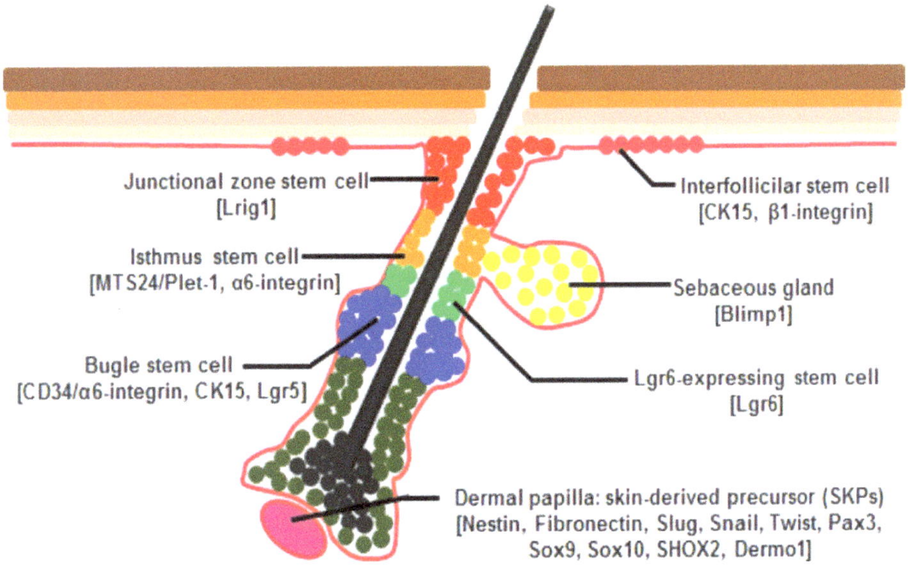

Figure 3: Location of SKPs and ESCs in the skin.

Therefore, oriented mitotic division of epithelial cells, especially ESCs, is particularly important for wound healing. The prolonged symmetric divisions of ESCs during early embryonic development generate large pools of stem cells for tissues repair in need. Perhaps the ability to switch back and forth between symmetric and asymmetric division, increases the capacity for repair and facilitates longer lifespan. When the cells undergo oriented mitotic division to repair the wound by natural process, the impaired skin can complete perfect repair; and when the cell division orientation is random, there will form fibre hyperplasia, abnormal tissue structure and eventually lead to scar formation. Up to now, the currently important questions are that the key factors decide the mitotic division of epithelial cells and how to regulate to achieve oriented cell division. As it should be, we can find some interesting clues from previous researches. But how to utilize these basic theories related cell mitotic division to promote wound healing is still a long way to go.

PERSPECTIVES

Oriented cell divisions are crucial for a variety of developmental and homoeostatic processes ranging from elongation to stratification. Both PCP and Par/Insc pathway mainly control the oriented cell division in many different systems. However, many questions need to be resolved, such as mechanism controlled the position of the spindle, cortical determinants transmitting external cues to the actin cytoskeleton and microtubules, roles of cell–cell and cell–ECM interactions in determining how cells divide. A key issue for the future is how stem cells are regulated to switch between asymmetric and symmetric divisions in the development of tissues and organs and skin wound healing.

Furthermore, as the epidermis must maintain a balance of between asymmetric and symmetric divisions, it must also regulate the two types of division to ensure proper development. Then, how is the machinery required for asymmetric or symmetric cell division localized to specific regions in the cell? Whereas there have been significant advances in uncovering proteins involved in spindle orientation, we still do not have the full complement of players. Developmental questions are also outstanding. How does the epidermis commit to stratification? What are the developmental cues? This information may also shed light on how the balance of symmetric and asymmetric divisions is regulated to yield a normal sized and shaped skin. Maybe that is the key question to achieve perfectly wound repair in skin.

REFERENCES

1. Hafen E. Interplay between growth factor and nutrient signaling: lessons from *Drosophila* TOR. Curr. Top. Microbiol. Immunol. 2004;279:153–167.
2. Lawrence P.A., Struhl G. Morphogens, compartments, and pattern: lessons from *Drosophila*? Cell.1996;85:951–961. doi: 10.1016/S0092-8674(00)81297-0.
3. Pan D. Hippo signaling in organ size control. Genes Dev. 2007;21:886–897. doi: 10.1101/gad.1536007.
4. Glotzer M. The 3Ms of central spindle assembly:

microtubules, motors and MAPs. Nat. Rev. Mol. Cell Biol. 2009;10:9–20. doi: 10.1038/nrm2609.

5. Tanaka K., Kitamura E., Tanaka T.U. Live-cell analysis of kinetochore-microtubule interaction in budding yeast. Methods. 2010;51:206–213. doi: 10.1016/j.ymeth.2010.01.017.

6. Park D.H., Rose L.S. Dynamic localization of LIN-5 and GPR-1/2 to cortical force generation domains during spindle positioning. Dev. Biol. 2008;315:42–54. doi: 10.1016/j.ydbio.2007.11.037.

7. Dujardin D.L., Vallee R.B. Dynein at the cortex. Curr. Opin. Cell Biol. 2002;14:44–49. doi: 10.1016/S0955-0674(01)00292-7.

8. Pearson C.G., Bloom K. Dynamic microtubules lead the way for spindle positioning. Nat. Rev. Mol. Cell Biol. 2004;5:481–492. doi: 10.1038/nrm1402.

9. Morrison S.J., Kimble J. Asymmetric and symmetric stem-cell divisions in development and cancer. Nature. 2006;441:1068–1074. doi: 10.1038/nature04956.

10. da Silva S.M., Vincent J.P. Oriented cell divisions in the extending germband of *Drosophila*. Development. 2007;134:3049–3054. doi: 10.1242/dev.004911.

11. Baena-Lopez L.A., Baonza A., Garcia-Bellido A. The orientation of cell divisions determines the shape of *Drosophila* organs. Curr. Biol. 2005;15:1640–1644. doi: 10.1016/j.cub.2005.07.062.

12. Mao Y., Tournier A.L., Bates P.A., Gale J.E., Tapon N., Thompson B.J. Planar polarization of the atypical myosin Dachs orients cell divisions in *Drosophila*. Genes Dev. 2011;25:131–136. doi: 10.1101/gad.610511.

13. Quesada-Hernandez E., Caneparo L., Schneider S., Winkler S., Liebling M., Fraser S.E., Heisenberg C.P. Stereotypical cell division orientation controls neural rod midline formation in zebrafish. Curr. Biol.2010;20:1966–1972. doi: 10.1016/j.cub.2010.10.009.

14. Gong Y., Mo C., Fraser S.E. Planar cell polarity signalling controls cell division orientation during zebrafish gastrulation. Nature. 2004;430:689–693. doi: 10.1038/nature02796.

15. Concha M.L., Adams R.J. Oriented cell divisions and cellular morphogenesis in the zebrafish gastrula and neurula: a time-lapse analysis. Development. 1998;125:983–994.

16. Lyons D.A., Guy A.T., Clarke J.D. Monitoring neural progenitor fate through multiple rounds of division in an intact vertebrate brain. Development. 2003;130:3427–3436. doi: 10.1242/dev.00569.

17. Geldmacher-Voss B., Reugels A.M., Pauls S., Campos-Ortega J.A. A 90-degree rotation of the mitotic spindle changes the orientation of mitoses of zebrafish neuroepithelial cells. Development. 2003;130:3767–3780. doi: 10.1242/dev.00603.

18. Tawk M., Araya C., Lyons D.A., Reugels A.M., Girdler G.C.,Bayley P.R.,Hyde D.R.,Tada M.,Clarke J.D.A mirror-symmetric cell division that orchestrates neuroepithelial morphogenesis. Nature.2007;446:797–800. doi: 10.1038/nature05722.

19. Ciruna B., Jenny A., Lee D., Mlodzik M., Schier A.F. Planar cell polarity signalling couples cell division and morphogenesis during neurulation. Nature. 2006;439:220–224. doi: 10.1038/nature04375.

20. Zigman M., Trinh le A, Fraser S.E., Moens C.B. Zebrafish neural tube morphogenesis requires scribble-dependent oriented cell divisions. Curr. Biol. 2011;21:79–86. doi: 10.1016/j.cub.2010.12.005.

21. Finley K.R., Bouchonville K.J., Quick A., Berman J. Dynein-dependent nuclear dynamics affect morphogenesis in *Candida albicans* by means of the Bub2p spindle checkpoint. J. Cell Sci. 2008;121(Pt 4):466–476. doi: 10.1242/jcs.015172.

22. Li C.R., Wang Y.M., De Zheng X., Liang H.Y., Tang J.C., Wang Y. The formin family protein CaBni1p has a role in cell polarity control during both yeast and hyphal growth in *Candida albicans*. J. Cell Sci.2005;118(Pt 12):2637–2648. doi: 10.1242/jcs.02393.

23. Knoblich J.A. Asymmetric cell division: recent developments and their implications for tumour biology. Nat. Rev. Mol. Cell Biol. 2010;11:849–860. doi: 10.1038/nrm3010.

24. Hyenne V., Chartier N.T., Labbe J.C. Understanding the role of asymmetric cell division in cancer using*C. elegans*. Dev. Dyn. 2010;239:1378–1387.

25. Munro E., Bowerman B. Cellular symmetry breaking during *Caenorhabditis elegans* development. Cold Spring Harb. Perspect. Biol. 2009;1:a003400. doi: 10.1101/cshperspect.a003400.

26. Gonczy P. Mechanisms of asymmetric cell division: flies and worms pave the way. Nat. Rev. Mol. Cell Biol. 2008;9:355–366. doi: 10.1038/nrm2388.

27. Furman D.P., Bukharina T.A. Drosophila mechanoreceptors as a model for studying asymmetric cell division. Int. J. Dev. Biol. 2011;55:133–141. doi: 10.1387/ijdb.103129df.

28. Doe C.Q. Neural stem cells: balancing self-renewal with differentiation. Development. 2008;135:1575–1587. doi: 10.1242/dev.014977.

29. Siegrist S.E., Doe C.Q. Extrinsic cues orient the cell division axis in *Drosophila* embryonic neuroblasts. Development. 2006;133:529–536. doi: 10.1242/dev.02211.

30. Siller K.H., Doe C.Q. Spindle orientation during asymmetric cell division. Nat. Cell Biol. 2009;11:365–374. doi: 10.1038/ncb0409-365.

31. Galli M., van den Heuvel S. Determination of the cleavage plane in early *C. elegans* embryos. Ann. Rev. Genet. 2008;42:389–411. doi: 10.1146/annurev.genet.40.110405.090523.

32. Strauss B., Adams R.J., Papalopulu N. A default mechanism of spindle orientation based on cell shape is sufficient to generate cell fate diversity in polarised *Xenopus blastomeres*. Development. 2006;133:3883–3893. doi: 10.1242/dev.02578.

33. Honda H. Geometrical models for cells in tissues. Int. Rev. Cytol. 1983;81:191–248. doi: 10.1016/S0074-

7696(08)62339-6.

34. Bellaiche Y., Radovic A., Woods D.F., Hough C.D., Parmentier M.L., O'Kane C.J., Bryant P.J., Schweisguth F. The Partner of Inscuteable/Discs-large complex is required to establish planar polarity during asymmetric cell division in *Drosophila*. Cell. 2001;106:355–366. doi: 10.1016/S0092-8674(01)00444-5.

35. Bellaiche Y., Beaudoin-Massiani O., Stuttem I., Schweisguth F. The planar cell polarity protein Strabismus promotes Pins anterior localization during asymmetric division of sensory organ precursor cells in*Drosophila*. Development. 2004;131:469–478. doi: 10.1242/dev.00928.

36. Bellaiche Y., Gho M., Kaltschmidt J.A., Brand A.H., Schweisguth F. Frizzled regulates localization of cell-fate determinants and mitotic spindle rotation during asymmetric cell division. Nat. Cell Biol. 2001;3:50–57. doi: 10.1038/35050558.

37. Segalen M., Johnston C.A., Martin C.A., Dumortier J.G., Prehoda K.E., David N.B., Doe C.Q., Bellaiche Y. The Fz-Dsh planar cell polarity pathway induces oriented cell division via Mud/NuMA in*Drosophila* and zebrafish. Dev. Cell. 2010;19:740–752. doi: 10.1016/j.devcel.2010.10.004.

38. Kimmel C.B., Warga R.M., Kane D.A. Cell cycles and clonal strings during formation of the zebrafish central nervous system. Development. 1994;120:265–276.

39. Kimmel C.B., Ballard W.W., Kimmel S.R., Ullmann B., Schilling T.F. Stages of embryonic development of the zebrafish. Dev. Dyn. 1995;203:253–310. doi: 10.1002/aja.1002030302.

40. Witzel S., Zimyanin V., Carreira-Barbosa F., Tada M., Heisenberg C.P. Wnt11 controls cell contact persistence by local accumulation of Frizzled 7 at the plasma membrane. J. Cell Biol. 2006;175:791–802. doi: 10.1083/jcb.200606017.

41. Simon M.A. Planar cell polarity in the *Drosophila* eye is directed by graded Four-jointed and Dachsous expression. Development. 2004;131:6175–6184. doi: 10.1242/dev.01550.

42. Rogulja D., Rauskolb C., Irvine K.D. Morphogen control of wing growth through the Fat signaling pathway. Dev. Cell. 2008;15:309–321. doi: 10.1016/j.devcel.2008.06.003.

43. Broadus J., Doe C.Q. Extrinsic cues, intrinsic cues and microfilaments regulate asymmetric protein localization in *Drosophila* neuroblasts. Curr. Biol. 1997;7:827–835. doi: 10.1016/S0960-9822(06)00370-8.

44. Betschinger J., Knoblich J.A. Dare to be different: asymmetric cell division in *Drosophila*, *C. elegans*and vertebrates. Curr. Biol. 2004;14:R674–R685. doi: 10.1016/j.cub.2004.08.017.

45. Knoblich J.A. Asymmetric cell division during animal development. Nat. Rev. Mol. Cell Biol.2001;2:11–20. doi: 10.1038/35048085.

46. Schober M., Schaefer M., Knoblich J.A. Bazooka recruits Inscuteable to orient asymmetric cell divisions in *Drosophila* neuroblasts. Nature. 1999;402:548–551. doi: 10.1038/990135.

47. Yu F., Morin X., Cai Y., Yang X., Chia W. Analysis of partner of inscuteable, a novel player of Drosophila asymmetric divisions, reveals two distinct steps in inscuteable apical localization. Cell.2000;100:399–409. doi: 10.1016/S0092-8674(00)80676-5.

48. Yu F., Ong C.T., Chia W., Yang X. Membrane targeting and asymmetric localization of Drosophila partner of inscuteable are discrete steps controlled by distinct regions of the protein. Mol. Cell. Biol.2002;22:4230–4240. doi: 10.1128/MCB.22.12.4230-4240.2002.

49. Du Q., Macara I.G. Mammalian Pins is a conformational switch that links NuMA to heterotrimeric G proteins. Cell. 2004;119:503–516. doi: 10.1016/j.cell.2004.10.028.

50. Smith N.R., Prehoda K.E. Robust spindle alignment in *Drosophila* neuroblasts by ultrasensitive activation of pins. Mol. Cell. 2011;43:540–549. doi: 10.1016/j.molcel.2011.06.030.

51. Bowman S.K., Neumuller R.A., Novatchkova M., Du Q., Knoblich J.A. The *Drosophila* NuMA homolog Mud regulates spindle orientation in asymmetric cell division. Dev. Cell. 2006;10:731–742. doi: 10.1016/j.devcel.2006.05.005.

52. Siller K.H., Cabernard C., Doe C.Q. The NuMA-related Mud protein binds Pins and regulates spindle orientation in *Drosophila* neuroblasts. Nat. Cell Biol. 2006;8:594–600. doi: 10.1038/ncb1412.

53. Izumi Y., Ohta N., Hisata K., Raabe T., Matsuzaki F. *Drosophila* Pins-binding protein Mud regulates spindle-polarity coupling and centrosome organization. Nat. Cell Biol. 2006;8:586–593. doi: 10.1038/ncb1409.

54. Zhu J., Wen W., Zheng Z., Shang Y., Wei Z., Xiao Z., Pan Z., Du Q., Wang W., Zhang M. LGN/mInsc and LGN/NuMA complex structures suggest distinct functions in asymmetric cell division for the Par3/mInsc/LGN and Galphai/LGN/NuMA pathways. Mol. Cell. 2011;43:418–431. doi: 10.1016/j.molcel.2011.07.011.

55. Nguyen-Ngoc T., Afshar K., Gonczy P. Coupling of cortical dynein and G alpha proteins mediates spindle positioning in Caenorhabditis elegans. Nat. Cell Biol. 2007;9:1294–1302. doi: 10.1038/ncb1649.

56. Labbe J.C., Maddox P.S., Salmon E.D., Goldstein B. PAR proteins regulate microtubule dynamics at the cell cortex in *C. elegans*. Curr. Biol. 2003;13:707–714. doi: 10.1016/S0960-9822(03)00251-3.

57. Gillies T.E., Cabernard C. Cell division orientation in animals. Curr. Biol. 2011;21:R599–R609. doi: 10.1016/j.cub.2011.06.055.

58. O'Connell C.B., Wang Y.L. Mammalian spindle orientation and position respond to changes in cell shape in a dynein-dependent fashion. Mol. Biol. Cell. 2000;11:1765–1774. doi: 10.1091/mbc.11.5.1765.

59. Thery M., Jimenez-Dalmaroni A., Racine V., Bornens M., Julicher F. Experimental and theoretical study of mitotic spindle orientation. Nature. 2007;447:493–496. doi: 10.1038/nature05786.

60. Thery M., Racine V., Pepin A., Piel M., Chen Y.,

Sibarita J.B., Bornens M. The extracellular matrix guides the orientation of the cell division axis. Nat. Cell Biol. 2005;7:947–953. doi: 10.1038/ncb1307.

61. Pugacheva E.N., Roegiers F., Golemis E.A. Interdependence of cell attachment and cell cycle signaling.Curr. Opin. Cell Biol. 2006;18:507–515. doi: 10.1016/j.ceb.2006.08.014.

62. Grill S.W., Howard J., Schaffer E., Stelzer E.H., Hyman A.A. The distribution of active force generators controls mitotic spindle position. Science. 2003;301:518–521. doi: 10.1126/science.1086560.

63. Grill S.W., Kruse K., Julicher F. Theory of mitotic spindle oscillations. Phys. Rev. Lett. 2005;94:108104. doi: 10.1103/PhysRevLett.94.108104.

64. Lechler T., Fuchs E. Asymmetric cell divisions promote stratification and differentiation of mammalian skin. Nature. 2005;437:275–280. doi: 10.1038/nature03922.

65. Williams S.E., Beronja S., Pasolli H.A., Fuchs E. Asymmetric cell divisions promote Notch-dependent epidermal differentiation. Nature. 2011;470:353–358. doi: 10.1038/nature09793.

66. Poulson N.D., Lechler T. Robust control of mitotic spindle orientation in the developing epidermis. J. Cell Biol. 2010;191:915–922. doi: 10.1083/jcb.201008001.

67. Clayton E., Doupe D.P., Klein A.M., Winton D.J., Simons B.D., Jones P.H. A single type of progenitor cell maintains normal epidermis. Nature. 2007;446:185–189. doi: 10.1038/nature05574.

68. Doupe D.P., Klein A.M., Simons B.D., Jones P.H. The ordered architecture of murine ear epidermis is maintained by progenitor cells with random fate. Dev. Cell. 2010;18:317–323. doi: 10.1016/j.devcel.2009.12.016.

69. Sanada K., Tsai L.H. G protein betagamma subunits and AGS3 control spindle orientation and asymmetric cell fate of cerebral cortical progenitors. Cell. 2005;122:119–131. doi: 10.1016/j.cell.2005.05.009.

70. Postiglione M.P., Juschke C., Xie Y., Haas G.A., Charalambous C., Knoblich J.A. Mouse inscuteable induces apical-basal spindle orientation to facilitate intermediate progenitor generation in the developing neocortex. Neuron. 2011;72:269–284. doi: 10.1016/j.neuron.2011.09.022.

71. Bultje R.S., Castaneda-Castellanos D.R., Jan L.Y., Jan Y.N., Kriegstein A.R., Shi S.H. Mammalian Par3 regulates progenitor cell asymmetric division via notch signaling in the developing neocortex. Neuron.2009;63:189–202. doi: 10.1016/j.neuron.2009.07.004.

72. Iden S., van Riel W.E., Schafer R., Song J.Y., Hirose T., Ohno S., Collard J.G. Tumor type-dependent function of the par3 polarity protein in skin tumorigenesis. Cancer Cell. 2012;22:389–403. doi: 10.1016/j.ccr.2012.08.004.

73. Roberts S., Delury C., Marsh E. The PDZ protein discs-large (DLG): the 'Jekyll and Hyde' of the epithelial polarity proteins. FEBS J. 2012;279:3549–3558. doi: 10.1111/j.1742-4658.2012.08729.x.

74. Roberts S., Calautti E., Vanderweil S., Nguyen H.O., Foley A., Baden H.P., Viel A. Changes in localization of human discs large (hDlg) during keratinocyte differentiation are [corrected] associated with expression of alternatively spliced hDlg variants. Exp. Cell Res. 2007;313:2521–2530. doi: 10.1016/j.yexcr.2007.05.017.

75. Rivera C., Simonson S.J., Yamben I.F., Shatadal S., Nguyen M.M., Beurg M., Lambert P.F., Griep A.E. Requirement for Dlgh-1 in planar cell polarity and skeletogenesis during vertebrate development. PLoS One.2013;8:e54410. doi: 10.1371/journal.pone.0054410.

76. Nechiporuk T., Klezovitch O., Nguyen L., Vasioukhin V. Dlg5 maintains apical aPKC and regulates progenitor differentiation during lung morphogenesis. Dev. Biol. 2013;377:375–384. doi: 10.1016/j.ydbio.2013.02.019.

77. Smart I.H. Variation in the plane of cell cleavage during the process of stratification in the mouse epidermis. Br. J. Dermatol. 1970;82:276–282. doi: 10.1111/j.1365-2133.1970.tb12437.x.

78. Blanpain C., Fuchs E. Epidermal homeostasis: a balancing act of stem cells in the skin. Nat. Rev. Mol. Cell Biol. 2009;10:207–217. doi: 10.1038/nrm2636.

79. Zhang Y.V., Cheong J., Ciapurin N., McDermitt D.J., Tumbar T. Distinct self-renewal and differentiation phases in the niche of infrequently dividing hair follicle stem cells. Cell Stem Cell. 2009;5:267–278. doi: 10.1016/j.stem.2009.06.004.

80. Frances D., Niemann C. Stem cell dynamics in sebaceous gland morphogenesis in mouse skin. Dev. Biol. 2012;363:138–146. doi: 10.1016/j.ydbio.2011.12.028.

81. Rebollo E., Sampaio P., Januschke J., Llamazares S., Varmark H., Gonzalez C. Functionally unequal centrosomes drive spindle orientation in asymmetrically dividing Drosophila neural stem cells. Dev. Cell.2007;12:467–474. doi: 10.1016/j.devcel.2007.01.021.

82. Rusan N.M., Peifer M. A role for a novel centrosome cycle in asymmetric cell division. J. Cell Biol.2007;177:13–20. doi: 10.1083/jcb.200612140.

83. Rebollo E., Roldan M., Gonzalez C. Spindle alignment is achieved without rotation after the first cell cycle in *Drosophila* embryonic neuroblasts. Development. 2009;136:3393–3397. doi: 10.1242/dev.041822.

84. Lechler T., Fuchs E. Desmoplakin: an unexpected regulator of microtubule organization in the epidermis.J. Cell Biol. 2007;176:147–154. doi: 10.1083/jcb.200609109.

85. Kirsner R.S., Eaglstein W.H. The wound healing process. Dermatol. Clin. 1993;11:629–640.

86. Boyce S.T., Warden G.D. Principles and practices for treatment of cutaneous wounds with cultured skin substitutes. Am. J. Surg. 2002;183:445–456. doi: 10.1016/S0002-9610(02)00813-9.

87. Phillips T.J. Current approaches to venous ulcers and compression. Dermatol. Surg. 2001;27:611–621.

88. Gomathysankar S., Halim A.S., Yaacob N.S. Proliferation of keratinocytes induced by adipose-derived stem cells on a chitosan scaffold and its role in wound healing, a review. Arch. Plastic Surg. 2014;41:452–457. doi: 10.5999/aps.2014.41.5.452.
89. Fu X. Wound care in China: from repair to regeneration. Int. J. Low. Extrem. Wounds. 2012;11:143–145. doi: 10.1177/1534734612457033.
90. Fu X. Wound care study and translation application: a team›s work in China. Int. J. Lo. Extrem. Wounds.2014;13:84–87. doi: 10.1177/1534734614535406.
91. Werner S., Grose R. Regulation of wound healing by growth factors and cytokines. Physiol. Rev.2003;83:835–870.
92. Rappolee D.A., Patel Y., Jacobson K. Expression of fibroblast growth factor receptors in peri-implantation mouse embryos. Mol. Reprod. Dev. 1998;51:254–264. doi: 10.1002/(SICI)1098-2795(199811)51:3<254::AID-MRD4>3.0.CO;2-O.
93. Martin P. Wound healing–aiming for perfect skin regeneration. Science. 1997;276:75–81. doi: 10.1126/science.276.5309.75.
94. Shaw T.J., Martin P. Wound repair at a glance. J. Cell Sci. 2009;122(Pt 18):3209–3213. doi: 10.1242/jcs.031187.
95. Baum C.L., Arpey C.J. Normal cutaneous wound healing: clinical correlation with cellular and molecular events. Dermatol. Surg. 2005;31:674–686. doi: 10.1097/00042728-200506000-00011.
96. Reinke J.M., Sorg H. Wound repair and regeneration. Eur. Surg. Res. 2012;49:35–43. doi: 10.1159/000339613.
97. Fernandes K.J., Toma J.G., Miller F.D. Multipotent skin-derived precursors: adult neural crest-related precursors with therapeutic potential. Philos. Trans. R. Soc. Lond. B Biol. Sci. 2008;363:185–198. doi: 10.1098/rstb.2006.2020.
98. Hunt D.P., Jahoda C., Chandran S. Multipotent skin-derived precursors: from biology to clinical translation. Curr. Opin. Biotechnol. 2009;20:522–530. doi: 10.1016/j.copbio.2009.10.004.
99. Cotsarelis G., Sun T.T., Lavker R.M. Label-retaining cells reside in the bulge area of pilosebaceous unit: implications for follicular stem cells, hair cycle, and skin carcinogenesis. Cell. 1990;61:1329–1337. doi: 10.1016/0092-8674(90)90696-C.
100. Tumbar T., Guasch G., Greco V., Blanpain C., Lowry W.E., Rendl M., Fuchs E. Defining the epithelial stem cell niche in skin. Science. 2004;303:359–363. doi: 10.1126/science.1092436.
101. Mackenzie I.C. Retroviral transduction of murine epidermal stem cells demonstrates clonal units of epidermal structure. J. Invest. Dermatol. 1997;109:377–383. doi: 10.1111/1523-1747.ep12336255.
102. Ito M., Yang Z., Andl T., Cui C., Kim N., Millar S.E., Cotsarelis G. Wnt-dependent *de novo* hair follicle regeneration in adult mouse skin after wounding. Nature. 2007;447:316–320. doi: 10.1038/nature05766.
103. Kaur P. Interfollicular epidermal stem cells: identification, challenges, potential. J. Invest. Dermatol.2006;126:1450–1458. doi: 10.1038/sj.jid.5700184.
104. Jensen U.B., Yan X., Triel C., Woo S.H., Christensen R., Owens D.M. A distinct population of clonogenic and multipotent murine follicular keratinocytes residing in the upper isthmus. J. Cell Sci.2008;121(Pt 5):609–617. doi: 10.1242/jcs.025502.
105. Horsley V., O›Carroll D., Tooze R., Ohinata Y., Saitou M., Obukhanych T., Nussenzweig M., Tarakhovsky A., Fuchs E. Blimp1 defines a progenitor population that governs cellular input to the sebaceous gland. Cell. 2006;126:597–609. doi: 10.1016/j.cell.2006.06.048.
106. Jensen K.B., Collins C.A., Nascimento E., Tan D.W., Frye M., Itami S., Watt F.M. Lrig1 expression defines a distinct multipotent stem cell population in mammalian epidermis. Cell Stem Cell. 2009;4:427–439. doi: 10.1016/j.stem.2009.04.014.
107. Snippert H.J., Haegebarth A., Kasper M., Jaks V., van Es J.H., Barker N., van de Wetering M., van den Born M., Begthel H., Vries R.G., et al. Lgr6 marks stem cells in the hair follicle that generate all cell lineages of the skin. Science. 2010;327:1385–1389. doi: 10.1126/science.1184733.
108. Uzarska M., Porowinska D., Bajek A., Drewa T. Epidermal stem cells–biology and potential applications in regenerative medicine. Postepy. Biochem. 2013;59:219–227.
109. Sun X., Fu X., Han W., Zhao M., Chalmers L. Epidermal stem cells: an update on their potential in regenerative medicine. Expert Opin. Biol. Ther. 2013;13:901–910. doi: 10.1517/14712598.2013.776036.
110. Dahl M.V. Stem cells and the skin. J. Cosmet. Dermatol. 2012;11:297–306. doi: 10.1111/jocd.12008.
111. Ghadially R. 25 years of epidermal stem cell research. J. Invest. Dermatol. 2012;132(3 Pt 2):797–810. doi: 10.1038/jid.2011.434.

CHAPTER 4

MOTILITY-DRIVEN GLASS AND JAMMING TRANSITIONS IN BIOLOGICAL TISSUES

Dapeng Bi[1,3], Xingbo Yang[1,4], M. Cristina Marchetti[1,2], and M. Lisa Manning[1,2]

[1]Department of Physics and Soft Matter Program, Syracuse University, Syracuse, New York 13244, USA

[2]Syracuse Biomaterials Institute, Syracuse University, Syracuse, New York 13244, USA

[3]Center for Studies in Physics and Biology, Rockefeller University, New York 10065, USA

[4]Department of Physics, Northwestern University, Evanston, Illinois 60208, USA

ABSTRACT

Cell motion inside dense tissues governs many biological processes, including embryonic development and cancer metastasis, and recent experiments suggest that these tissues exhibit collective glassy behavior. To make quantitative predictions about glass transitions in tissues, we study a self-propelled Voronoi model that simultaneously captures polarized cell motility and multibody cell-cell interactions in a confluent tissue, where there are no gaps between cells. We demonstrate that the model exhibits a jamming transition from a solidlike state to a fluidlike state that is controlled by three parameters: the single-cell motile speed, the persistence time of single-cell tracks, and a target shape index that characterizes the competition between cell-cell adhesion and cortical tension. In contrast to traditional particulate glasses, we are able to identify an experimentally accessible structural order parameter that specifies the entire jamming surface as a function of model parameters. We demonstrate that a continuum soft glassy rheology model precisely captures this transition in the limit of small persistence times and explain how it fails in the limit of large persistence times. These results provide a framework for understanding the collective solid-to-liquid transitions that have been observed in embryonic development and cancer progression, which may be associated with epithelial-to-

INTRODUCTION

Recent experiments have revealed that cells in dense biological tissues exhibit many of the signatures of glassy materials, including caging, dynamical heterogeneities, and viscoelastic behavior[1–5]. These dense tissues, where cells are touching one another with minimal spaces in between, are found in diverse biological processes including wound healing, embryonic development, and cancer metastasis.

Citation: Dapeng Bi, Xingbo Yang, M. Cristina Marchetti, and M. Lisa Manning, Motility-Driven Glass and Jamming Transitions in Biological Tissues, DOI: http://dx.doi.org/10.1103/PhysRevX.6.021011

In many of these processes, tissues undergo an epithelial-to-mesenchymal transition (EMT), where cells in a solidlike, well-ordered epithelial layer transition to a mesenchymal, migratory phenotype with less well-ordered cell-cell interactions [6,7], or an inverse process, the mesenchymal-to-epithelial transition (MET). Over many decades, detailed cell biology research has uncovered many of the signaling pathways involved in these transitions [8,9], which are important in developing treatments for cancer and congenital disease.

Most previous work on EMT or MET has focused, however, on properties and expression levels in single cells or pairs of cells, leaving open the interesting question of whether there is a collective aspect to these transitions: Are some features of EMT or MET generated by large numbers of interacting cells? Although there is no definitive answer to this question, several recent works have suggested that EMT might coincide with a collective solid-to-liquid jamming transition in biological tissues [3,10–12]. Therefore, our goal is to develop a framework for jamming and glass transitions in a minimal model that accounts for both cell shapes and cell motility, in order to make predictions that can quantitatively test this conjecture.

Jamming occurs in nonbiological particulate systems (such as granular materials, polymers, colloidal suspensions, and foams) when their packing density is increased above some critical threshold, and glass transitions occur when the fluid is cooled below a critical temperature. Over the past 20 years these phenomena have been unified by "jamming phase diagrams" [13,14].

Building on these successes, researchers have recently used self-propelled particle (SPP) models to describe dense biological tissues [15–20]. These models are similar to those for inert particulate matter—cells are represented as disks or spheres that interact with an isotropic soft repulsive potential—but unlike Brownian particles in a thermal bath, self-propelled particles exhibit persistent random walks. SPP models typically exhibit a glass transition from a diffusive fluid state to an arrested subdiffusive solid that is controlled by (1) the strength of self-propulsion [16,17,21] and (2) the packing density □ [17,21–24]. Just like in thermal systems, a jamming transition occurs at a critical packing density $□_G$, but this critical density is altered by the persistence time of the random walks [17,21–24].

During many biological processes, however, a tissue remains at confluence (packing fraction equal to unity) while it changes from a liquidlike to a solidlike state or vice versa. For example, in wound healing, cells collectively organize to form a "moving sheet" without any change in their packing density [25], and during vertebrate embryogenesis, mesendoderm tissues are more fluidlike than ectoderm tissues, despite both having packing fraction equal to unity [5].

Recently, Bi and co-workers [26] have demonstrated that the well-studied vertex model for 2D confluent tissues [27–32] exhibits a rigidity transition in the limit of zero cell motility. Specifically, the rigidity of the tissue vanishes at a critical balance between cortical tension and cell-cell adhesion. An important insight is that this transition depends sensitively on cell shapes, which are well defined in the vertex model. While promising, vertex models are difficult to compare to some aspects of experiments because they do not incorporate cell motility.

In this work, we bridge the gap between the confluent tissue mechanics and cell motility by studying a hybrid between the vertex model and the SPP model, which we name the self-propelled Voronoi (SPV) model. A similar model was introduced by Li and Sun [33], and cellular Potts models also bridge this gap [34,35], although glass transitions have not been carefully studied in any of these hybrid systems.

SPV MODEL

While the vertex model describes a confluent tissue as a polygonal tiling of space where the degrees of freedom are the vertices of the polygons, the SPV model identifies each cell using only the center (r_i) of Voronoi cells in a Voronoi tessellation of space (Dirichlet domains) [36]. The observation that Voronoi tessellations can describe cellular patterns in epithelial tissues was first proposed by Honda[37]. For a tissue containing N cells, the intercellular interactions are captured by a total energy that is the same as that in the vertex model. Since the tessellation is completely determined by the {ri}, the total tissue mechanical energy can be fully expressed as $E=E(\{r_i\})$:

$$E = \sum_{i=1}^{N} [K_A(A(r_i) - A_0)^2 + K_P(P(r_i) - P_0)^2]. \tag{1}$$

The term quadratic in cell area A(ri) results from a combination of cell volume incompressibility and the monolayer's resistance to height fluctuations [29]. The term involving the cell perimeter $P(r_i)$originates from active contractility of the actomyosin subcellular cortex (quadratic in perimeter) and effective cell membrane tension due to cell-cell adhesion and cortical tension (both linear in perimeter). This gives rise to an effective target shape index that is dimensionless: $p_0 = P_0/\sqrt{A_0}$. K_A and KP are the area and perimeter moduli, respectively. For the remainder of this article, we assume p0 is homogeneous across a tissue, although heterogeneous properties are also interesting to consider [38].

In the vertex model [26], a rigidity transition takes place at a critical value of $p_0 = p_0^* \approx 3.81$. When $p_0 < p_0^*$, cortical tension dominates over cell-cell adhesion and the energy barriers for local cell rearrangement and motion are finite. The tissue then behaves as an elastic solid with finite shear modulus. When $p_0 > p_0^*$, cell-cell adhesion dominates and the energy barriers for local rearrangements vanish, resulting in zero rigidity and fluidlike behavior. While the energy functional for cell-cell interactions is identical in the vertex and SPV models, the two are truly distinct: the local minimum energy states of the vertex model are not guaranteed to be similar to a Voronoi tessellation of cell centers, although we do find them to be very similar in practice. Therefore, we are also interested in whether a rigidity transition in the SPV model coincides with the rigidity transition of the vertex model.

We define the effective mechanical interaction force experienced by cell i as $F_i=-\nabla_i E$. In contrast to particle-based models, F_i is nonlocal and nonadditive: Ficannot be expressed as a sum of pairwise force between cells i and its neighboring cells. Nevertheless, one can show that momentum is still precisely conserved by this energy functional in the absence of the additional self-propulsion forces introduced below.

In addition to Fi, cells can also move due to self-propelled motility. Just as in SPP models, we assign a polarity vector $n\hat{}i=(\cos\theta_i,\sin\theta_i)$ to each cell; along n^i the cell exerts a self-propulsion force with constant magnitude v_0/μ, where μ is the mobility (the inverse of a frictional drag). Together these forces control the overdamped equation of motion of the cell centers r_i:

$$\frac{d\boldsymbol{r}_i}{dt} = \mu \boldsymbol{F}_i + v_0 \hat{\boldsymbol{n}}_i. \tag{2}$$

The polarity is a minimal representation of the front or rear characterization of a motile cell [35]. While the precise mechanism for polarization in cell motility is an area of intense study, here we model its dynamics as a unit vector that undergoes random rotational diffusion,

$$\partial_t\theta_i = \eta_i(t),\ \langle\eta_i(t)\eta_j(t')\rangle = 2D_r\delta(t-t')\delta_{ij}, \qquad (3)$$

where θ_i is the polarity angle that defines n^i, and $\eta_i(t)$ is a white-noise process with zero mean and variance 2Dr. The value of angular noise D_r determines the memory of stochastic noise in the system, giving rise to a persistence time scale $\tau=1/D_r$ for the polarization vector n^. For smallDr≪1, the dynamics of n^ is more persistent than the dynamics of the cell position. At large values of Dr, i.e., when $1/D_r$ becomes the shortest time scale in the model, Eq. (2) approaches simple Brownian motion.

The model can be nondimensionalized by expressing all lengths in units of $\sqrt{A_0}$ and time in units of$1/(\mu K_A A_0)$. There are three remaining independent model parameters: the self-propulsion speed v_0, the cell shape index p0, and the rotational noise strength Dr. We simulate a confluent tissue under periodic boundary conditions with a constant number of N=400 cells (no cell divisions or apoptosis) and assume that the average cell area coincides with the preferred cell area, i.e., $\langle A_i\rangle=A_0$. This approximates a large confluent tissue in the absence of strong confinement. We numerically simulate the model using molecular dynamics by performing 105 integration steps at step size Δt=10−1using Euler's method. A detailed description of the SPV implementation can be found.

CHARACTERIZING GLASSY BEHAVIOR

We first characterize the dynamics of cell motion within the tissue by analyzing the mean-squared displacement (MSD) of cell trajectories. In Fig. 1(a), we plot the MSD as function of time, for tissues at various values of p_0 and fixed $v_0=0.1$ and $D_r=1$. The MSD exhibits ballistic motion (slope close to 2 on a log-log plot) at short times and plateaus at intermediate time scales. The plateau is an indication that cells are becoming caged by their neighbors. For large values of p_0, the MSD eventually becomes diffusive (slope=1), but as p_0 is decreased, the plateau persists for increasingly longer times. This indicates dynamical arrest due to caging effects and broken ergodicity, which is a characteristic signature of glassy dynamics.

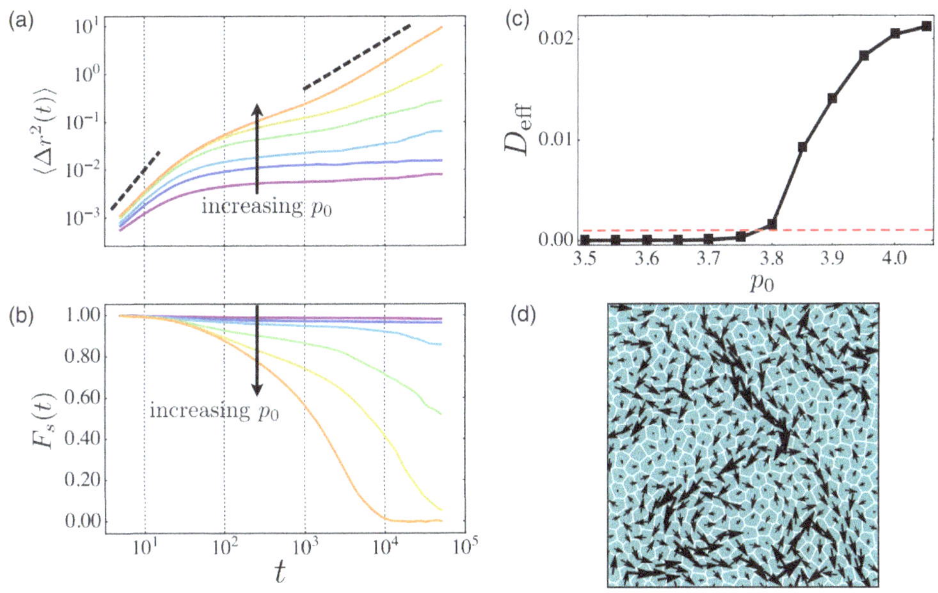

Figure 1: Analysis of glassy behavior. (a) The mean-squared displacement of cell centers for Dr = 1 and $v_0 = 0.1$ and various values of p_0 (bottom to top: $p_0 = 3.5, 3.65, 3.7, 3.75, 3.8, 3.85$) show the onset of dynamical arrest as p0 is changed indicating a glass transition. The dashed lines indicate a slope of 2 (ballistic) and 1 (diffusive) on a log-log plot. (b) The self-intermediate scattering function at the same values of p0 shown in (a) shows the emergence of caging behavior at the glass transition. (c) The effective self-diffusivity as a function of p0 at v0 ¼ 0.1. At the glass transition Deff becomes nonzero. (d) The cell displacement map in the SPV model for a fluid state very close to the glass transition ($p_0 = 3.8$, $v_0 = 0.1$, and $D_r = 1$) over a time window t = 10^4 corresponding to the structural relaxation at which FsðtÞ ≈ 1/2.

Another standard method for quantifying glassy dynamics is the self-intermediate scattering function[39]: $F_s(k,t) = \langle e^{i\vec{k}\cdot\Delta\vec{r}(t)}\rangle$. Glassy systems possess a broad range of relaxation time scales, which show up as a long plateau in Fs(t) when it is analyzed at a length scale q comparable to the nearest-neighbor distance. Figure 1(b) illustrates precisely this behavior in the SPV model, when $|\vec{k}| = \pi/r_0$, where r0 is the position of the first peak in the pair correlation function. The average ⟨···⟩ is taken temporally as well as over angles of $\vec{k}$. $F_s(t)$ also clearly indicates that there is a glass transition as a function of p0: at high p0 values, Fs approaches zero at long times, indicating that the structure is changing and the tissue behaves as a viscoelastic liquid. At lower values of p_0, F_s remains large at all time scales, indicating that the structure is arrested and the tissue is a glassy solid. Figure 1(d) demonstrates that at the structural relaxation time, the cell displacements show collective behavior across large length scales, suggesting strong dynamical heterogeneity. This is strongly reminiscent of the "swirl-like" collective motion seen in experiment in epithelial monolayers [1,2,16,40,41].

A. Dynamical Order Parameter for the Glass Transition

Although the phase space for this model is three dimensional, we now study the model at a fixed value of D_r=1.

We then search for a dynamical order parameter that distinguishes between the glassy and fluid states as a function of the two remaining model parameters (v_0,p_0). A candidate order parameter is the self-diffusivity $D_s = \lim_{t\to\infty}\langle\Delta r(t)^2\rangle/(4t)$. For practicality, we calculate Ds using simulation runs of 10^5 time steps, chosen to be much longer than the typical caging time scale in the fluid regime. We present the self-diffusivity in units of D0=$v_0^2/(2D_r)$, which is the free diffusion constant of an isolated cell. Deff=D_s/D_0 then serves as an accurate dynamical order parameter that distinguishes a fluid state from a solid (glassy) state in the space of (v_0,p_0), matching the regimes identified using the MSD and Fq. In Fig. 2, the fluid region is characterized by a finite value of Deffand Deff drops below a noise floor of $\sim 10^{-3}$ as the glass transition is approached. In practice, we label materials with Deff>10−3 as fluids indicated by an orange dot, and those with Deff≤10^{-3} as solids indicated by blue squares. Importantly, we find that the SPV model in the limit of zero cell motility shares a rigidity transition with the vertex model [26] at p0≈3.81, and that this rigidity transition controls a line of glass transitions at finite cell motilities. Typical cell tracks (Fig. 2) clearly show caging behavior in the glassy solid phase.

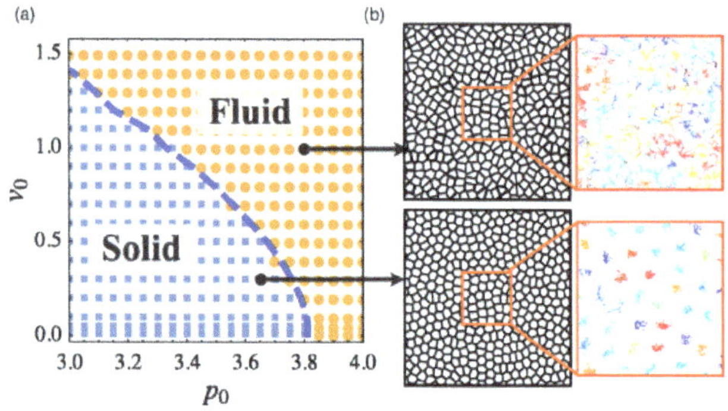

Figure 2: (A) Glassy phase diagram for confluent tissues as function of cell motility v_0 and target shape index p_0 at fixed D_r = 1. Blue data points correspond to solid-like tissue with vanishing D_{eff}; orange points correspond to flowing tissues (finite D_{eff}). The dynamical glass transition boundary also coincides with the locations in phase space where the structural order parameter $q = \langle p/\sqrt{a}\rangle = 3.81$ (dashed line). In the solid phase, q ≈ 3.81 and q > 3.81 in the fluid phase. (B) Instantaneous tissue snapshots show the difference in cell shape across the transition. Cell tracks also show dynamical arrest due to caging in the solid phase and diffusion in the fluid phase

B. Cell Shape is a Structural Order Parameter for the Glass Transition

In glassy systems it can be difficult to experimentally distinguish between a truly dynamically arrested state and a state with relaxation times longer than the experimental time window. Similarly, in tissues it is experimentally challenging to quantify a glass transition through the measurement of a dynamical quantity such as the diffusivity D_s.

Identifying a static quantity that directly probes the mechanical properties of a tissue would therefore be a powerful tool for experiments. Puliafito et al. have suggested that shape changes accompany dynamical arrest in proliferating tissues [41]. Similarly, a structural signature based on cell shapes – the shape index $q = \langle p/\sqrt{a}\rangle$ – was previously shown to be an excellent order parameter for the confluent tissue rigidity transition in the vertex model [38]. In a model where cells were not motile (v_0 = 0) we found that when p0 < 3.813, q is constant ~ 3.81 and when p_0 > 3.81 q grows linearly with p_0. Quite surprisingly, we found that the prediction of q = 3.813 works perfectly in identifying a jamming transition in in-vitro experiments involving primary human tissues, where cells are clearly motile (v_0 6= 0) [38]. At that time, we did not understand why the v_0 = 0 theory worked so well for these tissues. The prediction of a solid-liquid transition in the SPV model presented here provides an explanation for this observation. We find that q (which can be easily calculated in experiments or simulations from a snapshot) can be used as a structural order parameter for the glass transition for all values of v0, not just at v0 = 0. Specifically, the boundary defined by q = 3.813, shown by the blue dashed line in Fig. 2(A) coincides extremely well with the glass transition line obtained using the dynamical order parameter, shown by the round and square data points. The insets to Fig. 2 also illustrate typical cell shapes: cells are isotropic on average in the solid phase and anisotropic in the fluid phase. This highlights the fact that q can be used as a structural order parameter for the glass transition at all cell motilities, providing a powerful new tool for analyzing tissue mechanics.

A THREE-DIMENSIONAL JAMMING PHASE DIAGRAM FOR TISSUES

Having studied the glass transition as function of v0 and p0 at a large value of D_r, we next investigate the full threedimensional phase diagram by characterizing the effect of D_r on tissue mechanics and structure. Dr controls the persistence time τ = 1/D_r and persistence length or Peclet number ′ $P_e \sim v_0$/Dr of cell trajectories; smaller values of Dr correspond to more persistent motion.

In Fig. 3(A), we show several 2D slices of the three dimensional jamming boundary. Solid lines illustrate the phase transition line identified by the structural order parameter q = 3.813 as function of v_0 and p_0 for a large range of D_r values (from 10^{-2} to 10^3). In contrast to results for particulate matter [6], this figure illustrates that the glass transition lines meet at a single point (p_0 = 3.81) in the limit of vanishing cell motility, regardless of persistence

Fig. 3(B) shows an orthogonal set of slices of the jamming diagram, illustrating how the phase boundary shifts as function of p0 and Dr at various values of v_0. This highlights the interesting result that a solid-like material at high value of Dr can be made to flow simply by lowering its value of D_r . The crossover in behavior at large v_0 occurs when the persistence time 1/D_r is approximately equal to the viscous relaxation time 1/(μ$K_A A_0$) = 1.

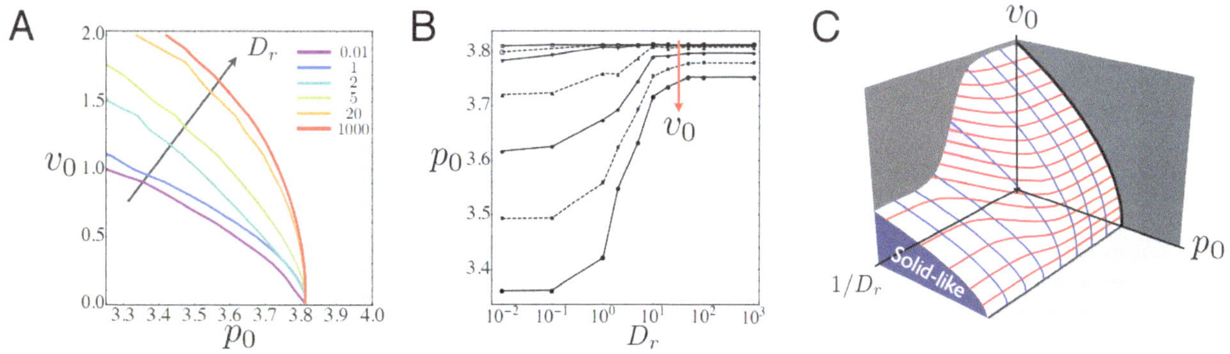

Figure 3. (A) The glass transition in v_0 − p0 phase space shifts as the persistence time changes. Lines represent the glass transition identified by the structural order parameter q = 3.81. The phase boundary collapse to a single point at p ∗ 0 = 3.81, regardless of D_r, in the limit $v_0 \rightarrow 0$. (B) The glass transition in p_0 −Dr phase space shifts as a function of v_0 (from top to bottom: v0 = 0.02,0.08,0.14,0.2,0.26) For large v_0 there is a crossover in the behavior at D_r ~ μKAA0 = 1, as discussed in the main text. (C) The phase boundary between solid and fluid as function of motility v_0, persistence $1/D_r$ and p_0 which is tuned by cell-cell adhesion can be organized into a schematic 3D phase diagram. Red lines on the surface correspond to iso-v_0 contours and blue lines correspond to iso-Dr contours.

These slices can be combined to generate a three dimensional jamming phase diagram for confluent biological tissues, shown in Fig. 3(C). This diagram provides a concrete, quantifiable prediction for how macroscopic tissue mechanics depends on single-cell properties such as motile force, persistence, and the interfacial tension generated by adhesion and cortical tension. We note that Fig. 3(C) is significantly different from the jamming phase diagram conjectured by Sadati et al [42], which was informed by results from adhesive particulate matter [55]. For example, in particulate matter adhesion enhances solidification, while in confluent models adhesion increases cell perimeters/surface area and enhances fluidization. In addition, we identify "persistence" as a new axis with a potentially significant impact on cell migration rates in dense tissue

To better understand why persistence is so important in dense tissues, we first have to characterize the transitions between different cellular structures. In the limit of zero cell motility, the system can be described by a potential energy landscape where each allowable arrangement of cell neighbors corresponds to a metastable minimum in the landscape. There are many possible pathways out of each metastable state: some of correspond to localized cell rearrangements, while others correspond to large-scale collective modes. The maximum energy required to transition out of a metastable state along each pathway is called an energy barrier [8]. We observe that tissue fluidity can increase drastically with increasing D_r at finite cell speeds. This suggests that different pathways (with lower energy barriers) must become dynamically accessible at higher values of D_r.

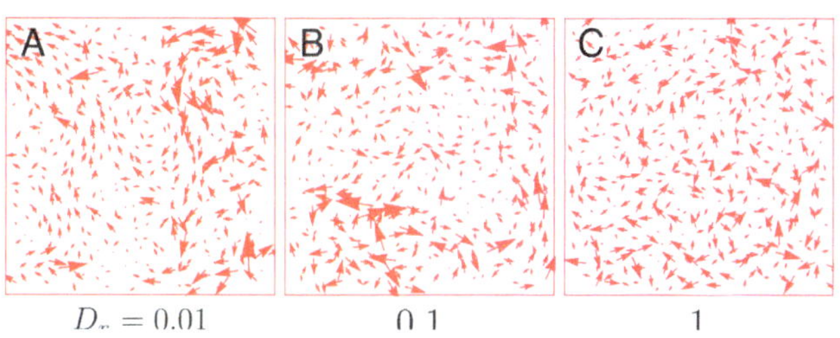

Figure 4: (A-C) Instantaneous cell displacements at p_0 = 3.65 and v_0 = 0.5. They are different from the displacements shown in Fig. 1(D) which are averaged over the structural relaxation timescale. (A) At the lowest value of D_r = 0.01, the cells are able to flow by collectively displacing along the 'soft' modes of the system. (B) At D_r = 0.1, collective displacements are less pronounced. (C) For D_r = 1 and larger, the displacements appear disordered and uncorrelated.

One hint about these pathways comes from the instantaneous cell displacements, shown for different values of D_r in Fig. 4. At high values of D_r , (p0 = 3.78, v_0 = 0.1) the instantaneous displacement field is essentially random and largely uncorrelated, as shown in Fig. 4, and the material is solidlike. There is no collective behavior among cells, and each cell 'rattles' independently near its equilibrium position. However, as D_r is lowered, the instantaneous displacement field becomes much more collective (Fig. 4) and the tissue begins to flow, presumably because these collective displacement fields correspond to pathways with lower transition energies.

Two obvious questions remain: How does a lower value of D_r generate more collective instantaneous displacements? Why should collective instantaneous displacements generically have lower energy barriers? The first question can be answered by extending ideas first proposed by Henkes, Fily and Marchetti [20] to explain why motion in self-propelled particle models seems to follow the 'soft modes' of a solid. This argument is based on a simple, yet powerful observation: in the limit of zero motility (v_0 = 0), a solid-like state will have a well-defined set of normal modes of vibration (with frequencies $\{\omega_v\}$), and a corresponding set of eigenvectors ($\{\hat{e}_v\}$) that forms a complete basis. At higher motilities (v_0 > 0) near the glass transition, the motion of particles in the system can be expanded in terms of the eigenvectors. As discussed one can use this observation to show that in the limit of $D_r \rightarrow 0$, motion along the lowest frequency eigenmodes is amplified – the amplitude along each mode is proportional to $1/\omega_v^2$. These low-frequency normal modes are precisely the collective displacements observed for low D_r .

The second question is more difficult to answer because it is impossible to enumerate all of the possible transition pathways and energy barriers in a disordered material. However, a partial answer comes from recent work in disordered particulate matter showing that low-frequency normal modes do have significantly lower energy barriers [30, 61] than higher frequency normal modes. A similar analysis could potentially be performed in vertex or SPV models.

A CONTINUUM MODEL FOR GLASS TRANSITIONS IN TISSUES

Although continuum hydrodynamic equations of motion have been developed by coarse graining SPP models in the dilute limit, there is no existing continuum model for a dense active matter system near a glass transition. Here we propose that a simple trap [32] or Soft Glassy Rheology (SGR) [47] model provides an excellent continuum approximation for the phase behavior in the large Dr Brownian regime, but fails in the small D_r limit.

For large Dr it is known that particle behave like Brownian particles with an effective temperature $T_{eff} = v_0^2/2\mu D_r$. This mapping becomes exact when $D_r \to \infty$ at fixed "effective inertia" $(\mu D_r)^{-1}$ [13]. In other words, like in granular systems [1, 9], the effective temperature in SPP is dominated by kinetic effects. Guided by this result we conjecture that in our model the temperature also scale quadratically with the velocity,

$$T_{eff} \propto c v_0^2.$$

Physically, this effective temperature gives the amount of energy available for individual cells to vibrate within their cage or 'trap'.

The next important question is how to characterize the 'trap depths', or energy barriers between metastable states. In the Brownian regime (large D_r) there is no dynamical mechanism for the cells to organize collectively, and therefore a reasonable assumption is that the rearrangements are small and localized.

In [8], some of us explicitly calculated the statistics of energy barriers for localized rearrangements in the equilibrium vertex model. In the 2D vertex model, one can show that localized rearrangements must occur via so-called T1 transitions [58].

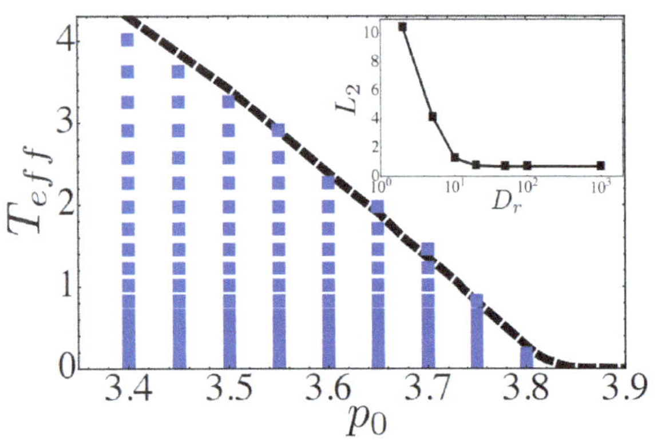

Figure 5: Comparison between SPV glass transition and an analytic prediction based on a Soft Glass Rheology (SGR) continuum model. The dashed line corresponds to an SGR prediction with no fit parameter based on previously measured vertex model trap depths [8]. Data points correspond to SPV simulations with Dr = 10^{-3} and where we have defined $T_{eff} = cv_2$ 0 with c = 0.1 as the best-fit normalization parameter. Blue points correspond to simulations which are solid-like, with De f f < 10^{-3}, and the boundary of these points define the observed SPV glass transition line. (Inset) L_2 difference between SPV glass transition line (at best-fit value of c) and the predicted SGR transition line at various values of D_r. The SGR prediction based on localized T1 trap depths works well in the high Dr limit, but not in the low Dr limit.

Using a trap model [32] or Soft Glassy Rheology(SGR) [47] framework, we were able to use these statistics to generate an analytic prediction, with no fit parameters, for the glass transition temperature T_g as function of p_0.

To see if the SGR prediction for the glass transition holds for the SPV model in the large D_r limit, we simply overlay the data points corresponding to glassy states from the SPV model with the glass transition T_g line predicted in [8]. There is one fitting parameter c that characterizes the proportionality constant in Eq. 4. Fig 5, shows that the SPV data for $D_r = 10^3$ is in excellent agreement with our previous SGR prediction. Because $T_{eff} \sim v_0^2$, and the glass transition line scales as $T_g \sim p_0^* - p_0$ for $p_0 \ll p_0^*$ and $D_r \to 0$, the glass transition line scales as $v_0 \sim (p_0^* - p_0)^{0.5}$ in those limits.

The reason the effective temperature SGR model works here is that, like in SPP models of spherical active Brownian colloids, the angular dynamics of each cell evolves independently of cell-cell interactions and of the angular dynamics of other cells. An additional alignment interaction that couples the angular and translational dynamics may therefore modify this behavior. To our knowledge, this is the first time that a SGR / trap model prediction has been precisely tested in any glassy system. This is because, unlike most glass models, we can enumerate all of the trap depths for localized transition paths in the vertex model.

However, for small values of D_r, we have shown that cell displacements are dominated by collective normal modes, and therefore the energy barriers for localized T_1 transitions are probably irrelevant in this regime. The inset to Fig 5 shows the deviation (L_2 -norm) between glass transition lines in the SPV model and T_1-based SGR prediction as a function of Dr . We see that the SGR prediction fails in the small D_r limit, as expected. A better understanding of the energy barriers associated with collective modes will be required to modify the theory at small D_r .

DISCUSSION AND CONCLUSIONS

We have shown that a minimal model for confluent tissues with cell motility exhibits glassy dynamics at constant density. This model allows us to make a quantitative prediction for how the fluid-to-solid/jamming transition in biological tissues depends on parameters such as the cell motile speed, the persistence time associated with directed cell motion, and the mechanical properties of the cell (governed by adhesion and cortical tension). We define a simple, experimentally accessible structural order parameter – the cell shape index – that robustly identifies the jamming transition, and we show that a simple analytic model based on localized T1 rearrangements precisely predicts the jamming transition in the large D_r limit. We also show that this prediction fails in the small Dr limit, because the instantaneous particle displacements are dominated by collective normal modes.

This model makes several experimentally-verifiable predictions for cell shape and tissue mechanics:

- The order parameter q = 3.81 is a structural signature

for the glass transition, even in tissues with significant cell motility or dynamics. This prediction has already been tested in epithelial lung tissue [38], but it should be much more broadly applicable. We have performed a rudimentary shape analysis of a small number of images from other systems that have been previously published, including proliferating MCDK monolayers [10] and convergent extension in fruit fly development [11] and found that the shapes are consistent with this prediction. A much more careful analysis with full data sets should be performed to further validate this prediction or understand where it breaks down.

- In the limit of vanishing cell motility, shape and pressure fluctuations should vanish when the jamming transition is approached from the solid side, and remain zero in the fluid. A finite motility v_0 will induce such fluctuations in the fluid phase, as confirmed by preliminary calculation of cellular stresses and pressure in the SPV model [62]. This could be studied by combining measurement of cell shape fluctuations with traction force microscopy (TFM) in wound healing assays. After locating the glass transition by imaging cell shape changes, it may be possible to extract information on cell motility v_0 from cellular stresses and pressure inferred from TFM in the fluid phase near the glass transition. This suggests that one may estimate cell motility by examining the changes in cellular stresses and pressure in the cell monolayer near the unjamming transition and assuming that the local velocity of the monolayer is very small just above the transition. The latter assumption can also be verified independently via particle image velocimetry (PIV).
- Cell proliferation, so far neglected in our model, causes an increase in cell number density in confluent tissues. Often this is accompanied by a reduction in individual cell motility v0, via contact inhibition of locomotion. In cases where this is the dominant effect and changes to the ratio between A0 and P0 are negligible, our work predicts that proliferation would drive the system towards jamming. This is consistent with existing reports in the literature [10], although more work is required to test the prediction carefully. In tissues where v0 remains low at all times [41], our model predicts that proliferation can either cause jamming or unjamming, depending on whether cell divisions are oriented in such a way to decrease or increase cell shape anisotropy.
- Spatial correlations and fluctuations of the cell displacement field, such as swirl sizes [2, 3, 15, 39, 40] , should grow as a tissue approaches the glass transition from the fluid side. Very recently, a similar prediction for displacements and correlation lengths based on a particle-based model has been verified in one cell type [15]. The SPV model, which makes predictions for cell shapes in addition to displacements and correlation lengths, could be tested simply by compiling detailed statistics about cell shapes and cell motion in epithelial monolayers.

Although all of the work presented here focuses on the SPV model, which tracks cell centers and therefore has only two degrees of freedom per cell, we found that in the limit of zero cell motility it exhibits the same rigidity transition as the vertex model which has two degrees of freedom per vertex. We have also checked that an "active vertex model", where active motile forces are added to the vertex model vertices, also exhibits a robust glass transition characterized by the shape order parameter q. The fact that two models with ostensibly different degrees of freedom share the same transition suggests that there is a deeper universality, perhaps generated by isostaticity, that remains to be understood.

Another result of this work is the surprising and unexpected differences between confluent models (such as the vertex and SPV models) and particle-based models (such as LennardJones glasses and SPP models). For example, works by Berthier [6], Fily and Marchetti [14] in SPP models suggest that the location of the zero motility glass transition packing density φ_G (defined as the density at which dynamics cease in the limit of $v_0 \rightarrow 0$) depends the value of noise, D_r . This is also related to the observation that the jamming and glass transition are not controlled by the same critical point in nonactive systems [23, 37]. We find this is not the case in the SPV model. Fig. 3(a & b) show that while the glass transition point p_0^* shifts with D_r at finite values of v_0, in the limit of vanishing motility, all glass transition lines merge on to a single point in the limit $v_0 \rightarrow 0$, namely $p_0^* = 3.81$.

Given these differences, it is important to ask which type of model is appropriate for a given system. We argue that SPV models are maybe more appropriate for many biological tissues. Whereas SPP models interact with two-body interactions that only depend on particle center positions, both SPV and vertex models naturally incorporate contractility as a key property of living cells and capture the inherently multibody nature of intercellular forces due to shape deformations. Unlike equilibrium vertex models, SPV models account for cell motility, and they are also much easier to simulate in 3D (which is nearly impossible in practice for the vertex model.) Recent work by Li and Sun [28] also models a confluent cell as a Voronoi tessellation of the plane. An important difference between this work and ours is that in Ref. [28] cellcell adhesion is captured via a potential that is quadratic in the distance between cell centers, just as in particle models. We might guess that stronger cell-cell adhesion in their model will result in stiffening of the tissue, which is common for particle based models, although that remains to be tested in active systems. In contrast, adhesion enters our model through the coupling of the shape energy to the cell perimeter. Increasing cell-cell adhesion (or decreasing cortical tension) yields a larger value of p_0, which leads to the tissue becoming softer. We expect that other shape-based models of confluent tissue dynamics will also yield the glass transition described here. For example, it has been reported in recent works based on the Cellular Potts model [24, 50] and in a modified SPP model [15] when the cell motile force is decreased beyond a certain threshold, the motion of cells transitions from diffusive to sub-diffusive. This is similar to crossing the glass transition line in the SPV model by decreasing the value of v_0

In this work and in previous work based on the vertex model, the cell volume is generally assumed to be fixed. While this is a good assumption in developmental systems such as drosopholia [16, 25] and zebrafish [31], epithelial tissues cells can show significant volume fluctuations, as reported recently [63, 64]. Therefore, it will be important to incorporate volume fluctuations in future iterations of the vertex model or the SPV model, as they introduce another source of active shape fluctuation and could therefore lead to jamming or unjamming of the tissue locally and potentially shift the location of the rigidity and glass transitions. In our version of the SPV model, we have assumed that cell polarity is controlled by simple rotational white noise. It is also possible to include more complex mechanisms. For example, external chemical or mechanical cues could be modeled by coupling v_0 and $\hat{n}_i$ to chemoattractant or mechanical gradients, allowing waves or other pattern formation mechanisms to interact with the jamming transition. Similarly, simple alignment rules (such as those in the Viscek model [57]) could lead to collective flocking modes that also affect glassy dynamics.

Another interesting extension of the SPV model would be to study the role of cell-cell friction, which has already been shown to be important in controlling collective dynamics in particle-based tissue models [15]. Our current model includes viscous frictional coupling of cell to the 2D substrate and cellcell adhesion enters as a negative line tension on interfaces. However, it would be possible to add a frictional force between cells proportional to the length of the edge shared between two cells, and we know from previous work on particulate glasses that these localized frictions can change the location of jamming/glass transition and the nature of spatial correlations in a glass [19, 46].

It is also tempting to speculate about the relationship between the unjamming transition captured by our model and the epithelial-mesenchimal transition (EMT) that precedes cell escape from a solid tumor mass. The EMT involves significant changes in cell-cell adhesion and cytoskeletal composition, with associated changes in cell shape and motility. This suggests that escape from the tumor mass is controlled not just by the chemical breakdown of the basement membrane, but also by specific changes in mechanical properties of both individual cells and the surrounding tissue [60]. One could then hypothesize that the collective unjamming described here may provide the first necessary step towards the mechanical changes needed for cell escape from primary tumors. In particular, recent work suggests that cancer tumors are mechanically heterogeneous, with mixtures of stiff and soft cells that have varying degrees of active contractility [59]. Our jamming phase diagram suggests that the soft cells, which often exhibit mesenchymal markers and presumably correspond to higher values of p_0, might unjam and move towards the boundary of a primary tumor more easily than their stiff counterparts. Examining the effects of tissue heterogeneity on tissue rigidity and patterns of cell motility is therefore a very promising avenue for developing predictive theories for tumor invasiveness and metastasis.

REFERENCES

1. T.E. Angelini, E. Hannezo, X. Trepat, J.J. Fredberg, and D.A. Weitz, Cell Migration Driven by Cooperative Substrate Deformation Patterns, Phys. Rev. Lett. **104**, 168104 (2010).
2. T.E. Angelini, E. Hannezo, X. Trepat, M. Marquez, J.J. Fredberg, and D.A. Weitz, Glass-like Dynamics of Collective Cell Migration, Proc. Natl. Acad. Sci. U.S.A. **108**, 4714 (2011).
3. K.D. Nnetu, M. Knorr, J. Käs, and M. Zink, The Impact of Jamming on Boundaries of Collectively Moving Weak-Interacting Cells, New J. Phys. **14**, 115012 (2012).
4. E.-M. Schoetz, R.D. Burdine, F. Julicher, M.S. Steinberg, C.-P. Heisenberg, and R.A. Foty, Quantitative Differences in Tissue Surface Tension Influence Zebrafish Germ Layer Positioning,HFSP J. **2**, 1 (2008).
5. E.-M. Schoetz, M. Lanio, J.A. Talbot, and M.L. Manning, Glassy Dynamics in Three-Dimensional Embryonic Tissues, J. R. Soc. Interface **10**, 20130726 (2013).
6. J.P. Thiery, Epithelial-Mesenchymal Transitions in Tumour Progression, Nat. Rev. Cancer **2**, 442 (2002).
7. E.W. Thompson and D.F. Newgreen, Carcinoma Invasion and Metastasis: A Role for Epithelial-Mesenchymal Transition?, Cancer Res. **65**, 5991 (2005).
8. N.P.A.D. Gunasinghe, A. Wells, E.W. Thompson, and H.J. Hugo, Mesenchymalepithelial Transition (Met) as a Mechanism for Metastatic Colonisation in Breast Cancer, Cancer Metastasis Rev. **31**, 469 (2012).
9. Y. Nakaya, S. Kuroda, Y.T. Katagiri, K. Kaibuchi, and Y. Takahashi, Mesenchymal-Epithelial Transition during Somitic Segmentation is Regulated by Differential Roles of Cdc42 and Rac1,Dev. Cell **7**, 425 (2004).
10. A. Haeger, M. Krause, K. Wolf, and P. Friedl, Cell Jamming: Collective Invasion of Mesenchymal Tumor Cells Imposed by Tissue Confinement, Biochim. Biophys. Acta **1840**, 2386 (2014); N.K. Karamanos, Matrix-mediated cell behaviour and properties, **1840**, 2385 (2014).
11. J.-A. Park *et al.*, Unjamming and Cell Shape in the Asthmatic Airway Epithelium, Nat. Mater.**14**, 1040 (2015).
12. M. Sadati, N.T. Qazvini, R. Krishnan, C.Y. Park, and J.J. Fredberg, Collective Migration and Cell Jamming., Differentiation (Berlin) **86**, 121 (2013).
13. A.J. Liu and S.R. Nagel, The Jamming Transition and the Marginally Jammed Solid, Annu. Rev. Condens. Matter Phys. **1**, 347 (2010).
14. V. Trappe, V. Prasad, L. Cipelletti, P. Segre, and D.A. Weitz, Jamming Phase Diagram for Attractive Particles, Nature (London) **411**, 772 (2001).
15. J.M. Belmonte, G.L. Thomas, L.G. Brunnet, R.M.C. de Almeida, and H. Chaté, Self-Propelled Particle Model for Cell-Sorting Phenomena, Phys. Rev. Lett. **100**, 248702 (2008).
16. S. Garcia, E. Hannezo, J. Elgeti, J.-F. Joanny, P. Silberzan, and N.S. Gov, Physics of Active Jamming during Collective Cellular Motion in a Monolayer, Proc. Natl. Acad. Sci. U.S.A. **112**, 15314 (2015).
17. S. Henkes, Y. Fily, and M.C. Marchetti, Active Jamming: Self-Propelled Soft Particles at High Density, Phys. Rev. E **84**, 040301 (2011).
18. N. Sepulveda, L. Petitjean, O. Cochet, E. Grasland-Mongrain, P. Silberzan, and V. Hakim, Collective Cell Motion in an Epithelial Sheet Can be Quantitatively Described by a Stochastic Interacting Particle Model, PLoS Comput. Biol. **9**, e1002944 (2013).

19. S.S. Soumya, A. Gupta, A. Cugno, L. Deseri, K. Dayal, D. Das, S. Sen, and M.M. Inamdar, Coherent Motion of Monolayer Sheets under Confinement and Its Pathological Implications, PLoS Comput. Biol. **11**, e1004670 (2015).
20. B. Szabo, G.J. Szöllösi, B. Gönci, Zs. Jurányi, D. Selmeczi, and T. Vicsek, Phase Transition in the Collective Migration of Tissue Cells: Experiment and Model, Phys. Rev. E **74**, 061908 (2006).
21. R. Ni, M.A.C. Stuart, and M. Dijkstra, Pushing the Glass Transition towards Random Close Packing Using Self-Propelled Hard Spheres, Nat. Commun. **4**, 2704 (2013).
22. L. Berthier, Nonequilibrium Glassy Dynamics of Self-Propelled Hard Disks, Phys. Rev. Lett.**112**, 220602 (2014).
23. Y. Fily, S. Henkes, and M.C. Marchetti, Freezing and Phase Separation of Self-Propelled Disks, Soft Matter **10**, 2132 (2014).
24. Y. Fily and M.C. Marchetti, Athermal Phase Separation of Self-Propelled Particles with No Alignment, Phys. Rev. Lett. **108**, 235702 (2012).
25. J.H. Kim, X. Serra-Picamal, Dhananjay T. Tambe, E.H. Zhou, C.Y. Park, M. Sadati, J.-A. Park, R. Krishnan, B. Gweon, E. Millet, J.P. Butler, X. Trepat, and J.J. Fredberg, Propulsion and Navigation within the Advancing Monolayer Sheet, Nat. Mater. **12**, 856 (2013).
26. D. Bi, J.H. Lopez, J.M. Schwarz, and M.L. Manning, A Density-Independent Rigidity Transition in Biological Tissues, Nat. Phys. **11**, 1074 (2015).
27. D. Bi, J.H. Lopez, J.M. Schwarz, and M.L. Manning, Energy Barriers and Cell Migration in Densely Packed Tissues, Soft Matter **10**, 1885 (2014).
28. R. Farhadifar, J.-C. Roeper, B. Aigouy, S. Eaton, and F. Julicher, The Influence of Cell Mechanics, Cell-Cell Interactions, and Proliferation on Epithelial Packing, Curr. Biol. **17**, 2095 (2007).
29. L. Hufnagel, A.A. Teleman, H. Rouault, S.M. Cohen, and B.I. Shraiman, On the Mechanism of Wing Size Determination in Fly Development, Proc. Natl. Acad. Sci. U.S.A. **104**, 3835 (2007).
30. M.L. Manning, R.A. Foty, M.S. Steinberg, and E.-M. Schoetz, Coaction of Intercellular Adhesion and Cortical Tension Specifies Tissue Surface Tension, Proc. Natl. Acad. Sci. U.S.A.**107**, 12517 (2010).
31. T. Nagai and H. Honda, A Dynamic Cell Model for the Formation of Epithelial Tissues, Philos. Mag. B **81**, 699 (2001).
32. D.B. Staple, R. Farhadifar, J.C. Roeper, B. Aigouy, S. Eaton, and F. Julicher, Mechanics and Remodelling of Cell Packings in Epithelia, Eur. Phys. J. E **33**, 117 (2010).
33. B. Li and S.X. Sun, Coherent Motions in Confluent Cell Monolayer Sheets, Biophys. J. **107**, 1532 (2014).
34. A.J. Kabla, Collective Cell Migration: Leadership, Invasion and Segregation, J. R. Soc. Interface **9**, 3268 (2012).
35. A. Szabo, R. Unnep, E Mehes, W.O. Twal, W.S. Argraves, Y. Cao, and A. Czirok, Collective Cell Motion in Endothelial Monolayers, Phys. Biol. **7**, 046007 (2010).
36. G.L. Dirichlet, On the Reduction of Positive Quadratic Forms with Three Undetermined Integers, J. Reine Angew. Math. **1850**, 209 (1850).
37. H. Honda, Description of Cellular Patterns by Dirichlet Domains: The Two-Dimensional Case,J. Theor. Biol. **72**, 523 (1978).
38. F. Wetzel, A. Fritsch, D. Bi, R. Stange, S. Pawlizak, T. Kieβling, L.-C. Horn, K. Bendrat, M. Oktay, M. Zink, A. Niendorf, J. Condeelis, M. Höckel, M.C. Marchetti, M. L. Manning, and J.K. Kas (unpublished).
39. L. Van Hove, Correlations in Space and Time and Born Approximation Scattering in Systems of Interacting Particles, Phys. Rev. **95**, 249 (1954).
40. L. Petitjean, M. Reffay, E. Grasland-Mongrain, M. Poujade, B. Ladoux, A. Buguin, and P. Silberzan, Velocity Fields in a Collectively Migrating Epithelium, Biophys. J. **98**, 1790 (2010).
41. M. Poujade, E. Grasland-Mongrain, A. Hertzog, J. Jouanneau, P. Chavrier, B. Ladoux, A. Buguin, and P. Silberzan, Collective Migration of an Epithelial Monolayer in Response to a Model Wound, Proc. Natl. Acad. Sci. U.S.A. **104**, 15988 (2007).
42. See Supplemental Material at http://link.aps.org/supplemental/10.1103/PhysRevX.6.021011for simulations videos of typical fluid and solid phases.
43. A. Puliafito, L. Hufnagel, P. Neveu, S. Streichan, A. Sigal, D.K. Fygenson, and B.I. Shraiman, Collective and Single Cell Behavior in Epithelial Contact Inhibition, Proc. Natl. Acad. Sci. U.S.A.**109**, 739 (2012).
44. M.L. Manning and A.J. Liu, Vibrational Modes Identify Soft Spots in a Sheared Disordered Packing, Phys. Rev. Lett. **107**, 108302 (2011).
45. N. Xu, V. Vitelli, A.J. Liu, and S.R. Nagel, Anharmonic and Quasi-Localized Vibrations in Jammed Solids—Modes for Mechanical Failure, Europhys. Lett. **90**, 56001 (2010).
46. C. Monthus and J.-P. Bouchaud, Models of Traps and Glass Phenomenology, J. Phys. A **29**, 3847 (1996).
47. P. Sollich, Rheological Constitutive Equation for a Model of Soft Glassy Materials, Phys. Rev. E**58**, 738 (1998).
48. A.R. Abate and D.J. Durian, Effective Temperatures and Activated Dynamics for a Two-Dimensional Air-Driven Granular System on Two Approaches to Jamming, Phys. Rev. Lett. **101**, 245701 (2008).
49. F. Cecconi, A. Puglisi, U.M.B. Marconi, and A. Vulpiani, Noise Activated Granular Dynamics,Phys. Rev. Lett. **90**, 064301 (2003).
50. D.L. Weaire and S. Hutzler, *The Physics of Foams* (Oxford University Press, New York, 1999).
51. M. Deforet, V. Hakim, H.G. Yevick, G. Duclos, and P. Silberzan, Emergence of Collective Modes and Tri-Dimensional Structures from Epithelial Confinement, Nat. Commun. **5**, 05 (2014).
52. R. Etournay, M. Popovi, M. Merkel, A. Nandi, C. Blasse, B. Aigouy, H. Brandl, G. Myers, G. Salbreux, F. Jlicher, and S. Eaton, Interplay of Cell Dynamics and Epithelial Tension during Morphogenesis of the *Drosophila* Pupal

Wing, eLife, **4**, e07090 (2015).

53. X.B. Yang, D. Bi, M.C. Marchetti, and M.L. Manning (unpublished).

54. A. Ikeda, L. Berthier, and P. Sollich, Unified Study of Glass and Jamming Rheology in Soft Particle Systems, Phys. Rev. Lett. **109**, 018301 (2012).

55. P. Olsson and S. Teitel, Athermal Jamming versus Thermalized Glassiness in Sheared Frictionless Particles, Phys. Rev. E **88**, 010301 (2013).

56. M.A. Gelbart, B. He, A.C. Martin, S.Y. Thiberge, E.F. Wieschaus, and M. Kaschube, Volume Conservation Principle Involved in Cell Lengthening and Nucleus Movement during Tissue Morphogenesis, Proc. Natl. Acad. Sci. U.S.A. **109**, 19298 (2012).

57. K.E. Kasza, D.L. Farrell, and J.A. Zallen, Spatiotemporal Control of Epithelial Remodeling by Regulated Myosin Phosphorylation, Proc. Natl. Acad. Sci. U.S.A. **111**, 11732 (2014).

58. S.M. Zehnder, M. Suaris, M.M. Bellaire, and T.E. Angelini, Cell Volume Fluctuations in MDCK Monolayers, Biophys. J. **108**, 247 (2015).

59. S.M. Zehnder, M.K. Wiatt, J.M. Uruena, A.C. Dunn, W.G. Sawyer, and T.E. Angelini, Multicellular Density Fluctuations in Epithelial Monolayers, Phys. Rev. E **92**, 032729 (2015).

60. T. Vicsek, A. Czirók, E. Ben-Jacob, I. Cohen, and O. Shochet, Novel Type of Phase Transition in a System of Self-Driven Particles, Phys. Rev. Lett. **75**, 1226 (1995).

61. S. Henkes, M. van Hecke, and W. van Saarloos, Critical Jamming of Frictional Grains in the Generalized Isostaticity Picture, Europhys. Lett. **90**, 14003 (2010).

62. L.E. Silbert, Jamming of Frictional Spheres and Random Loose Packing, Soft Matter **6**, 2918 (2010).

63. I.Y. Wong, S. Javaid, E.A. Wong, S. Perk, D.A. Haber, M. Toner, and D. Irimia, Collective and Individual Migration Following the Epithelial-Mesenchymal Transition, Nat. Mater. **13**, 1063 (2014).

64. S. Torquato and H.W. Haslach, Random Heterogeneous Materials: Microstructure and Macroscopic Properties., Appl. Mech. Rev. **55**, B62 (2002).

65. E. Ben-Isaac, É. Fodor, P. Visco, F. van Wijland, and N.S. Gov, Modeling the Dynamics of a Tracer Particle in an Elastic Active Gel, Phys. Rev. E **92**, 012716 (2015).

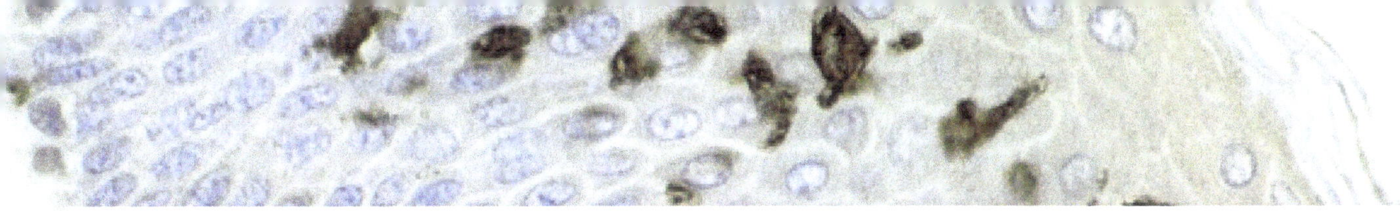

CHAPTER 5 DENDRITIC CELLS IN AUTOIMMUNE DISEASES

Mohan S. Maddur[1,2,3], Janakiraman Vani[1,2,3], Jordan D. Dimitrov[1,2,3], Kithiganahalli N. Balaji[4], Sébastien Lacroix-Desmazes[1,2,3], Srini V. Kaveri[1,2,3], Jagadeesh Bayry[1,2,3]

[1]Institut National de la Santé et de la Recherche Médicale, Unité 872, 15 rue de l'Ecole de Médicine, Paris, F-75006, France

[2]Centre de Recherche des Cordeliers, Equipe 16- Immunopathology and Therapeutic Immunointervention, Université Pierre et Marie Curie – Paris 6, UMR S 872, 15 rue de l'Ecole de médicine, Paris, F-75006, France

[3]Université Paris Descartes, UMR S 872, 15 rue de l'Ecole de médicine, Paris, F-75006, France

[4]Department of Microbiology and Cell Biology, Indian Institute of Science, Bangalore 560012, India

ABSTRACT

Dendritic cells (DCs) are professional antigen presenting cells, which play a crucial role both in maintaining immune tolerance and in inducing adaptive immune responses. Therefore, DCs are the most sought in inciting autoimmunity and its sustenance to autoimmune diseases. The emerging knowledge of the importance of DCs, DC-derived cytokines, and intracellular signal transduction pathways in mediating autoimmune diseases and in creating an inflammatory environment provides a rationale for pursuing strategies to block these inflammatory pathways for therapeutic purposes.

Dendritic cells (DCs) are the prototype professional antigen presenting cells (APCs) derived from hemopoietic cells and are essential link between innate and adaptive arms of the immune system. They are sparse, but widely distributed in different types of tissues, serving as sentinels of the immune system [1]. DCs are capable of stimulating naïve T lymphocytes and hence play a central role in initiating primary immune responses [2]. In addition, DCs also play a critical role in maintaining immune tolerance through cross talk with T and B cells [1,3].

SUBSETS OF DENDRITIC CELLS

Despite similar basic function of antigen presentation, DCs differ in their developmental pathways, location, migratory properties, phenotypic markers and immunological functions. Thus, DCs are heterogeneous and several subtypes with distinct features have been identified in lymphoid and non-lymphoid tissues of human and mouse [1,3]. The major subtypes of human DCs are Langerhans cells, interstitial dermal DCs, monocyte-derived inflammatory DCs (Mo-DCs), BDCA-1$^+$and BDCA-3$^+$myeloid DC subsets, and plasmacytoid DCs (pDCs). The major subsets of mouse DCs include CD4$^-$CD8α^{hi}CD205^{hi}CD11b$^-$DCs, CD4$^+$CD8α^-CD205$^-$CD11b$^+$DCs, CD4$^-$CD8α^-CD205$^-$CD11b$^+$DCs, CD4$^-$CD8α^-CD205$^+$CD11b$^-$ DCs, plasmacytoid DCs, and Langerhans DCs [3]. However, as depicted in Table1, human and mouse DC subsets vary in their expression of CD markers (Table1) [4].

Citation: Mohan S. Maddur, Janakiraman Vani, Jordan D. Dimitrov, Kithiganahalli N. Balaji, Sébastien Lacroix-Desmazes, Srini V. Kaveri, Jagadeesh Bayry, Dendritic Cells in Autoimmune Diseases, DOI: 10.2174/1876539401003010001.

Table 1: Differentiation Markers on Different Subsets of DCs in Human and Mice

Human DC Subsets	Epidermal Langerhans Cells	Dermal Interstitial DCs	Blood Myeloid DCs	Blood Plasmacytoid DCs
C-type lectins	Langerins	DC-SIGN, Mannose receptor	DC-SIGN, Mannose receptor	BDCA-2
Specific Molecules	CD11c, CD1a, E-cadherin	CD11c, CD1a/CD14, CD11b, CD36, FactorXIIIa	CD11c	CD123, ILT7

Mouse DC subsets	$CD8\alpha^+CD4^-$	$CD8\alpha^-CD4^-$	$CD8\alpha^-CD4^+$	$CD8\alpha^-$ $CD4^-$	'Langerhans' DCs $CD8\alpha^{low}CD4^-$	Plasmacytoid DCs
C-type lectins	$DEC205^{high}$	$DEC205^-$	DCIR	$DEC205^+$	$DEC205^{high}$, Langerin	$DEC205^-$
Specific Molecules	CD8, $CD11c^{high}$,	CD11b, $CD11c^{high}$	CD4, CD11b, $CD11c^{high}$	$CD11c^{high}$	CD11b	$CD11c^{low}$, $CD4^{low}$, B220 (CD45R)

DCIR= Dendritic cell immunoreceptor. Markers that are useful for segregating mouse DC subtypes include CD11b and the interdigitating DC marker CD205 (the multilectin domain moleculeDEC205, originally known as NLDC-145). The upper table is reproduced from Immunol Rev 2007; 219: 118-142 (ref. [4]) with permission from Wiley-Blackwell.

MATURATION AND ACTIVATION OF DENDRITIC CELLS

Phenotypically, DCs are identified in "immature" and "mature" stage, based on the expression of surface markers [2]. Resting or immature DCs have high capacity to capture and internalize antigens due to surface expression of receptors that enables the recognition and endocytosis of potential antigens. These receptors include members of pattern recognition receptors (PRRs) family (such as Toll-like receptors (TLRs), C-type lectin receptors (CLRs), intracytoplasmic nucleotide oligomerization domain (NOD)-like receptors (NLRs)), Fc receptors (FcR), complement receptors, scavenger receptors and receptors involved in uptake of apoptotic bodies such as phosphatidylserine receptor. The expression of surface receptors differs with DC subsets and hence enables identification of subtypes [4]. DCs also possess numerous cytokine receptors that enable them to respond to the appropriate stimuli. Thus, immature DCs act as immunological sensors that receive stimuli from environment to alert the immune system. However, immature DCs are poor stimulators of T cells and must undergo a maturation and activation process.

Upon antigen capture by immature DCs, a complex process of maturation and activation is initiated based on the nature of the stimuli. The molecules that provide maturation-associated signals include pathogen-derived molecules (eg., dsRNA, LPS, CpG), dying cell components, tissue factors such as hyaluronan fragments, heparin sulphate, immune complexes and inflammatory cytokines [5]. The components of the dying cells include heat shock proteins (HSPs), high-mobility group box1 proteins (HMGB1), β-defensins, ATP and uric acid, which are collectively called damage-associated molecular pattern molecules (DAMPS) [6]. NF-κB and mitogen-activated protein kinase are the two major signaling pathways that are activated by these stimuli during maturation of DCs [7,8].

Maturation is defined as the series of phenotypic changes that enables DCs to initiate immunity as professional APCs [9]. This process involves similar morphological and functional changes in both human and mouse DCs [9] that include (i) loss of endocytic/phagocytic receptors; (ii) morphological changes such as the loss of adhesive structures, cytoskeleton reorganization, and the acquisition of high cellular motility; (iii) secretion of immunoregulatory cytokines and chemokines such as IL-1β, IL-6, IL-10, IL-12, IL-23, TNF, CCL17, CCL19 and CCL22; (iv) up-regulation of co-stimulatory and adhesion molecules CD80, CD86, CD40, CD54, CD58 critical in mediating clustering with and activation of T cells; (v) translocation of MHC class I/II compartments to the cell surface; and (vi) migration of mature DCs into regional/draining lymphoid tissue. As a result, antigen acquired in peripheral sites is retained, processed and presented to T cells in lymph nodes.

INDUCTION OF ADAPTIVE IMMUNE RESPONSES BY DENDRITIC CELLS

The process of T-cell activation is regulated primarily by signaling events derived from the T-cell receptor TCR (signal 1) and from co-stimulatory and adhesion molecules (signal 2) CD80, CD86, CD40, CD54, CD58 on APCs and CD28, CD154, CD11a and CD2 on T cells. The CD80 and/or CD86–CD28 interactions provide signals important for T-cell activation and survival. CD40–CD154 interactions are crucial for the development of $CD4^+$T-dependent effector functions such as help for B-cell differentiation and class switch. The interactions of adhesion molecules CD54 and CD11a, and CD58 and CD2 play a central part in the clustering of DCs and $CD4^+$T lymphocytes and facilitate TCR triggering by increasing T-cell–DC interactions at low antigen densities. An additional signal has also been proposed in the form of immunomodulatory cytokines (signal 3) produced by activated DCs that dictate the polarization of $CD4^+$T cells towards Th1, Th2, Th17 or regulatory T cells (Tregs) (Fig.1) [10]. Interleukin (IL)-12 has an important role in the differentiation of Th1 cells, and IL-4 is crucial for Th2 differentiation. IL-1β and IL-6 promote IL-17 and IL-21 production from activated and memory human T cells. IL-21 promotes the differentiation of human Th17 cells from naïve CD4+ T cells in the presence of transforming growth factor (TGF)-β, whereas IL-23 might further expand or stabilize differentiated Th17 cells [11-14].

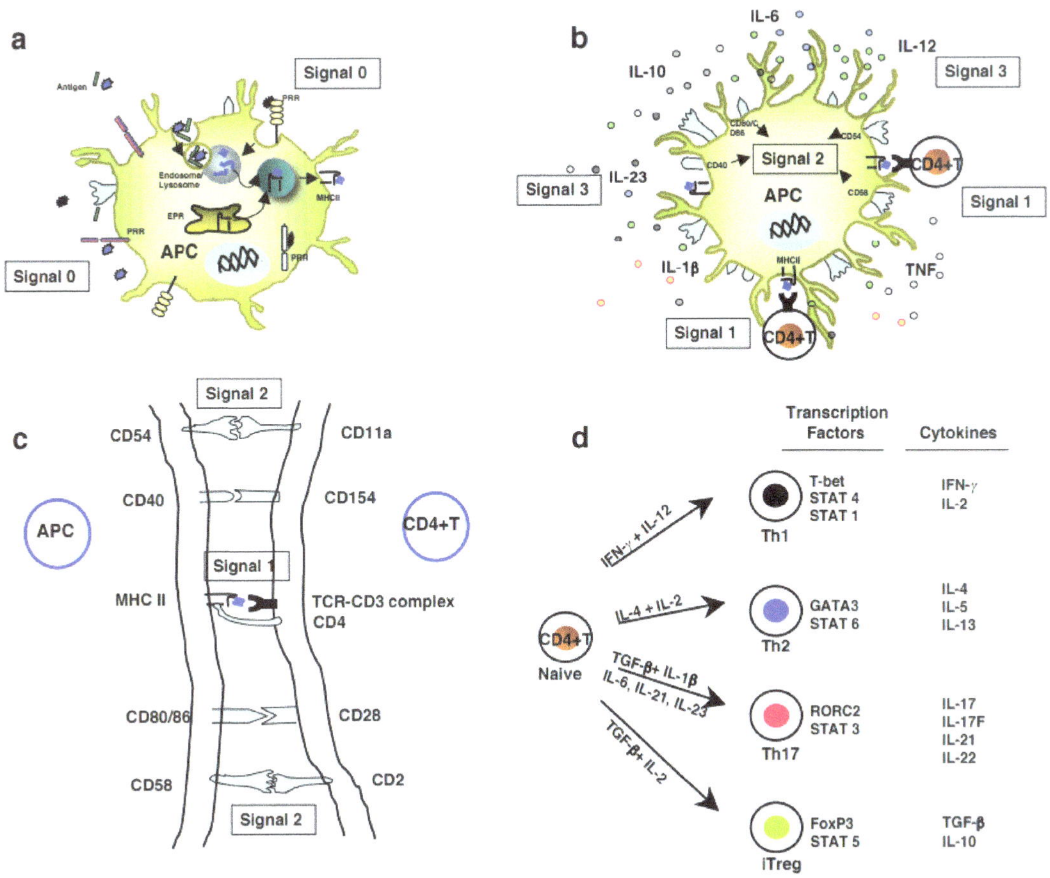

Figure (1): (a) Immature DCs present with high capacity to take up exogenous antigens. The recognition of antigen-associated PAMPs by PRRs (signal 0) leads to maturation and activation of DCs. In endosomes/lysosmes, the internalized antigens are processed into smaller peptides. These antigenic peptides bind to MHC II molecules and transported to cell surface. EPR, endoplasmic reticulum (b) The 'antigen-educated' DCs present antigenic peptides in the context of MHC II to TCR of naïve CD4+T cells (Signal 1). These mature and activated DCs express high levels of MHC, co-stimulatory (CD40, CD80 and CD86) and adhesion molecules (CD54 and CD58) (signal 2) and secrete high amounts of cytokines IL-1α, IL-6, IL-10, IL-12, IL-23 and TNF (Signal 3) that play a critical role in inflammation and differentiation of CD4+T cells. (c) The interaction of signaling molecules at APC-T cell interface (d) CD4+T cells under the influence of signaling by DCs and DC-derived cytokines can be differentiated into 4 distinct subclasses expressing distinct transcription factors: Th1, Th2, Th17 and induced Tregs (iTregs). The figure is reprinted from Trends Pharmacol Sci 2009, 30:287-295 (ref. [10]) with permission from Elsevier.

In addition to T cell stimulation, DCs contribute to the stimulation of B cells, both in the lymph node T cell areas and in germinal centers. Antigens in the immune complex captured by inhibitory Fc receptor (FcRγIIB) can be retained in non-degradative intracellular vesicles and presented in native form to B cells [15]. DCs can also modulate the functions of non-specific effectors such as natural killer cells and natural killer T cells and influence the outcome of immune response [16,17].

DENDRITIC CELLS IN IMMUNE TOLERANCE: GENERATION OF REGULATORY T CELLS

In addition to their central role in initiating adaptive immunity, DCs are also involved in the induction of immune tolerance in both central and peripheral compartments by different regulatory mechanisms that include, clonal deletion, clonal anergy, receptor editing of T/B cells and generation of Tregs [16,17]. Role of DCs in generation of regulatory T cells is of prime importance [15]. In human thymus, the thymic stromal lymhopoietin (TSLP) derived from the hassall's corpuscles instruct DCs to switch from negative selection of autoreactive T cells to positive selection of natural Tregs (nTregs). Interestingly, the preformed immature myeloid tolerogenic DCs carrying peripheral antigens migrate into the thymus and induce the generation of nTregs*in vivo*[15]. In addition, immature/tolerogenic DCs or DCs stimulated to produce IL-10/ TGF-β induce Tregs in peripheral tissues referred to as induced regulatory cells (iTregs) [18,19]. Tregs play a crucial role in maintaining immune tolerance by acting on several cellular targets [20,21].

DENDRITIC CELLS: INCITING AND SUSTAINING AUTOIMMUNITY

Autoimmunity refers to inappropriate immune response against self-components of the host that might result in pathological conditions. A low level of autoreactivity is acknowledged as physiologic and crucial for normal immune function [22]. Autoimmune diseases are clinical syndromes that occur in selected genetic backgrounds characterised by the abnormal activation of T cells, B cells or both in the absence of on going infection or other discernible cause [23]. Autoimmune diseases affect approximately 5% of the population based on genetic characters and environmental factors. They can be clinically classified either systemic (eg., systemic lupus erythematosus (SLE) or organ specific (eg., type 1 diabetes mellitus) [23].

A breakdown in immune tolerance is central to the initiation and progression of an autoimmune disease. Given the central role in induction of immune response and tolerance, DCs are the most sought after cells in inciting autoimmunity and its sustenance to autoimmune diseases [18,24-28]. In human leukocyte antigen-restricted autoimmune diseases, DC priming of the immune response in lymphoid organs is a requisite first step that leads to the activation of T cells and the production of pathogenic autoantibodies and the ensuing chronic inflammatory process [29-31]. Individuals with autoimmune disease show a high number of aberrantly activated DCs either in circulation or in the autoimmune lesions secreting large amounts of proinflammatory cytokines that mediate inflammation and differentiation of pathogenic Th1 and Th17 cells [26]. In the mouse model, altered DC homeostasis has been demonstrated to result in an apparent autoimmune response, thus signifying a central role of DCs in maintaining immune self-tolerance and potential to incite autoimmunity [32,33]. Further, defects in the apoptosis of the immune cells and defective clearance of apoptotic cell debris due to an impaired uptake can result in development of autoimmunity. The priming of autoreactive lymphocytes by activated DCs that have taken up apoptotic materials may lead to break down of self-tolerance [34]. The role of TLRs expressed on different DC subsets has been increasingly evidenced in the pathogenesis of many autoimmune diseases [35,36]. An important sequel of continued antigenic stimulation*via*DCs is the formation of lymphoid structures at the site of inflammation. By coordinating the recruitment and/or activation of other immune cells, DCs can drive the generation of ectopic lymphoid tissues, as in the case of inflamed synovia in rheumatoid arthritis (RA) and SLE [18,19].

Although Tregs can suppress the maturation and activation and the production of inflammatory cytokines by recently activated DCs (Fig.2) [31,37-40], chronically activated DCs can escape suppression by Tregs and generate activated T cells that are refractory to the suppression by Tregs. High amounts of pro-inflammatory cytokines that are released by DCs due to constant and prolonged activation can render Tregs functionally defective. For example, prolonged exposure to TNF can inhibit the expression of FoxP3 and the function of Tregs by signaling through TNF receptor II [41]. Further, IL-6, IL-12, and IL-1β may also promote autoimmunity by interfering with the regulatory functions of Tregs and promoting activation and expansion of effector T cells [31]. In addition to directly suppressing the functions of Tregs, DC-derived cytokines such as IL-6, IL-1β, and IL-23 can reciprocally regulate the populations of iTregs and IL-17-producing pathogenic Th17 cells [42].

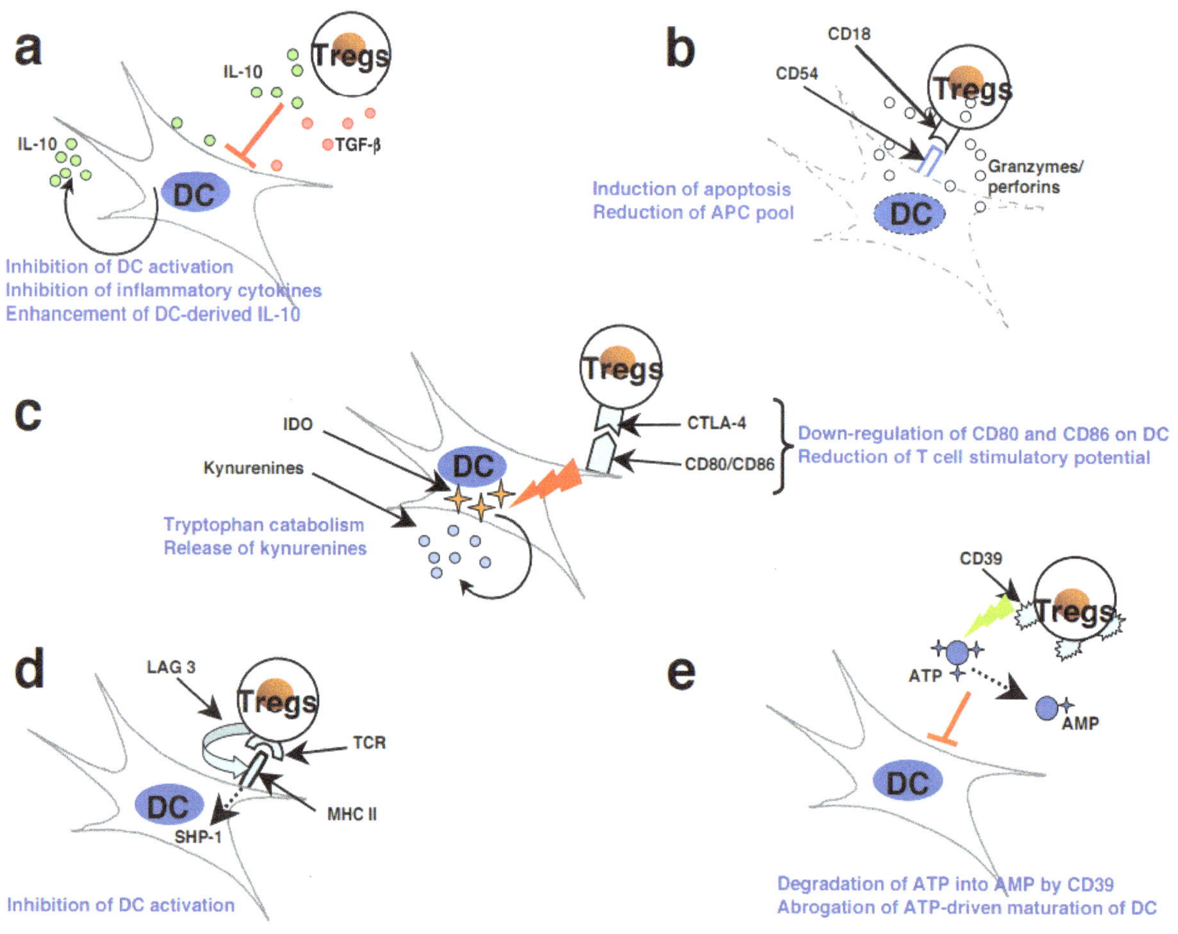

Figure (2): Several mutually nonexclusive mechanisms have been proposed to account for the modulation of DCs by Tregs. The suppression of DCs by Tregs is mediated by both soluble factors and cell-associated molecules. However, recent reports suggest that CTLA-4 molecule has a key role in Treg-mediated suppressive functions. (a) Treg-derived IL-10 and TGF-β induce a suppressive phenotype in DCs and inhibit inflammatory cytokine expression by DCs while enhancing the secretion of the anti-inflammatory cytokine IL-10. (b) Activated Tregs express high levels of granzyme A and perforin and exert CD18/CD54 adhesive interaction-dependent cytotoxicity against both immature and mature DCs, thus reducing the antigen-presenting and co-stimulatory pool. (c) CTLA-4 interaction with B7 molecules transduces a negative signal for DCs CTLA-4 diminishes the co-stimulatory molecules CD80 and CD86 and converts DCs into less potent APCs with reduced T cell stimulatory potential. CTLA-4 can also induce indoleamine 2,3-dioxygenase (IDO) expression in DCs. IDO catalyzes the conversion of tryptophan into kynurenines, which are potent immunosuppressive metabolites. (d) Tregs can interact with MHC class II on DCs via lymphocyte activation gene-3 (LAG-3), a CD4-related transmembrane protein that inhibits DC activation by a mechanism involving ERK-mediated recruitment of SHP-1. (e) Tregs express CD39 (nucleoside triphosphate diphosphohydrolase-1), an ectoenzyme that degrades ATP to AMP. Thus, Tregs reduce ATP-mediated activation of DCs and the adjuvant activities of ATP. Some of these mechanisms are also reported to be critical in Treg-mediated regulation of functions of macrophages and B cells. The figure is reprinted from Am J Pathol 2009, 174:1575-1587 (Ref. [31]) with permission from the American Society for Investigative Pathology.

Type 1 interferons (IFN-α and IFN-β) are secreted in large amount by pDCs not only following viral infection, but also in autoimmune diseases [1]. In psoriasis, pDCs accumulate in the inflamed skin at an early stage of the disease, and these cells secrete type I IFN [43]. In insulin-dependent diabetes mellitus, elevated expression of IFN was found in the pancreas of recently diagnosed patients, though the role of pDCs is unclear [44]. In idiopathic inflammatory myopathies, DCs present in inclusion body myositis and polymyositis are primarily myeloid DCs, whereas in dermatomyositis, type I IFN producing pDCs have been found abundantly in muscle tissue [45]. In Sjogren's syndrome, an autoimmune disease affecting mainly salivary and lacrimal glands, pDCs infiltrate salivary glands, where IFN-inducible genes are over-expressed [46]. In immune thrombocytopenic pupura patients, DCs are found to be highly efficient in presenting apoptotic platelets and stimulating autologous T-cell proliferation [47].

DENDRITIC CELLS IN AUTOIMMUNE DISEASES: FEW SPECIFIC EXAMPLES

Rheumatoid Arthritis (RA)

Spontaneous arthritis animal models resembling RA have revealed defects related to interaction of DCs and thymocytes and as a consequence escape of pathogenic autoreactive T cells. These T cells upon stimulation by activated DCs in the presence of TNF-α and IL-1 are shown to drive autoantibody production and proinflammatory arthritogenic response [48]. Further, DCs can present altered citrullinated collagens and fibrin to autoreactive T cells (Table2) [49]. DCs enter the inflamed synovium through blood vessels and are found to have activated NF-κB pathway. Mature myeloid DCs are in perivascular T cell rich areas of synovial tissue and immature DCs are abundant in the synovial lining and sublining layers of synovium [28,49]. Myeloid DCs can enhance inflammation in the synovium by secreting IL-23 that induces Th17 cells. pDCs are also observed in the inflamed synovium that might respond to immunostimulatory nucleic acid sequences complexed to rheumatoid factor, by producing IFN-α [50]. TNF-α is a dominant inflammatory cytokine that is over-produced in the joints by macrophages and mature myeloid DCs [51].

Table 2: The Role of DC Subsets in Human Autoimmune Diseases

Disease	Role of Dendritic Cells
Rheumatoid arthritis	mDCs present citrullinated collagens and fibrin to autoreactive T cells and drive autoantibody and TNF-a production
	pDCs produce IFN-a in response to nucleic acid sequences complexed to rheumatoid factor
Systemic lupus erythematosus	pDCs produce IFN-a in response to nucleic acid-containing immune complexes which signal through TLR7 and TLR9
	pDC-produced IFNs drive autoantibody production
Multiple Sclerosis	mDCs present self-antigens of myelin to pathogenic T cells
	Intra-cerebral DCs modulate encephalitogenic versus regulatory immune responses in CNS
Psoriasis	pDCs in inflammed skin secrete IFN-a
Insulin-dependent diabetes mellitus	pDCs might be involved in IFN-a secretion in pancreas
Myositis and Polymyositis	mDCs are mainly found at the site of lesions
Dermatomyositis	Type I IFN producing pDCs are abundant
Sjogren's syndrome	pDCs in salivary glands might favour IFN-inducible genes

Systemic Lupus Erythematosus (SLE)

IFN-α is the cytokine implicated in pathogenesis of SLE. pDCs are the main source of IFN-α, produced in response to nucleic acid-containing immune complexes which are detected by TLR7 and TLR9 (Table2) [52]. Unabated induction of DC maturation by IFN-α has been proposed to drive the autoimmune responses in systemic SLE [53]. Thus, IFN-α induces differentiation and activation of DCs from monocytes in SLE individuals, which can activate autoreactive T and B cells [53,54]. IFN-α/β also enhances B-cell response to nucleic acid-containing immune complexes by upregulating TLR7 and promoting cell death to release increased amounts of nucleic acid. pDC-produced IFNs initiate a self-perpetuating feedback loop to drive autoantibody production in SLE [52]. Aberrant phenotype and function of DCs in SLE is also observed [55-57].

Multiple Sclerosis (MS)

MS is a T cell-mediated disease, where Th1 and Th17 cells predominate resulting in chronic inflammation, demyelination and progressive paralysis [58]. DCs are recruited to MS lesions, where uptake and presentation of self-antigens which are made available by continuous myelin destruction, may contribute to the local activation and expansion of presumably pathogenic T cells (Table2) [59,60]. DCs found within MS lesions have been shown to be functionally abnormal [61]. Further, intra-cerebral DCs were found to critically modulate encephalitogenic versus regulatory immune responses in CNS of EAE mice, an animal model of MS and hence considered to be rate-limiting factor for neuro-inflammation [62].

CONCLUSION

DCs are the prime initiators and mediators of autoimmune diseases in an altered host microenvironment (cytokine imbalance) arising out of environmental and genetic perturbations. DCs act in concert with adaptive and other innate immune cells in moulding the autoimmune response. The emerging knowledge of the importance of DCs, DC-derived cytokines, and intracellular signal transduction pathways in mediating and creating an inflammatory environment in autoimmune diseases provides a rationale for pursuing strategies to block these inflammatory pathways for the therapeutic purposes.

ACKNOWLEDGEMENTS

Supported by grants Institut National de la Santé et de la Recherche Médicale (INSERM), Centre National de la Recherche Scientifique (CNRS), Université Paris Descartes-Paris 5, Université Pierre et Marie Curie-Paris 6, Agence Nationale de la Recherche (ANR-07-JCJC-0100-01), Coopération Institut National de la Santé et de la Recherche Médicale–Indian Council for Medical Research (INSERM-ICMR-AO 2009/2010, eS-PIN (European Scientific Progress - Immunoglobulins in Neurology) award 2009 and Talents research grant from Talecris Biotherapeutics.

REFERENCES

1. Shortman K, Naik SH. Steady-state and inflammatory dendriticcell development. Nat Rev Immunol 2007; 7: 19-30.
2. Banchereau J, Steinman RM. Dendritic cells and the control of immunity. Nature 1998; 392: 245-52.
3. Shortman K, Liu YJ. Mouse and human dendritic cell subtypes. Nat Rev Immunol 2002; 2: 151-61.
4. Ueno H, Klechevsky E, Morita R, et al. Dendritic cell subsets in health and disease. Immunol Rev 2007; 219: 118-42.
5. Steinman RM, Turley S, Mellman I, Inaba K. The induction of tolerance by dendritic cells that have captured apoptotic cells. J Exp Med 2000; 191: 411-6.
6. Seong SY, Matzinger P. Hydrophobicity: an ancient damageassociated molecular pattern that initiates innate immune responses. Nat Rev Immunol 2004; 4: 469-78.
7. Ghosh S, May MJ, Kopp EB. NF-kappa B and Rel proteins: evolutionarily conserved mediators of immune responses. Annu Rev Immunol 1998; 16: 225-60.
8. O'Sullivan B, Thompson A, Thomas R. NF-kappa B as a therapeutic target in autoimmune disease. Expert Opin Ther Targets 2007; 11: 111-22.
9. Villadangos JA, Schnorrer P. Intrinsic and cooperative antigenpresenting functions of dendritic-cell subsets in vivo. Nat Rev Immunol 2007; 7: 543-55.
10. Aimanianda V, Haensler J, Lacroix-Desmazes S, Kaveri SV, Bayry J. Novel cellular and molecular mechanisms of induction of immune responses by aluminum adjuvants. Trends Pharmacol Sci 2009; 30: 287-95.
11. Sporri R, Reis e Sousa C. Inflammatory mediators are insufficient for full dendritic cell activation and promote expansion of CD4+ T cell populations lacking helper function. Nat Immunol 2005; 6: 163-70.
12. Blander JM, Medzhitov R. Toll-dependent selection of microbial antigens for presentation by dendritic cells. Nature 2006; 440: 808- 12.
13. van Kooyk Y, Geijtenbeek TB. Toll-like receptors keep antigen sorting on the right track. Immunity 2006; 25: 525-7.
14. Zhu J, Paul WE. CD4 T cells: fates, functions, and faults. Blood 2008; 112: 1557-69.
15. Bergtold A, Desai DD, Gavhane A, Clynes R. Cell surface recycling of internalized antigen permits dendritic cell priming of B cells. Immunity 2005; 23: 503-14.
16. Kitamura H, Iwakabe K, Yahata T, et al. The natural killer T (NKT) cell ligand alpha-galactosylceramide demonstrates its immunopotentiating effect by inducing interleukin (IL)-12 production by dendritic cells and IL-12 receptor expression on NKT cells. J Exp Med 1999; 189: 1121-8.
17. Munz C, Steinman RM, Fujii S. Dendritic cell maturation by innate lymphocytes: coordinated stimulation of innate and adaptive immunity. J Exp Med 2005; 202: 203-7.
18. Ludewig B, Odermatt B, Ochsenbein AF, Zinkernagel RM, Hengartner H. Role of dendritic cells in the

induction and maintenance of autoimmune diseases. Immunol Rev 1999; 169: 45-54.

19. Aloisi F, Pujol-Borrell R. Lymphoid neogenesis in chronic inflammatory diseases. Nat Rev Immunol 2006; 6: 205-17.

20. Sakaguchi S, Yamaguchi T, Nomura T, Ono M. Regulatory T cells and immune tolerance. Cell 2008; 133: 775-87.

21. Tang Q, Bluestone JA. The Foxp3+ regulatory T cell: a jack of all trades, master of regulation. Nat Immunol 2008; 9: 239-44.

22. Steinman RM, Banchereau J. Taking dendritic cells into medicine. Nature 2007; 449: 419-26.

23. Davidson A, Diamond B. Autoimmune diseases. N Engl J Med 2001; 345: 340-50.

24. Drakesmith H, Chain B, Beverley P. How can dendritic cells cause autoimmune disease? Immunol Today 2000; 21: 214-7.

25. Marrack P, Kappler J, Kotzin BL. Autoimmune disease: why and where it occurs. Nat Med 2001; 7: 899-905.

26. Turley SJ. Dendritic cells: inciting and inhibiting autoimmunity. Curr Opin Immunol 2002; 14: 765-70.

27. Bayry J, Thirion M, Delignat S, et al. Dendritic cells and autoimmunity. Autoimmun Rev 2004; 3: 183-7.

28. Lebre MC, Tak PP. Dendritic cells in rheumatoid arthritis: Which subset should be used as a tool to induce tolerance? Hum Immunol 2009.

29. Ludewig B, Odermatt B, Landmann S, Hengartner H, Zinkernagel RM. Dendritic cells induce autoimmune diabetes and maintain disease via de novo formation of local lymphoid tissue. J Exp Med 1998; 188: 1493-501.

30. Boyman O, Conrad C, Tonel G, Gilliet M, Nestle FO. The pathogenic role of tissue-resident immune cells in psoriasis. Trends Immunol 2007; 28: 51-7.

31. Andre S, Tough DF, Lacroix-Desmazes S, Kaveri SV, Bayry J. Surveillance of antigen-presenting cells by CD4+CD25+ regulatory T cells in autoimmunity. immunopathogenesis and therapeutic implications. Am J Pathol 2009; 174: 1575-87.

32. Chen M, Wang YH, Wang Y, et al. Dendritic cell apoptosis in the maintenance of immune tolerance. Science 2006; 311: 1160-4.

33. Ohnmacht C, Pullner A, King SB, et al. Constitutive ablation of dendritic cells breaks self-tolerance of CD4 T cells and results in spontaneous fatal autoimmunity. J Exp Med 2009; 206: 549-59.

34. Lleo A, Selmi C, Invernizzi P, Podda M, Gershwin ME. The consequences of apoptosis in autoimmunity. J Autoimmun 2008; 31: 257-62.

35. Fischer M, Ehlers M. Toll-like receptors in autoimmunity. Ann N Y Acad Sci 2008; 1143: 21-34.

36. Hurst J, von Landenberg P. Toll-like receptors and autoimmunity. Autoimmun Rev 2008; 7: 204-8.

37. Fallarino F, Grohmann U, Hwang KW, et al. Modulation of tryptophan catabolism by regulatory T cells. Nat Immunol 2003; 4: 1206- 12.

38. Houot R, Perrot I, Garcia E, Durand I, Lebecque S. Human CD4+CD25high regulatory T cells modulate myeloid but not plasmacytoid dendritic cells activation. J Immunol 2006; 176: 5293-8.

39. Bayry J, Triebel F, Kaveri SV, Tough DF. Human dendritic cells acquire a semimature phenotype and lymph node homing potential through interaction with CD4+CD25+ regulatory T cells. J Immunol 2007; 178: 4184-93.

40. Wing K, Onishi Y, Prieto-Martin P, et al. CTLA-4 control over Foxp3+ regulatory T cell function. Science 2008; 322: 271-5.

41. Valencia X, Stephens G, Goldbach-Mansky R, Wilson M, Shevach EM, Lipsky PE. TNF downmodulates the function of human CD4+CD25hi T-regulatory cells. Blood 2006; 108: 253-61.

42. Bettelli E, Korn T, Oukka M, Kuchroo VK. Induction and effector functions of T(H)17 cells. Nature 2008; 453: 1051-7.

43. Nestle FO, Conrad C, Tun-Kyi A, et al. Plasmacytoid predendritic cells initiate psoriasis through interferon-alpha production. J Exp Med 2005; 202: 135-43.

44. Tisch R, Wang B. Role of plasmacytoid dendritic cells in type 1 diabetes: friend or foe? Diabetes 2009; 58: 12-3.

45. de Padilla CM, Reed AM. Dendritic cells and the immunopathogenesis of idiopathic inflammatory myopathies. Curr Opin Rheumatol 2008; 20: 669-74.

46. Gottenberg JE, Cagnard N, Lucchesi C, et al. Activation of IFN pathways and plasmacytoid dendritic cell recruitment in target organs of primary Sjogren's syndrome. Proc Natl Acad Sci USA 2006; 103: 2770-5.

47. Catani L, Fagioli ME, Tazzari PL, et al. Dendritic cells of immune thrombocytopenic purpura (ITP) show increased capacity to present apoptotic platelets to T lymphocytes. Exp Hematol 2006; 34: 879-87.

48. Sakaguchi N, Takahashi T, Hata H, et al. Altered thymic T-cell selection due to a mutation of the ZAP-70 gene causes autoimmune arthritis in mice. Nature 2003; 426: 454-60.

49. Lutzky V, Hannawi S, Thomas R. Cells of the synovium in rheumatoid arthritis. Dendritic cells. Arthritis Res Ther 2007; 9: 219.

50. Cavanagh LL, Boyce A, Smith L, et al. Rheumatoid arthritis synovium contains plasmacytoid dendritic cells. Arthritis Res Ther 2005; 7: R230-40.

51. Banchereau J, Pascual V, Palucka AK. Autoimmunity through cytokine-induced dendritic cell activation. Immunity 2004; 20: 539-50.

52. Marshak-Rothstein A. Toll-like receptors in systemic autoimmune disease. Nat Rev Immunol 2006; 6: 823-35.

53. Blanco P, Palucka AK, Gill M, Pascual V, Banchereau J. Induction of dendritic cell differentiation by IFN-alpha in systemic lupus erythematosus. Science 2001; 294: 1540-3.

54. Ronnblom L, Pascual V. The innate immune system in SLE: type I interferons and dendritic cells. Lupus 2008; 17: 394-9.

55. Ding D, Mehta H, McCune WJ, Kaplan MJ. Aberrant

phenotype and function of myeloid dendritic cells in systemic lupus erythematosus. J Immunol 2006; 177: 5878-89.

56. Jin O, Kavikondala S, Sun L, et al. Systemic lupus erythematosus patients have increased number of circulating plasmacytoid dendritic cells, but decreased myeloid dendritic cells with deficient CD83 expression. Lupus 2008; 17: 654-62.

57. Crispin JC, Alcocer-Varela J. The role myeloid dendritic cells play in the pathogenesis of systemic lupus erythematosus. Autoimmun Rev 2007; 6: 450-6.

58. Korn T. Pathophysiology of multiple sclerosis. J Neurol 2008; 255 (Suppl 6): 2-6.

59. Serafini B, Rosicarelli B, Magliozzi R, et al. Dendritic cells in multiple sclerosis lesions: maturation stage, myelin uptake, and interaction with proliferating T cells. J Neuropathol Exp Neurol 2006; 65: 124-41.

60. Miller SD, McMahon EJ, Schreiner B, Bailey SL. Antigen presentation in the CNS by myeloid dendritic cells drives progression of relapsing experimental autoimmune encephalomyelitis. Ann NY Acad Sci 2007; 1103: 179-91.

61. Wu GF, Laufer TM. The role of dendritic cells in multiple sclerosis. Curr Neurol Neurosci Rep 2007; 7: 245-52.

62. Zozulya AL, Ortler S, Lee J, et al. Intracerebral dendritic cells critically modulate encephalitogenic versus regulatory immune responses in the CNS. J Neurosci 2009; 29: 140-52.

CHAPTER 6

THE LONGEVITY OF THE HUMORAL IMMUNE RESPONSE: SURVIVAL OF LONG-LIVED PLASMA CELLS

Pei Xiong Liew

Snyder Institute for Chronic Diseases, University of Calgary, Alberta, Canada

ABSTRACT

The production of antibodies by terminally differentiated plasma cells is the central humoral response that mediates host immunity to infections. Many plasma cells are short-lived and die by apoptosis after several days. Long-term production of antigen-specific antibodies is usually attributed to memory B cells. However, the central paradigm of plasma cell biology is changing with the discovery of long-lived plasma cells which contribute to life-long humoral immunity. Long-lived plasma cells have only been recognized as a part of immunological memory in the last decade – there is much to learn about the complexities of their development and survival. In this article, I review the extrinsic signals involved in long-lived plasma cell longevity, with an emphasis on how the bone marrow microenvironment contributes to the survival of long-lived plasma cells. Insights into these mechanisms will have a profound impact on plasma cell biology and provide novel therapeutic targets in autoimmune diseases and plasma cell malignancies.

INTRODUCTION

The main function of the immune response is to generate an appropriate and robust response to defend against foreign stimuli, i.e., infection by viruses or bacterial pathogens. As the main producers and secretors of antibodies, terminally differentiated B-cells, also called plasma cells (PCs), represent the humoral arm of this response [1, 2]. Antibodies can directly neutralize microbes by blocking factors essential for their survival and also mark the pathogen for clearance by other cells of the immune system [3, 4]. A remarkable property of humoral immunity is the long-lasting 'memory' of the response, providing decades and sometimes lifelong protection against previously encountered pathogens or antigens [5, 6, 7, 8, 9]. The longevity of humoral immunity is provided by two different types of B-cells: memory B cells (B_{mem}) and long-lived plasma cells (LPCs). The role of B_{mem} in long-lived humoral immunity has been extensively reviewed elsewhere [10, 11]. Here, I will review recent developments which have led to a renewed understanding of the longevity of PCs and provide a detailed mechanistic overview of factors in the bone marrow microenvironment that sustain their survival. LPCs persist after the resolution of an immune reaction and contribute to

Citation: Pei Xiong Liew, The Longevity of the Humoral Immune Response: Survival of Long-lived Plasma Cells, http://www.akademeia.ca/index.php/main/article/view/ea0116/50.

an enhanced protective state during re-infection. They are substantially different from B_{mem} cells as they remain non-dividing in the bone marrow and secrete abundant amounts of antibodies for extended periods of time [8, 9]. LPC persistence do not require presence of the antigen [12].

Before the discovery of LPCs, PCs were considered to be differentiated B-cells which survive for less than a week [1, 10, 13] and antibody levels were presumed to be maintained by continuous generation of PCs from activated B cells or B_{mem} after antigenic challenge [14, 15]. This principal dogma was derived from several lines of evidence. Very early in vivo studies suggested PCs were short-lived [16, 17, 18] and these findings were substantiated by later studies that observed that PCs rapidly die after ex vivo isolation as PCs fail to survive after three days of in vitro cell culture [19, 20]. PC numbers were also observed to decline to very low numbers in secondary lymphoid organs after their initial peak response to an antigen [21, 22].

This textbook paradigm was re-examined when two different research groups, using mouse models, found evidence that PCs can be long-lived. In 1997, Manz et al. [8] used proliferation markers to differentiate between short- and long-lived antiovalbumin-secreting PCs in the bone marrow of mice secondarily immunized with ovalbumin. Seventy percent of these anti-ovalbumin-secreting PCs migrated to the bone marrow within the first month after immunization, and persisted for at least 2 months without proliferation (characterized by a lack of cell division and DNA synthesis) while constantly secreting antibodies. The same group then further demonstrated that the persistence of these cells did not require the presence of antigen [12]. The second group, Slifka et al. [9], employed a different approach. Using lymphocytic choriomeningitis (LCMV)-infected mice, they demonstrated that these mice continued to produce specific LCMV-antibodies for 250 days after irradiation. Irradiation depletes memory B cells and prevents generation of new PCs. Adoptive transfer of these plasma cells into naïve non-immunized mice also resulted in consistent antiLCMV antibody titres.

Direct evidence for LPCs in humans has not been demonstrated thus far, but several studies point to their existence. The first observation was obtained from patients with autoimmune diseases who were treated with the therapeutic antibody rituximab, which depletes all B-lineage cells except those that secrete antibodies [23]. Total serum antibody titres of these patients were found to remain within normal range and stable in the long-term after rituximab treatment even though naïve and memory B cells were confirmed to be absent [24, 25, 26]. In other cases, specific serum antibody titres remain stable for decades in humans even without repeated antigenic challenge for more than 70 years [6]. This phenomenon was also independent of the number of vaccine administrations [6, 27]. In addition, recent findings from Mesin et al. [28] suggests that LPCs are present in the human small intestine. As the evidence for LPCs contributing to long-lived humoral immunity continues to increase, these new cells provide novel targets and new therapeutic opportunities for autoimmunity and PC malignancies such as multiple myeloma.

The importance of antibodies and long-lived humoral immunity has been acknowledged for decades [29, 30]. Indeed, the ability of naïve B cells to differentiate into antibody-producing cells is the basis of using vaccines [31, 32]. The emergence of a new player (long-lived plasma cells) contributing to this process brings about a need to understand its role and importance in immunological memory. Importantly, we need to appreciate how the regulation and formation of LPCs is controlled, as failure to balance between a continued production of antibodies targeting foreign antigens and choosing when to turn off this response may result in collateral damage to host cells [33]. In severe cases, dysregulation of this response may result in the development of multiple myeloma [34, 35] or autoimmunity diseases [36, 37, 38]. There is increasing evidence that the survival of LPCs is influenced by extracellular conditions which regulates their lifespan. Below, I will briefly summarise the developmental pathway of LPCs and review the current findings and interpretation of the extrinsic factors which regulate the formation and survival of LPCs. The bone marrow microenvironment as a survival niche for LPCs will also be discussed.

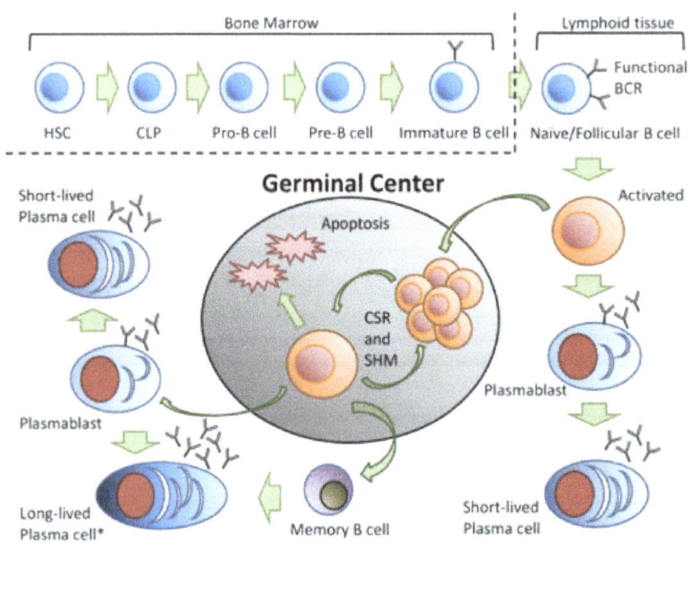

Figure 1: Development of B cells and generation of short- and long-lived plasma cells in mice and humans. B cells develop from hematopoietic stem cells (HSC) present in bone marrow and migrate to lymphoid tissues upon maturation (naïve/follicular B cell). Upon encountering

an antigen, naïve B cells can differentiate into plasmablasts and short-lived PCs which provide a first-line defense against infections. Some of these activated naïve B cells can enter germinal centre (GC) reactions. In GCs, B cells first undergo several rounds of expansion before beginning somatic hypermutation (SHM) and class switch recombination (CSR) (for further details, see [39, 50, 51, 52]). These cells are in a state of activated apoptosis and will die unless they obtain a survival cue from dendritic cells or T cells in the GC environment. In this GC selection process, B cells with BCRs that have a higher affinity for the unique antigenic epitope are selected to survive whereas B cells with low affinity BCRs are out-competed and die. With a survival signal, these B cells then leave the GC where they develop into plasmablasts and PCs and begin to secrete antibodies. PCs become LPCs if they are able to migrate into specific survival niches in the bone marrow and receive survival signals from cells in the microenvironment. CLP: Common lymphoid progenitor.

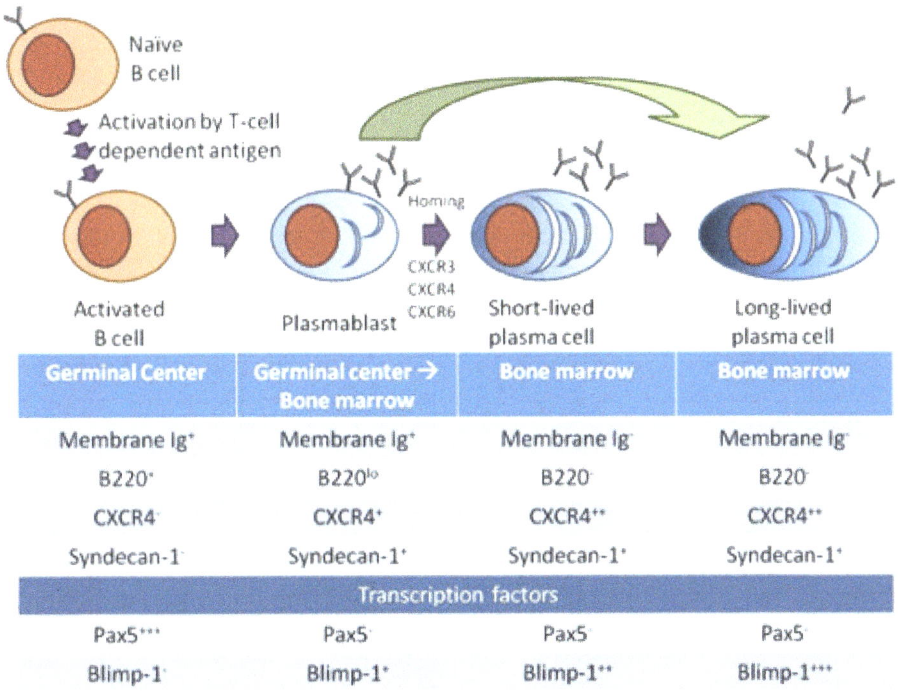

Figure 2: Key surface markers and transcription factors during PC development after activation in the germinal centre. Following activation in the germinal centre, PC development can be identified by the presence of several surface markers and transcription factors [1, 14, 89, 90, 91]. B220 is a surface antigen that defines B-cell lineage and is down-regulated upon terminal differentiation into a PC. As activated B cells differentiate into PCs, surface immunoglobulins (membrane Ig) are secreted as antibodies. CXCR4 and its ligand CXCR12 are important for PCs migration to bone marrow. Syndecan-1, a traditional PC marker, is a cellassociated proteoglycan which binds to the extracellular matrix and is upregulated as activated B cells differentiate into PCs. Pax5 is a transcription factor which mediates B-cell identity and is expressed throughout B lineage and all mature Bcell subsets. Blimp-1 is a crucial transcription factor required for full PC differentiation. Pax5 and Blimp- 1 are mutually exclusive transcriptions factors where Pax5 must be eliminated to permit successful PC development.

DEVELOPMENT OF LONG-LIVED PLASMA CELLS

Plasma cells are the differentiated result of a series of cellular maturation steps beginning from a bone marrow hematopoietic multipotent stem cell (fig. 1). This multistep developmental pathway yields a naïve B cell expressing a functional B cell receptor (BCR) and is antigen-independent [1, 39]. Activation of B cells via the BCR induces naïve B cells to differentiate into short-lived antibody-secreting plasmablasts (fig. 2) [1, 14]. Plasmablasts are larger cellular precursors to mature plasma cells which retain the ability to proliferate and migrate into the bone marrow or during autoimmune and inflammatory conditions, immigrate to inflamed tissues [10, 14, 40]. The migratory ability of plasmablasts is achieved by alterations in the expression of chemokine receptors. Plasmablasts upregulate $CXCR_4$ and $CXCR_3$ in the presence of interferon-γ, which enables them to be guided by chemotactic molecules $CXCL_9$, $CXCL_{10}$, $CXCL_{11}$ and $CXCL_{12}$ [21, 41, 42]. Chemotactic molecules are specific chemical signals that are upregulated or secreted in tissues to attract and recruit specific immune cells which express the matching chemotactic receptor [43]. Upon arrival at tissue niches, the plasmablasts terminally differentiate into non-migratory PCs which either reside there for several days before dying by apoptosis or persist as LPCs over years [14, 40, 44, 45]. Few intrinsic differences have been found between PCs and LPCs, as genes involved in B cell to PC differentiation, adhesion and survival have been found to be essential for all PCs [40]. (cf. a recent manuscript demonstrates that LPCs in the bone marrow were able to respond to signalling via the receptor molecule CD28 which was not observed in shortlived PCs from the spleen [46]). Instead, current data suggests they are regulated by extrinsic factors. The extrinsic determinants which govern the survival of PCs are discussed in further detail in the next few sections.

Although it is clear that PCs are differentiated end-products of activated B cells, the origin from which PCs arise can differ. The route from which PCs arise depends on the type, dose and form of antigen as well as the current location of the B cell [39, 47]. This, in turn, affects the strength and timing of signals propagated by the BCR. Further, PCs can emerge from naïve or memory B cells upon encountering antigen and be dependent or independent of T cell help [1, 39]. Although both dependent and independent T cell help can induce naïve B cells to become plasmablasts, T celldependent help induces B cells to seed germinal centres in lymphoid follicles. Germinal centres are specific, defined structures in secondary lymphoid tissues where mature B cells proliferate [33, 48, 49]. In addition, antibodies at germinal centres undergo affinity maturation (a situation whereby antibodies that are produced against a specific antigen display increasingly greater affinity for the antigen over the time of the immune response) by somatic hypermutation (SHM) [50, 51] as well as antibody class switching [52]. The development of the germinal centre has now been well-defined and is outside the scope of this review. For further reading, I direct readers to an excellent review by Klein and Dalla-Favera [48].

Why is the route of PC differentiation important? The current literature suggests that PCs generated outside of germinal centres (i.e., extrafollicularly) are thought to be mostly short-lived [1, 10]. They provide a rapid initial response to pathogens but undergo apoptosis in situ within a few days [13, 53]. In contrast, PCs generated from germinal centres with T cell-dependent help mostly migrate to the bone marrow [10, 54] with a small proportion remaining in the spleen [53]. Additional studies have also demonstrated that PCs can migrate and persist in inflamed tissues under chronic inflammatory conditions [38, 55]. This migration to bone marrow (and/or other sites) plays an important role in determining whether a PC becomes long-lived or not, as little evidence exists to demonstrate that LPC generation is dependent on intrinsic or internal cell signals. Consistent with this idea, LPCs harvested from bone marrow exhibit affinity-matured antibodies [8, 9, 40] suggesting that they emerged from germinal centre reactions. At the moment, current evidence suggests that the lifespan of a PC is dependent on multiple extrinsic survival factors and cell-cell interactions. Therefore, being a LPC is conditional upon locating a dedicated plasma cell survival niche such as the bone marrow - if a plasma cell arrives at the right place, it can become longlived.

EXTRINSIC FACTORS FOR SURVIVAL SECRETED BY THE BONE MARROW MICROENVIRONMENT

Bone marrow hosts the majority of LPCs and plays an important role in their survival. Increasing experimental data points towards the idea that the bone marrow functions as an ecological 'niche' where the microenvironment provides survival signals to LPCs through 'cellular survival factors'. Here, I will first review the extrinsic factors involved in the regulation of LPC survival, before discussing how the bone marrow microenvironment provides a survival niche to maintain a long-lived plasma cell. The removal of PCs from bone marrow for in vitro cell culture led investigators to observe that PCs had a poor survival rate [19]. This brought about studies which attempted to isolate soluble factors that could sustain PCs and contribute to the longevity of LPCs. The cytokines IL-5, IL-6, TNFα and CXC-chemokine ligand 12 (CXCL12, also known as stromal cellderived factor 1) were identified as individual factors which improve LPC survival in vitro, promoting the survival of seeded LPCs for at least 3 days [19]. Intriguingly, the role of IL-6 as an essential survival signal for LPCs was initially contradictory. In vitro studies indicated that it is a crucial factor for the survival of bone marrow derived plasma cells [19, 56, 57]. In addition, IL-6 had the most pronounced effect on PC survival, as addition of this cytokine rescued 70% of cultured PCs [19].

However, LPC numbers and persistence were similar in IL-6-deficient mice as compared to wild-type mice [19] which indicates that IL-6 was not essential for LPC survival in vivo. Interestingly, the presence of PCs alone could induce the production of IL-6 from bone marrow stromal cells (BMSCs), which suggests that PCs themselves could contribute to the formation of a survival niche [56]. BMSCs are a collection of adherent cells (endothelial cells, reticular cells, adipocytes, osteoblasts, stromal fibroblasts and smooth muscle cells) in bone marrow with a variety of immunological roles including the maintenance of immunological memory and LPC survival [56, 58, 59]. The observation that PCs themselves contribute to a survival niche is not unlike the tumorigenesis of multiple myeloma (MM), whereby BMSCs stimulated by MM cells secrete abundant amounts of IL-6 which then promotes the growth of malignant MM cells [35, 60]. Thus, the production and effector function of IL-6 appears to result in a positive feedback loop for PC survival.

However, even when these factors were collectively provided in vitro, their synergistic operation did not result in a complete recovery of plasma cells [19]. Even the co-incubation of these cells with bone marrow supernatant did not result in a prolonged PC survival rate [19]. These observations suggest that non-soluble factors (i.e., cell-cell interactions) most likely provided additional signals for LPCs survival. To determine which bone marrow cell types could provide those interactions, Roldan et al. [61] showed that adherent cells (but not non-adherent cells) fractionated from bone marrow cell suspensions were able to support PC survival in vitro. To investigate if cellular interactions between LPCs and BMSCs play a role in LPC survival, BMSCs were used as feeder cells for LPCs in vitro, and significantly increased the survival of LPCs for up to 4 weeks [56]. Further investigations revealed that VLA-4 and VCAM-1 interactions between LPCs and BMSCs provided the supportive effect [56, 62, 63]. Signalling via CD44 was also shown to promote a modest survival response [19]. CD44 is a cell surface glycoprotein involved in cell-cell interactions, cell adhesion and migration [64]. Its principal ligand is hyaluronic acid (HA) which is a common component of the extracellular matrix [64, 65].

Another group of important survival factors is the tumour necrosis factor subfamily of ligands and receptors. The receptors in this subfamily include the transmembrane activator and calcium modulator ligand interactor (TACI), B-cell maturation antigen (BCMA) and B-cell activating factor receptor (BAFFR). The ligands are B-cell activating factor (BAFF) and a proliferation-inducing ligand (APRIL). The precise interactions between each ligand and receptor have been difficult to identify [66, 67, 68, 69] because the ligands are able to adopt monomeric or oligomeric forms. These forms or complexes can also be soluble or insoluble. As a result, the different ligand forms greatly affect their in vivo interactions with the receptors (reviewed in [70]). Nevertheless, it is clear that BAFF-BAFFR interactions do not play a role in LPC survival [70, 71]. In contrast, interactions between the ligands (BAFF, APRIL) with the receptors (BCMA, TACI) have been shown to be important for the survival of LPCs. In vivo studies demonstrated impaired LPC survival in BCMA-/- mice

[72], and that blockage of TACI with TACIImmunoglobulin caused a decay in the LPC population [71, 72]. TACI was later shown to be necessary for the sustained expression of a B-cell gene that regulates the differentiation of long-lived antibody secreting cells [73]. Noelle's group [71] also demonstrated that eradication of the LPC population from bone marrow required the neutralization of both BAFF and APRIL ligands. The presence of either BAFF or APRIL is sufficient for long-term PC survival which indicates that these two ligands have redundant supportive roles.

THE BONE MARROW MICROENVIRONMENT AS A SURVIVAL NICHE FOR LPCS

How do these soluble and insoluble factors contribute to a specific survival niche in bone marrow? With knowledge of the factors involved and their role in LPC survival, a representation of the interactions occurring within the bone marrow microenvironment can be envisioned (fig. 3). Following activation in the GC and differentiation, plasmablasts migrate to the bone marrow [49]. As with most immune effector cells, the homing of the PC to bone marrow is controlled by a potpourri of adhesion molecules, chemokines and receptors [74]. Plasma cell homing to the bone marrow is dependent on three chemokine receptors: CXCR3 [21, 75], CXCR4 [41, 76] and CXCR6 [77] which bind to CXCL9, CXCL12 and CXCL16 respectively. Following the migration of plasmablasts to bone marrow , these cells bind to BMSCs which express vascular cell-adhesion molecule (VCAM-1) via VLA- 4 [62, 63]. Endothelial-cell selectin (E-selectin) on the surface of BMSCs will also retain these PCs in bone marrow by binding to polysaccharides on the PC's surface [78]. The interaction of PCs with BMSCs will signal these cells to produce and release IL-6 as a survival signal [56]. BMSCs will also secrete other soluble factors such as IL-5 and TNFα, which also act as survival signals [19]. In addition, interaction of the PC with HA in the extracellular matrix activates CD44 signalling which provides another survival stimulus for PCs [19]. Further interactions between PCs and BMSCs via BAFF/APRIL with BMCA/TACI would also provide crucial survival signals to LPCs [71, 72]. Therefore, current data suggests that LPCs are regulated by a survival mechanism, i.e., anti-apoptotic signals. Accordingly, they show a pro-survival phenotype and gene expression profile (expression of anti-apoptotic Bcl-2, no CD95 FasR) [10, 11, 47].

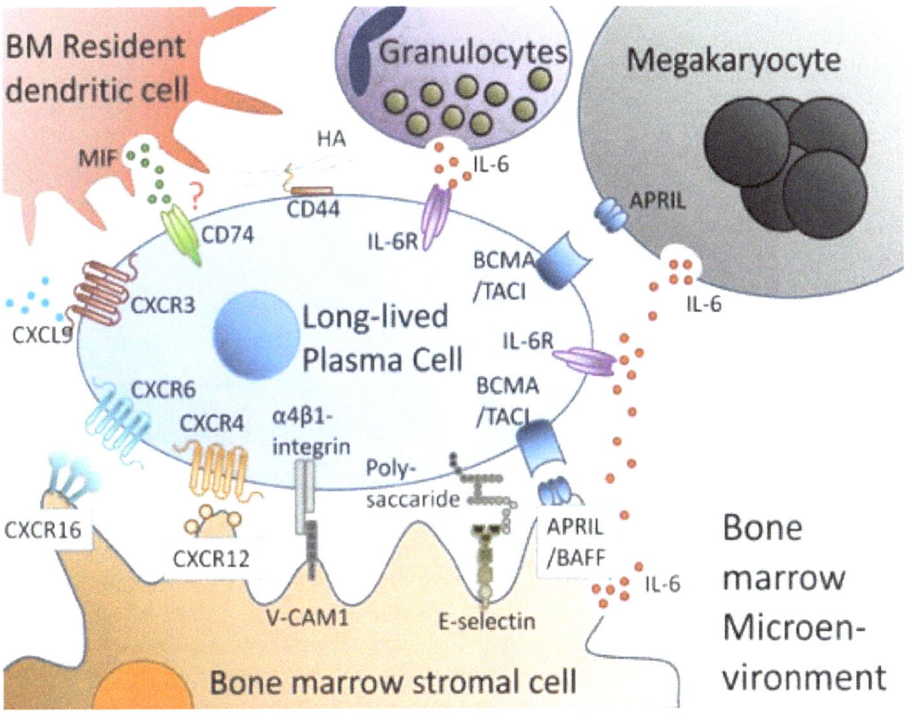

Figure 3: Survival factors which contribute to the persistence of long-lived PCs in the bone marrow microenvironment. Following the GC response, PCs with somatically-mutated, class-switched immunoglobulin leave the germinal centre and home to bone marrow with the help of CXCR3, CXCR12 and CXCR16 produced by bone marrow stromal cells. The expression of V-CAM1 and E-selectin on the surface of stromal cells retains PCs in bone marrow through interactions with VLA-4 and polysaccharides on the surface of the PC. Stromal cells, megakaryocytes or granulocytes (basophils, eosinophils) provides crucial survival signals to the PC through IL-6 and the ligands, B-cell activating factor (BAFF) or a proliferationinducing ligand (APRIL). BAFF/APRIL activates the Bcell maturation antigen (BCMA) or transmembrane activator and calcium modulator ligand interactor (TACI) receptors on the PC. CD44 interactions with hyaluronic acid provide a modest survival signal. Resident bone marrow dendritic cells may provide a survival signal via macrophage migratory inhibitory factor (MIF). The '?' denotes a survival pathway which requires further experiments for confirmation.

A large variety of cells comprise the bone marrow environment and it is likely that other cell types besides BMSCs play a role in the persistence of LPCs. Indeed, a recent series of studies has shown that a variety of non-stromal cells in the bone marrow - megakaryocytes (platelet precursors) [79], basophils [80]and eosinophils [81] all contribute to LPC survival. The contribution of these cells to LPC survival was demonstrated by several in vivo and in vitro approaches. In these studies, the authors first showed that these cells could interact with LPCs. Megakaryocytes were shown to co-localize and interact with LPCs and that megakaryocytes are an important source of two LPC survival factors - IL-6 and APRIL [79]. Co-culture of mouse PCs with either basophils or eosinophils also increases PC survival and antibody secretion which was dependent on IL-4 and IL-6 (from basophils) [80] or IL-4, IL-6, IL-10, TNFα and APRIL (from eosinophils) [81, 82]. Next, depletion of any of these three cell types in mice in vivo reduced the number of LPCs and significantly impaired PC responses [79, 80, 81, 82]. Finally, increasing the number of megakaryocytes by stimulating megakaryopoiesis in mice (a process tightly regulated by thrombopoietin and its receptor c-mpl) resulted in an increased number of PCs and increased PC persistence.

Finally, two studies have implicated that bone marrow dendritic cells (bmDCs) could play a role in the survival of PCs. The first study by Geffroy-Luseau et al. [83] involved co-culturing plasmablasts with bmDCs or osteoclasts (a stromal cell-type) in vitro. The authors observed that dendritic cells only supported the survival of plasmablasts but were unable to support PC survival. In contrast, osteoclasts (a stromal cell-type) could support the survival of both plasmablasts and plasma cells. By inhibiting BCMA with antibodies, they demonstrated that secreted BAFF and APRIL ligands were not involved. Instead, PC survival was dependent on cell-cell contact. In another study, Sapoznikov et al. [84] demonstrated that bmDCs could contribute to the survival of recirculating mature B cells in bone marrow through the secretion of macrophage migration-inhibitory factor (MIF). The authors used CD11c-diphtheria toxin receptor transgenic mice which when treated with diphtheria toxin results in the conditional ablation of bmDCs [85]. Interestingly, the authors also noted that bmDCs were restricted to specific perivascular clusters and speculated that these clusters localized to the same ecological niche as LPCs. However, neither study characterized the role of bmDCs in LPC survival nor further experiments are needed to determine if bmDCs can contribute to the LPC persistence.

The current data suggest that microenvironments within bone marrow function as ecological niches to provide specific survival signals promoting the persistence of LPCs. Referred to as the 'plasma cell competition model', this model describes how the survival of LPCs depends on their ability to access a survival niche [47]. This 'search' is suggested to occur in a completely stochastic weighted-random manner. The model predicts that only a limited number of these ecological niches exist. The number of BMSCs which express VCAM-1 are estimated to be 17% of all stromal cells [86]. This will limit the total pool of LPCs which can be sustained in bone marrow. Indeed, the number of PCs in human has been estimated to be 0.1%-1.0% of bone marrow cells [87] and mouse bone marrow has been estimated to be able to sustain 10^6 LPCs [88]. This raises the question of how the number of LPCs would be maintained in a stable state over years of infections, considering that in a single year, an individual may generate 10^4 -10^5 new PCs [8]. Thus, the competition model indicates that LPCs may persist for long periods of time when in a survival niche, but may be displaced through competition with newly formed migratory plasmablasts [47]. These new cells compete with older resident LPCs for the finite space available and any LPCs which are displaced are believed to be unable to regain localization and subsequently die due to ER stress. Multiple myeloma could represent a dysfunction of this system where available ecological niches to support PC survival are overtaken by MM cells [35]. Indeed, a common symptom in MM patients is the decreased production and increased destruction of normal antibodies resulting in diffuse hypogammaglobulinemia [34].

CONCLUSION

The importance of long-lived plasma cells in humoral immunity has only been recognized recently [8]. This has opened a new direction of humoral immunity research. Our knowledge of LPCs and how they play a role in humoral immunity has increased greatly in the past decade. Many extrinsic factors (cytokines, chemokines, surface cell-cell molecules) and neighbouring cells within the survival niche microenvironment have now been identified. As described above, the principal factors which govern the immortality of LPCs are the cytokine IL-6 and the TNF-subfamily ligands, APRIL and BAFF. Targeting these molecules to impair LPC survival could provide novel opportunities for therapeutic intervention in diseases such as autoimmunity or PC cancers such as multiple myeloma. However, the ideal therapeutic option would be the identification of novel LPC-specific molecules which will allow specific depletion of pathogenic LPCs in autoimmune diseases. This remains an important challenge of the future and a central step towards the development of curative therapy. Although great strides have been made in elucidating the regulation of survival of LPCs in the bone marrow microenvironment, detailed characteristics of these BM niches (e.g., cell types, size, accessibility, ontogeny, dynamics) are still not well defined and require further attention. However, as paradigms become known, new data and conclusions emerge to challenge them. No doubt, these are extremely exciting times for research in B cell humoral memory as we await further groundbreaking publications in this area.

ACKNOWLEDGEMENTS

The author wishes to thank Dr. Julie Deans and Dr. Yan Shi for their helpful comments on the initial draft of this manuscript as well as Dr. Paul Kubes for being a great mentor. The author would also like to acknowledge Alberta Innovates – Health Solutions (AIHS) for providing the Graduate Studentship funding support for his graduate studies.

REFERENCES

1. Shapiro-Shelef M & Calame K (2005) Regulation of plasmacell development. Nat Rev Immunol 5: 230-242.
2. Tangye SG (2011) Staying alive: regulation of plasma cell survival. Trends Immunol 32: 595-602.
3. Lamm ME (1997) Interaction of antigens and antibodies at mucosal surfaces. Annu Rev Microbiol 51: 311-340.
4. Casadevall A, Dadachova E, Pirofski LA (2004) Passive antibody therapy for infectious diseases. Nat Rev Microbiol 2: 695-703.
5. Crotty S, Felgner P, Davies H, Glidewell J, Villarreal L, Ahmed R (2003) Cutting edge: longterm B cell memory in humans after smallpox vaccination. J Immunol 171: 4969-4973.

6. Hammarlund E, Lewis MW, Hansen SG, Strelow LI, Nelson JA, Sexton GJ, Hanifin JM, Slifka MK (2003) Duration of antiviral immunity after smallpox vaccination. Nat Med 9: 1131-1137.
7. Amanna IJ, Carlson NE, Slifka MK (2007) Duration of humoral immunity to common viral and vaccine antigens. N Engl J Med 357: 1903-1915.
8. Manz RA, Thiel A, Radbruch A (1997) Lifetime of plasma cells in the bone marrow. Nature 388: 133-134.
9. Slifka MK, Antia R, Whitmire JK, Ahmed R (1998) Humoral immunity due to long-lived plasma cells. Immunity 8: 363- 372.
10. Yoshida T, Mei H, Dörner T, Hiepe F, Radbruch A, Fillatreau S, Hoyer BF (2010) Memory B and memory plasma cells. Immunol Rev 237: 117-139.
11. Elgueta R, de Vries VC, Noelle RJ (2010) The immortality of humoral immunity. Immunol Rev 236: 139-150.
12. Manz RA, Lohning M, Cassese G, Thiel A, Radbruch A (1998) Survival of long-lived plasma cells is independent of antigen. Int Immunol 10: 1703-1711.
13. Smith KG, Hewitson TD, Nossal GJ, Tarlinton DM (1996) The phenotype and fate of the antibody-forming cells of the splenic foci. Eur J Immunol 26: 444-448.
14. Radbruch A, Muehlinghaus G, Luger EO, Inamine A, Smith KG, Dorner T, Hiepe F (2006) Competence and competition: the challenge of becoming a long-lived plasma cell. Nat Rev Immunol 6: 741-750.
15. Zinkernagel RM, Bachmann MF, Kundig TM, Oehen S, Pirchet H, Hengartner H (1996) On immunological memory. Annu Rev Immunol 14: 333-367.
16. Schooley JC (1961) Autoradiographic observations of plasma cell formation. J Immunol 86: 331-337.
17. Nossal GJ & Makela O (1962) Autoradiographic studies on the immune response.I. The kinetics of plasma cell proliferation. J Exp Med 115: 209-230.
18. Makela O & Nossal GJ (1962) Autoradiographic studies on the immune response. II. DNA synthesis amongst single antibody-producing cells. J Exp Med 115: 231-244.
19. Cassese G, Arce S, Hauser AE, Lehnert K, Moewes B, Mostarac M, Muehlinghaus G, Szyska M, Radbruch A, Manz RA (2003) Plasma cell survival is mediated by synergistic effects of cytokines and adhesion-dependent signals. J Immunol 171: 1684-1690.
20. Auner HW, Beham-Schmid C, Dillon N, Sabbattini P (2010) The life span of short-lived plasma cells is partly determined by a block on activation of apoptotic caspases acting in combination with endoplasmic reticulum stress. Blood 116: 3445-3455.
21. Hauser AE, Debes GF, Arce S, Cassese G, Hamann A, Radbruch A, Manz RA (2002) Chemotactic responsiveness toward ligands for CXCR3 and CXCR4 is regulated on plasma blasts during the time course of a memory immune response. J Immunol 169: 1277-1282.
22. Miller JJr (1964) An autoradiographic study of plasma cell and lymphocyte survival in rat popliteal lymph nodes. J Immunol 92: 673-681.
23. Rastetter W, Molina A, White CA (2004) Rituximab: expanding role in therapy for lymphomas and autoimmune diseases. Annu Rev Med 55: 477-503.
24. Edwards JC, Szczepanski L, Szechinski J, FilipowiczSosnowska A, Emery P, Close DR, Stevens RM, Shaw T (2004) Efficacy of B-cell-targeted therapy with rituximab in patients with rheumatoid arthritis. N Engl J Med 350: 2572- 2581.
25. Cambridge G, Leandro MJ, Edwards JC, Ehrenstein MR, Salden M, Bodman-Smith M, Webster AD (2003) Serologic changes following B lymphocyte depletion therapy for rheumatoid arthritis. Arthritis Rheum 48: 2146-2154.
26. Vallerskog T, Gunnarsson I, Widhe M, Risselada A, Klareskog L, van Vollenhoven R, Malmstrom V, Trollmo C (2007) Treatment with rituximab affects both the cellular and the humoral arm of the immune system in patients with SLE. Clin Immunol 122: 62-74.
27. Amanna IJ, Carlson NE, Slifka MK (2007) Duration of humoral immunity to common viral and vaccine antigens. N Engl J Med 357: 1903-1915.
28. Mesin L, Di Niro R, Thompson KM, Lundin KE, Sollid LM (2011) Long-lived plasma cells from human small intestine biopsies secrete immunoglobulins for many weeks in vitro. J Immunol 187: 2867-2874.
29. O'Connor BP, Gleeson MW, Noelle RJ, Erickson LD (2003) The rise and fall of long-lived humoral immunity: terminal differentiation of plasma cells in health and disease. Immunol Rev 194: 61-76.
30. Strugnell RA & Wijburg OL (2010) The role of secretory antibodies in infection immunity. Nat Rev Microbiol 8: 656- 667.
31. Burton DR (2002) Antibodies, viruses and vaccines. Nat Rev Immunol 2: 706-713.
32. Detmer A & Glenting J (2006) Live bacterial vaccines--a review and identification of potential hazards. Microb Cell Fact 5: 23.
33. Goodnow CC, Vinuesa CG, Randall KL, Mackay F, Brink R (2010) Control systems and decision making for antibody production. Nat Immunol 11: 681-688.
34. Hartley-Brown MA, Sullivan DM, Baz R. 2010. State-of-the-Art Management of Complications of Myeloma and Its Treatment. In Adv Hematol. 1-7.
35. Basak GW, Srivastava AS, Malhotra R, Carrier E (2009) Multiple myeloma bone marrow niche. Curr Pharm Biotechnol 10: 335-346.
36. Dörner T, Jacobi AM, Lipsky PE (2009) B cells in autoimmunity. Arthritis Res Ther 11: 247.
37. Yu D & Vinuesa CG (2010) Multiple checkpoints keep follicular helper T cells under control to prevent autoimmunity. Cell Mol Immunol 7: 198-203.
38. Starke C, Frey S, Wellmann U, Urbonaviciute V, Herrmann M, Amann K, Schett G, Winkler T, Voll RE (2011) High frequency of autoantibody-secreting cells and long-lived plasma cells within inflamed kidneys of NZB/W F1 lupus mice. Eur J Immunol 41: 2107-2112.
39. LeBien TW & Tedder TF (2008) B lymphocytes: how they develop and function. Blood 112: 1570-1580. 40.

40. Hiepe F, Dorner T, Hauser AE, Hoyer BF, Mei H, Radbruch A (2011) Long-lived autoreactive plasma cells drive persistent autoimmune inflammation. Nat Rev Rheumatol 7: 170-178.

41. Hargreaves DC, Hyman PL, Lu TT, Ngo VN, Bidgol A, Suzuki G, Zou YR, Littman DR, Cyster JG (2001) A coordinated change in chemokine responsiveness guides plasma cell movements. J Exp Med 194: 45-56.

42. Muehlinghaus G, Cigliano L, Huehn S, Peddinghaus A, Leyendeckers H, Hauser AE, Hiepe F, Radbruch A, Arce S, Manz RA (2005) Regulation of CXCR3 and CXCR4 expression during terminal differentiation of memory B cells into plasma cells. Blood 105: 3965-3971.

43. Comerford I & McColl SR (2011) Mini-review series: focus on chemokines. Immunol Cell Biol 89: 183-184.

44. Hoyer BF, Moser K, Hauser AE, Peddinghaus A, Voigt C, Eilat D, Radbruch A, Hiepe F, Manz RA (2004) Short-lived plasmablasts and long-lived plasma cells contribute to chronic humoral autoimmunity in NZB/W mice. J Exp Med 199: 1577-1584.

45. Ho F, Lortan JE, MacLennan IC, Khan M (1986) Distinct short-lived and long-lived antibody-producing cell populations. Eur J Immunol 16: 1297-1301.

46. Rozanski CH, Arens R, Carlson LM, Nair J, Boise LH, Chanan-Khan AA, Schoenberger SP, Lee KP (2011) Sustained antibody responses depend on CD28 function in bone marrow-resident plasma cells. J Exp Med 208: 1435-1446.

47. Oracki SA, Walker JA, Hibbs ML, Corcoran LM, Tarlinton DM (2010) Plasma cell development and survival. Immunol Rev 237: 140-159.

48. Klein U & Dalla-Favera R (2008) Germinal centres: role in Bcell physiology and malignancy. Nat Rev Immunol 8: 22-33.

49. Moser K, Tokoyoda K, Radbruch A, MacLennan I, Manz RA (2006) Stromal niches, plasma cell differentiation and survival. Curr Opin Immunol 18: 265-270.

50. Teng G & Papavasiliou FN (2007) Immunoglobulin somatic hypermutation. Annu Rev Genet 41: 107-120.

51. Peled JU, Kuang FL, Iglesias-Ussel MD, Roa S, Kalis SL, Goodman MF, Scharff MD (2008) The biochemistry of somatic hypermutation. Annu Rev Immunol 26: 481-511.

52. Stavnezer J, Guikema JE, Schrader CE (2008) Mechanism and regulation of class switch recombination. Annu Rev Immunol 26: 261-292.

53. Sze DM, Toellner KM, Garcia de Vinuesa C, Taylor DR, MacLennan IC (2000) Intrinsic constraint on plasmablast growth and extrinsic limits of plasma cell survival. J Exp Med 192: 813-821.

54. Fooksman DR, Schwickert TA, Victora GD, Dustin ML, Nussenzweig MC, Skokos D (2010) Development and migration of plasma cells in the mouse lymph node. Immunity 33: 118-127.

55. Cassese G, Lindenau S, de Boer B, Arce S, Hauser A, Riemekasten G, Berek C, Hiepe F, Krenn V, Radbruch A, Manz RA (2001) Inflamed kidneys of NZB / W mice are a major site for the homeostasis of plasma cells. Eur J Immunol 31: 2726-2732.

56. Minges Wols HA, Underhill GH, Kansas GS, Witte PL (2002) The role of bone marrow-derived stromal cells in the maintenance of plasma cell longevity. J Immunol 169: 4213- 4221.

57. Ellyard JI, Avery DT, Phan TG, Hare NJ, Hodgkin PD, Tangye SG (2004) Antigen-selected, immunoglobulinsecreting cells persist in human spleen and bone marrow. Blood 103: 3805–3812.

58. Tokoyoda K, Hauser AE, Nakayama T, Radbruch A (2010) Organization of immunological memory by bone marrow stroma. Nat Rev Immunol 10: 193-200.

59. Dilosa RM, Maeda K, Masuda A, Szakal AK, Tew JG (1991) Germnal center B cells and antibody production in bone marrow. J Immunol 146: 4071-4077.

60. Xie JY, Li MX, Xiang DB, Mou JH, Qing Y, Zeng LL, Yang ZZ, Guan W, Wang D (2010) Elevated Expression of APE1/Ref-1 and its Regulation on IL-6 and IL-8 in Bone Marrow Stromal Cells of Multiple Myeloma. Clin Lymphoma Myeloma Leuk 10: 385-393.

61. Roldan E & Brieva JA (1991) Terminal differentiation of human bone marrow cells capable of spontaneous and highrate immunoglobulin secretion: role of bone marrow stromal cells and interleukin 6. Eur J Immunol 21: 2671-2677.

62. DiLillo DJ, Hamaguchi Y, Ueda Y, Yang K, Uchida J, Haas KM, Kelsoe G, Tedder TF (2008) Maintenance of long-lived plasma cells and serological memory despite mature and memory B cell depletion during CD20 immunotherapy in mice. J Immunol 180: 361-371.

63. Schaumann DH, Tuischer J, Ebell W, Manz RA, Lauster R (2007) VCAM-1-positive stromal cells from human bone marrow producing cytokines for B lineage progenitors and for plasma cells: SDF-1, flt3L, and BAFF. Mol Immunol 44: 1606- 1612.

64. Ponta H, Sherman L, Herrlich PA (2003) CD44: from adhesion molecules to signalling regulators. Nat Rev Mol Cell Biol 4: 33-45.

65. Miyake K, Underhill CB, Lesley J, Kincade PW (1990) Hyaluronate can function as a cell adhesion molecule and CD44 participates in hyaluronate recognition. J Exp Med 172: 69-75.

66. Liu Y, Hong X, Kappler J, Jiang L, Zhang R, Xu L, Pan CH, Martin WE, Murphy RC, Shu HB, Dai S, Zhang G (2003) Ligand-receptor binding revealed by the TNF family member TALL-1. Nature 423: 49-56.

67. Hendriks J, Planelles L, de Jong-Odding J, Hardenberg G, Pals ST, Hahne M, Spaargaren M, Medema JP (2005) Heparan sulfate proteoglycan binding promotes APRIL-induced tumor cell proliferation. Cell Death Differ 12: 637-648.

68. Bossen C, Cachero TG, Tardivel A, Ingold K, Willen L, Dobles M, Scott ML, Maquelin A, Belnoue E, Siegrist CA, Chevrier S, Acha-Orbea H, Leung H, Mackay F, Tschopp J, Schneider P (2008) TACI, unlike BAFF-R, is solely activated by oligomeric BAFF and APRIL to support survival of activated B cells and plasmablasts. Blood 111: 1004-1012.

69. Kimberley FC, van Bostelen L, Cameron K, Hardenberg G, Marquart JA, Hahne M, Medema JP (2009) The proteoglycan (heparan sulfate proteoglycan) binding domain of APRIL serves as a platform for ligand multimerization and crosslinking. FASEB J 23: 1584-1595.

70. Mackay F & Schneider P (2009) Cracking the BAFF code. Nat Rev Immunol 9: 491-502.

71. Benson MJ, Dillon SR, Castigli E, Geha RS, Xu S, Lam KP, Noelle RJ (2008) Cutting edge: the dependence of plasma cells and independence of memory B cells on BAFF and APRIL. J Immunol 180: 3655-3659.

72. O'Connor BP, Raman VS, Erickson LD, Cook WJ, Weaver LK, Ahonen C, Lin LL, Mantchev GT, Bram RJ, Noelle RJ (2004) BCMA is essential for the survival of long-lived bone marrow plasma cells. J Exp Med 199: 91-98.

73. Tsuji S, Cortesao C, Bram RJ, Platt JL, Cascalho M (2011) TACI deficiency impairs sustained Blimp-1 expression in B cells decreasing long-lived plasma cells in the bone marrow. Blood 118: 5832-5839.

74. Kunkel EJ & Butcher EC (2003) Plasma-cell homing. Nat Rev Immunol 3: 822-829.

75. Odendahl M, Mei H, Hoyer BF, Jacobi AM, Hansen A, Muehlinghaus G, Berek C, Hiepe F, Manz R, Radbruch A, Dörner T (2005) Generation of migratory antigen-specific plasma blasts and mobilization of resident plasma cells in a secondary immune response. Blood 105: 1614-1621.

76. Wehrli N, Legler DF, Finke D, Toellner KM, Loetscher P, Baggiolini M, MacLennan IC, Acha-Orbea H (2001) Changing responsiveness to chemokines allows medullary plasmablasts to leave lymph nodes. Eur J Immunol 31: 609-616.

77. Nakayama T, Hieshima K, Izawa D, Tatsumi Y, Kanamaru A, Yoshie O (2003) Cutting edge: profile of chemokine receptor expression on human plasma cells accounts for their efficient recruitment to target tissues. J Immunol 170: 1136- 1140.

78. Underhill GH, H.A. MW, Fornek JL, Witte PL, Kansas GS (2002) IgG plasma cells display a unique spectrum of leukocyte adhesion and homing molecules. Blood 99: 2905- 2912.

79. Winter O, Moser K, Mohr E, Zotos D, Kaminski H, Szyska M, Roth K, Wong DM, Dame C, Tarlinton DM, Schulze H, MacLennan IC, Manz RA (2010) Megakaryocytes constitute a functional component of a plasma cell niche in the bone marrow. Blood 116: 1867-1875.

80. Rodriguez Gomez M, Talke Y, Goebel N, Hermann F, Reich B, Mack M (2010) Basophils support the survival of plasma cells in mice. J Immunol 185: 7180-7185.

81. Chu VT & Berek C (2012) Immunization induces activation of bone marrow eosinophils required for plasma cell survival. Eur J Immunol 42: 130-137.

82. Chu VT, Frohlich A, Steinhauser G, Scheel T, Roch T, Fillatreau S, Lee JJ, Lohning M, Berek C (2011) Eosinophils are required for the maintenance of plasma cells in the bone marrow. Nat Immunol 12: 151-159.

83. Geffroy-Luseau A, Jégo G, Bataille R, Campion L, PellatDeceunynck C (2008) Osteoclasts support the survival of human plasma cells in vitro. Int Immunol 20: 775-782.

84. Sapoznikov A, Pewzner-Jung Y, Kalchenko V, Krauthgamer R, Shachar I, Jung S (2008) Perivascular clusters of dendritic cells provide critical survival signals to B cells in bone marrow niches. Nat Immunol 9: 388-395.

85. Jung S, Unutmaz D, Wong P, Sano G, De los Santos K, Sparwasser T, Wu S, Vuthoori S, Ko K, Zavala F, Pamer EG, Littman DR, Lang RA (2002) In vivo depletion of CD11c+ dendritic cells abrogates priming of CD8+ T cells by exogenous cell-associated antigens. Immunity 17: 211-220.

86. Tokoyoda K, Egawa T, Sugiyama T, Choi BI, Nagasawa T (2004) Cellular niches controlling B lymphocyte behavior within bone marrow during development. Immunity 20: 707- 718.

87. Terstappen LW, Johnsen S, Segers-Nolten IM, Loken MR (1990) Identification and characterization of plasma cells in normal human bone marrow by high-resolution flow cytometry. Blood 76: 1739-1747.

88. Haaijman JJ, Schuit HR, Hijmans W (1977) Immunoglobulincontaining cells in different lymphoid organs of the CBA mouse during its life-span. Immunology 32: 427-434.

89. Good-Jacobson KL & Shlomchik MJ (2010) Plasticity and heterogeneity in the generation of memory B cells and longlived plasma cells: the influence of germinal center interactions and dynamics. J Immunol 185: 3117-3125.

90. Delogu A, Schebesta A, Sun Q, Aschenbrenner K, Perlot T, Busslinger M (2006) Gene repression by Pax5 in B cells is essential for blood cell homeostasis and is reversed in plasma cells. Immunity 24: 269-281.

91. Cobaleda C, Schebesta A, Delogu A, Busslinger M (2007) Pax5: the guardian of B cell identity and function. Nat Immunol 8: 463-470.

CHAPTER 7

REVIEW OF QUANTITATIVE PHASE-DIGITAL HOLOGRAPHIC MICROSCOPY: PROMISING NOVEL IMAGING TECHNIQUE TO RESOLVE NEURONAL NETWORK ACTIVITY AND IDENTIFY CELLULAR BIOMARKERS OF PSYCHIATRIC DISORDERS

Pierre Marquet[a,b], Christian Depeursinge[b,c] and Pierre J. Magistretti[a,b,c]

[a]Centre Hospitalier Universitaire Vaudois (CHUV), Centre de Neurosciences Psychiatriques, Département de Psychiatrie, Site de Cery, Prilly/ Lausanne CH-1008, Switzerland

[b]Brain Mind Institute, École Polytechnique Fédérale de Lausanne, Lausanne CH-1015, Switzerland

[c]King Abdullah University of Science and Technology (KAUST), Division of Biological and Environmental Sciences and Engineering, Thuwal 23955-6900, Kingdom of Saudi Arabia

ABSTRACT

Quantitative phase microscopy (QPM) has recently emerged as a new powerful quantitative imaging technique well suited to noninvasively explore a transparent specimen with a nanometric axial sensitivity. In this review, we expose the recent developments of quantitative phase-digital holographic microscopy (QP-DHM). Quantitative phase-digital holographic microscopy (QP-DHM) represents an important and efficient quantitative phase method to explore cell structure and dynamics. In a second part, the most relevant QPM applications in the field of cell biology are summarized. A particular emphasis is placed on the original biological information, which can be derived from the quantitative phase signal. In a third part, recent applications obtained, with QP-DHM in the field of cellular neuroscience, namely the possibility to optically resolve neuronal network activity and spine dynamics, are presented. Furthermore, potential applications of QPM related to psychiatry through the identification of new and original cell biomarkers that, when combined with a range of other biomarkers, could significantly contribute to the determination of high risk developmental trajectories for psychiatric disorders, are discussed.

INTRODUCTION

Elucidating the mechanisms underlying physiological and pathophysiological cellular processes requires the possibility to quantitatively and noninvasively observe cell structure and dynamics down to cellular and subcellular levels.1 In addition, cell processes are usually multifaceted and cannot be efficiently grasped with techniques providing a view of a single aspect. Thus, the development of high resolution imaging approaches providing simultaneous measurements of multiple relevant biophysical cell parameters is a prerequisite to dissect complex cellular processes. Considering that most biological cells are transparent, i.e., they differ only slightly from their surroundings in terms of optical properties (including absorbance, reflectance, etc.), it is obvious that obtaining a

Citation: Pierre Marquet ; Christian Depeursinge and Pierre J. Magistretti, "Review of quantitative phase-digital holographic microscopy: promising novel imaging technique to resolve neuronal network activity and identify cellular biomarkers of psychiatric disorders", Neurophoton. 1(2), 020901 (Sep 22, 2014). ; http://dx.doi.org/10.1117/1.NPh.1.2.020901.

comprehensive understanding of cellular physiological and pathophysiological processes with optical approaches remains a difficult challenge. Consequently, a wide range of optical approaches have been developed in cell biology.2 Specifically, the development of both new modes of contrast generation and new forms of high resolution microcopy techniques which allow us to break the so-called "diffraction limit" of resolution, which limits the smallest observable details in the image to around 200–500 nm, represent a very active area of research in the field of cell biology.3

In contrast to fluorescence microscopy, which today represents a powerful and widely used cell imaging technique,4 contrast-generating modes, based on wavefront phase information, represent an intrinsic contrast of transparent specimens, which have already in the mid-20th century demonstrated their relevance for noninvasive visualization of cell structure, in particular because the Zernicke's invention of phase contrast (PhC)5 and the DIC techniques developed by Smith6 and Nomarski.7 Currently, PhC and Normarski's differential interference contrast (DIC) are widely used contrast-generating techniques available for high-resolution light microscopy. Without using any staining, PhC and DIC allow the visualization of transparent specimens making visible, in particular, the fine subcellular structural organization. Basically, these two noninvasive contrast-generating techniques are based on their capacity for transforming in detectable intensity modulation the minute relative phase shift that a transparent microscopic object induces either between the transmitted wave light and the undeviated background wave (PhC) or between two orthogonally polarized transmitted waves (DIC).

In contrast to PhC and DIC, the invention of interference microscopy with separate object and reference beams at the end of the 19th century provides a direct and quantitative measurement of the phase difference introduced by the observed object. Practically, due to the seminal work of Barer, interference microscopy applied to cell imaging has permitted the quantitative measurement of important biophysical cells parameters including dry mass8 and intracellular RI9. However, because of temporal phase shifting, interferometric techniques require very demanding and costly opto-mechanical designs and phase shifts are very sensitive to experimental artifacts and noise, only a few attempts to dynamically image live cells in biology have been reported.10,11

Nevertheless, scientific advances lowering the cost of lasers and data acquisition equipment as well as the development of computing facilities, the large spread of personal computers and digital signal processors have brought fundamentally new perspectives in the field of interferometry and holography. Consequently, during the last decades, numerous quantitative phase microscopy (QPM) techniques based on an interferometer or holographic approaches have been developed. These approaches are much simpler to implement than classical interference microscopy and provide reliable and quantitative phase mapping of the observed specimen, therefore, they have started to be efficiently used in the field of cell biology.12,13

In a first part of this review, we present the current state of the art of quantitative phase measurement based on digital holographic microscopy (DHM) with a special focus on the advantages provided by the possibility to propagate the whole object wave (amplitude and phase) diffracted by the observed specimen during the numerical reconstruction of the digitally recorded holograms. A paragraph describing quantitative phase tomographic approaches providing a full three-dimensional (3-D) image—in particular a 3-D RI distribution—of the observed specimen are also presented. In a second part, we briefly describe some relevant applications related to cell biology provided by different QPM techniques. Special attention is paid to the biological information provided by the quantitative phase signal (QPS) through the derivation of specific cell biophysical parameters. In the third part, we describe how quantitative phase (QP)-DHM and quantitative phase tomography (QPT), due to their capacity for quantitatively monitoring cell structure and dynamics through the measurements of different cell biophysical parameters, can begin to address specific questions in the field of cellular neuroscience and psychiatric research. Specifically, the possibility to noninvasively visualize spine dynamics and resolve neuronal network activity by performing a simultaneous multiple sites optical recording of transmembrane currents as well as identifying new early cell biomarkers of vulnerability, neuroprogession, and diagnosis of mental disorders are discussed. Finally, some perspectives related to the development of QPM and/or QPT in high-throughput multimodality approaches are presented.

PRINCIPLES OF QUANTITATIVE PHASE MEASUREMENTS BASED ON DIGITAL HOLOGRAPHIC MICROSCOPY

Digital Holography

Holography techniques were developed by Gabor in 1948 with the aim of improving the detection of spatial resolution in the X-ray wavelength by exploiting its lensless imaging capabilities. This resulted in the possibility of generating an exact replica during the illumination of the recorded hologram (reconstruction process) with a specific magnification of the full object wavefront created by the observed specimen.14 The use of digital means in holography gradually occurred at the end of the 1960s, when Goodman used a vidicon detector to encode a hologram which could be reconstructed on a computer.15 However, the interest in digital holography rose with the availability of cheaper digital detectors and charge-coupled device cameras.16,17 The two practical main approaches to recover the object wave from the recorded hologram are namely temporal decoding, i.e., phase shifting and spatial decoding i.e., off-axis methods. Phase-shifting reconstruction methods are based on the combination of several frames, and enable the suppression of the zero order and one of the cross terms through temporal sampling.18,19The most well-known phase-shifting algorithm, proposed by Yamaguchi,20 is based on the recording of four frames separated by a phase shift of a quarter of a wavelength. Various combinations of frames derived from interferometry have been considered19,21and many different approaches have been developed to produce the phase shift including high precision piezo-electric transducers, which move a mirror in the reference wave or acousto-optics modulators using the light frequency shift, etc. One of the main issues in phase-shifting methods is the requirement of several frames for reconstruction in interferometric setups, which are commonly very sensitive to vibrations. Thus, it could be difficult to ensure stable phase shifts and an invariant sample state during acquisition. In addition, the requirements on the accuracy of the phase shifts are rather high in regard to displacements on the order of magnitude of hundreds of nanometers, implying the use of high-precision transducers. Consequently, several attempts were made to either reduce the required number of frames for reconstruction, which led to two-frame reconstruction,22·23 or to simultaneously enable the recoding of the various phase-shift frames, by employing, for example, multiplexing methods.24 On the other hand, more refined algorithms were developed to relax the accuracy requirements of phase-shift methods.25-27 It should also be mentioned that the development of a fast temporal phase-shifting procedure within the framework of tomography28 as well as approaches based on the spatial analysis of fringe pattern,29·30 sometimes referred as spatial phase shifting, has permitted to successfully reconstruct the object wave in an off-axis configuration.31·32

The second main approach for recovering the object wave is based on an off-axis configuration, so that the different diffraction terms encoded in the hologram (zero-order wave, real and virtual images) are propagating in different directions, enabling their separation for reconstruction.33 This configuration was the one employed for the first demonstration of a fully numerical recording and reconstruction holography.16·17

In practice, reconstruction methods based on an off-axis configuration usually rely on Fourier methods to filter one of the diffraction terms. This concept was first proposed by Takeda et al.34 in the context of interferometric topography. The method was later extended for smooth topographic measurements for phase recovery35 and generalized for use in DHM with amplitude and phase recovery.36 As discussed in the following paragraphs, the main characteristic of this approach is its capability for recovering the complex object wave through only one acquisition, thus greatly reducing the influence of vibrations. However, as the diffraction terms are spatially encoded in the hologram, this one shot capability potentially comes at the cost of the usable bandwidth. In addition, the frequency modulation, induced by the angle between the reference and the object wave, has to guarantee the separability of the information contained in the different diffraction terms that are encoded in the hologram while carrying a frequency compatible with the sampling capacity of the digital detectors.

Quantitative Phase Digital Holographic Microscopy

A digital holographic microscope37 comprises both an optical setup devoted to hologram formation and software specially developed to numerically process the digitized hologram. The hologram results from the interference of the object beam with a reference wave that can be kept separate from the object beam. The goal is to precisely estimate the propagated wavefront corresponding to either the virtual or the real image of the specimen. Different optical setups have been proposed to perform microscopy with holography. Some lensless setups have been proposed to image small fields38-40 by using strongly diverging wavefields, such as spherical waves emitted by punctual sources. Optical setups that make use of a microscope objective (MO) offer the largest numerical apertures.36·41 A typical DHM arrangement we have developed and used to explore transparent specimens including living cells, is the quantitative monitoring of the phase retardation of the transmitted wavefront [Fig. 1(a)].41

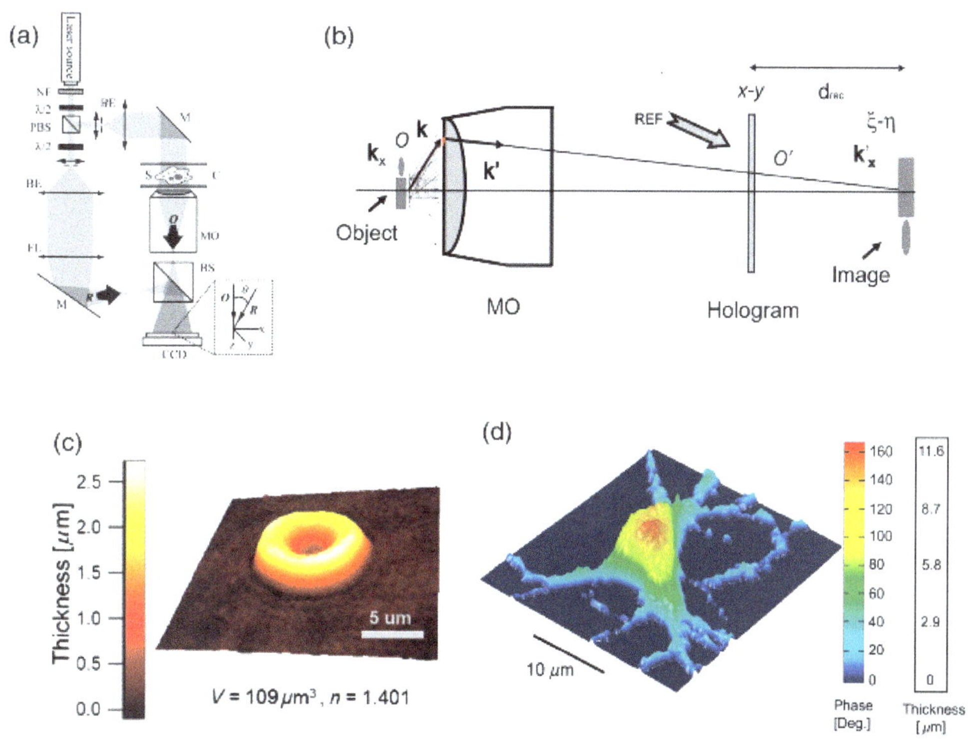

Figure 1: (a) Transmission optical setup for QP-DHM (b) role of the microscope objective (MO): magnifying the image of the specimen but also, in terms of wavevectors, decreasing the lateral wavevector component k_x or y by a factor proportional to the magnification (M) of the MO, so that the hologram can be sampled by an electronic camera according to the Shannon's rules (c,d) three-di-

mensional (3-D) perspective images in false colors of a living human red blood cell (RBC) and a mouse cortical neuron in culture, respectively. Each pixel represents a quantitative measurement of the phase retardation or cellular optical path length (OPL) induced by the cell with a sensitivity corresponding to a few tens of nanometers. By taking into account the measured mean values of the RBC and neuronal cell body refractive indices (RIs) resulting from the decoupling procedure (cf. below), scales (at right), which relate OPL (deg) to morphology in the zz-axis (μmμm), can be constructed. Abbreviations: BS, beam splitter; CCD, camera; M1 and M2, mirrors; MO, microscope objective; O, object beam; R, reference beam; RL, reference lens and OC, object condenser; object or specimen.

Other configurations are possible depending on the targeted application,12·42 but will not be reviewed in detail here. An important issue is the need for a reference beam that is controllable in both in intensity and polarization, thus enabling one to improve contrast and signal. The holographic principle also permits other valuable concepts which particularly building on the possibility of superimposing several holograms. Holograms with several reference waves corresponding to several polarization states can be generated in order to analyze the birefringence properties of specimens including strained dielectrics or biological molecules.43·44 Reference waves corresponding to different wavelengths can also be generated, permitting the use of synthetic wavelengths from a single hologram.45

As in other developments in digital holography, the hologram reconstruction method in DHM is based on the theory of diffraction. When an MO is used and the specimen can be maintained in the MO's focal plane, the image is formed at infinity. The hologram at infinity is simply the Fourier transform of the specimen. Inverse Fourier transformation straightforwardly provides the reconstructed wavefront in the focal plane of the MO. However, specimens are generally not fully contained in the focal plane of the MO, and thus a real and/or virtual image is formed at a finite distance from the MO [Fig. 1(b)]. Intercepting the wavefront at a finite distance from the specimen image gives rise to holograms in the Fresnel zone. Thus, the Huygens-Fresnel expression of diffraction can be used to calculate the propagation of the reconstructed wave. It should be noted that the MO introduced in the interferometer allows one to adapt the wavefield to the sampling capacity of the camera. Indeed, the lateral components of the wavevector k_x or y can be divided by the magnification factor M of the MO. Taking advantage of a full digital approach, a computer can perform the reconstruction by simulating the reference wave and multiplying it by the hologram intensity distribution. Then, the distribution of the wavefront in space is obtained by simulating the propagation of the wavefront reconstructed in the plane of the hologram. As previously mentioned, the wavefront gives rise to several propagated beams corresponding to the zero,+1 and −1 diffracted orders. When an MO is used, the latter two orders give rise to real and virtual images, respectively, that appear as twin images which are the reflection of one another on a virtual mirror that takes the place of the hologram plane.

The off-axis geometry introduces a spatial carrier frequency, and demodulation restores the full spatial frequency content of the wavefront. The main advantage of this approach is that all the information for reconstructing the complex wavefield comes from a single hologram.36 In microscopy, the full bandwidth of the beam delivered by the MO can be acquired without limitation. Scale adjustments can made to preserve the correct sampling of the MO's full bandwidth so that no loss of resolution is conceded. Finally, selection of a signal corresponding to a diffraction order of +1 in the Fourier domain of the hologram allows for the full restoration of the object wavefront transmitted by the MO. Therefore, reconstructing the wavefront in 3-D is simply done by propagating the wavefront generated in the hologram plane x-y to the image plane ξ-η, which is situated at a distance d_{rec}. This could be efficiently achieved by computing the Fresnel transform of the wavefield. Thus, DHM brings quantitative data simultaneously derived from the amplitude and the phase of the complex reconstructed wavefront that is diffracted by the object. Our method requires the adjustment of several reconstruction parameters, which can be done using a computer-aided method we developed. Some image processing is also needed to improve the accuracy of the reconstructed phase.46 Moreover any type of aberrations can also be corrected during such a numerical reconstruction process.47·48 Using a high numerical aperture, submicron transverse resolution has been achieved to a 300-nm lateral resolution. Accuracies of approximately 0.1° have been estimated for phase measurements. In reflection geometry, this corresponds to a vertical resolution of less than 1 nm at a wavelength of 632 nm.49 In transmission geometry, the resolution is limited to a few nanometers as far as living cells are concerned. It is also worthwhile to note that when the experimental configuration of the hologram recording requires distances of reconstruction for which the Fresnel approximation does not properly work, the propagation of the wavefront generated by the hologram can be efficiently performed with the angular spectrum approach.50·51 On the other hand, the characterization of the effect of noise on the formation of the hologram and on the reconstruction of the image results in an improved signal-to-noise ratio (SNR), using the coherent detection of scattered light.52·53 This improvement is often described as "coherent amplification" of the signal.

Quantitative Phase Tomographic Microscopy: Full Three-Dimensional Imaging Technique

In general, a single wavefront determination obtained from a single hologram is not sufficient to obtain a full 3-D image of an object. An exception concerns sparse media, similar to those found in particle image velocimetry (PIV), where each particle can be considered as an isolated punctual object and its position determined in 3-D by the methods of "in-line" holography.54·55 Thus, the combination of data reconstructed from several holograms, obtained at either several wavelengths (variable k-vector amplitude) or at several incidence angles (variable k-vector direction), is needed to achieve true 3-D imaging. Several articles report results obtained by changing the wavelength (variable k-vector amplitude) as shown by Marron.56 However, in this study, the range of the wavelength scan remained very small and consequently the resolution was weak. Arons et al. using "Fourier synthesis holography"57 have also discussed a similar approach. In digital holography, multiple wavelengths have been used to reconstruct 3-D structures.58 In the field of microscopy, such multiwavelength tomographic imaging techniques, based on reconstruction from multiple holograms, have yielded very accurate images of cells, erythrocytes in particular.59·60 An alternative, but somehow similar approach, is to use a wide bandwidth source and form a hologram in the plane where the mutual coherence between the object and reference wave is nonzero: this concept introduces coherence gating in the space domain. It has been proven to perform well.61·62 On the other hand, the variable k-vector direction approach more exactly meets the concept found in the literature under the name of "diffraction tomography"63 for reconstruction of the scattering potential associated with the structure of the diffracting object. A diffracted wave can be collected and reconstructed from the holograms at various incidences. Practically, the phase and amplitude of the diffracted wave are directly reconstructed from the hologram and can be used to compute the scattering potential at every point of the specimen. From these data, the 3-D distribution of the RI can be established providing a 3-D tomographic image of the cell constituents. Tomography of cells based on DHM is new and original. In 2006, a first approach consisting of the rotation of the specimen,64·65 has been developed. It was demonstrated that the RI of the cell body could be measured in 3-D with a spatial resolution better than 3 μm in all directions. Other approaches consisting of rotating the beam rather than the object have been proposed.28·66‑69 Kou70 and Sheppard71 have compared both approaches in their principles and shown consistent differences in the performance of each modality. Different reconstruction algorithms have been used to reconstruct objects from several holograms obtained at variable kk-vector directions or amplitudes including projection28·64 and diffraction algorithms.72‑74

CELL IMAGING WITH QUANTITATIVE PHASE MICROSCOPY

Biological specimens, such as living cells and tissues, are usually phase objects, that is they are transparent and made visible most often by PhC as explained above. The phase signal originates in the RI difference generated by the presence of organic molecules in cells, including proteins, DNA, organelles, and nuclei. Consequently, QPM visualizes cells by quantitatively providing the phase retardation that they induce on the transmitted wavefront.41 3-D perspective images of such a phase retardation corresponding to a living human red blood cell and a mouse cortical neuron in culture are shown in Figs. 1(c) and1(d), respectively. This QPS signal is given by the following equation:

$$\varphi = \frac{2\pi}{\lambda}(\bar{n}_c - n_m)d, \qquad (1)$$

where d is the cellular thickness, $\bar{n}_c$ is the intracellular RI averaged over the optical path length of optical rays crossing the specimen, and n_m is the RI of the surrounding medium. The cellular quantitative phase contrast results from the fact that the intracellular RI is different from the RI of the perfusion solution nmnm.

Cell Counting, Classification Analysis, and Cytotoxic Screening Assay

Practically, the QPS is particularly well suited to the development of algorithms, allowing automated cell counting,75‑79 recognition, and classification.80‑89 As far as red blood cells (RBCs) are concerned, QPM allows one to achieve a morphometry-based RBCs' analysis. Indeed, the intracellular content of mature nucleus-and organelle-free RBCs, which can be assimilated from an optical point of view to a homogeneous solution mainly containing hemoglobin, induces a quantitative phase shift directly proportional to the RBC thickness.78·85·87·88·90‑96

Since the QPS is also sensitive to intracellular RI, it specifically provides some information about the intracellular content. Based on this information, interesting applications allowing discrimination between physiological and pathophysiological states have been achieved, particularly in the fields of assisted reproduction86·97·98 and cancer research.99‑102 In addition, *in vitro* label-free cytotoxicity screening assays based on QP-DHM have been recently validated due to the flexibility offered by numerical reconstruction.103‑106

Exploration of Cell Movements and Dynamics

QPM approaches as nonscanning techniques are usually well suited to monitor dynamic cell processes. Dynamic motions during the migrations of human dendritic cells107 and subcellular contraction of embryonic cardiomyocytes108·109 have been successfully measured. On the other hand, due to the numerical propagation of the reconstructed object wavefront, QP-DHM allowing, one to apply autofocusing32·110·111‑114·115 and extended depth of focus116‑122 has opened the possibility of efficiently monitoring cell migration in 3-D.110·112·123‑130 In addition, these numerical possibilities offer an alternative to the shallow depth of field of conventional microscopy, which hampers any fast 3-D tracking of cells in their environment specifically when microfluidic devices are considered.131·132 Coherently, autofocusing and extended depth of focus abilities also facilitate the 3-D tracking of micro- or nano-particles,126·133‑135 particularly in combination with approaches capable of second harmonic generation,136·137 the use of which is highly promising in medical fields including drug carrier, tumor detection and treatment, gene therapy.

Dry Mass, Cell Growth, and Cell Cycle

As described by Eq. (1), the measured quantitative phase shift induced by an observed cell on the transmitted light wavefront is proportional to the intracellular RI, which mainly depends on protein content. Therefore, this measure can be used to directly monitor protein production, owing to a relation established more than 50 years ago by Barer.8Practically, the phase shift induced by a cell is related to its dry mass (DM) by the following equation (converted to the International System of Units):

$$\mathrm{DM} = \frac{10\lambda}{2\pi\alpha}\int_{S_c} \Delta\varphi \mathrm{d}s = \frac{10\lambda}{2\pi\alpha}\Delta\bar{\varphi}S_c. \qquad (2)$$

where $\Delta\bar{\varphi}$ is the mean phase shift induced by the whole cell, λ is the wavelength of the illuminating light source, S_c is the projected cell surface, and α is a constant known as the specific refraction increment (in cubic meters per kilogram) and is related to the intracellular content. α is approximated by $1.8–2.1\times10^{-3}$ m^{3}Kg–11.8–2.1×10^{-3} m^{3}Kg^{-1}, when considering a mixture of all the components of a typical cell.

Recently, several groups using various QPM techniques have begun to exploit this phase-DM relationship to study the dynamics of cell growth as well as to the characterization of the cell cycle.[138-143] As far as RBCs are concerned, this relationship has also been explicitly and implicitly used to calculate hemoglobin content, which corresponds to the RBC DM.[92,144-147]

Cell Membrane Fluctuations and Biomechanical Properties

RBCs, while being carried throughout the vascular tree to deliver oxygen from the lungs to the tissues, are squeezed as they pass capillaries which are often smaller than the cell diameter. This ability can be attributed to the remarkable elastic properties of the membrane structure. As a consequence of these elastic properties, RBCs show cell membrane fluctuations (CMFs) at the nanometric scale, often called flickering. Owing to their high sensitivity allowing one to quantitatively measure RBC membrane fluctuations over the whole cell surface, different QPM techniques have shed new light on these CMFs by providing quantitative information about the biomechanical properties of the RBC membrane.[145,148-153] We should also mention the integration of QPM with optical tweezers, which represents a very promising tool, especially with respect to monitoring trapped objects along the axial direction including nanoparticles,[154,155] as well as manipulating and testing the biomechanical properties of cells).[81,156-160]

Dynamics of Absolute Cell Volume and Transmembrane Water Movements

Membranes of animal cells are highly permeable to water; therefore, movement of water across membranes is dictated in large part by osmotic pressure gradients.[161,162] Any imbalance in intracellular and extracellular osmolarity is paralleled by water movements across cell membranes affecting the cell volume along with the concentration of intracellular compounds.[163] However, even at constant extracellular osmolarity, the volume constancy of any mammalian cell is permanently challenged by the normal activity of the cells.[164] Thus, the maintenance of a constant volume is critical for cell homeostasis and requires the continued operation of cell volume regulatory mechanisms.[161] Water crosses cell membranes through several routes (simple diffusion through the lipid bilayer, transmembrane proteins such as specialized water channels, aquaporins, etc.).[162]Consequently, the accurate measurement of both the absolute cell volume and water membrane permeability represents indices which can provide highly relevant information about the mechanisms involved in the cell volume regulatory processes. However, the noninvasive measurement of these two parameters at the single cell level is not a trivial issue.[165] Below, we present how these parameters can be calculated from the QPS although, as indicated by Eq. (1), information concerning the intracellular content related to $\bar{n}_c$ is intrinsically mixed with morphological information relating to a thickness dd. Practically, different strategies have been developed to separately measure cell morphology and RI. Some authors[166,167] measured the intracellular RI by trapping cells between two cover slips separated by a known distance. However, this approach which prevents cell movement, precludes the possibility of exploring dynamic cellular processes, including cell volume regulation. Recently, spectroscopy phase microscopy approaches[145,147,168] have addressed this limitation, at least as far as cells with high intrinsic dispersion properties are considered, including red blood cells (owing to presence of hemoglobin pigment). However, such spectroscopy approaches are only applicable to a very limited variety of cell types, most of which have intrinsic dispersion almost identical to that of water. We have developed another approach called the decoupling procedure, to separately measure the parameters $\bar{n}_c$ and dd from the phase signal φ, based on a modification of the extracellular RI, n_m. Basically, this method consists of performing a slight alteration of the extracellular RI n_m and recording two holograms corresponding to the two different values of n_m, providing two quantitative phase images from which $\bar{n}_c$ and d can be calculated. We have considered two different approaches to modify nmnm: the first approach requires sequentially perfusing a standard cell perfusion solution and a second solution with a different RI, but with the same osmolarity to record the two corresponding holograms at a single wavelength.[169] Practically, this procedure has permitted us to quantitatively measure some highly relevant RBC parameters, including mean corpuscular volume and mean corpuscular hemoglobin concentration.[92] However, owing to the solution exchange time, this approach precludes the possibility of monitoring dynamic changes of cell morphology and RI that occur during fast biological processes. To overcome these drawbacks, we have developed a second approach based on a dual-wavelength DHM (DW-DHM)[45] which exploits the dispersion of the extracellular medium, enhanced by the utilization of an extracellular dye, to achieve separate measurements of the intracellular RI and the absolute cell volume in real time.[170] These approaches have been successfully applied to study the osmotic water membrane permeability P_f—representing the water volume flux per unit of time per unit of membrane surface for a given applied osmotic gradient—by monitoring cell volume changes while retaining the cell functionality.[165] In addition, resulting from the linear relationship between the intracellular RI and the DM as well as from a DM balance equation, it has been possible to determine the RI of the transmembrane flux n_f.[165] The high precision with which n_f can be determined reveals if the transmembrane water flux is accompanied by solute transport and provides some important information about the underlying mechanisms involved in the transmembrane water movements.

RESOLVING NEURONAL NETWORK ACTIVITY AND VISUALIZING SPINE DYNAMICS

Background

Within the brain, water is distributed between blood, cerebrospinal fluid, and interstitial and intracellular compartments. Practically, water moves across these compartments following the differences in osmotic and hydrostatic pressures. At constant hydrostatic pressure gradients, water movements largely depend on the osmotic gradients, which are created by the concentration of osmolytes, including ions, amino acids, and derivatives, in particular, some neurotransmitters, in the extracellular and intracellular compartments. At the microscopic level, water transport is involved in cell volume regulation and in controlling the dimensions of the extracellular space (ECS).171·172 However, neuronal activity challenges the hemostasis of these intra and extra cellular compartments. Indeed, considering a low intracellular concentration of Na^+, resulting from both a neuronal membrane with a low permeability to Na^+ and sodium-potassium transporters continuously pumping out Na^+, if the Na^+ conductances suddenly increase it will cause Na^+ to rush into the neuronal cell. This movement of Na^+ ions along their concentration gradient is also enhanced by the transmembrane electrostatic gradient corresponding to a negative resting membrane potential (−70 mV−70 mV) mainly due to a large quantity of negatively charged intracellular proteins. This inward flow of Na^+ ions changes the electrochemical gradient, which in turn produces a further rise in the membrane potential called a depolarization. This mechanism is precisely what initiates the action potential. Otherwise, the Na^+ currents flowing inward at a point on the axon during an action potential spread out along the axon and depolarize the adjacent sections of its membrane. When actions potentials are conducted down an axon—their size remains undiminished as they travel—they finally induces a depolarization in a presynaptic terminal button, which is the end of axonal branches which generally face the postsynaptic membrane of the neuron that receives the message. The two membranes face each other across the synaptic cleft which is a gap of around 20-nm wide. Practically, when the depolarization reaches the terminal button, small vesicles containing molecules of the neurotransmitter and located just inside the presynaptic membrane, fuse with the membrane and then break down, spilling their contents into the synaptic cleft. Then, the neurotransmitter diffuses across the synaptic clef and by attaching postsynaptic membrane opens neurotransmitter–receptor ion channels, which permit the passage of specific ions into or out the cell, thus changing the local membrane potential of the postsynaptic neuron, which could, in turn, induces or inhibits the generation of action potentials.173

Consequently, these transmembrane ions movements and neurotransmitter releases including glutamate, glycine, and GABA (gamma-aminobutyric acid) related to the neuronal activity are accompanied by transmembrane water movements for osmotic reasons. This results in the regulation of both the ECS and the cell volume being affected by the level of neural activity.

On the other hand, a distinct feature of nervous tissue is the intricate network of synaptic connections among neurons with diverse morphologies. Although initial connections are formed largely through molecular mechanisms, it has been clearly established that electrical activities influence not only the structure and efficacy of existing synapses, but can also contribute to the production of new synapses, in particular through changes in the size and shape of dendritic spines and the formation on new ones. Indeed, synapses can mainly occur in three places of the postsynpatic neuron: on the soma, on other axons, and on the smooth surface of dendrites—branched projections which compose most of the receptive surface of a neuron- or on dendritic spines—small membranous protrusions, with typical sizes of a few hundred nanometers, that stud the dendrites. The axodendritic synapses represent the largest proportion of the excitatory synapses in the mammalian brain and the glutamate is the most common excitatory neurotransmitter, inducing a depolarization of the postsynaptic neurons particularly by opening the (NMDA), the 2-amino-3-(3-hydroxy-5-methylisoxazole-4-yl) propionate (AMPA), and the kainate ionotropic receptors, representing the three main classes of glutamate-receptor ion channels. All these abilities of synapses to strengthen, weaken or to be created over time in response to increases or decreases to the neuronal network activity, called synaptic plasticity, that have been intensively studied in recent years, have been significantly stimulated by the development of new optical methods, which have permitted the resolution of the morphology and the dynamics of dendritic spines.174 Furthermore, plasticity mechanisms affect information transfer on multiple time scales and levels of specificity and play a pivotal role in integrative function and learning. However, with a few exceptions, including activity-induced LTP and LTD (long-term potentiation and long-term depression) and some specific activity-dependent mechanisms that can modulate the density of dendritic spines onto which the majority of fast excitatory synaptic transmission operates, our understanding of how electrical activity affects the structure and function of neuronal networks is very limited. Within this framework, the development of techniques that provide a noninvasive resolution of both local neuronal network activity and dendritic spine dynamics is highly relevant. In light of this, we present how QP-DHM can start to originally address both of these issues due to the possibilities of high resolution and of monitoring cell volume and transmembrane water.

Imaging Neuronal Activity by Measuring Transmembrane Water Movements with Quantitative Phase Digital Holographic Microscopy

With respect to the study of neuronal activity, electrophysiological approaches, in particular patch-clamp techniques, have permitted major breakthroughs leading, in particular, to the discovery of LTP and LTD, by setting a voltage across the neuronal membrane and directly measuring the current flowing through a single ion channel. Basically, a patch clamp is the best standard for assessing ion channel function because it allows the discrimination of ionic currents in the femto-ampere (10–15 fA) range with microsecond time resolution. Practically, the most commonly used patch-clamp mode is the whole-cell mode. In short, in this configuration, an open tip thin glass micropipette, which contains both an ionic solution mimicking the intracellular solution and chlorided silver electrode connected to a highly sensitive electronic amplifier, is pressed against a cell membrane. A brief strong suction is applied to both assist in the formation of a high-resistance seal between the glass and the cell membrane and to disrupt the membrane patch. The interior of the pipette becomes continuous with the cytoplasm and a bath electrode is used to set the zero level. In whole-cell measurements, the two main configurations are the (1) voltage-clamp mode and the (2) current-clamp mode. In the voltage-clamp mode, the cell membrane potential is kept constant via a negative feedback loop injecting a compensating current into the cell that resembles the current that is flowing through the whole-cell membrane. Practically, the recording of this compensating current provides information about the whole-cell membrane conductance. In addition, the current-voltage relationships can be investigated since the membrane potential can be manipulated independently of ionic currents. In contrast, in the current-clamp mode, the injected current is kept constant and changes in the membrane potential can be measured. However, a patch clamp is still a highly invasive laborious process requiring precise micromanipulations and a high degree of operations, which generally impose the recording of voltages on a limited number of cells that form a neuronal network. Optical techniques, however, seem to be an ideal solution for measuring membrane potentials since they are relatively noninvasive and could work both at low and high magnifications. For instance, calcium indicators used in combination with high resolution two-photon microscopy allow the performing measurements of the spiking activity from hundreds to several thousand neurons in mammalian circuits while still keeping track of the activity of each neuron individually.175·176 However, calcium imaging has its shortcomings and cannot substitute for voltage imaging.176 Practically, voltage imaging methods have lagged behind calcium imaging due to important challenges related to physical constraints of the measurements themselves, including an electrical field highly located in the thin membrane region, an essentially two-dimension plasma membrane which cannot contain an arbitrary number of voltage chromophores without disrupting its properties, as well as the fact that the plasma membrane represents only a small proportion of the total membrane surface in the neuron on which the chromophores are attached. Finally, the relatively high speed of the electrical responses of mammalian neurons also represents a serious challenge for voltage measurements. Consequently, despite some promising perspectives, the different voltage imaging methods suffer from poor SNR and secondary side effects, and they have up to now fallen short of providing single-cell resolution when imaging the activity of neuronal populations.176

Otherwise, it is well known that neuronal activity induces modifications of the intrinsic optical properties at the subcellular,177-180 cellular181·182 and tissue level.183-185Consequently, we have attempted to establish an accurate relationship between QPS and the transmembrane current. Practically, we set out experiments combining electrophysiology in a whole-cell configuration and QP-DHM, performed in a well-established biological model involving Cl^- conductance, a human embryonic kidney (HEK) cell line, and transfected to express the neuronal ligand-gated chloride ion channel called $GABA_A$, the acronym of its endogenous ligand, the gamma-aminobutyric acid which is an inhibitory neurotransmitter. These experiments have stressed the fact that a net ionic influx corresponds to a phase decrease due to the osmotic water entry while a net ionic efflux induces a phase increase reflecting the osmotic exit of water. Under the following two assumptions—(1) the n_c value linearly depends on the concentration of the different intracellular components according to the seminal work of Davies and Wilkins186and (2) the volume variations induced by the transmembrane water movements accompanying the ionic movements are proportional to the number of ions having passed through the membrane or channels—it has been possible to establish the following relation between the QP signal φ(t) and transmembrane current I(t)

$$I(t) = \frac{V_0}{\varepsilon}\frac{d}{dt}\left[\frac{\varphi_0}{\varphi(t)}\right]^{1/s-r} - \frac{1}{\varepsilon}\frac{\mathrm{d}\Delta V(t)}{\mathrm{d}t}, \tag{3}$$

where V_0 is the cell volume measured at the beginning of the experiments at a time $t=t_0$, ΔV are the nonelectrogenic volume changes corresponding to any cell volume variation except those directly related to a transmembrane net current, ε is a constant representing the volume variation associated with the net charge movement through the cell membrane [ml/C], r is a parameter related to the cell deformation associated with the volume variation and the parameter s which takes cell processes inducing volume variations or not but leading to intracellular RI changes by mechanisms other than dilution or concentration of the intracellular content by ions and water fluxes.187 Practically, r∈[0,1] with r=1/3 and r=1 correspond to isotropic volume changes and piston-like cell movements in the z direction, i.e., along the optical axis, respectively. Usually, the adhesion of cultured cells on a coverslip facilitates a movement along the z-axis resulting in r∈[0.5,0.8] typically. In contrast, the case s=1 corresponds to a simple dilution or concentration of the cell content, while s<1 reflects modification of intracellular RI by processes mediating the release or cellular internalization of substrates, including protein release mechanisms or cotransport activation involved in the volume regulatory processes. Practically, the parameters V_0, r, and s can be determined by the decoupling procedure. Consequently, two parameters remains to be determined, namely ε and ΔV, to calculate the current.

Experiments with HEK cells have validated Eq. (3) regarding Cl^- currents mediated by GABA applications [Fig 2(a1)] with εGABA lies within the range of 90–110 μm3/nC90–110 μm3/nC, and have permitted us to construct a phase response as a function of the clamped membrane potential (φ/Vφ/V curve) allowing the determination of the equilibrium potential for Cl^- with the same precision as an electrophysiological approach measuring the classical I/V curves [Fig. 2(a2)].187 Practically, εGABA has been estimated for a membrane potential clamped at-40 mV-40 mV by performing a least-squares fit to adjust the phase current (Pred. Curr.) derived from Eq. (3) with the measured current (electrophysiological recordings in a voltage-clamp mode). Using this εGABA determination, Pred. Curr. has been calculated for other values of the membrane potential by using Eq. (3) [Fig. 2(a1)].

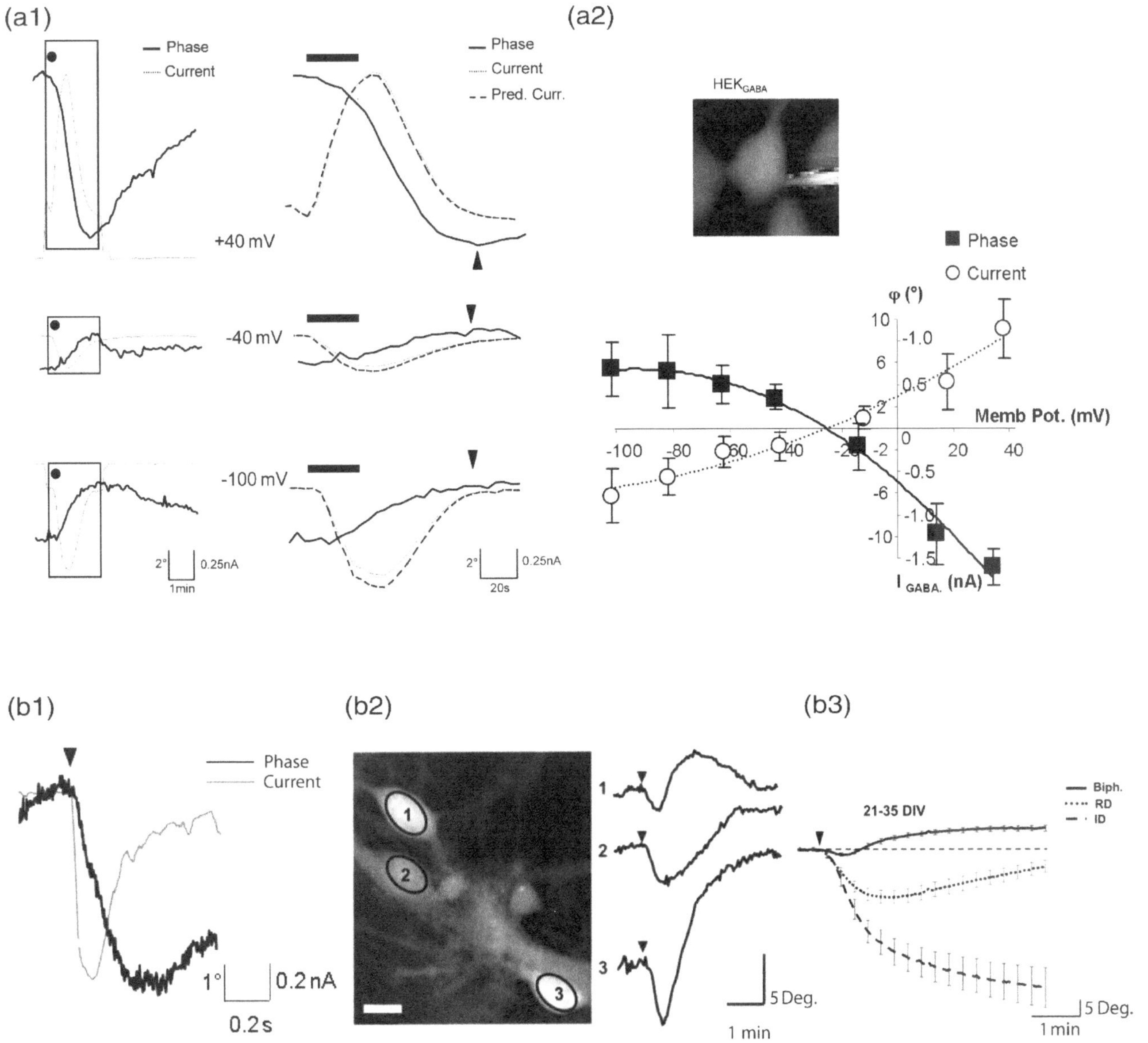

Figure 2: (A) HEK cells and GABA applications, (B) neurons in response to glutamate. (a1): (left) Example of three simultaneous traces of current (gray line) and phase shift (black line) recorded with 44 mM of $[Cl^-]_{intrapip}$ on the same HEK_{GABA} after application of GABA (3 μM; 30 s; dot). At −100 mV, application of GABA triggered an inward current concomitantly to an increase in the phase signal. Conversely, at +40 mV, the same applications of GABA triggered an outward current accompanied by a decrease of the phase signal. Note that for −40 mV (close to the resting potential for Cl), the current and the phase shift were very small. (a1) (right) Expansion of traces visualized in a1 (parts defined by rectangles). For each level of membrane potential, the peak of phase shift (indicated by the arrow head) was reached when the I_{GABA} was terminated. According to Eq. (3) the phase signal can predict the current (Pred. Curr.: dashed line) superimposed on the recorded current: (a2) The φ/Vφ/V curve (full square and thick line) and the I/V curve (empty circle and thin line) obtained with GABA (n=7) indicated an ECl of −26 and −26 mV, respectively. (a2) inset representative quantitative phase image of patched a Hek cell. (b1): Local application of glutamate (500 μM, 200 ms; arrow head) on a patched neuron induces a strong transient decrease of the phase signal associated to an inward current. Phase is expressed in degrees (b2) application of glutamate triggers three main types of phase shifts, Left, representative quantitative phase image of three cultured cortical neurons visualized in DHM. Ovals in the middle of the cells corresponded to the region of interest for the phase signal (scale bar, 10 μm). Right, application of glutamate (30 μM, 30 s; arrowhead) triggered different types of shapes of optical signals from these three recorded neurons. (b3), averaged traces of the three characteristic optical signals induced by glutamate application (30_M, 30 s, arrowhead) and those recorded from 21–35 DIV neurons. These phase signals are classified into three categories: a biphasic response (Biph.; n=137), a reversible decrease of the phase signal (RD, n=80), and an irreversible decrease of the phase signal (ID, n=35). All these experiments were performed at room temperature.

Subsequently, Eq. (3) has been used to optically monitor neuronal network activity mediated by glutamate application, the main excitatory neurotransmitter in the brain released at 80% of the synapses. Practically, a multimodality microscope, QP-DHM, and electrophysiology setup has been developed to study the early stage of neuronal responses induced by glutamate on primary culture of mouse cortical neurons. This study involving pharmacological experiments revealed that glutamate produces three distinct optical responses, predominantly mediated by NMDA receptors: biphasic (Biph.), reversible decrease (RD), and irreversible decrease (ID) responses [Fig. 2(b3)]. The shape and amplitude of the optical signal were not associated with a particular cellular phenotype, but reflected the physio-pathological status of neurons linked to the degree of NMDA activity. Thus, the "Biph.", "RD," and "ID" responses respectively indicated low-, a high-, and "excito-toxic" levels of NMDA activation. For these three distinct responses, a strong inward current associated with the early phase decrease, whose amplitude is proportional to the concentration of glutamate and to the duration of the application, has been measured. This inward current is consistent with glutamate-mediated activation of specific ionotropic receptors including the NMDA, AMPA, and the kainate ionotropic receptors, whose opening induces influxes of Ca^{2+} and Na^{+} down their electrochemical gradient. Consistently, the phase decrease results from water entrance accompanying the influxes of Ca^{2+} and Na^{+} for osmotic reasons. Moreover, furosemide and bumetanide, two inhibitors of sodium and/or potassium-coupled chloride movement, strongly modify the phase shift suggesting an involvement of two neuronal co-transporters, Na-K-Cl (NKCC1) and K-Cl (KCC2) in the genesis of the optical signal. In addition, the phase responses can be decomposed into two components, a rapid one accompanying the glutamate-mediated current (IGLUTIGLUT)—the phase decrease on Fig. 2(b1), and a slow one generally corresponding to a phase recovery while IGLUT=0IGLUT=0. The phase recovery, which is much slower than the fast one, is likely to correspond to a nonelectrogenic neuronal volume regulation involving several mechanisms. Interestingly, the time course presented inFig. 2(b1) shows that the water movements are not significantly delayed—at the tenth of second scale at least—relative to the recorded current. The measurements of I_{GLUT} as well as the corresponding intracellular RI and neuronal volume changes allowed us to estimate the parameter "ε_{GLUT}" [mL/C][mL/C]. Practically, the values of ε_{GLUT} lie within the range of 60–120 µm3/nC equivalent to 340–620 water molecules transported per ion having crossed the membrane. Interestingly, the typical intracellular RI change induced by a glutamate pulse (500 uM 0.2 s) is around 0.002–0.003. The associated neuronal swelling is around 100 femtoliter (fL) for a typical neuronal cell body of 1500 fL corresponding to a 6–7% cell volume variation. However, these orders of magnitude correspond to an exogenous glutamate application during a few tenths of seconds. Physiological release of endogenous glutamate is likely to induce smaller water movements and intracellular RI changes. Finally, it is important to note that, although we do not yet have a comprehensive understanding of the underlying mechanisms inducing these phase responses, these consequent transmembrane movements of water accompanying the neuronal activity are in good agreement with the typical amount of water co-transported by various cotransports including NKCC1.188

Furthermore, information concerning transmembrane current obtained from electrophysiological recordings combined with QP-DHM signal allows us to pave the way for developing a simultaneous multiple sites optical recording of transmembrane currents capable of resolving local neuronal network activity. In addition, appropriate mathematical treatment of the optical signal affords the possibility to quantitatively determine the dynamics of the current triggered by the ionotropic receptor activity, making this technique amenable to use for pharmacological screenings of modulators developed for the management of human pathologies involving dysfunctions of specific ionotropic channel including chloride channels associated with cystic fibrosis.189

Three-Dimensional Visualization of Dendritic Spine Dynamics with Quantitative Phase Tomographic Microscopy

As previously mentioned, dendritic spines are the postsynaptic components of most excitatory synapses in the mammalian brain and are the loci of structural and functional plasticity. In past decades, studies have revealed that the number and size of dendritic spines are regulated by a variety of gene products and environmental factors, underscoring the dynamic nature of spines and their importance to brain plasticity. Because dendritic spines are the key elements for information acquisition and retention, understanding how spines are formed and maintained in relation to neuronal network activity will likely provide fundamental insights into how the brain possesses the extraordinary capacity to learn and to remember.190 However, dendritic spines are tiny transparent processes whose visualization and resolution remain a challenge for optical approaches. However, the advent of two-photon (2P) excitation laser scanning microscopy, despite limitations inherent to the use of fluorescence, has greatly contributed to the understanding of dendritic spine function and plasticity.191·192 On the other hand, the diffraction tomography, which is particularly efficient when the wavelength of the irradiating beam is comparable to the dimensions of the diffracting parts of the object, represents a promising approach to dynamically and noninvasively resolve subcellular structures including dendritic spines. Indeed, appealing applications to cell biology have already been obtained and published. Practically, the investigation of the RI 3-D map has been published for pollen grains (Charrière, Marian et al. 2006), amoeba,65human RBCs parasitized by plasmodium falciparum,193 human influenza virus infected cells, epithelial cheek cells, and granulocytes.194 Images of Hela cells at 0 deg and up to 50 deg illumination incidences have been published195 and DIC has been derived from synthetic aperture phase contrast microscopy. Interestingly, an approach based on the rotation of the sample using a hollow optical fiber as a live-cell imaging chamber has permitted one to measure the RI 3-D maps of cell clusters.196 Recently, Cotte et al. firmly established the benefit of synthetic aperture elevated to quasi 2π by pushing the occupation of the Fourier space up to a diameter equal to the diameter of the Ewald sphere.197 Super-resolution could be obtained and details as small as 100 nm could be observed. Moreover, dynamic images showing the detailed evolution of dendrite spines (Fig. 3) could be obtained.

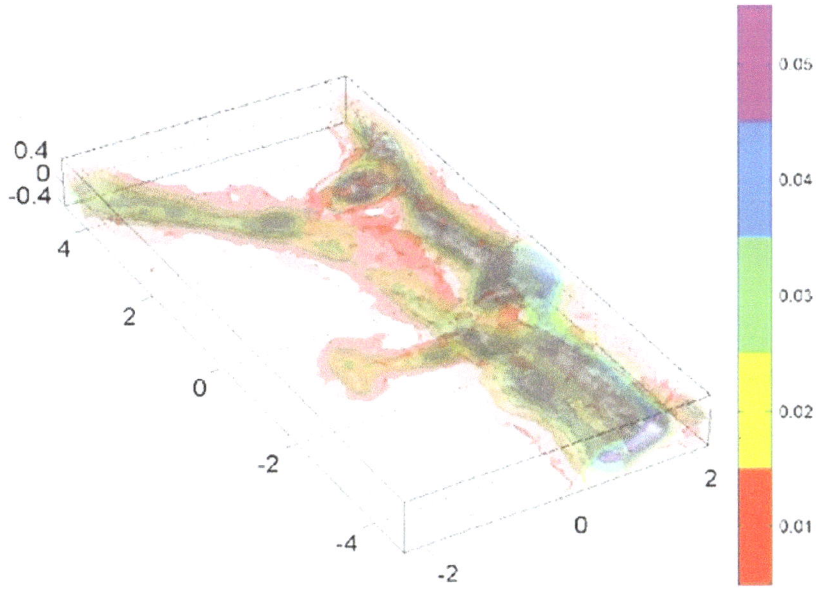

Figure 3: Three-dimensional (3-D) representation of a tomographic image of a neuronal process obtained by synthetic aperture quantitative phase image. The details of protruding dendritic spines can be visualized. Scale is in μm and the color code represents the difference of refractive index (RI) [from 0.04 (red) to 0.16 (cyan)] compared with the extracellular medium (Yann Cotte). These experiments were performed at room temperature.

QUANTITATIVE PHASE DIGITAL HOLOGRAPHIC MICROSCOPY: TOOL TO SCREEN AND IDENTIFY NEW CELL BIOMARKERS OF PSYCHIATRIC DISORDERS

Background

Psychiatric disorders represent major public health problems and, while significant improvements of the conditions of the patients have been achieved due to the discovery of neuroleptics and antidepressants, the available treatments remain largely insufficient and unsatisfactory. A major paradigm shift is needed to move the field forward. Advances in the field are impaired in large part due to the lack of proper tools to conduct adequate biological measurements that can guide diagnosis and treatment. This is particularly critical in light of growing evidence showing that early treatment of psychiatric disorders, already in their prodromal phase and before the onset of debilitating symptoms, leads to significantly improved clinical outcomes.198 It is therefore important to identify in children of parents affected by the disease, biomarkers or endophenotypes indicating a high risk of developing the disease later in life However, to date there are insufficient data to support any of these biomarkers as vulnerability or diagnostic or prognostic factors, notably regarding bipolar disorders.199 In this field of research, the most relevant investigations are longitudinal studies conducted on cohorts consisting of patients with a specific psychiatric disorder and their children (high-risk children). Indeed, epidemiological studies, particularly in regard to mood disorders,200·201 have clearly shown that children of patients with a psychiatric disorder usually have a significantly higher risk of developing the disease. In addition, psychiatric disorders are frequently comorbid with other mental illnesses as well as endocrine, metabolic (e.g., diabetes mellitus and obesity) and cardiovascular diseases. Thus, they must be considered as a multisystem condition, and as a consequence, their underlying neurobiological bases are complex. Coherently, only studies combining several biomarkers have been successful in separating acute mood states (mania and depression) from controls202 and providing a high accuracy for the diagnosis of major depression.203Consequently, the determination of high risk developmental trajectories for psychiatric disorders requires to identify and to longitudinally study, in high risk children, various sets of biomarkers very diverse in nature including, e.g., neuropsychological and neuroimaging (structural and functional MRI), as well as peripheral biomarkers or cellular biomarkers.

Practically, the study of cell biomarkers can be achieved from somatic cells (fibroblasts) obtained from patients as well as from high-risk children (skin biopsy). However, recent years have seen a renewed interest for the cell biomarkers due to the possibility, through induced pluripotent stem cells, to convert somatic cells (fibroblast) obtained from patients into neurons (grown in culture) with the genetic makeup necessary for the development of certain phenotypes of the disease. Although this approach currently has several limitations,204 it allows one to consider the particularly attractive hypothesis that some phenotypes are already found *in vitro* at the cell level or in a network of cells collected from patients or their offspring.205

Within this framework, a multimodality imaging approach developed around QP-DHM has the capacity to explore new, original optical cellular biomarkers down at the nanoscale that can then be put in contrast to other more conventional markers as mentioned above. Indeed, QP-DHM provides the unique monitoring of fine cell structure and dynamics due to its ability to yield quantitatively information on a wide variety of cellular parameters including volume, morphology, intracellular protein content, dry weight, nanoscale membrane fluctuations, membrane water permeability, transmembrane water movements, and volume regulation. The value of such an approach can be illustrated by a study we are conducting in a cohort of diabetic patients aimed at correlating energy distribution among vibrational eigenmodes of RBC membranes with the levels of glycosylated hemoglobin (Fig. 4).

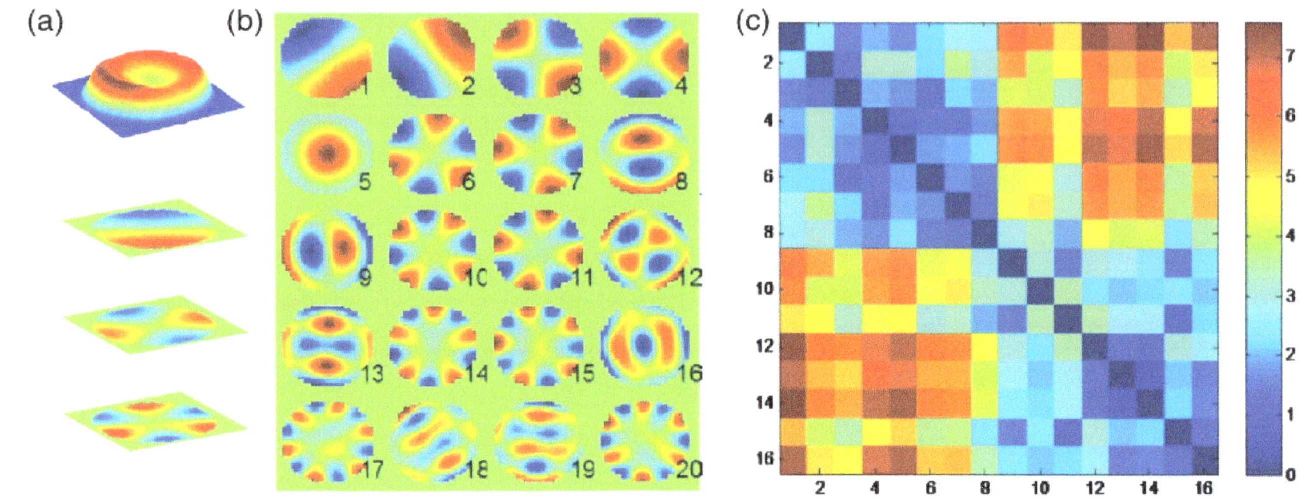

Figure 4: (a) Quantitative phase image of a red blood cell (RBC) in a perspective color coded representation and projections of three principal component analysis (PCA) modes of the RBC cell membrane fluctuations. (b) Degeneracy free representation of 20 PCA modes, measured experimentally. (c) Each t-value (i,j) in the matrix indicates how similar the energy distributions of the PCA modes between subject ii and subject j are. At lower t-values, the energy distribution becomes more similar. Healthy (1–8) and diabetic (9–16) samples can be discriminated.

Building on these results, we are conducting a study monitoring the RBC membrane eigenmode energy distribution in a sample of the prodromal phase of first-episode psychosis patients before initial treatment, aimed at identifying specific biomarkers for early detection and intervention.

PERSPECTIVES

As illustrated through these different applications, QPM and specifically DH-QPM because of the unique possibilities offered by the numerical reconstruction and propagation of the object wavefront including the extended depth of focus, *a posteriori* numerical autofocusing, allows one to quantitatively and noninvasively study cell structure and dynamics as well as the content at different scales of time ranging from a few milliseconds to several days. However, although QPS is highly sensitive and contains a wealth of information, its interpretation in terms of specific biological processes remains an issue. Practically, addressing this interpretation issue beyond the well-established calculation of the cell DM, involves the ability to calculate relevant biophysical cell parameters including absolute volume, membrane fluctuations at the nanoscale, biomechanical properties, transmembrane water permeability, current, and so on from the phase signal. Concretely, the derivation of these cell biophysical parameters often requires one to separately obtain cell morphology (thickness) and content (intracellular RI). Within this framework, the development of DHM-based optical diffraction tomography, providing a direct access to the 3-D map of the intracellular RI, could allow us to revisit this difficulty of interpretation. Future developments leading to the real time and high resolution 3-D maps of the intracellular RI could provide, in addition to the possibility of noninvasively resolving both neuronal network activity and dendritic spine dynamics as previously explained, valuable information about cytoachitecture and compartmentalization of cytoplasm, which plays a critical role in several fundamental cell mechanisms including proteins' synthesis. On the other hand, information about transmembrane water movements measured with DH-QPM allows us to pave the way for studying the activity of specific ionotropic receptors and/or co-transporters including nonelectrogenic ones as well as their modulation by specific pharmacological agents.

In addition, QP-DHM and QP-TM in multimodality approaches combining, e.g., fluorescence,206 spectroscopy, coherent anti-Stokes Raman scattering (CARS), second-(SHG) and third-harmonic generation (THG), should allow simultaneous measurements of a large number of relevant and specific parameters yielding an extended cell profiling approach. Specifically, SHG207 and THG208 convey highly relevant information, particularly concerning membrane proteins and intracellular constituents. Due to the highly nonlinear character of these light generating processes, local properties of the cytoskeleton and membrane as well as 3-D cellular arrangements are obtainable by taking advantage of the remarkable penetration power of the SHG or, more particularly, the THG processes. Within this framework, it is worthwhile mentioning one of the first attempts to combine Raman information with quantitative phase information paving thus the way to simultaneously explore the cell's molecular composition and dynamics.209 Otherwise, used in conjunction with AFM, QP-DHM would provide an image of the local deformations, and, therefore, a more detailed and informative insight of biophysical properties of the cell. In addition, the use of free wandering particles would also precisely provide information on the nature of binding forces and properties of the macromolecules. This detection of multiple biophysical cell parameters could even be scaled into a high-throughput cellular screening assay105 allowing the concurrent monitoring of several cellular processes as well as their modulations as a function of the transcriptome and metabolome of different cell types derived from control patients as well as high-risk subjects. Thus, within the framework of longitudinal studies conducted on cohorts of patients and their children as previously mentioned, such high-throughput multimodality approaches would be highly promising to identify new and original early cell biomarkers or endophenotypes of mental disorders that will become invaluable tools for early diagnosis, improved stratification, a personalized medicine approach, and assessment of disease progression as well as of treatment outcome.

ACKNOWLEDGMENTS

The work presented in this review is the result of a close collaboration between the microvision and microdiagnosis group (SCI/STI/CHD group) of EPFL with Florian Charrière, Jonas Kühn, Nicolas Pavillon, Etienne Shafer, Fatih Toy, Yann Cotte, the Laboratory of Neuroenergetics and Cellular Dynamics, Brain and Mind Institute of EPFL with Benjamin Rappaz, Pascal Jourdain, the Center for Psychiatric neuroscience, CHUV with Daniel Boss, Kaspar Rothefusser and the company Lyncée Tec SA, Lausanne (http://www.lynceetec.com). We thank the Swiss National Science Foundation (SNSF) Grant No. CR3213_132993 the National Center of competence in research Synapsy (http://www.nccr-synapsy.ch/) as well as the fondation de Préfargier.

REFERENCES

1. Kumar V., Abbas A. K., Aster J., Robbins and Cotran Pathologic Basis of Disease. , Elsevier Saunders , Philadelphia, PA (2014).
2. Inoué S., Spring R. K., Video Microscopy, The Fundamentals. , Plenum , New York and London (1997).
3. Han R. et al., "Recent advances in super-resolution fluorescence imaging and its applications in biology," J. Genet. Genomics. 40, (12), 583 –595 (2013). 1673-8527
4. Cornea A., Conn P. M., Fluorescence Microscopy: Super-Resolution and other Novel Techniques. , Academic Press , USA (2014).
5. Zernike F., "Phase contrast, a new method for the microsopic observation of transparent objects," Physica. 9, , 686 –698 (1942). 0031-8914
6. Smith F. H., "Microscopic interferometry," Research. 8, , 385 –395 (1955).
7. Nomarski G., "*Dispositif Oculaire a Contraste De phase pour microscope," J. De Physique Et Le Radium. 11, (12), S9 –S10 (1950). 0368-3842
8. Barer R., "Interference microscopy and mass determination," Nature. 169, (4296), 366 –367 (1952). 0028-0836
9. Barer R., "Refractometry and interferometry of living cells," J. Opt. Soc. Am.. 47, (6), 545 –556 (1957). 0030-3941
10. Dunn G. A., "Transmitted-light interference microscopy: a technique born before its time," Proc. R. Microsc. Soc.. 33, , 189 –196 (1998).
11. Dunn G. A., Zicha D., "Cell behaviour: adhesion and motility," Symp. Soc. Exp. Biol.. 47, , 91 –106 (1993). 0081-1386
12. Lee K. et al., "Quantitative phase imaging techniques for the study of cell pathophysiology: from principles to applications," Sensors. 13, (4), 4170 –4191 (2013). 0746-9462
13. Marquet P., Depeursinge C., Magistretti P. J., "Exploring neural cell dynamics with digital holographic microscopy," Annu. Rev. Biomed. Eng.. 15, , 407 –431 (2013). 1523-9829
14. Gabor D., "A new microscopic principle," Nature. 161, (4098), 777 –778 (1948). 0028-0836
15. Goodman J. W., Lawrence R. W., "Digital image formation from electronically detected holograms," Appl. Phys. Lett.. 11, (3), 77 –79 (1967). 0003-6951
16. Schnars U., Juptner W., "Direct recording of holograms by a CCD target and numerical reconstruction," Appl. Opt.. 33, (2), 179 –181 (1994). 0003-6935
17. Coquoz O. et al., "Performances of endoscopic holography with a multicore optical- fiber," Appl. Opt.. 34, (31), 7186 –7193 (1995). 0003-6935
18. Rastogi P., Holographic Interferometry: Principles and Methods. , Springer-Verlag , New York (1994).
19. Kreis T., Handbook of Holographic Interferometry: Optical and Digital Methods. , Wiley-VCH Verlag GmbH & Co. KGaA , Weinheim, FRG (2005).
20. Yamaguchi I., Zhang T., "Phase-shifting digital holography," Opt. Lett.. 22, (16), 1268 –1270 (1997). 0146-9592
21. Carré P., "Installation et utilisation du comparateur photoélectrique et interférentiel du Bureau International des Poids et Mesures," Metrologia. 2, (1), 13 –33 (1966). 0026-1394
22. Guo P., Devaney A. J., "Digital microscopy using phase-shifting digital holography with two reference waves," Opt. Lett.. 29, (8), 857 –859 (2004). 0146-9592
23. Liu J. P., Poon T. C., "Two-step-only quadrature phase-shifting digital holography," Opt. Lett.. 34, (3), 250 –252 (2009). 0146-9592
24. Awatsuji Y., Sasada M., Kubota T., "Parallel quasi-phase-shifting digital holography," Appl. Phys. Lett.. 85, (6), 1069 –1071 (2004). 0003-6951
25. Guo C. S. et al., "Phase-shifting error and its elimination in phase-shifting digital holography," Opt. Lett.. 27, (19), 1687 –1689 (2002). 0146-9592
26. Wang Z., Han B., "Advanced iterative algorithm for phase extraction of randomly phase-shifted interferograms," Opt. Lett.. 29, (14), 1671 –1673 (2004). 0146-9592
27. Xu X. F. et al., "Simple direct extraction of unknown phase shift and wavefront reconstruction in generalized phase-shifting interferometry: algorithm and experiments," Opt. Lett.. 33, (8), 776 –778 (2008). 0146-9592
28. Choi W. et al., "Tomographic phase microscopy," Nat. Methods. 4, (9), 717 –719 (2007). 1548-7091
29. Ichioka Y., Inuiya M., "Direct phase detecting system," Appl. Opt.. 11, (7), 1507 –1514 (1972). 0003-6935
30. Mertz L., "Real-time fringe-pattern analysis," Appl. Opt.. 22, (10), 1535 –1539 (1983). 0003-6935
31. Carl D. et al., "Parameter-optimized digital holographic microscope for high-resolution living-cell analysis," Appl. Opt.. 43, (36), 6536 –6544 (2004). 0003-6935
32. Liebling M., Blu T., Unser M., "Complex-wave retrieval from a single off-axis hologram," J. Opt. Soc. Am.. 21, (3), 367 –377 (2004). 0030-3941
33. Leith E. N., Upatniek J., "Reconstructed wavefronts and communication theory," J. Opt. Soc. Am.. 52, (10), 1123 (1962). 0030-3941
34. Takeda M., Ina H., Kobayashi S., "Fourier-transform

method of fringe-pattern analysis for computer-based topography and interferometry," J. Opt. Soc. Am.. 72, (1), 156 –160 (1982). 0030-3941

35. Kreis T., "Digital holographic interference-phase measurement using the Fourier-transform method," J. Opt. Soc. Am. A. 3, (6), 847 –855 (1986). 0740-3232
36. Cuche E., Marquet P., Depeursinge C., "Simultaneous amplitude-contrast and quantitative phase-contrast microscopy by numerical reconstruction of Fresnel off-axis holograms," Appl. Opt.. 38, (34), 6994 –7001 (1999). 0003-6935
37. Cuche E., Bevilacqua F., Depeursinge C. "Digital holography for quantitative phase-contrast imaging," Opt. Lett.. 24, (5), 291 –293 (1999). 0146-9592
38. Haddad W. S. et al., "Fourier-transform holographic microscope," Appl. Opt.. 31, (24), 4973 –4978 (1992). 0003-6935
39. Takaki Y., Kawai H., Ohzu H., "Hybrid holographic microscopy free of conjugate and zero-order images," Appl. Opt.. 38, (23), 4990 –4996 (1999). 0003-6935
40. Xu W. B. et al., "Digital in-line holography for biological applications," Proc. Natl. Acad. Sci. U. S. A.. 98, (20), 11301 –11305 (2001). 0027-8424
41. Marquet P. et al., "Digital holographic microscopy: a noninvasive contrast imaging technique allowing quantitative visualization of living cells with subwavelength axial accuracy," Opt. Lett.. 30, (5), 468 –470 (2005). 0146-9592
42. Popescu G., "Quantitative phase imaging of nanoscale cell structure and dynamics," Methods Cell Biol.. 90, , 87 –115 (2008). 0091-679X
43. Colomb T. et al., "Polarization imaging by use of digital holography," Appl. Opt.. 41, (1), 27 –37 (2002). 0003-6935
44. Colomb T. et al., "Polarization microscopy by use of digital holography: application to optical fiber birefringence measurements," Appl. Opt.. 44, (21), 4461 –4469 (2005). 0003-6935
45. Kuhn J. et al., "Real-time dual-wavelength digital holographic microscopy with a single hologram acquisition," Opt. Express. 15, (12), 7231 –7242 (2007). 1094-4087
46. Cuche E., Marquet P., Depeursinge C., "Spatial filtering for zero-order and twin-image elimination in digital off-axis holography," Appl. Opt.. 39, (23), 4070 –4075 (2000). 0003-6935
47. Colomb T. et al., "Automatic procedure for aberration compensation in digital holographic microscopy and application to specimen shape compensation," Appl. Opt.. 45, , 851 –863 (2006). 0003-6935
48. Colomb T. et al., "Total aberrations compensation in digital holographic microscopy with a reference conjugated hologram," Opt. Express. 14, (10), 4300 –4306 (2006). 1094-4087
49. Kuhn J. et al., "Axial sub-nanometer accuracy in digital holographic microscopy," Meas. Sci. Technol.. 19, (7), 074007 (2008). 0957-0233
50. Kemper B. et al., "Investigation of living pancreas tumor cells by digital holographic microscopy," J. Biomed. Opt.. 11, (3), 034005 (2006). 1083-3668
51. Kim M. K., Yu L., Mann C. J., "Interference techniques in digital holography," J. Opt.. 8, (7), S518 –S523 (2006). 0150-536X
52. Charriere F. et al., "Shot-noise influence on the reconstructed phase image signal-to-noise ratio in digital holographic microscopy," Appl. Opt.. 45, (29), 7667 –7673 (2006). 0003-6935
53. Charriere F. et al., "Influence of shot noise on phase measurement accuracy in digital holographic microscopy," Opt. Express. 15, (14), 8818 –8831 (2007). 1094-4087
54. Liebling M., Unser M., "Autofocus for digital Fresnel holograms by use of a Fresnelet-sparsity criterion," J. Opt. Soc. Am.. 21, (12), 2424 –2430 (2004). 0030-3941
55. Dixon L., Cheong F. C., Grier D. G., "Holographic deconvolution microscopy for high-resolution particle tracking," Opt. Express. 19, (17), 16410 –16417 (2011). 1094-4087
56. Marron J. C., Schroeder K. S., "Holographic laser-radar," Opt. Lett.. 18, (5), 385 –387 (1993). 0146-9592
57. Arons E., Dilworth D., "Lensless imaging by spatial Fourier synthesis holography," Appl. Opt.. 35, (5), 777 –781 (1996). 0003-6935
58. Kim M. K., "Wavelength-scanning digital interference holography for optical section imaging," Opt. Lett.. 24, (23), 1693 –1695 (1999). 0146-9592
59. Montfort F. et al., "Submicrometer optical tomography by multiple-wavelength digital holographic microscopy," Appl. Opt.. 45, (32), 8209 –8217 (2006). 0003-6935
60. Kuhn J. et al., "Submicrometer tomography of cells by multiple-wavelength digital holographic microscopy in reflection," Opt. Lett.. 34, (5), 653 –655 (2009). 0146-9592
61. Cuche E., Poscio P., Depeursinge C., "Optical tomography by means of a numerical low-coherence holographic technique," J. Opt.. 28, (6), 260 –264 (1997). 0150-536X
62. Massatsch P. et al., "Time-domain optical coherence tomography with digital holographic microscopy," Appl. Opt.. 44, (10), 1806 –1812 (2005). 0003-6935
63. Wolf E., "Three dimensional structure determination of semi-transparent object from holographic data," Opt. Commun.. 1, (4), 153 –156 (1969).
64. Charriere F. et al., "Cell refractive index tomography by digital holographic microscopy," Opt. Lett.. 31, (2), 178 –180 (2006). 0146-9592
65. Charriere F. et al., "Living specimen tomography by digital holographic microscopy: morphometry of testate amoeba," Opt. Express. 14, (16), 7005 –7013 (2006). 1094-4087
66. Debailleul M. et al., "High-resolution three-dimensional tomographic diffractive microscopy of transparent inorganic and biological samples," Opt. Lett.. 34, (1), 79 –81 (2009). 0146-9592
67. Sung Y. J. et al., "Optical diffraction tomography for

high resolution live cell imaging," Opt. Express. 17, (1), 266 –277 (2009). 1094-4087

68. Cotte Y., Toy M. F., Depeursinge C., "Beyond the lateral resolution limit by phase imaging," J. Biomed. Opt.. 16, (10), 106007 (2011). 1083-3668
69. Kim M. et al., "High-speed synthetic aperture microscopy for live cell imaging," Opt. Lett.. 36, (2), 148 –150 (2011). 0146-9592
70. Kou S. S., Sheppard C. J. R., "Comparison of three dimensional transfer function analysis of alternative phase imaging methods," Proc. SPIE. 6443, , 64430Q (2007). 0277-786X
71. Sheppard C. J. R., Kou S. S., "3D Imaging with holographic tomography," in Proc. Int. Conf. on Advanced Phase Measurement Methods in Optics an Imaging , Vol. 1236, pp. 65 –69, AIP (2010).
72. Wolf E., "Three-dimensional structure determination of semi-transparent object from holographic data," Opt. Commun.. 1, (4), 153 –156 (1969). 0030-4018
73. Lauer V., "New approach to optical diffraction tomography yielding a vector equation of diffraction tomography and a novel tomographic microscope," J. Microsc-Oxford. 205, , 165 –176 (2002). 0022-2720
74. Devaney A., Cheng J., "Optical diffraction tomography in an inhomogeneous background medium," Meas. Sci. Technol.. 19, (8), 085505 (2008). 0957-0233
75. Milgram J. H., Li W. C., "Computational reconstruction of images from holograms," Appl. Opt.. 41, (5), 853 –864 (2002). 0003-6935
76. Molder A. et al., "Non-invasive, label-free cell counting and quantitative analysis of adherent cells using digital holography," J. Microsc.. 232, (2), 240 –247 (2008). 0022-2720
77. Seo S. et al., "High-throughput lens-free blood analysis on a chip," Anal. Chem.. 82, (11), 4621 –4627 (2010). 0003-2700
78. Mihailescu M. et al., "Automated imaging, identification, and counting of similar cells from digital hologram reconstructions," Appl. Opt.. 50, (20), 3589 –3597 (2011). 0003-6935
79. Bianco V. et al., "Imaging through scattering microfluidic channels by digital holography for information recovery in lab on chip," Opt. Express. 21, (20), 23985 –23996 (2013). 1094-4087
80. Moon I., Javidi B., "Three-dimensional identification of stem cells by computational holographic imaging," J. R. Soc. Interface. 4, (13), 305 –313 (2007). 1742-5689
81. DaneshPanah M. et al., "3-D holographic imaging and trapping for non-invasive cell identification and tracking," J. Disp. Technol.. 6, (10), 490 –499 (2010). 0193-2691
82. Javidi B., Daneshpanah M., Moon I., "Three-dimensional holographic imaging for identification of biological micro/nanoorganisms," IEEE Photon. J.. 2, (2), 256 –259 (2010). 1943-0655
83. Moon I., Yi F., Javidi B., "Automated three-dimensional microbial sensing and recognition using digital holography and statistical sampling," Sensors. 10, (9), 8437 –8451 (2010). 0746-9462
84. Anand A., Chhaniwal V. K., Javidi B., "Imaging embryonic stem cell dynamics using quantitative 3-D digital holographic microscopy," IEEE Photon. J.. 3, (3), 546 –554 (2011). 1943-0655
85. Liu R. et al., "Recognition and classification of red blood cells using digital holographic microscopy and data clustering with discriminant analysis," J. Opt. Soc. Am.. 28, (6), 1204 –1210 (2011). 0030-3941
86. Memmolo P. et al., "Identification of bovine sperm head for morphometry analysis in quantitative phase-contrast holographic microscopy," Opt. Express. 19, (23), 23215 –23226 (2011). 1094-4087
87. Moon I. et al., "Automated statistical quantification of three-dimensional morphology and mean corpuscular hemoglobin of multiple red blood cells," Opt. Express. 20, (9), 10295 –10309 (2012). 1094-4087
88. Yi F., Lee C. G., Moon I. K., "Statistical analysis of 3D volume of red blood cells with different shapes via digital holographic microscopy," J. Opt. Soc. Korea. 16, (2), 115 –120 (2012). 1226-4776
89. Yi F. et al., "Automated segmentation of multiple red blood cells with digital holographic microscopy," J. Biomed. Opt.. 18, (2), 026006 (2013). 1083-3668
90. Popescu G. et al., "Erythrocyte structure and dynamics quantified by Hilbert phase microscopy," J. Biomed. Opt.. 10, (6), 060503 (2005). 1083-3668
91. Bernhardt I. et al., "Application of digital holographic microscopy to investigate the sedimentation of intact red blood cells and their interaction with artificial surfaces," Bioelectrochemistry. 73, (2), 92 –96 (2008). 0302-4598
92. Rappaz B. et al., "Comparative study of human erythrocytes by digital holographic microscopy, confocal microscopy, and impedance volume analyzer," Cytometry Part A. 73A, (10), 895 –903 (2008). 1552-4922
93. Mir M., Tangella K., Popescu G., "Blood testing at the single cell level using quantitative phase and amplitude microscopy," Biomed. Opt. Express. 2, (12), 3259 –3266 (2011). 2156-7085
94. Mehta D. S., Srivastava V., "Quantitative phase imaging of human red blood cells using phase-shifting white light interference microscopy with colour fringe analysis," Appl. Phys. Lett.. 101, (20), 203701 (2012). 0003-6951
95. Shaked N. T. et al., "Dynamic quantitative microscopy and nanoscopy of red blood cells in sickle cell disease," Proc. SPIE. 8227, , 822711 (2012). 0277-786X
96. Moon I. et al., "Automated quantitative analysis of 3D morphology and mean corpuscular hemoglobin in human red blood cells stored in different periods," Opt. Express. 21, (25), 30947 –30957 (2013). 1094-4087
97. Crha I. et al., "Digital holographic microscopy in human sperm imaging," J. Assist. Reprod. Genet.. 28, (8), 725 –729 (2011). 1058-0468
98. Merola F. et al., "Digital holography as a method for 3D imaging and estimating the biovolume of motile cells," Lab Chip. 13, (23), 4512 –4516 (2013). 1473-0197
99. Mann C. J. et al., "High-resolution quantitative phase-

contrast microscopy by digital holography," Opt. Express. 13, (22), 8693 –8698 (2005). 1094-4087

100. Janeckova H., Vesely P., Chmelik R., "Proving tumour cells by acute nutritional/energy deprivation as a survival threat: a task for microscopy," Anticancer Res.. 29, (6), 2339 –2345 (2009). 0250-7005

101. Wang P. et al., "Nanoscale nuclear architecture for cancer diagnosis by spatial-domain low-coherence quantitative phase microscopy," Proc. SPIE. 7907, , 790704 (2011). 0277-786X

102. Wang P. et al., "Nanoscale nuclear architecture for cancer diagnosis beyond pathology via spatial-domain low-coherence quantitative phase microscopy," J. Biomed. Opt.. 15, (6), 066028 (2010). 1083-3668

103. Bauwens A. et al., "Differential cytotoxic actions of Shiga toxin 1 and Shiga toxin 2 on microvascular and macrovascular endothelial cells," Thromb. Haemost.. 105, (3), 515 –528 (2011). 0340-6245

104. Pavillon N. et al., "Early cell death detection with digital holographic microscopy," PLoS One. 7, (1), e30912 (2012). 1932-6203

105. Kuhn J. et al., "Label-free cytotoxicity screening assay by digital holographic microscopy," Assay Drug Dev. Technol.. 11, (2), 101 –107 (2013). 1540-658X

106. Rappaz B. et al., "Digital holographic microscopy: a quantitative label-free microscopy technique for phenotypic screening," Comb. Chem. High Throughput Screen. 17, (1), 80 –88 (2014). 1386-2073

107. Garay J. et al., "Crosstalk between PKA and Epac regulates the phenotypic maturation and function of human dendritic cells," J. Immunol.. 185, (6), 3227 –3238 (2010). 0022-1767

108. Ellerbee A. K., Creazzo T. L., Izatt J. A., "Investigating nanoscale cellular dynamics with cross-sectional spectral domain phase microscopy," Opt. Express. 15, (13), 8115 –8124 (2007). 1094-4087

109. Ansari R. et al., "Quantitative analysis of cardiomyocyte dynamics with optical coherence phase microscopy," Proc. SPIE. 8213, , 821338 (2012). 0277-786X

110. Choi Y. S., Lee S. J., "Three-dimensional volumetric measurement of red blood cell motion using digital holographic microscopy," Appl. Opt.. 48, (16), 2983 –2990 (2009). 0003-6935

111. Kim T., Poon T. C., "Autofocusing in optical scanning holography," Appl. Opt.. 48, (34), H153 –H159 (2009). 0003-6935

112. Langehanenberg P. et al., "Automated three-dimensional tracking of living cells by digital holographic microscopy," J. Biomed. Opt.. 14, (1), 014018 (2009). 1083-3668

113. Wang H. Y. et al., "photonics and optoelectronics," in SOPO Symposium on Photonics and Optoelectronics , pp. 1 –4, IEEE (2009).

114. Zhao J. et al., "Autofocusing on pure phase object for living cell imaging in lensless Fourier transform digital holography," Proc. SPIE. 7790, , 779017 (2010). 0277-786X

115. Toy M. F. et al., "Digital holographic microscopy for the cytomorphological imaging of cells under zero gravity," Proc. SPIE. 8227, , 82270S (2012). 0277-786X

116. Ferraro P. et al., "3-D imaging with large focus extension by a coherent optical microscope," in Information Optics. , Cristobal G., Javidi B., Vallmitjana S., Eds., 860, pp. 52 –62 (2006).

117. Ferraro P. et al., "Extended focused image in microscopy by digital holography," Opt. Express. 13, (18), 6738 –6749 (2005). 1094-4087

118. Antkowiak M. et al., "Accurate three-dimensional detection of micro-particles by means of digital holographic microscopy," Proc. SPIE. 6995, , 699514 (2008). 0277-786X

119. McElhinney C., Hennelly B., Naughton T., "Extended focused imaging for digital holograms of macroscopic three-dimensional objects," Appl. Opt.. 47, (19), D71 –D79 (2008). 0003-6935

120. Bergoend I. et al., "Depth-of-field extension and 3-D reconstruction in digital holographic microscopy," Proc. SPIE. 7390, , 73901C (2009). 0277-786X

121. Colomb T. et al., "Extended depth-of-focus by digital holographic microscopy," Opt. Lett.. 35, (11), 1840 –1842 (2010). 0146-9592

122. Pan W. Q., "Multiplane imaging and depth-of-focus extending in digital holography by a single-shot digital hologram," Opt. Commun.. 286, , 117 –122 (2013). 0030-4018

123. Mann C. J., Yu L. F., Kim M. K., "Movies of cellular and sub-cellular motion by digital holographic microscopy," Biomed. Eng. Online. 5, , 21 (2006). 1475-925X

124. DaneshPanah M., Javidi B., "Tracking biological microorganisms in sequence of 3D holographic microscopy images," Opt. Express. 15, (17), 10761 –10766 (2007). 1094-4087

125. Sun H. Y. et al., "Visualization of fast-moving cells in vivo using digital holographic video microscopy," J. Biomed. Opt.. 13, (1), 014007 (2008). 1083-3668

126. Bae Y. et al., "Three-dimensional single particle tracking using off-axis digital holographic microscopy," Proc. SPIE. 7574, , 757408 (2010). 0277-786X

127. Javidi B., Moon I., Daneshpanaha M., "Detection, identification and tracking of biological micro/nano organisms by computational 3-D optical imaging," Proc. SPIE. 7759, , 77590R (2010). 0277-786X

128. Bohm M. et al., "PACE4, a member of the prohormone convertase family, mediates increased proliferation, migration and invasiveness of melanoma cells in vitro and enhanced subcutaneous tumor growth in vivo," J. Invest. Dermatol.. 131, , S108 –S108 (2011). 0022-202X

129. Memmolo P. et al., "On the holographic 3D tracking of in vitro cells characterized by a highly-morphological change," Opt. Express. 20, (27), 28485 –28493 (2012). 1094-4087

130. Merola F. et al., "Simultaneous optical manipulation, 3-D tracking, and imaging of micro-objects by digital holography in microfluidics," IEEE Photon. J.. 4, (2), 451 –454 (2012). 1943-0655

131. Miccio L. et al., "Detection and visualization improvement

of spermatozoa cells by digital holography," Proc. SPIE. 8089, , 80890C (2011). 0277-786X

132. Paturzo M. et al., "Microscopy imaging and quantitative phase contrast mapping in turbid microfluidic channels by digital holography," Lab Chip. 12, (17), 3073 –3076 (2012). 1473-0197

133. Xu W. et al., "Tracking particles in four dimensions with in-line holographic microscopy," Opt. Lett.. 28, (3), 164 –166 (2003). 0146-9592

134. Antkowiak M. et al., "Extended focused imaging of a microparticle field with digital holographic microscopy," Opt. Lett.. 33, (14), 1626 –1628 (2008). 0146-9592

135. Warnasooriya N. et al., "Imaging gold nanoparticles in living cell environments using heterodyne digital holographic microscopy," Opt. Express. 18, (4), 3264 –3273 (2010). 1094-4087

136. Hsieh C. L. et al., "Three-dimensional harmonic holographic microcopy using nanoparticles as probes for cell imaging," Opt. Express. 17, (4), 2880 –2891 (2009). 1094-4087

137. Shaffer E., Depeursinge C., "Digital holography for second harmonic microscopy: application to 3D-tracking of nanoparticles," Proc. SPIE. 7715, , 771505 (2010). 0277-786X

138. Zicha D., Dunn G. A., "An image-processing system for cell behavior studies in subconfluent cultures," J. Microsc-Oxford. 179, , 11 –21 (1995). 0022-2720

139. Popescu G. et al., "Imaging red blood cell dynamics by quantitative phase microscopy," Blood Cells Mol. Dis.. 41, (1), 10 –16 (2008). 1079-9796

140. Rappaz B. et al., "Noninvasive characterization of the fission yeast cell cycle by monitoring dry mass with digital holographic microscopy," J. Biomed. Opt.. 14, (3), 034049 (2009). 1083-3668

141. Kemper B. et al., "Label-free quantitative cell division monitoring of endothelial cells by digital holographic microscopy," J. Biomed. Opt.. 15, (3), 036009 (2010). 1083-3668

142. Mir M. et al., "Optical measurement of cycle-dependent cell growth," Proc. Natl. Acad. Sci. U. S. A.. 108, (32), 13124 –13129 (2011). 0027-8424

143. Byun H. et al., "Optical measurement of biomechanical properties of individual erythrocytes from a sickle cell patient," Acta Biomater.. 8, (11), 4130 –4138 (2012). 1742-7061

144. Yusipovich A. I. et al., "Laser interference microscopy in erythrocyte study," J. Appl. Phys.. 105, (10), 102037 (2009). 0021-8979

145. Jang Y., Jang J., Park Y., "Dynamic spectroscopic phase microscopy for quantifying hemoglobin concentration and dynamic membrane fluctuation in red blood cells," Opt. Express. 20, (9), 9673 –9681 (2012). 1094-4087

146. Pham H. et al., "Spectroscopic diffraction phase microscopy," Opt. Lett.. 37, (16), 3438 –3440 (2012). 0146-9592

147. Rinehart M., Zhu Y. Z., Wax A., "Quantitative phase spectroscopy," Biomed. Opt. Express. 3, (5), 958 –965 (2012). 2156-7085

148. Popescu G. et al., "Optical imaging of cell mass and growth dynamics," Am. J. Physiol. Cell Physiol.. 295, (2), C538 –C544 (2008). 0363-6143

149. Rappaz B. et al., "Spatial analysis of erythrocyte membrane fluctuations by digital holographic microscopy," Blood Cells Mol. Dis.. 42, (3), 228 –232 (2009). 1079-9796

150. Park Y. K. et al., "Metabolic remodeling of the human red blood cell membrane," Proc. Natl. Acad. Sci. U. S. A.. 107, (4): 1289 –1294 (2010). 0027-8424

151. Lee S. et al., "Detrended fluctuation analysis of membrane flickering in discocyte and spherocyte red blood cells using quantitative phase microscopy," J. Biomed. Opt.. 16, (7), 076009 (2011). 1083-3668

152. Park Y. et al., "Measurement of the nonlinear elasticity of red blood cell membranes," Phys. Rev. E. 83, (5), 051925 (2011). 1063-651X

153. Boss D. et al., "Spatially-resolved eigenmode decomposition of red blood cells membrane fluctuations questions the role of ATP in flickering," PLoS One. 7, (8), e40667 (2012). 1932-6203

154. Higuchi T. et al., "Three-dimensional positioning of optically trapped nanoparticles," Appl. Opt.. 50, (34), H183 –H188 (2011). 0003-6935

155. Hayasaki Y., Sato A., "Three-dimensional position measurement of a gold nanoparticle trapped in optical tweezers using in-line low-coherence digital holography," in Proc. 2012 11th Euro-American Workshop on Information Optics (Wio) , IEEE , Québec (2012).

156. Mauritz J. M. A. et al., "Biophotonic techniques for the study of malaria-infected red blood cells," Med. Biol. Eng. Comput.. 48, (10), 1055 –1063 (2010). 0140-0118

157. Cardenas N., Yu L. F., Mohanty S. K., "Probing orientation and rotation of red blood cells in optical tweezers by digital holographic microscopy," Proc. SPIE. 7906, , 790613 (2011). 0277-786X

158. Cardenas N., Yu L. F., Mohanty S. K., "Stretching of red blood cells by optical tweezers quantified by digital holographic microscopy," Proc. SPIE. 7897, , 78971J (2011). 0277-786X

159. Esseling M. et al., "Multimodal biophotonic workstation for live cell analysis," J. Biophotonics. 5, (1), 9 –13 (2012). 1864-063X

160. Kemper B. et al., "Towards 3D modelling and imaging of infection scenarios at the single cell level using holographic optical tweezers and digital holographic microscopy," J. Biophotonics. 6, (3), 260 –266 (2013). 1864-063X

161. Friedrich B., Matskevich I., Lang F., "Cell volume regulatory mechanisms," Contrib. Nephrol.. 152, , 1 –8 (2006). 0302-5144

162. Hoffmann E. K., Lambert I. H., Pedersen S. F., "Physiology of cell volume regulation in vertebrates," Physiol. Rev.. 89, (1), 193 –277 (2009). 0031-9333

163. Pasantes-Morales H., Cruz-Rangel S., "Brain volume regulation: osmolytes and aquaporin perspectives," Neuroscience. 168, (4), 871 –884 (2010). 0735-2743

164. Strange K., "Cellular volume homeostasis," Adv.

Physiol. Educ.. 28, (4), 155 –159 (2004). 1043-4046

165. Boss D. et al., "Measurement of absolute cell volume, osmotic membrane water permeability, and refractive index of transmembrane water and solute flux by digital holographic microscopy," J. Biomed. Opt.. 18, (3), 036007 (2013). 1083-3668

166. Lue N. et al., "Live cell refractometry using microfluidic devices," Opt. Lett.. 31, (18), 2759 –2761 (2006). 0146-9592

167. Kemper B. et al., "Integral refractive index determination of living suspension cells by multifocus digital holographic phase contrast microscopy," J. Biomed. Opt.. 12, (5), 054009 (2007). 1083-3668

168. Park Y. et al., "Spectroscopic phase microscopy for quantifying hemoglobin concentrations in intact red blood cells," Opt. Lett.. 34, (23), 3668 –3670 (2009). 0146-9592

169. Rappaz B. et al., "Measurement of the integral refractive index and dynamic cell morphometry of living cells with digital holographic microscopy," Opt. Express. 13, (23), 9361 –9373 (2005). 1094-4087

170. Rappaz B. et al., "Simultaneous cell morphometry and refractive index measurement with dual-wavelength digital holographic microscopy and dye-enhanced dispersion of perfusion medium," Opt. Lett.. 33, (7), 744 –746 (2008). 0146-9592

171. Sykovà E. et al., "The relationship between changes in intrinsic optical signals and cell swelling in rat spinal cord slices," Neuroimage. 18, , 214 –230 (2003).

172. Nagelhus E. A., Mathiisen T. M., Ottersen O. P., "Aquaporin-4 in the central nervous system: cellular and subcellular distribution and coexpression with KIR4.1," Neuroscience. 129, (4), 905 –913 (2004). 0735-2743

173. Carlson N., Physiology of Behavior. , Pearson , Harlow (2014).

174. Padamsey Z., Emptage N. J., "Imaging synaptic plasticity," Mol Brain. 4, , 36 (2011). 1756-6606

175. Cossart R., Aronov D., Yuste R., "Attractor dynamics of network UP states in the neocortex," Nature. 423, (6937), 283 –288 (2003). 0028-0836

176. Peterka D. S., Takahashi H., Yuste R., "Imaging voltage in neurons," Neuron. 69, (1), 9 –21 (2011). 0896-6273

177. Hill D. K., Keynes R. D., "Opacity changes in stimulated nerve," J. Physiol.. 108, (3), 278 –281 (1949). 0022-3751

178. Tasaki I. et al., "Changes in fluorescence, turbidity, and birefringence associated with nerve excitation," Proc. Natl. Acad. Sci. U. S. A.. 61, (3), 883 –888 (1968). 0027-8424

179. Cohen L. B., "Changes in neuron structure during action potential propagation and synaptic transmission," Physiol. Rev.. 53, (2), 373 –418 (1973). 0031-9333

180. Carter K. M., George J. S., Rector D. M., "Simultaneous birefringence and scattered light measurements reveal anatomical features in isolated crustacean nerve," J. Neurosci. Methods. 135, (1–2), 9 –16 (2004). 0165-0270

181. Stepnoski R. A. et al., "Noninvasive detection of changes in membrane potential in cultured neurons by light scattering," Proc. Natl. Acad. Sci. U. S. A.. 88, (21), 9382 –9386 (1991). 0027-8424

182. Tasaki I., Byrne P. M., "Rapid structural changes in nerve fibers evoked by electric current pulses," Biochem. Biophys. Res. Commun.. 188, (2), 559 –564 (1992). 0006-291X

183. MacVicar B. A., Hochman D., "Imaging of synaptically evoked intrinsic optical signals in hippocampal slices," J. Neurosci.. 11, (5), 1458 –1469 (1991). 0270-6474

184. Andrew R. D., Adams J. R., Polischuk T. M., "Imaging NMDA- and kainate-induced intrinsic optical signals from the hippocampal slice," J. Neurophysiol.. 76, (4), 2707 –2717 (1996). 0022-3077

185. Holthoff K., Witte O. W., "Intrinsic optical signals in rat neocortical slices measured with near-infrared dark-field microscopy reveal changes in extracellular space," J. Neurosci.. 16, (8), 2740 –2749 (1996). 0270-6474

186. Davies H. G., Wilkins M. H. F., "Interference microscopy and mass determination," Nature. 169, (4300), 541 (1952).

187. Jourdain P. et al., "Simultaneous optical recording in multiple cells by digital holographic microscopy of chloride current associated to activation of the ligand-gated chloride channel GABA(A) receptor," PLoS One. 7, (12), e51041 (2012). 1932-6203

188. Hamann S. et al., "Water permeability of Na+-K+-2Cl− cotransporters in mammalian epithelial cells," J. Physiol.. 568, (1), 123 –135 (2005). 0022-3751

189. Jourdain P. et al., "The human CFTR protein expressed in CHO cells activates aquaporin-3 in a cAMP-dependent pathway: study by digital holographic microscopy," J. Cell Sci.. 127, (3), 546 –556 (2014). 0021-9533

190. Bhatt D. H., Zhang S., Gan W. B., "Dendritic spine dynamics," Annu. Rev. Physiol.. 71, , 261 –282 (2009). 0066-4278

191. Svoboda K., Yasuda R., "Principles of two-photon excitation microscopy and its applications to neuroscience," Neuron. 50, (6), 823 –839 (2006). 0896-6273

192. Lee K. F. C., Soares C., Beique J. C., "Examining form and function of dendritic spines," Neural Plast.. 2012, , 704103 (2012). 0792-8483

193. Park Y. et al., "Static and dynamic light scattering of healthy and malaria-parasite invaded red blood cells," J. Biomed. Opt.. 15, (2), 020506 (2010). 1083-3668

194. Simon B. et al., "High-resolution tomographic diffractive microscopy of biological samples," J. Biophotonics. 3, (7), 462 –467 (2010). 1864-063X

195. Kim M. et al., "Three-dimensional differential interference contrast microscopy using synthetic aperture imaging," J. Biomed. Opt.. 17, (2), 026003 (2012). 1083-3668

196. Kus A. et al., "Tomographic phase microscopy of living three-dimensional cell cultures," J. Biomed. Opt.. 19, (4), 046009 (2014). 1083-3668

197. Cotte Y. et al., "Marker-free phase nanoscopy (vol 7, pg

113, 2013)," Nat. Photonics. 7, (5), 418 –418 (2013). 1749-4885

198. Berk M. et al., "From neuroprogression to neuroprotection: implications for clinical care," Med. J. Aust.. 193, (4 Suppl), S36 –S40 (2010). 0025-729X

199. Luby J. L., Navsaria N., "Pediatric bipolar disorder: evidence for prodromal states and early markers," J. Child Psychol. Psychiatry. 51, (4), 459 –471 (2010).

200. Vandeleur C. et al., "Mental disorders in offspring of parents with bipolar and major depressive disorders," Bipolar Disord.. 14, (6), 641 –653 (2012).

201. Mesman E. et al., "The Dutch bipolar offspring study: 12-year follow-up," Am. J. Psychiatry. 170, (5), 542 –549 (2013). 0002-953X

202. Kapczinski F. et al., "Peripheral biomarkers and illness activity in bipolar disorder," J. Psychiatry Res.. 45, (2), 156 –161 (2011). 0022-3956

203. Papakostas G. I. et al., "Assessment of a multi-assay, serum-based biological diagnostic test for major depressive disorder: a pilot and replication study," Mol. Psychiatry. 18, (3), 332 –339 (2013). 1359-4184

204. Wu S. M., Hochedlinger K., "Harnessing the potential of induced pluripotent stem cells for regenerative medicine (vol 13, pg 497, 2011)," Nat. Cell Biol.. 13, (6), 734 –734 (2011). 1465-7392

205. Brennand K. J. et al., "Modeling psychiatric disorders at the cellular and network levels," Mol. Psychiatry. 17, (12), 1239 –1253 (2012). 1359-4184

206. Pavillon N. et al., "Cell morphology and intracellular ionic homeostasis explored with a multimodal approach combining epifluorescence and digital holographic microscopy," J. Biophotonics. 3, (7), 432 –436 (2010). 1864-063X

207. Campagnola P., "Second harmonic generation imaging microscopy: applications to diseases diagnostics," Anal. Chem.. 83, (9), 3224 –3231 (2011). 0003-2700

208. Fujita K., Smith N. I., "Label-free molecular imaging of living cells," Mol. Cells. 26, (6), 530 –535 (2008). 1016-8478

209. Pavillon N., Hobro A. J., Smith N. I., "Cell optical density and molecular composition revealed by simultaneous multimodal label-free imaging," Biophys. J.. 105, (5), 1123 –1132 (2013). 0006-3495

CHAPTER 8

TRANSCRIPTOMICS RESOURCES OF HUMAN TISSUES AND ORGANS

Mathias Uhlén[1,2,3] Björn M Hallström[1,2], Cecilia Lindskog[4], Adil Mardinoglu[1,5], Fredrik Pontén[4] and Jens Nielsen[1,3,5]

[1]Science for Life Laboratory, KTH - Royal Institute of Technology, Stockholm, Sweden

[2]Department of Proteomics, KTH - Royal Institute of Technology, Stockholm, Sweden

[3]Novo Nordisk Foundation Center for Biosustainability, Technical University of Denmark, Hørsholm, Denmark

[4]Department of Immunology, Genetics and Pathology, Science for Life Laboratory, Uppsala University, Uppsala, Sweden

[5]Department of Biology and Biological Engineering, Chalmers University of Technology, Gothenburg, Sweden

ABSTRACT

Quantifying the differential expression of genes in various human organs, tissues, and cell types is vital to understand human physiology and disease. Recently, several large-scale transcriptomics studies have analyzed the expression of protein-coding genes across tissues. These datasets provide a framework for defining the molecular constituents of the human body as well as for generating comprehensive lists of proteins expressed across tissues or in a tissue-restricted manner. Here, we review publicly available human transcriptome resources and discuss body-wide data from independent genome-wide transcriptome analyses of different tissues. Gene expression measurements from these independent datasets, generated using samples from fresh frozen surgical specimens and postmortem tissues, are consistent. Overall, the different genome-wide analyses support a distribution in which many proteins are found in all tissues and relatively few in a tissue-restricted manner. Moreover, we discuss the applications of publicly available omics data for building genome-scale metabolic models, used for analyzing cell and tissue functions both in physiological and in disease contexts.

INTRODUCTION

The global classification of the human proteome with regard to its spatiotemporal expression patterns and its functions represents one of the major challenges for studying human biology and disease (Lamond *et al*, 2012). Recently, genomic, transcriptomic, and proteomic technologies have been employed to analyze the human proteome on a genome-wide level. Genome annotation efforts, such as Ensembl (Cunningham *et al*, 2015) and Gencode consortium (Harrow *et al*, 2012) in the Encode project (Nilsson *et al*, 2015), have identified approximately 20,000 genes coding for proteins, and the UniProt consortium (UniProt, 2015) has manually annotated the majority of these genes. On the transcript level, expression levels of human genes have been monitored to study the effects of diseases, treatments, and developmental stages using microarray-based gene expression profiling (Brawand *et al*, 2011; Petryszak *et al*, 2015). Recently, several efforts have been published with the quantitative analysis of RNA levels based on next-generation sequencing in samples representing most of the major organs and tissues in the human body (Fig 1), including the Fantom consortium (Yu *et al*, 2015), the Human Protein Atlas (HPA) consortium (Uhlen *et al*, 2015), and the

Citation: Mathias Uhlén, Björn M Hallström, Cecilia Lindskog, Adil Mardinoglu, Fredrik Pontén andJens Nielsen, Transcriptomics resources of human tissues and organs, DOI: 10.15252/msb.20155865

genome-based tissue expression (GTEx) consortium (Keen & Moore, 2015). On the protein level, several large-scale studies based on mass spectrometry analysis have also been published (Kim *et al*, 2014; Wilhelm *et al*, 2014), and these studies have been complemented with antibody-based protein profiling using tissue microarrays containing samples representing most major tissues and organs in the human body (Fagerberg *et al*, 2014; Uhlen *et al*, 2015). Most of the quantitative data on the expression of protein-coding genes are based on recent transcriptomics studies based on RNA-seq. Here, we review some of the publicly available human transcriptome resources and discuss tissue data from independent research groups.

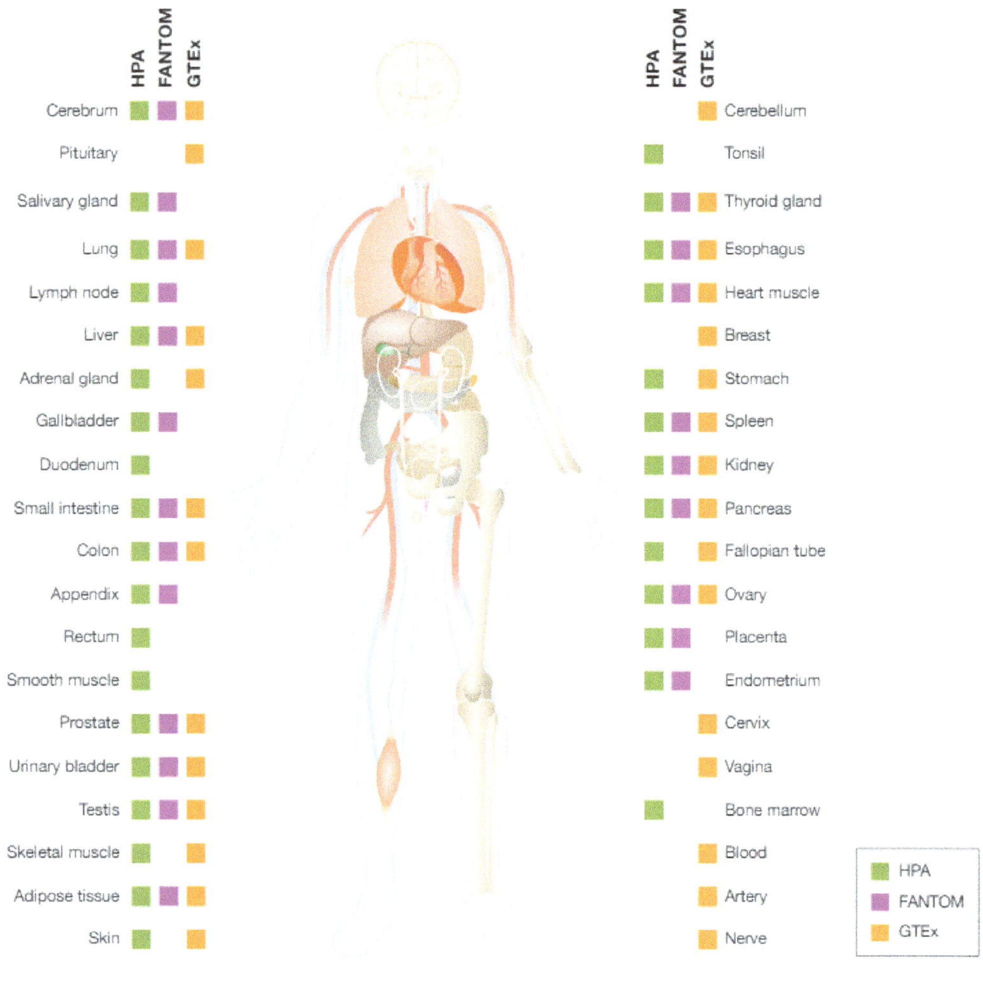

Figure 1. Global transcriptomics analysis of human tissues and organs.

Overview of the tissues and organs analyzed using RNA-seq by the Human Protein Atlas consortium (HPA, green), tissues studied with cap analysis gene expression (CAGE) within the FANTOM consortium (purple), and tissues analyzed using RNA-seq by the genome-based tissue expression consortium (GTEx, orange). Altogether, 22 tissues and organs were studied with both the HPA and FANTOM datasets, while 21 tissues overlapped between the HPA and GTEx datasets.

An interesting aspect of the integration of omics technologies is the sampling that depends on the sensitivity and the resolution of each technology. Consequently, the analysis of tissue samples is normally performed on a mixture of cell types using transcriptomics and mass spectrometry-based proteomics, whereas *in situ* hybridization techniques, successfully applied for mapping the distribution of transcripts in the brain (Hawrylycz *et al*, 2012), and more qualitative approaches involving antibody-based profiling allow analyzing single cells in their natural environment to reveal the differences in protein expression levels between neighboring cells. Thus, antibody-based protein profiling complements quantitative transcriptomics and proteomics, based on a mixture of cell types, to reach single-cell resolution in the analysis of gene expression in complex tissues. Recently, single-cell technologies have been developed for transcriptomics (Tang *et al*, 2009), but no global analysis across many tissues has yet been published using such methodology. In the near future, the possibility to move to single-cell analyses of tissues for transcriptomics and proteomics will allow analyzing cell-to-cell variability, which is particular interesting, that is, in the context of cancer. Moreover, the development of more quantitative assays for immunohistochemistry using fluorescent probes will be valuable for providing quantitative data for whole-cell modeling (Stadler *et al*, 2013).

The correlation between mRNA levels and the corresponding protein levels is an important issue for the comparability of the different omics-based technologies, and the presence or absence of such correlation on an individual gene/protein level has been debated in the literature for many years (Anderson & Seilhamer, 1997; Tian *et al*, 2004; Gry *et al*, 2009; Maier *et al*, 2009, 2011; Lundberg *et al*, 2010; Schwanhausser*et al*, 2011). However, a comparison of steady-state levels of mRNA and proteins in human cell lines using RNA-seq and quantitative triple-SILAC analysis showed good genome-wide correlations when the mRNA and protein levels of an individual gene were compared in three separate cell lines (Lundberg *et al*, 2010). These observations were supported by Kuster and coworkers (Wilhelm *et al*, 2014) comparing mass spectrometry data from different tissues with RNA-seq data obtained from the HPA consortium (Uhlen *et al*, 2015). Again, the steady-state levels for individual genes correlated across several tissues. Overall, these studies suggest that the amount of a given protein in a cell or tissue is, in general, reflected by the corresponding mRNA level, although this gene-/protein-specific ratio may vary greatly between different gene products depending on various factors, mainly translational rates and protein half-lives (Eden *et al*, 2011), and the transcript level for a given gene might therefore be used to predict the corresponding protein level. This hypothesis needs to be confirmed by more in-depth studies using quantitative measurements at both the RNA and protein levels. However, it suggests that next-generation sequencing of the transcriptome is a powerful tool for indirect measurements of protein expression levels, thus forming an attractive link between the field of genomics and proteomics.

ANALYSES OF THE HUMAN TRANSCRIPTOME BY DIFFERENT METHODS

Several genome-wide transcriptome profiling methods have been used for identifying and quantifying global gene expression patterns, each allowing for a quantitative analysis of RNA transcripts. Whole-body maps based on microarray analyses have been launched, such as BioGPS (Wu *et al*, 2009), and these have recently been followed by several tissue-based data portals based on RNA-seq. Some of these publicly available repositories for transcriptomics data are summarized in Table 1 with a focus on datasets from RNA-seq experiments. The resources include repositories for external data, such as Expression Atlas from the European Bioinformatics Institute (EBI) and Gene Expression Omnibus from the National Center for Biotechnology Information (NCBI), as well as repositories with internally generated transcriptome data, such as the GTEx, the Human Protein Atlas, and the Allen Brain Atlas. In a recent study (Danielsson *et al*, 2015), the concordance of RNA-seq data from four large-scale efforts was compared based on gene expression measurements for ostensibly similar samples (specifically, human brain, heart, and kidney samples). The authors concluded that human tissue RNA-seq expression measurements appear quite consistent, considering that samples cluster by tissue rather than laboratory of origin given simple preprocessing transformations.

Table 1: Data resources for RNA expression data with relevance for human protein-encoding genes

Resource	Affiliation	Description	Link (URL)	References
Human Protein Atlas	Science for Life Lab (Sweden)	Tissue-based RNA data based on surgically removed tissues (RNA-Seq)	http://www.proteinatlas.org/	(Uhlen *et al*, 2015)
GTEx	Broad Institute (USA)	Tissue-based RNA data based on postmortem samples (RNA-Seq)	http://gtexportal.org/home/	(Keen & Moore, 2015)
FANTOM	Riken Institute (Japan)	Tissue-based RNA data based on CAGE	http://fantom.gsc.riken.jp/	(Yu *et al*, 2015)
RNA-Seq Atlas	J. Gutenberg University (Germany)	A reference database for gene expression profiling in normal tissue by next-generation sequencing	http://medicalgenomics.org/rna_seq_atlas	(Krupp *et al*, 2012)
Allen Brain Atlas	Allen Institute (USA)	An anatomically comprehensive atlas of the adult human brain transcriptome	http://human.brain-map.org/	(Hawrylycz *et al*, 2012)
Evolution of gene expression	University of Lausanne (Switzerland)	The evolution of gene expression levels in mammalian organs	http://www.ncbi.nlm.nih.gov/geo/query/acc.cgi?acc=GSE30352	(Brawand *et al*, 2011)
AltIso	MIT (USA)	Alternative isoform regulation in human tissue transcriptomes.	http://www.ncbi.nlm.nih.gov/geo/query/acc.cgi?acc=GSE12946	(Wang *et al*, 2008)
Expression Atlas	EBI (UK)	Repository for RNA expression data (both microarray and RNA-Seq)	https://www.ebi.ac.uk/gxa	(Petryszak *et al*, 2015)
ArrayExpress	EBI (UK)	International functional genomics public data repositories	http://www.ebi.ac.uk/arrayexpress/	(Rustici *et al*, 2013)
Illumina Body Map	Illumina (USA)	RNA-Seq of 16 human individual tissues	http://www.ebi.ac.uk/arrayexpress/experiments/E-MTAB-513/	(Rustici *et al*, 2013)
Gene Expression Omnibus	NCBI (USA)	Repository for RNA expression data (both microarray and RNA-Seq)	http://www.ncbi.nlm.nih.gov/geo/	(Barrett *et al*, 2013)

An alternative approach to RNA-seq, named cap analysis gene expression (CAGE), has been described by the Fantom consortia (Yu *et al*,2015) and allows for quantitative measurements of transcripts based on sequencing the 5′-end of capped mRNA molecules. The correlation between RNA-seq and CAGE for transcriptome analysis was recently investigated (Yu *et al*, 2015), and the transcriptome of 22 tissues was analyzed using both methods (Fig 1) based on 79 RNA-seq (HPA) and 27 CAGE (FANTOM) samples. Tissue-to-tissue comparisons showed a high genome-wide correlation between the two datasets (Yu *et al*, 2015). Interestingly, discrepancies between the two datasets can largely be explained by gene model annotation issues or technical artifacts inherent in the respective methodologies. As an example, the HPA data excluded mRNA without poly-adenylation tails and it is therefore not surprising that many histone genes were lacking in the RNA-seq data, but are present in the CAGE data. Conversely, CAGE peaks mapping more than 500 base pairs from the transcriptional start site are lacking in the CAGE dataset, as well as CAGE peaks mapping to two or more locations on the genome, which are removed from the dataset. Thus, the two methods are complementary and it would be attractive to integrate data obtained by these two approaches to refine gene models and to improve the interpretation of gene expression values.

CLASSIFICATION OF ALL HUMAN PROTEIN-CODING GENES BASED ON TISSUE PROFILING

The different omics-based analyses of the human proteome have allowed the classification of protein-coding genes with regard to tissue-restricted expression. In the analysis performed by the HPA consortium (Fagerberg *et al*, 2014; Uhlen *et al*, 2015), a cutoff of 1 FPKM (Hebenstreit *et al*, 2011) was used to indicate the presence or absence of transcripts for a particular gene in a tissue. Based on this definition, all human protein-coding genes were classified into (i) genes with an elevated expression in one or several tissues, (ii) genes expressed in all analyzed tissues, (iii) genes with mixed expression found in several, but not all tissues, and (iv) genes not detected in any tissues. The elevated genes were further stratified into "tissue enriched", "group enriched", or "tissue enhanced". The term "tissue specific" was avoided as it depends on the definition of cutoff values, and only few genes, including well-known proteins such as insulin, PSA, and troponin, were found to be exclusively expressed in a single tissue type (Uhlen *et al*, 2015). A classification of all protein-coding genes is shown in Table 2(cutoff of 0.5 FPKM).

Table 2: Classification of all human protein-coding genes based on transcript expression levels in tissues and organs. The columns HPA and GTEx indicate the number of genes identified in the different categories using the datasets (Keen & Moore, 2015; Uhlen *et al*, 2015) from these two consortia

Category	Definition	HPA	GTEx
Tissue enriched	At least fivefold higher mRNA levels (FPKM) in a particular tissue as compared to all other tissues	2,359	2,289
Group enriched	At least fivefold higher mRNA levels in a group of tissues (2-7)	1,208	1,307
Enhanced	At least fivefold higher mRNA levels in a particular tissue as compared to the average levels in all tissues	3,227	3,077
Expressed in all	Detected in all tissues	8,385	8,459
Mixed	Detected in at least two tissues, but not in all, and not part of any of the categories above	2,484	2,537
Not detected	Not present in any of the analyzed tissues (under cutoff)	1,021	1,015
Total	Total number of genes analyzed	18,684	18,684
Total elevated	Total number of tissue-enriched, group-enriched, and tissue-enhanced genes	6,794	6,673

The classification of the human protein-coding transcriptome showed that almost half of the genes were detected in all tissues (45%), while 13% showed a mixed expression (Fig 2A). Approximately one-third of the genes showed a tissue elevated expression with 13% of the genes enriched in one of the analyzed tissues. Only 5% of the genes were not detected in any of the analyzed tissues. A further analysis of the number of genes with a tissue elevated expression (Fig 2B) showed that the testis has by far the highest number of tissue-enriched genes followed by the brain (cerebral cortex) and liver.

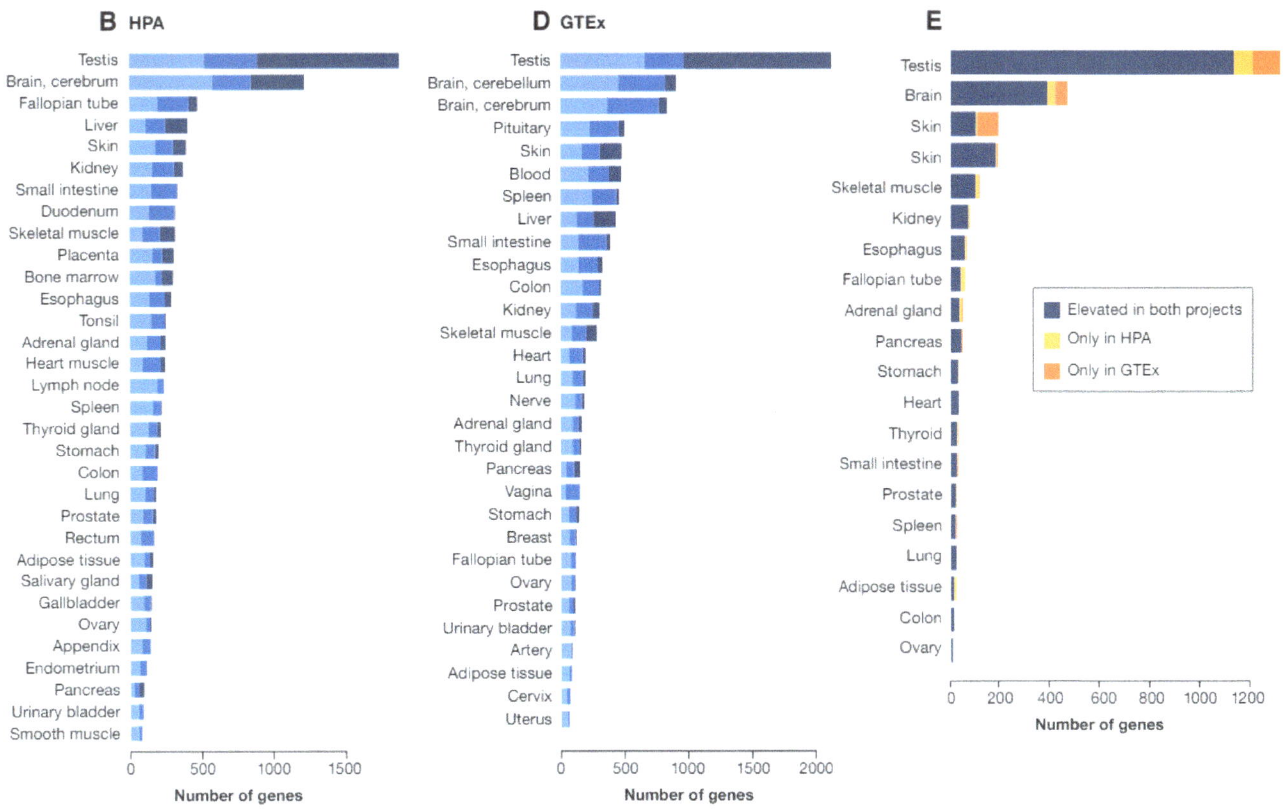

Figure 2. Classification of all protein-coding genes using transcriptomics data.

(A) Pie chart showing the number of genes that fall into each expression specificity category, based on the classifications of HPA (32 tissues, 137 samples) (with a cutoff of 0.5 FPKM). (B) The number of protein-coding genes classified as tissue enriched (dark blue), group enriched (medium blue), and tissue enhanced (light blue) based on the HPA dataset. (C) Pie chart showing the number of genes that fall into each expression specificity category, based on the classifications of GTEx (30 tissues, 2,510 samples) (Keen & Moore,2015) (with a cutoff of 0.5 FPKM). (D) The number of protein-coding genes classified as above based on GTEx dataset. (E). Barplot showing the overlap of tissue-enriched genes between the two datasets. All genes that are tissue enriched in either dataset are depicted. Genes classified as tissue enriched/group enriched/tissue enhanced in the same tissue in both datasets are shown in blue; genes only enriched in one of the datasets are shown in yellow (only HPA) or orange (only GTEx).

The recently published RNA-Seq data generated by the GTEx consortium (Bahcall, 2015; GTEx Consortium, 2015; Gibson, 2015) allow for an independent tissue-based classification of the human proteome. The GTEx dataset includes more than 1,600 postmortem samples from mostly overlapping, but in some cases unique, tissues compared to the HPA consortium (Fig 1). For example, the GTEx dataset includes more tissue samples from the brain, blood, and nerves, which are not included in the HPA dataset. As illustrated in Fig 2C, the overall tissue-based classification based on the GTEx dataset and an identical cutoff of 0.5 FPKM is similar to that of the HPA with 45% of the genes expressed in all tissues, 14% showing a mixed expression, 12% being tissue-enriched expression, and 5% of the genes not detected in any of the analyzed tissues. With respect to tissue-elevated genes (Fig 2B and D) the testis is again observed to contain the largest number of tissue-enriched genes, followed by the brain (cerebellum, cortex, and pituitary), skin, and liver. The values for the HPA and the GTEx datasets can be found in Table EV1.

TISSUE-ENRICHED GENES VS. UBIQUITOUSLY EXPRESSED GENES

Tissue-enriched genes identified by the analysis of the HPA data (Uhlen *et al*, 2015) based on the definitions shown in Table 2 can be found for all tissues in the interactive HPA database (www.proteinatlas.org/humanproteome/tissue+specific). A functional Gene Ontology analysis of the tissue-enriched genes in the HPA dataset has been performed and the results are consistent with the function of each tissue (Uhlen *et al*,2015). As an example, genes elevated in liver encode secreted plasma and bile proteins, detoxification proteins, and proteins associated with metabolic processes and glycogen storage (Kampf *et al*, 2014), whereas genes elevated in adipose tissue encode proteins involved in lipid metabolic processes (Mardinoglu *et al*, 2014b) and genes elevated in skin encode proteins associated with functions related to the barrier function (squamous cell differentiation and cornification), skin pigmentation, and hair development (Edqvist *et al*, 2015). In order to further validate these lists, we have compared the overlap of tissue-enriched genes identified using the independent HPA and GTEx datasets. The number of tissue-enriched genes in the different tissues and the overlap between the two datasets are shown in Figs 2E and 3A, andTable EV2. Overall, it is reassuring that there is a significant overlap in the tissue classification of the genes based on the two independent datasets. The fact that similar results are obtained when using fresh frozen tissue (HPA) and postmortem tissue (GTEx) suggests negligible effects of the sampling procedures used by the GTEx consortium on RNA degradation. In the comparison, note that in the HPA dataset, the brain contains only one tissue (cerebral cortex), while the corresponding GTEx dataset is based on three different tissues (cerebellum, cortex, and pituitary). The large discrepancy for skin can be explained by the fact that the sampling of skin in the HPA was based on shave biopsies including mainly epidermis (Uhlen *et al*, 2015), while the GTEx consortium also included the underlying dermis, most likely containing skin adnexal structures such as hair follicles and sweat glands.

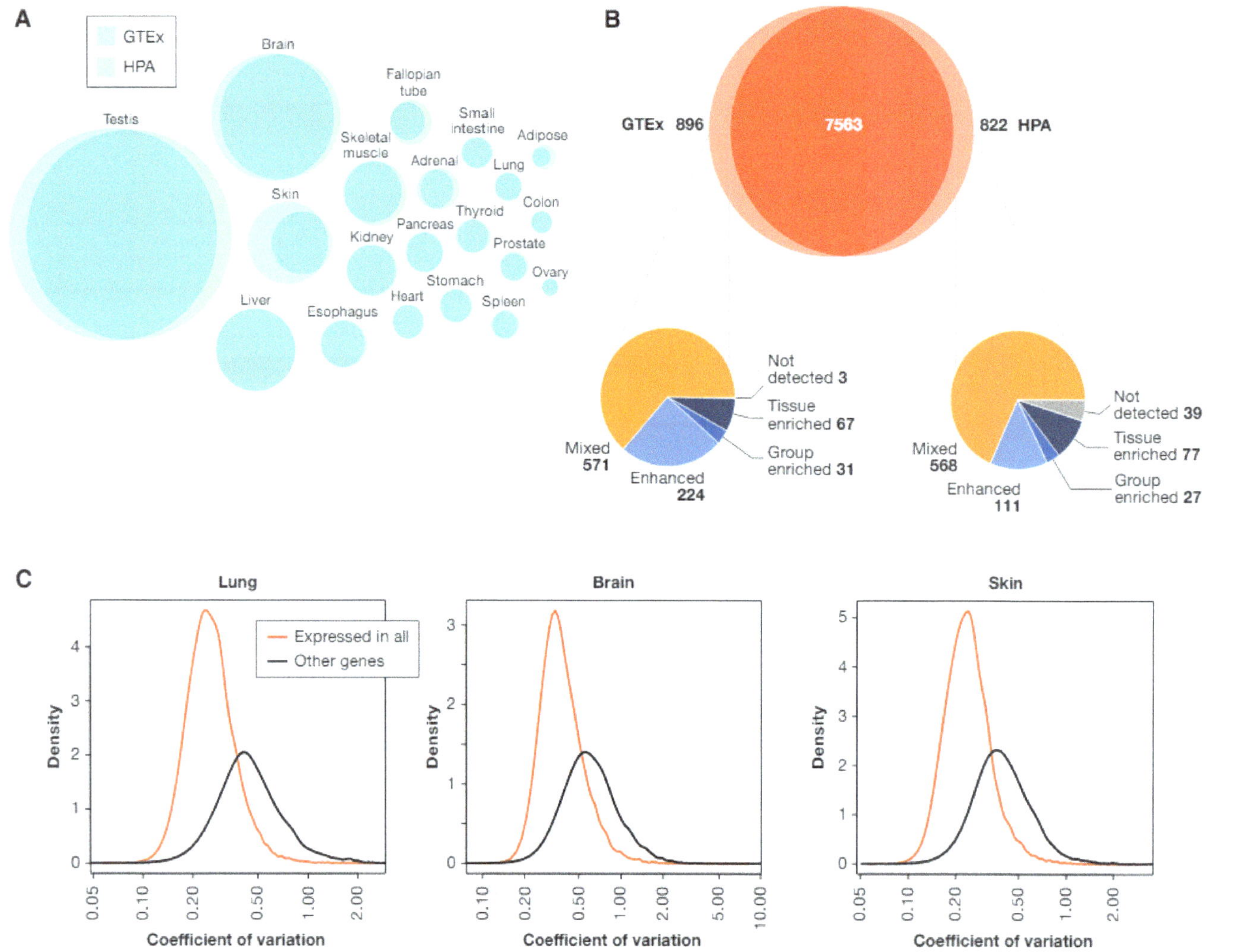

Figure 3: Protein classification and interindividual variations.

(A) Venn diagrams showing the overlap between tissue-elevated genes between the two datasets, HPA in light green and GTEx in light blue. (B) Venn diagram showing the overlap between genes classified as "expressed in all tissues" between the two datasets. The pie charts show the classification of the non-overlapping genes in the dataset where the gene was not detected in all tissues. (C) Comparison of interindividual variation between genes that are annotated as "expressed in all tissues" and all other genes, in lung, brain, and skin (these tissues were selected because they have a large number of biological replicates). The plots illustrate the distribution of the coefficient of variation (CV) within the tissue for all genes in each of the two classes (red: expressed in all, black: other). The CV is shifted toward the lower side in the "expressed in all" category ($P \ll 0.001$), suggesting that genes that are expressed in all tissues have lower variation between individuals.

Genes identified as "expressed in all tissues" are expected to be either "housekeeping" genes for which the protein product is needed in every cell, such as proteins involved in transcription, translation, and energy metabolism, or genes expressed in cell types that are present in all tissue types, such as lymphocytes, macrophages, fibroblasts, and endothelial cells. HPA and GTEx data largely overlap for this category as well, with 7,563 genes identified as "expressed in all" by both resources (Fig 3B). Between 800 and 900 genes were only identified in one of the two datasets, and a more detailed analysis shows that the vast majority of these genes were identified as "mixed" by the other dataset (Fig 3B). This suggests that these genes move between categories based on the relatively arbitrary FPKM cutoff and indicates that expression in a single tissue below the detection threshold makes a gene move from "expressed in all" to "mixed".

THE VARIATION IN PROTEIN PROFILES BETWEEN INDIVIDUALS

A relevant question arising is the level of interindividual variation in gene expression levels. A comparison of individual variation for "housekeeping" genes (defined as expressed in all analyzed tissues) and genes with a more tissue-restricted expression using GTEx data is presented in Fig 3C for three different tissues (lung, brain, and skin) that are represented by a large number of biological replicates. For all three tissues, the coefficient of variation (CV) in the "expressed in all" category shifted toward the lower side, suggesting that genes expressed in all tissues seem to vary less between individuals for a particular tissue as compared to genes with a tissue-restricted expression pattern. This illustrates that the proteins found in all tissues are expressed at relatively similar levels across the analyzed tissues, suggesting, as perhaps expected, that these proteins that are involved in "basic functions" are required at similar concentrations in the various tissue types.

BUILDING GENOME-SCALE METABOLIC MODELS FOR HUMAN TISSUES

High-quality genome-wide proteomics and transcriptomics data can be used for generating and improving context-specific biological networks including protein–protein interaction (PPI), regulatory, signaling, and metabolic networks (Papin *et al*, 2005; Qian *et al*, 2005; Bossi & Lehner,2009) in order to gain further insights into the differences in cellular functions across tissues. Genome-scale metabolic models (GEMs) that can be reconstructed directly from proteomics or transcriptomics data are particularly well suited for the analysis of biological functions, since they can be applied to examine the metabolic functions associated with a given cell type. Several studies have recently reported the use of proteomics data to reconstruct GEMs for analyzing metabolic processes across different cell and tissue types in humans (Mardinoglu & Nielsen, 2015; O'Brien *et al*, 2015; Yizhak *et al*, 2015; Bjornson *et al*, 2016) and mice (Mardinoglu *et al*, 2015b). GEMs contain thousands of biochemical reactions and their catalyzing protein-coding genes in a cell/tissue, which generate a complex network of molecular interactions capturing the metabolic functions of this cell/tissue (Fig 4A). This reaction network is converted into a computational model using a stoichiometric (S) matrix and can be applied for the analysis of physiological data collected from both healthy and diseased states (Mardinoglu & Nielsen, 2012; Mardinoglu *et al*, 2013b)

Table 3: List of generic and cell-/tissue-specific human GEMs

Model name	Application	References
Generic human GEMs		
Recon1	Integration of genomic and bibliomic data	(Duarte *et al*, 2007)
EHMN	Integration of genomic and bibliomic data	(Ma *et al*, 2007)
HMR	Integration of previous generic human GEMs and publicly available databases	(Agren *et al*, 2012)
Recon2	Community-based reconstruction of human metabolism	(Thiele *et al*, 2013)
HMR2	Incorporation of extensive lipid metabolism into the generic human GEM	(Mardinoglu *et al*, 2014a)
Cell-/tissue-specific GEMs		
Red blood cell	Analysis of the metabolic loads in red blood cells	(Wiback & Palsson, 2002)
Mitochondria	Study of the human mitochondrial metabolism	(Vo *et al*, 2004)
Fibroblasts	Metabolic alterations in Leigh syndrome	(Vo *et al*, 2007)
HepatoNet1	Investigation of hepatic enzyme deficiencies	(Gille *et al*, 2010)
Computational liver model	Discovery of biomarkers of liver disorders including hyperammonemia and hyperglutaminemia	(Jerby *et al*, 2010)
Kidney	Prediction of causal drug off-targets that impact kidney function	(Chang *et al*, 2010)
Brain (three neuron types and astrocytes)	Revealing the metabolic alterations in Alzheimer's disease	(Lewis *et al*, 2010)
IAB-AMQ-1410	Analysis of the host pathogen interactions with *Mycobacterium tuberculosis*	(Bordbar *et al*, 2010)
Multitissue (hepatocytes, myocytes, and adipocytes)	Revealing the metabolic alterations in T2D	(Bordbar *et al*, 2011a)
Erythrocyte (iAB-RBC-283)	Revealing the complexity in the functional capabilities of human erythrocyte metabolism	(Bordbar *et al*, 2011b)
69 cell-specific GEMs	Studying the metabolic differences between healthy cells and cancers	(Agren *et al*, 2012)
126 tissue-specific GEMs	Comparative analysis between healthy tissues and tumor	(Wang *et al*, 2012)
CardioNet	The effect of oxygen and substrate supply on the efficiency of selected metabolic functions of cardiomyocytes	(Karlstaedt *et al*, 2012)
iAdipocytes1809	Revealing the metabolic differences in obese subjects	(Mardinoglu *et al*, 2013a)
Tissue-specific GEMs	Studying the metabolic differences between healthy tissues and cancers	(Nam *et al*, 2014)
Liver GEM	Studying urea metabolism in liver tissue	(Vlassis *et al*, 2014)
83 cell-specific GEMs	Defining the major metabolic functions in human cell types	(Agren *et al*, 2014)
iHepatocytes2322	Revealing the metabolic alterations in response to NAFLD	(Mardinoglu *et al*, 2014a)
iMyocyte2419	Revealing the metabolic alterations in response to T2D	(Varemo *et al*, 2015)
32 tissue-specific GEMs	Global analysis of the metabolic functions in major human tissues	(Uhlen *et al*, 2015)

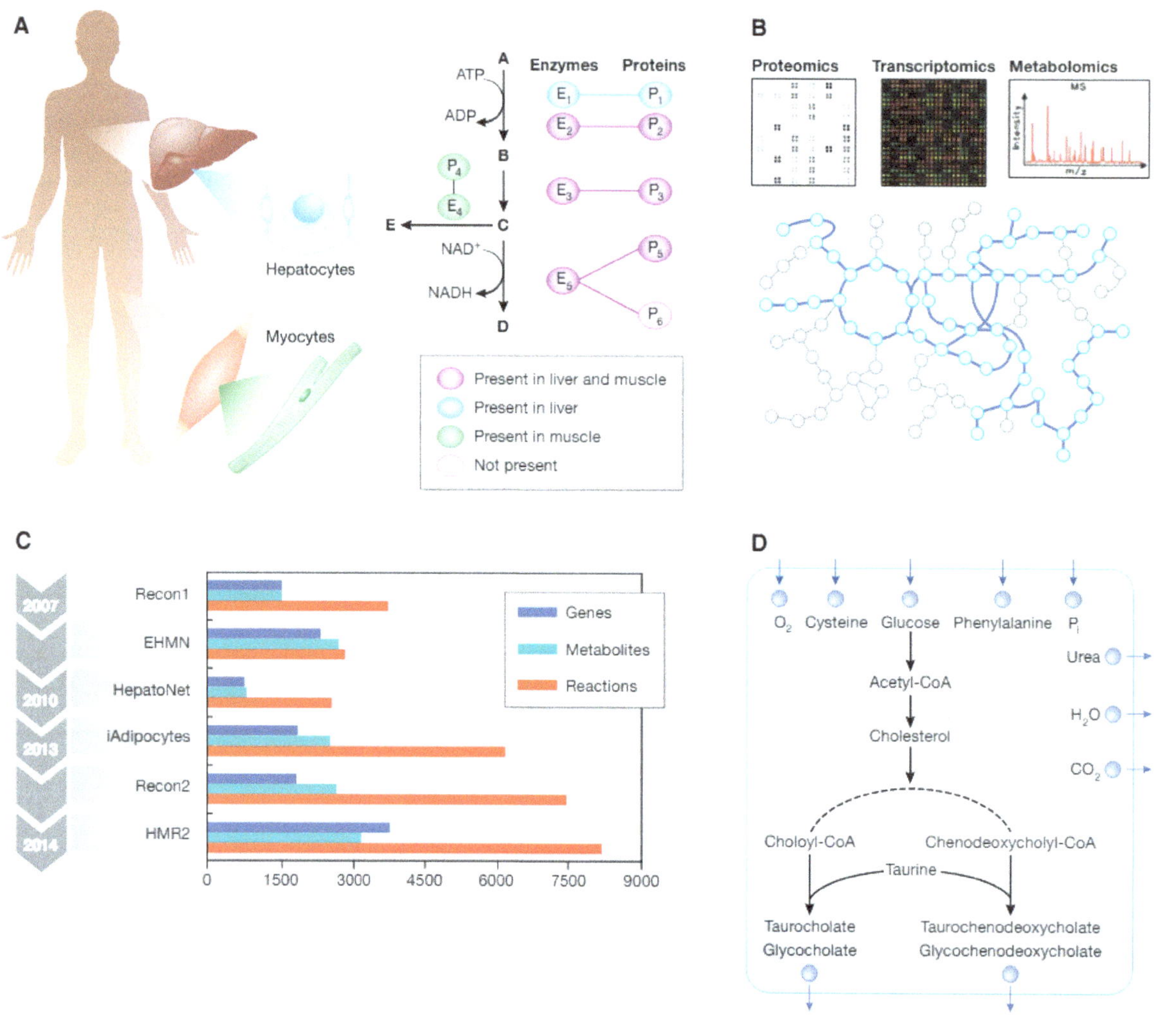

Figure 4: Genome-scale metabolic models for human cells/tissues.

(A) GEMs incorporate the known biochemical reactions and their catalyzing enzymes in a particular cell/tissue type. The information related to the reaction–gene association is used for the reconstruction of context-specific GEMs. (B) The continuously increasing number of reactions, metabolites, and genes included in generic human GEMs and manually curated cell-/tissue-specific GEMs generated in the recent years is shown. (C) High-throughput omics data including proteomics, transcriptomics, and metabolomics have been used for reconstructing cell-/tissue-specific GEMs based on generic human GEMs. (D) The metabolic tasks that are known to occur in a given human cell/tissue need to be defined to generate functional cell-/tissue-specific GEMs. The definition of the metabolic task related to bile acid synthesis in the liver is presented. Glucose, cysteine, phenylalanine, oxygen (O_2), and phosphate (Pi) must be taken up, whereas urea, water (H_2O), and carbon dioxide (CO_2) must be secreted in order to successfully simulate bile acid synthesis in liver GEM.

The first human GEMs, Recon1 (Duarte *et al*, 2007) and EHMN (Ma *et al*, 2007), were developed nearly 10 years ago and have now grown to the level where they can be used for predicting the metabolic response of cell/tissue to a given perturbation. These integrative models allowed the identification of new drug target candidates by theoretical analyses (Frezza *et al*, 2011), and many enzymes have already been proposed as drug targets for cancer treatment (Yizhak *et al*, 2015). Recently, more comprehensive generic human GEMs including Recon2 (Thiele *et al*, 2013) and HMR2 (Mardinoglu *et al*, 2014a) were constructed by integrating the components of the first generic human GEMs with manually reconstructed context-specific GEMs. Recon2 covers the content of the HepatoNet, a manually reconstructed GEM for hepatocytes (Gille *et al*, 2010), whereas HMR2 covers the content of both HepatoNet and *iAdipocytes1809*, a manually reconstructed GEM for adipocytes (Mardinoglu *et al*, 2013a). HMR2 also includes the extensive description of lipid metabolism present in human adipocytes and hepatocytes. The number of reactions, metabolites, and genes incorporated in each model as well as the conceptual evolution of global reconstructions of human metabolism is presented in Fig 4B. As illustrated in Fig 4B, HMR2 is the most comprehensive global reconstruction of human metabolism and this model together with other generic models of human metabolism has served as a basis for the reconstruction of context-specific GEMs (Fig 4C).

Context-specific GEMs were generated by manually curating the existing literature as well as by using various algorithms that have been reviewed elsewhere (Machado & Herrgard, 2014). For instance, the recently developed tINIT algorithm enables the reconstruction of simulation-ready GEMs based on proteomics data and metabolic functions that are known to occur in the cell/tissue of interest (Agren *et al*,2014). The implementation of a metabolic function related to bile acid synthesis into the liver-specific GEM is shown as an example in Fig 4D. Recently, 32 tissue-specific GEMs for healthy human tissues were generated by integrating RNA-seq data from the HPA in combination with the tINIT algorithm and they were used to compare the metabolic differences between these tissues (Uhlen *et al*, 2015). GEMs reconstructed based on RNA-seq data successfully predicted tissue-specific functions. For instance, the liver GEM was the only model that could successfully perform metabolic functions related to bile acid synthesis. Moreover, the liver GEM was able to perform all defined human metabolic functions and it was the largest GEM in terms of incorporated reactions, metabolites, and genes, reflecting its high metabolic activity compared to the other analyzed tissues. A list of the various cell-/tissue-specific GEMs that have been generated so far, either in physiological or in disease states, is presented in Table 3.

APPLYING CONTEXT-DEPENDENT GEMS FOR ANALYZING HUMAN DISEASES

Context-specific GEMs in combination with omics data obtained in disease-specific contexts have been used to elucidate the metabolic capabilities of cells/tissues involved in metabolism-related disorders including obesity (Mardinoglu *et al*, 2013a, 2014b, 2015a), non-alcoholic fatty liver disease (NAFLD) (Mardinoglu *et al*, 2014a; Hyötyläinen *et al*, 2016), type 2 diabetes (T2D) (Varemo *et al*, 2015), and aging (Yizhak*et al*, 2013), as well as to determine unique metabolic properties of cancer cells (Agren *et al*, 2012; Gatto *et al*, 2014; Nam *et al*, 2014) and even individual cell lines (Yizhak *et al*, 2014a,b; Gatto *et al*, 2015; Ghaffari *et al*, 2015) and tumors (Agren *et al*, 2014). Each of these studies advanced our understanding of the molecular mechanisms underlying these diseases and allowed the discovery of drug targets or biomarkers that can be used for designing effective treatment strategies.

Recently, a GEM for skeletal myocytes was reconstructed using cell type-specific RNA-seq data and incorporating cell type-specific proteomics data from the HPA. First, the presence/absence of each enzyme in myocytes was determined and based on this information the corresponding metabolic reaction was incorporated into the myocyte-specific GEM (Varemo *et al*, 2015). The model was employed for characterizing the metabolic alterations in skeletal muscle in response to T2D based on the meta-analysis of six published datasets on T2D muscle gene expression. The metabolic alterations observed in the skeletal muscle T2D patients involved differences in pyruvate oxidation, tetrahydrofolate metabolism, and branched-chain amino acid catabolism.

The interplay between a large number of biological pathways and the significant variation between patients makes it extremely difficult to identify effective drug targets and biomarkers for metabolic diseases. Personalized GEMs that account for interindividual differences as well as for the unique characteristics of disease progression in each individual (Agren *et al*, 2014) present a potential solution to these issues. In a recent study, personalized cancer GEMs for six hepatocellular carcinoma (HCC) patients as well as 83 healthy cell-specific GEMs were reconstructed using the tINIT algorithm to integrate proteomics data from the HPA and metabolic functions that are present in human cells (Agren *et al*, 2014). Based on these personalized GEMs, anticancer drug targets that can be used for inhibiting the HCC tumor growth in each patient were identified. One of the targets, predicted to be effective in all patients, was experimentally validated in human HCC cancer cell lines. Overall, the observation that fat oxidation was increased in the analyzed HCC tumors indicated that targeting this metabolic process could be used for developing treatment strategies for HCC.

Another recent application of GEMs in the context of HCC is presented in the study of Björnson *et al* (2015). In this case, gene expression data from approximately 360 HCC tumors and 50 non-cancerous liver samples were analyzed using a HCC-specific GEM. Interestingly, a group of patients showed an increased fat oxidation, whereas another group showed a decreased fat oxidation. The fact that HCC tumors from different patients may have completely opposite metabolic programming highlights that careful stratification of HCC patients and personalized medicine approaches are highly advantageous for developing effective treatment strategies. Overall, these studies provide valuable insights into inter- and intratumor heterogeneity and point out that it might be extremely difficult to treat all different HCC patients with a single drug. This drug can be effective in the right context, that is, in a given patient or patient group. Therefore, personalized GEMs and their predictions of a patient›s response to different drugs can be extremely useful for guiding precision medicine approaches.

CONCLUDING REMARKS

Here, we reviewed some of the publicly available human transcriptomics data resources with a focus on the expression data for protein-coding genes. Tissue-restricted and tissue-enriched genes can be consistently defined in a genome-wide manner by two independent datasets generated using either fresh surgically removed tissues or postmortem tissues taken within 24 hours after the death of the individual. Thus, comprehensive lists of protein-coding genes can be compiled for all the major tissues of the human body (see Table EV1), with their quantitative expression profiles generated by deep sequencing of the transcriptome.

The use of high-quality proteomics and transcriptomics data in combination with metabolic modeling allows for functional analyses in the context of different pathologies, for example, by comparing GEMs reconstructed using data from healthy and diseased subjects. On the one hand, a comparison of the healthy vs. diseased GEM topology can provide insights into how cancer metabolism differs from metabolism of the healthy tissue (Gatto *et al*, 2014; Björnson *et al*, 2015; Zhang *et al*, 2015). Furthermore, GEMs can be used for identifying drug targets (Agren *et al*, 2014), and therefore, their integration with omics data generated in a clinical setup can be applied to guide precision medicine in different disease types. Further improvement and expansion of GEMs to cover other biological processes, for example, protein secretion pathways and protein synthesis (Feizi *et al*, 2013), will allow this modeling framework to capture dysfunction of key cellular pathways in a range of different pathologies, potentially leading to the identification of new treatment strategies and biomarkers.

The transcriptomics data can be complemented with immunohistochemistry to define protein localization in the subcompartments of each tissue and organ down to the single-cell level (www.proteinatlas.org). Moreover, extending these tissue profiles to include splice variants and protein modifications is important for improving our understanding of the role of the isoform proteome and post-translational modifications in human physiology and disease. Finally, spatial proteomics using fluorescent-based antibody profiling (Marx, 2015) can provide even higher resolution with precise localizations of the corresponding proteins down to subcellular compartments and various substructures. The integration of transcriptomics data with other large-scale data, such as mass spectrometry-based proteomics, antibody-based profiling, and metabolomics, can thus generate an important molecular knowledge base for systems biology of human health and disease.

ACKNOWLEDGEMENTS

We acknowledge the entire staff of the Human Protein Atlas program and the Science for Life Laboratory for valuable contributions. Funding was provided by the Knut and Alice Wallenberg Foundation.

REFERENCES

1. Agren R, Bordel S, Mardinoglu A, Pornputtapong N, Nookaew I, Nielsen J (2012) Reconstruction of genome-scale active metabolic networks for 69 human cell types and 16 cancer types using INIT. *PLoS Comput Biol* 8: e1002518
2. Agren R, Mardinoglu A, Asplund A, Kampf C, Uhlen M, Nielsen J (2014) Identification of anticancer drugs for hepatocellular carcinoma through personalized genome-scale metabolic modeling. *Mol Syst Biol* 10: 721
3. Anderson L, Seilhamer J (1997) A comparison of selected mRNA and protein abundances in human liver. *Electrophoresis* 18: 533–537
4. Bahcall OG (2015) Human genetics: GTEx pilot quantifies eQTL variation across tissues and individuals. *Nat Rev Genet* 16: 375
5. Barrett T, Wilhite SE, Ledoux P, Evangelista C, Kim IF, Tomashevsky M, Marshall KA, Phillippy KH, Sherman PM, Holko M, Yefanov A,Lee H, Zhang N, Robertson CL, Serova N, Davis S, Soboleva A (2013) NCBI GEO: archive for functional genomics data sets–update. *Nucleic Acids Res* 41: D991–D995
6. Bjornson E, Boren J, Mardinoglu A (2016) Personalized Cardiovascular Disease Prediction and Treatment-A Review of Existing Strategies and Novel Systems Medicine Tools. *Front Physiol* 7: 2
7. Björnson E, Mukhopadhyay B, Asplund A, Pristovsek N, Cinar R, Romeo S, Uhlen M, Kunos G, Nielsen J, Mardinoglu A (2015)Stratification of hepatocellular carcinoma patients based on acetate utilization. *Cell Rep* 13: 2014–2026
8. Bordbar A, Lewis NE, Schellenberger J, Palsson BO, Jamshidi N (2010) Insight into human alveolar macrophage and M. tuberculosis interactions via metabolic reconstructions. *Mol Syst Biol* 6: 422
9. Bordbar A, Feist AM, Usaite-Black R, Woodcock J, Palsson BO, Famili I (2011a) A multi-tissue type genome-scale metabolic network for analysis of whole-body systems physiology. *BMC Syst Biol* 5: 180
10. Bordbar A, Jamshidi N, Palsson BO (2011b) iAB-RBC-283: a proteomically derived knowledge-base of erythrocyte metabolism that can be used to simulate its physiological and patho-physiological states. *BMC Syst Biol* 5: 110
11. Bossi A, Lehner B (2009) Tissue specificity and the human protein interaction network. *Mol Syst Biol* 5: 260
12. Brawand D, Soumillon M, Necsulea A, Julien P, Csardi G, Harrigan P, Weier M, Liechti A, Aximu-Petri A, Kircher M, Albert FW, Zeller U,Khaitovich P, Grutzner F, Bergmann S, Nielsen R, Paabo S, Kaessmann H (2011) The evolution of gene expression levels in mammalian organs. *Nature* 478: 343–348
13. Chang RL, Xie L, Bourne PE, Palsson BO (2010) Drug off-target effects predicted using structural analysis in the context of a metabolic network model. *PLoS Comput Biol* 6: e1000938
14. GTEx Consortium (2015) Human genomics. The Genotype-Tissue Expression (GTEx) pilot analysis: multitissue gene regulation in humans. *Science* 348: 648–660
15. Cunningham F, Amode MR, Barrell D, Beal K, Billis K, Brent S, Carvalho-Silva D, Clapham P, Coates G, Fitzgerald S, Gil L, Giron CG,Gordon L, Hourlier T, Hunt SE, Janacek SH, Johnson N, Juettemann T, Kahari AK, Keenan S *et al* (2015) Ensembl 2015. *Nucleic Acids Res* 43: D662–D669
16. Danielsson F, James T, Gomez-Cabrero D, Huss M (2015) Assessing the consistency of public human tissue RNA-seq data sets. *Brief Bioinform* 16: 941–949
17. Duarte NC, Becker SA, Jamshidi N, Thiele I, Mo ML, Vo TD, Srivas R, Palsson BO (2007) Global reconstruction of the human metabolic network based on genomic and bibliomic data. *Proc Natl Acad Sci U S A* 104: 1777–1782
18. Eden E, Geva-Zatorsky N, Issaeva I, Cohen A, Dekel E, Danon T, Cohen L, Mayo A, Alon U (2011) Proteome half-life dynamics in living human cells. *Science* 331: 764–768
19. Edqvist PH, Fagerberg L, Hallstrom BM, Danielsson A, Edlund K, Uhlen M, Ponten F (2015) Expression of human skin-specific genes defined by transcriptomics and antibody-based profiling. *J Histochem Cytochem* 63: 129–141
20. Fagerberg L, Hallstrom BM, Oksvold P, Kampf C, Djureinovic D, Odeberg J, Habuka M, Tahmasebpoor S, Danielsson A, Edlund K,Asplund A, Sjostedt E, Lundberg E, Szigyarto CA, Skogs M, Takanen JO, Berling H, Tegel H, Mulder J, Nilsson P *et al* (2014)Analysis of the human tissue-specific expression by genome-wide integration of transcriptomics and antibody-based proteomics. *Mol Cell Proteomics* 13: 397–406
21. Feizi A, Osterlund T, Petranovic D, Bordel S, Nielsen J (2013) Genome-scale modeling of the protein secretory machinery in yeast.*PLoS ONE* 8: e63284

22. Frezza C, Zheng L, Folger O, Rajagopalan KN, MacKenzie ED, Jerby L, Micaroni M, Chaneton B, Adam J, Hedley A, Kalna G,Tomlinson IP, Pollard PJ, Watson DG, Deberardinis RJ, Shlomi T, Ruppin E, Gottlieb E (2011) Haem oxygenase is synthetically lethal with the tumour suppressor fumarate hydratase. *Nature* 477: 225–228

23. Gatto F, Nookaew I, Nielsen J (2014) Chromosome 3p loss of heterozygosity is associated with a unique metabolic network in clear cell renal carcinoma. *Proc Natl Acad Sci USA* 111: E866–E875

24. Gatto F, Miess H, Schulze A, Nielsen J (2015) Flux balance analysis predicts essential genes in clear cell renal cell carcinoma metabolism. *Sci Rep* 5: 10738

25. Ghaffari P, Mardinoglu A, Asplund A, Shoaie S, Kampf C, Uhlen M, Nielsen J (2015) Identifying anti-growth factors for human cancer cell lines through genome-scale metabolic modeling. *Sci Rep* 5: 8183

26. Gibson G (2015) Human genetics. GTEx detects genetic effects. *Science* 348: 640–641

27. Gille C, Bolling C, Hoppe A, Bulik S, Hoffmann S, Hubner K, Karlstadt A, Ganeshan R, Konig M, Rother K, Weidlich M, Behre J,Holzhutter HG (2010) HepatoNet1: a comprehensive metabolic reconstruction of the human hepatocyte for the analysis of liver physiology. *Mol Syst Biol* 6: 411

28. Gry M, Rimini R, Stromberg S, Asplund A, Ponten F, Uhlen M, Nilsson P (2009) Correlations between RNA and protein expression profiles in 23 human cell lines. *BMC Genom* 10: 365

29. Harrow J, Frankish A, Gonzalez JM, Tapanari E, Diekhans M, Kokocinski F, Aken BL, Barrell D, Zadissa A, Searle S, Barnes I, Bignell A, Boychenko V, Hunt T, Kay M, Mukherjee G, Rajan J, Despacio-Reyes G, Saunders G, Steward C *et al* (2012) GENCODE: the reference human genome annotation for The ENCODE Project. *Genome Res* 22: 1760–1774

30. Hawrylycz MJ, Lein ES, Guillozet-Bongaarts AL, Shen EH, Ng L, Miller JA, van de Lagemaat LN, Smith KA, Ebbert A, Riley ZL, Abajian C, Beckmann CF, Bernard A, Bertagnolli D, Boe AF, Cartagena PM, Chakravarty MM, Chapin M, Chong J, Dalley RA *et al* (2012) An anatomically comprehensive atlas of the adult human brain transcriptome. *Nature* 489: 391–399

31. Hebenstreit D, Fang M, Gu M, Charoensawan V, van Oudenaarden A, Teichmann SA (2011) RNA sequencing reveals two major classes of gene expression levels in metazoan cells. *Mol Syst Biol* 7: 497

32. Hyötyläinen T, Jerby L, Petäjä EM, Mattila I, Jäntti S, Auvinen P, Gastaldelli A, Yki-Järvinen H, Ruppin E, Orešič M (2016) Genome-scale study reveals reduced metabolic adaptability in patients with non-alcoholic fatty liver disease. *Nat Commun* 7: 8994.

33. Jerby L, Shlomi T, Ruppin E (2010) Computational reconstruction of tissue-specific metabolic models: application to human liver metabolism. *Mol Syst Biol* 6: 401

34. Kampf C, Mardinoglu A, Fagerberg L, Hallström B, Edlund K, Nielsen J, Uhlen M (2014) The human liver-specific proteome defined by transcriptomics and antibody-based profiling. *FASEB J* 28: 2901–2914

35. Karlstaedt A, Fliegner D, Kararigas G, Ruderisch HS, Regitz-Zagrosek V, Holzhutter HG (2012) CardioNet: a human metabolic network suited for the study of cardiomyocyte metabolism. *BMC Syst Biol* 6: 114

36. Keen JC, Moore HM (2015) The Genotype-Tissue Expression (GTEx) Project: Linking Clinical Data with Molecular Analysis to Advance Personalized Medicine. *J Pers Med* 5: 22–29

37. Kim MS, Pinto SM, Getnet D, Nirujogi RS, Manda SS, Chaerkady R, Madugundu AK, Kelkar DS, Isserlin R, Jain S, Thomas JK,Muthusamy B, Leal-Rojas P, Kumar P, Sahasrabuddhe NA, Balakrishnan L, Advani J, George B, Renuse S, Selvan LD *et al* (2014)A draft map of the human proteome. *Nature* 509: 575–581

38. Krupp M, Marquardt JU, Sahin U, Galle PR, Castle J, Teufel A (2012) RNA-Seq Atlas–a reference database for gene expression profiling in normal tissue by next-generation sequencing. *Bioinformatics* 28: 1184–1185

39. Lamond AI, Uhlen M, Horning S, Makarov A, Robinson CV, Serrano L, Hartl FU, Baumeister W, Werenskiold AK, Andersen JS, Vorm O,Linial M, Aebersold R, Mann M (2012) Advancing cell biology through proteomics in space and time (PROSPECTS). *Mol Cell Proteomics* 11: O112. 017731

40. Lewis NE, Schramm G, Bordbar A, Schellenberger J, Andersen MP, Cheng JK, Patel N, Yee A, Lewis RA, Eils R, Konig R, Palsson BO(2010) Large-scale in silico modeling of metabolic interactions between cell types in the human brain. *Nat Biotechnol* 28: 1279–1285

41. Lundberg E, Fagerberg L, Klevebring D, Matic I, Geiger T, Cox J, Algenas C, Lundeberg J, Mann M, Uhlen M (2010) Defining the transcriptome and proteome in three functionally different human cell lines. *Mol Syst Biol* 6: 450

42. Ma H, Sorokin A, Mazein A, Selkov A, Selkov E, Demin O, Goryanin I (2007) The Edinburgh human metabolic network reconstruction and its functional analysis. *Mol Syst Biol* 3: 135

43. Machado D, Herrgard M (2014) Systematic evaluation of methods for integration of transcriptomic data into constraint-based models of metabolism. *PLoS Comput Biol* 10: e1003580

44. Maier T, Guell M, Serrano L (2009) Correlation of mRNA and protein in complex biological samples. *FEBS Lett* 583: 3966–3973

45. Maier T, Schmidt A, Guell M, Kuhner S, Gavin AC, Aebersold R, Serrano L (2011) Quantification of mRNA and protein and integration with protein turnover in a bacterium. *Mol Syst Biol* 7: 511

46. Mardinoglu A, Nielsen J (2012) Systems medicine and metabolic modelling. *J Intern Med* 271: 142–154

47. Mardinoglu A, Agren R, Kampf C, Asplund A, Nookaew I, Jacobson P, Walley AJ, Froguel P, Carlsson LM, Uhlen M, Nielsen J (2013a)Integration of clinical data with a genome-scale metabolic model of the human adipocyte. *Mol Syst Biol* 9: 649

48. Mardinoglu A, Gatto F, Nielsen J (2013b) Genome-scale modeling of human metabolism - a systems biology approach. *Biotechnol J* 8:985–996

49. Mardinoglu A, Agren R, Kampf C, Asplund A, Uhlen M, Nielsen J (2014a) Genome-scale metabolic modelling of hepatocytes reveals serine deficiency in patients with non-alcoholic fatty liver disease. *Nat Commun* 5: 3083

50. Mardinoglu A, Kampf C, Asplund A, Fagerberg L, Hallstrom BM, Edlund K, Bluher M, Ponten F, Uhlen M, Nielsen J (2014b) Defining the human adipose tissue proteome to reveal metabolic alterations in obesity. *J Proteome Res* 13: 5106–5119

51. Mardinoglu A, Nielsen J (2015) New paradigms for metabolic modeling of human cells. *Curr Opin Biotech* 34: 91–97

52. Mardinoglu A, Heiker JT, Gärtner D, Björnson E, Schön MR, Flehmig G, Klöting N, Krohn K, Fasshauer M, Stumvoll M, Nielsen J, Blüher M (2015a) Extensive weight loss reveals distinct gene expression changes in human subcutaneous and visceral adipose tissue. *Sci Rep* 5: 14841

53. Mardinoglu A, Shoaie S, Bergentall M, Ghaffari P, Zhang C, Larsson E, Bäckhed F, Nielsen J (2015b) The gut microbiota modulates host amino acid and glutathione metabolism in mice. *Mol Syst Biol* 11: 834

54. Marx V (2015) Mapping proteins with spatial proteomics. *Nat Methods* 12: 815–819

55. Nam H, Campodonico M, Bordbar A, Hyduke DR, Kim S, Zielinski DC, Palsson BO (2014) A systems approach to predict oncometabolites via context-specific genome-scale metabolic networks. *PLoS Comput Biol* 10: e1003837

56. Nilsson CL, Mostovenko E, Lichti CF, Ruggles K, Fenyo D, Rosenbloom KR, Hancock WS, Paik YK, Omenn GS, LaBaer J, Kroes RA,Uhlen M, Hober S, Vegvari A, Andren PE, Sulman EP, Lang FF, Fuentes M, Carlsohn E, Emmett MR *et al* (2015) Use of ENCODE resources to characterize novel proteoforms and missing proteins in the human proteome. *J Proteome Res* 14: 603–608

57. O›Brien EJ, Monk JM, Palsson BO (2015) Using Genome-scale Models to Predict Biological Capabilities. *Cell* 161: 971–987

58. Papin JA, Hunter T, Palsson BO, Subramaniam S (2005) Reconstruction of cellular signalling networks and analysis of their properties.*Nat Rev Mol Cell Biol* 6: 99–111

59. Petryszak R, Keays M, Tang YA, Fonseca NA, Barrera E, Burdett T, Fullgrabe A, Fuentes AM, Jupp S, Koskinen S, Mannion O, Huerta L, Megy K, Snow C, Williams E, Barzine M, Hastings E, Weisser H, Wright J, Jaiswal P *et al* (2015) Expression Atlas update-an integrated database of gene and protein expression in humans, animals and plants. *Nucleic Acids Res* 44: D746–D752

60. Qian J, Esumi N, Chen Y, Wang Q, Chowers I, Zack DJ (2005) Identification of regulatory targets of tissue-specific transcription factors: application to retina-specific gene regulation. *Nucleic Acids Res* 33: 3479–3491

61. Rustici G, Kolesnikov N, Brandizi M, Burdett T, Dylag M, Emam I, Farne A, Hastings E, Ison J, Keays M, Kurbatova N, Malone J, Mani R, Mupo A, Pedro Pereira R, Pilicheva E, Rung J, Sharma A, Tang YA, Ternent T *et al* (2013) ArrayExpress update–trends in database growth and links to data analysis tools. *Nucleic Acids Res* 41: D987–D990

62. Schwanhausser B, Busse D, Li N, Dittmar G, Schuchhardt J, Wolf J, Chen W, Selbach M (2011) Global quantification of mammalian gene expression control. *Nature* 473: 337–342

63. Stadler C, Rexhepaj E, Singan VR, Murphy RF, Pepperkok R, Uhlen M, Simpson JC, Lundberg E (2013) Immunofluorescence and fluorescent-protein tagging show high correlation for protein localization in mammalian cells. *Nat Methods* 10: 315–323

64. Tang F, Barbacioru C, Wang Y, Nordman E, Lee C, Xu N, Wang X, Bodeau J, Tuch BB, Siddiqui A, Lao K, Surani MA (2009) mRNA-Seq whole-transcriptome analysis of a single cell. *Nat Methods* 6: 377–382

65. Thiele I, Swainston N, Fleming RM, Hoppe A, Sahoo S, Aurich MK, Haraldsdottir H, Mo ML, Rolfsson O, Stobbe MD, Thorleifsson SG,Agren R, Bolling C, Bordel S, Chavali AK, Dobson P, Dunn WB, Endler L, Hala D, Hucka M *et al* (2013) A community-driven global reconstruction of human metabolism. *Nat Biotechnol* 31: 419–425

66. Tian Q, Stepaniants SB, Mao M, Weng L, Feetham MC, Doyle MJ, Yi EC, Dai H, Thorsson V, Eng J, Goodlett D, Berger JP, Gunter B,Linseley PS, Stoughton RB, Aebersold R, Collins SJ, Hanlon WA, Hood LE (2004) Integrated genomic and proteomic analyses of gene expression in Mammalian cells. *Mol Cell Proteomics* 3: 960–969

67. Uhlen M, Fagerberg L, Hallstrom BM, Lindskog C, Oksvold P, Mardinoglu A, Sivertsson A, Kampf C, Sjostedt E, Asplund A, Olsson I,Edlund K, Lundberg E, Navani S, Szigyarto CA, Odeberg J, Djureinovic D, Takanen JO, Hober S, Alm T *et al* (2015) Proteomics. Tissue-based map of the human proteome. *Science* 347: 1260419

68. UniProt C (2015) UniProt: a hub for protein information. *Nucleic Acids Res* 43: D204–D212

69. Varemo L, Scheele C, Broholm C, Mardinoglu A, Kampf C, Asplund A, Nookaew I, Uhlen M, Pedersen BK, Nielsen J (2015) Proteome- and transcriptome-driven reconstruction of the human myocyte metabolic network and its use for identification of markers for diabetes. *Cell Rep* 11: 921–933

70. Vlassis N, Pacheco MP, Sauter T (2014) Fast reconstruction of compact context-specific metabolic network models. *PLoS Comput Biol*10: e1003424

71. Vo TD, Greenberg HJ, Palsson BO (2004) Reconstruction and functional characterization of the human mitochondrial metabolic network based on proteomic and biochemical data. *J Biol Chem* 279: 39532–39540

72. Vo TD, Paul Lee WN, Palsson BO (2007) Systems analysis of energy metabolism elucidates the affected respiratory chain complex in Leigh›s syndrome. *Mol*

Genet Metab 91: 15–22

73. Wang ET, Sandberg R, Luo S, Khrebtukova I, Zhang L, Mayr C, Kingsmore SF, Schroth GP, Burge CB (2008) Alternative isoform regulation in human tissue transcriptomes. *Nature* 456: 470–476

74. Wang YL, Eddy JA, Price ND (2012) Reconstruction of genome-scale metabolic models for 126 human tissues using mCADRE. *BMC Syst Biol* 6: 153

75. Wiback SJ, Palsson BO (2002) Extreme pathway analysis of human red blood cell metabolism. *Biophys J* 83: 808–818

76. Wilhelm M, Schlegl J, Hahne H, Moghaddas Gholami A, Lieberenz M, Savitski MM, Ziegler E, Butzmann L, Gessulat S, Marx H,Mathieson T, Lemeer S, Schnatbaum K, Reimer U, Wenschuh H, Mollenhauer M, Slotta-Huspenina J, Boese JH, Bantscheff M,Gerstmair A *et al* (2014) Mass-spectrometry-based draft of the human proteome. *Nature* 509: 582–587

77. Wu C, Orozco C, Boyer J, Leglise M, Goodale J, Batalov S, Hodge CL, Haase J, Janes J, Huss JW 3rd, Su AI (2009) BioGPS: an extensible and customizable portal for querying and organizing gene annotation resources. *Genome Biol* 10: R130

78. Yizhak K, Gabay O, Cohen H, Ruppin E (2013) Model-based identification of drug targets that revert disrupted metabolism and its application to ageing. *Nat Commun* 4: 2632

79. Yizhak K, Gaude E, Le Devedec S, Waldman YY, Stein GY, van de Water B, Frezza C, Ruppin E (2014a) Phenotype-based cell-specific metabolic modeling reveals metabolic liabilities of cancer. *eLife* 3: e03641.

80. Yizhak K, Le Devedec SE, Rogkoti VM, Baenke F, de Boer VC, Frezza C, Schulze A, van de Water B, Ruppin E (2014b) A computational study of the Warburg effect identifies metabolic targets inhibiting cancer migration. *Mol Syst Biol* 10: 744.

81. Yizhak K, Chaneton B, Gottlieb E, Ruppin E (2015) Modeling cancer metabolism on a genome scale. *Mol Syst Biol* 11: 817

82. Yu NY, Hallstrom BM, Fagerberg L, Ponten F, Kawaji H, Carninci P, Forrest AR, Fantom CT, Hayashizaki Y, Uhlen M, Daub CO (2015)Complementing tissue characterization by integrating transcriptome profiling from the Human Protein Atlas and from the FANTOM5 consortium. *Nucleic Acids Res* 43: 6787–6798

83. Zhang C, Ji B, Mardinoglu A, Nielsen J, Hua Q (2015) Logical transformation of genome-scale metabolic models for gene level applications and analysis. *Bioinformatics* 31: 2324–2331

CHAPTER 9

BIOLOGY, DETECTION, AND CLINICAL IMPLICATIONS OF CIRCULATING TUMOR CELLS

Simon A Joosse, Tobias M Gorges and Klaus Pantel

Department of Tumor Biology, Center of Experimental Medicine, University Medical Center Hamburg-Eppendorf, Hamburg, Germany

ABSTRACT

Cancer metastasis is the main cause of cancer-related death, and dissemination of tumor cells through the blood circulation is an important intermediate step that also exemplifies the switch from localized to systemic disease. Early detection and characterization of circulating tumor cells (CTCs) is therefore important as a general strategy to monitor and prevent the development of overt metastatic disease. Furthermore, sequential analysis of CTCs can provide clinically relevant information on the effectiveness and progression of systemic therapies (e.g., chemo-, hormonal, or targeted therapies with antibodies or small inhibitors). Although many advances have been made regarding the detection and molecular characterization of CTCs, several challenges still exist that limit the current use of this important diagnostic approach. In this review, we discuss the biology of tumor cell dissemination, technical advances, as well as the challenges and potential clinical implications of CTC detection and characterization.

BIOLOGY OF CIRCULATING TUMOR CELLS

Introduction

The ability of a malignant tumor to become metastatic begins with the hallmarks of motility and invasiveness (Hanahan & Weinberg, 2011). When individual cells or clusters of cancer cells acquire the ability to separate and move away from the primary tumor mass, migrate through the surrounding tissue, and enter the lymphatic system and/or blood circulation, the metastatic trait of a given tumor becomes manifest (Fig 1). Tumor cells that escape their primary microenvironment need to escape anoikis and might be challenged by the host's immunologic defenses. Although the rate of tumor cell release in cancer patients is unknown, experimental models indicate that millions of tumor cells are continuously dispersed through the body, although only few of these cells might reach a distant organ, survive in a dormant state, evade the immune system and systemic therapy, and eventually grow into an overt metastasis (Chambers *et al*, 2002; Kang & Pantel, 2013). Adaption to a new microenvironment and proliferation of a single tumor cell in a distant site requires special traits (plasticity) that the cell must already have or acquire in order to develop to an overt

Citation: Simon A Joosse, Tobias M Gorges and Klaus Pantel, Biology, detection, and clinical implications of circulating tumor cells, DOI: 10.15252/emmm.201303698

metastasis detectable by clinical procedures. Tumor cells may stop cell division processes after dissemination but may persist in a quiescent status until environmental conditions provide appropriate signals to start proliferation again. This quiescent state, for which the underlying mechanisms are largely unknown, is called clinical cancer dormancy (Uhr & Pantel, 2011; Kang & Pantel, 2013) and may last more than 10 years. There is now increasing evidence that the microenvironment plays a critical role in this process (e.g., osteoclasts as regulators of osteolytic bone metastases (Lu *et al*, 2011; Ell *et al*, 2013)), in addition to changes in the tumor cells themselves (e.g., acquisition of new mutations).

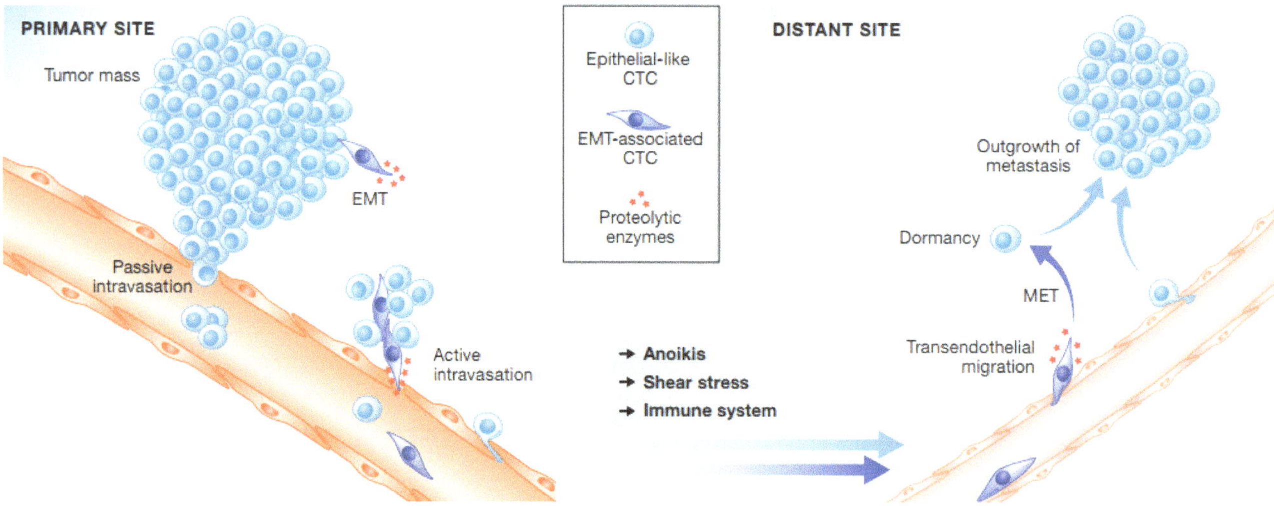

Figure 1: Metastatic cascade

Tumor cells may enter the bloodstream passively or actively via biological events such as EMT or centrosomes amplification. Disseminating tumor cells must overcome several hurdles including anoikis, shear stress in the bloodstream, and the immune system in- and outside of the blood circulation. Once at a distant site, tumor cells may extravasate, undergo MET, and grow locally to become a metastasis or remain in dormancy.

CTC migration: mobility and motility

Cancer metastasis has been correlated with specific genetic biomarkers such as mutations, chromosomal aberrations, and gene expression patterns (Nguyen & Massague, 2007; Wrage *et al*, 2009; Wikman *et al*, 2012; Hohensee *et al*, 2013). Nevertheless, whether CTCs enter the blood circulation through an active migration process, by passive means, or both, remains clinically and scientifically an unresolved question (Joosse & Pantel, 2013). In this review, we employ the terms *motile* and *mobile cells* to describe the differences between active and passive migration of the tumor cells.

Motile cancer cells are cells that are able to move on their own accord and that has thus gained the ability to move through the extracellular matrix and penetrate basement membranes and endothelial walls upon intravasation and extravasation. These active migration mechanisms imply modification of cell morphology, position, and surrounding tissue (Friedl & Alexander, 2011). Furthermore, cancer cells may infiltrate as single entities, in clusters, in strands, or in single (Indian) files as observed in lobular breast carcinoma. Single cells must weaken or completely lose their adhesive bonds with neighboring tumor cells for infiltration, whereas collective migration requires stable cell–cell adhesion and multicellular coordinated movement (Friedl & Gilmour, 2009). These clusters frequently comprise of different cell morphologies, that is, both epithelial- and mesenchymal-like. Collective migration may require a leader cell with mesenchymal features, able to create a path for the trailing tumor cells through the surrounding tissue (Friedl & Wolf, 2009).

Mobile cancer cells are moved by external forces such as growth of the tumor, mechanical forces, or friction which cause them to be dragged or pushed out of place (Camara *et al*, 2006; Fornvik *et al*, 2013). Although a vast amount of literature is available about the processes involved in active migration of cancer cells through the extracellular matrix, less is known concerning the passive dissemination of cancer cells and how they may be forced into the blood circulation. Carcinomas induce the formation of new blood vessels by the secretion of the vascular endothelial growth factor (VEGF) to facilitate the supply of nutrients and oxygen for growth. This process called angiogenesis, often results in leaky vessels caused by weak interconnections of the vessel's endothelial cells and intercellular openings (McDonald & Baluk, 2002). Due to outward pushing of the tumor during growth, single or clusters of tumor cells can be forced through the leaky vessels, thus ending up in the blood circulation as 'accidental CTCs'. Furthermore, tumor cells may be passively moved through the micro-tracks created by other tumor cells by proteolysis or through other pre-existing tissue structures (Friedl & Wolf, 2009).

Because the 'accidental CTCs' from epithelial malignancies are more likely to have retained their original phenotype, detection with epithelial-specific markers such as the epithelial cell adhesion molecule (EpCAM) would be feasible. On the other hand, tumor cells that have transformed to a more mesenchymal-like phenotype, and thus exhibit greater plasticity, may prove to be less easily detected by conventional EpCAM-based detection methods (Brabletz, 2012; Joosse & Pantel, 2013). Nevertheless, new evidence shows that a phenotypical transformation is not always required for tumor cell motility (Aceto *et al*, 2014; Godinho *et al*, 2014).

Epithelial–mesenchymal plasticity

Epithelial cells adhere to their neighboring cells through, among others, adherens junctions via binding of cadherins. Carcinoma cells are of epithelial origin and experience the same cell–cell interaction through adhesion molecules such as cadherins, claudins, or plakoglobin; these interactions prevent them from spreading in the first place (Pantel *et al*, 1998).

Normal epithelial cells show remarkable plasticity as they are able to reprogram and undergo dynamic and reversible transitions between the epithelial and mesenchymal cell phenotype (Tam & Weinberg, 2013). This de-differentiation process known as epithelial-to-mesenchymal transition (EMT) occurs frequently throughout embryogenesis because cells continuously migrate to form tissues and organs. EMT also plays an important role in wound healing and tissue regeneration (Nieto, 2013). Not surprisingly, carcinoma cells also make use of EMT as they become invasive. Indeed, they typically feature loss of cell–cell adherence proteins such as cadherin, followed by loss of apico-basal polarity, and finally, gaining the ability to migrate and invade (Thiery, 2002). EMT can be triggered by paracrine signaling of TGF-beta, WNT, platelet-derived growth factors, or interleukin-6 (IL-6) but can also be induced by nicotine, alcohol, and ultraviolet light (Thiery *et al*, 2009; Tam & Weinberg, 2013; Kishi*et al*, 2014). These triggers activate transcription factors such as Snail, Twist, and Zeb that maintain the mesenchymal phenotype by autocrine signaling (Tam & Weinberg, 2013). Because of the loss of tight and adherens junctions, as well as cytoskeletal changes, typical epithelial markers such as EpCAM and E-cadherin are down-regulated, keratin expression is altered, and finally, up-regulation of mesenchymal markers such as vimentin is observed during EMT (Joosse *et al*, 2012). Consequently, mesenchymal-like CTC subpopulations are difficult to identify in the hematopoietic cell environment which is also of mesenchymal origin (Joosse & Pantel, 2013). Of note, single tumor cells found in the blood of breast cancer patients exhibit EMT-associated changes, while cell clusters appear to require a partial EMT so that these cells possess the migratory abilities of a mesenchymal cell but retain the cell–cell interaction profile of an epithelial cell (Yu *et al*, 2013).

Non-EMT-associated motility

Centrosome amplification has also been linked to de-differentiation and recently been shown to induce invasion in cancer (Ghadimi *et al*, 2000; Lingle *et al*, 2002; Godinho *et al*, 2014). The centrosomes form the main microtubule organizing center in mammalian cells and facilitate chromosomal separation during mitosis via the meiotic spindle apparatus but also participate in the organization of flagella and cilia. EMT is thought not to play a role in the invasion of tumors with centrosome amplification since expression of E-cadherin is retained. Cell–cell adhesion appears instead reduced downstream of Rac1 signaling by increased Arp2/3-dependent actin polymerization (Godinho *et al*, 2014). Another form of non-EMT-associated dissemination, which is thought to substantially contribute to metastasis, is the release of circulating tumor cell clusters (Aceto *et al*, 2014). These clusters are constituted of 2–50 tumor cells held together by plakoglobin-dependent intercellular adhesion. Clustered cells are less likely to undergo anoikis and have an increased likelihood of being trapped in narrow blood vessels, thus favoring extravasation into distant organs.

Therefore, since not all metastatic tumor cells seem to lose their epithelial-like characteristics by EMT, it is important to consider CTCs with epithelial properties too in order to detect the metastasis-initiating cells (MICs).

CTCs in circulation

Once in the bloodstream, CTCs face several natural obstacles that hinder the metastatic process. First are the enormous shearing forces and collisions with blood cells, generated by blood flow. Although shear stresses decrease the number of viable cancer cells dramatically, tumor cells that underwent EMT seem to be more resistant against these forces compared to epithelial tumor cells (Mitchell & King, 2013). Second, CTCs must survive in the bloodstream without their cell–matrix interactions, an occurrence that would normally trigger apoptosis through a process called anoikis. Resistance against anoikis, however, is made possible in CTCs by activated tropomyosin-related kinase B (TrkB) that suppresses caspase-associated apoptosis and enables the cells to survive in liquid suspension (Douma *et al*, 2004). The third obstacle that CTCs face in the blood is the activity of the immune system. In colorectal cancer, immune escape is obtained by up-regulation of CD47 that prevents CTCs from macrophage and dendritic cell attack (Steinert *et al*, 2014). Finally, cancer cells must eventually leave the blood circulation, which requires binding to the endothelium lining the vessels. Because platelets enhance this binding, inhibition of platelet aggregation by for instance aspirin can decrease stable tumor cell binding to activated platelets (Uppal *et al*, 2014).

Tumor cell homing and dormancy

Extravasation starts when CTCs slow down in small capillaries, attach to the endothelium, and finally undergo transendothelial migration (Reymond *et al*, 2013). One of the most frequent CTCs homing sites are the bones, including for primary malignancies such as colorectal and lung cancer that less typically metastasize to the bone (Pantel & Brakenhoff, 2004; Braun *et al*, 2005; Riethdorf *et al*, 2008). It is therefore thought that the bone marrow functions as a reservoir for disseminated tumor cells (DTCs) (Wikman *et al*, 2008; Kang & Pantel, 2013). In breast cancer, it may take many years up to decades after surgery of the primary tumor until metastases become evident (Goss & Chambers, 2010; Uhr & Pantel, 2011). During this time, bone marrow DTCs (Janni *et al*, 2011) as well as CTCs derived from the DTCs (Meng *et al*, 2004) may be found in these patients. DTCs in bone marrow may linger in a dormant state, thus evading systemic therapy while waiting for the appropriate trigger to resume proliferation (Bragado *et al*, 2013; El Touny *et al*, 2014). The basis for tumor cell dormancy may be the initial EMT process itself. In squamous cell carcinoma, spatiotemporal regulation of the epithelial–mesenchymal transition is essential for the dissemination and eventual metastasis (Tsai *et al*, 2012). The mesenchymal phenotype of CTCs that underwent EMT promotes motility but does not favor growth (Celia-Terrassa *et al*, 2012). Indeed, cancer cells must undergo a reverse mesenchymal-to-epithelial transition (MET) to acquire the ability to proliferate and thus form a metastatic tumor. It has therefore been suggested that tumor cells with an intermediate phenotype can most efficiently disseminate and grow at the distant sites (Bednarz-Knoll & Weinberg, 2012; Tam & Weinberg, 2013).

ENRICHMENT AND DETECTION OF CTCS

Enrichment of CTCs

CTCs are infrequent and appear at an estimated level of one against the background of millions (10^6–10^7) of surrounding normal peripheral mononuclear blood cells (PBMCs) (Alix-Panabieres *et al*, 2012). The occurrence rate might even be lower in cancer patients without obvious metastases (Rink *et al*, 2012; Rack *et al*, 2014). Hence, selective enrichment of tumor cells and/or systematic removal of PBMCs and red blood cells (RBCs) is required to detect CTCs in the blood of a cancer patient. To date, more than 40 techniques have been developed for CTC detection (Parkinson *et al*, 2012) and novel strategies are published continuously. CTC enrichment and detection methods have been classified based on whether they exploit the physical or biological properties of the cells (Alix-Panabieres & Pantel, 2014). However, as many enrichment strategies rely on positive selection (usually targeting EpCAM), CTC assays are more commonly grouped into label-dependent and label-independent approaches (Fig 2). Among the numerous EpCAM-based CTC detection technologies, the semi-automated CellSearch® system is the most frequently used system and at present the only one cleared by the U.S. FDA. Through this approach, CTC counts have been associated with an independent prognostic power on progression-free survival (PFS) and overall survival (OS) in primary and metastatic disease (Cristofanilli *et al*, 2004; de Bono *et al*, 2008; Cohen *et al*, 2009; Bidard *et al*, 2014; Rack *et al*, 2014). Due to its clinical relevance, high reproducibility, and FDA clearance, CellSearch® can be considered as a benchmark for all other CTC detection methods appearing on the market. The AdnaTest® is another epithelial marker-based research tool that positively enriches CTCs from a blood sample. Using this approach, the presence or disappearance of CTCs was shown to be a prognostic and predictive marker in a study on metastatic breast cancer (Teweset *al*, 2009). Microfluidic device platforms like the CTC- or the Herringbone (HB)-Chip seem to be promising alternatives to selectively capture EpCAM-positive CTCs in cancer patients (Nagrath *et al*, 2007; Maheswaran *et al*, 2008). A combination of anti-EpCAM and anti-MUC1 capture in a single microfluidic device may further result in an improved capture performance of CTCs (Thege *et al*, 2014). Another chip-based platform combines a size-based filtration with an affinity-based enrichment strategy, thus enhancing the chance of systematic removal of PBMCs and RBCs (CTC-iChip) (Ozkumur *et al*, 2013). Fluxion Biosciences has lately introduced a commercially available microfluidic technology (IsoFlux®) with magnetic isolation zones to isolate EpCAM-positive tumor cells from biological samples. Using the IsoFlux®, recovery rates between 74 and 85% of $EpCAM^{low}$-(MDA-MB-231) and $EpCAM^{high}$ (SKBR3)-expressing tumor cells could be obtained (Harb *et al*, 2013). To circumvent sample volume limitations, GILUPI GmbH has designed an EpCAM-coated wire (CellCollector™) to capture CTCs *in vivo* (Saucedo-Zeni *et al*, 2012). This device is positioned through a cannula into the vein of a cancer patient. It is estimated that during the 30-min application time, up to 1.5 l of blood flows over the detector, thus increasing the yield of detectable CTCs.

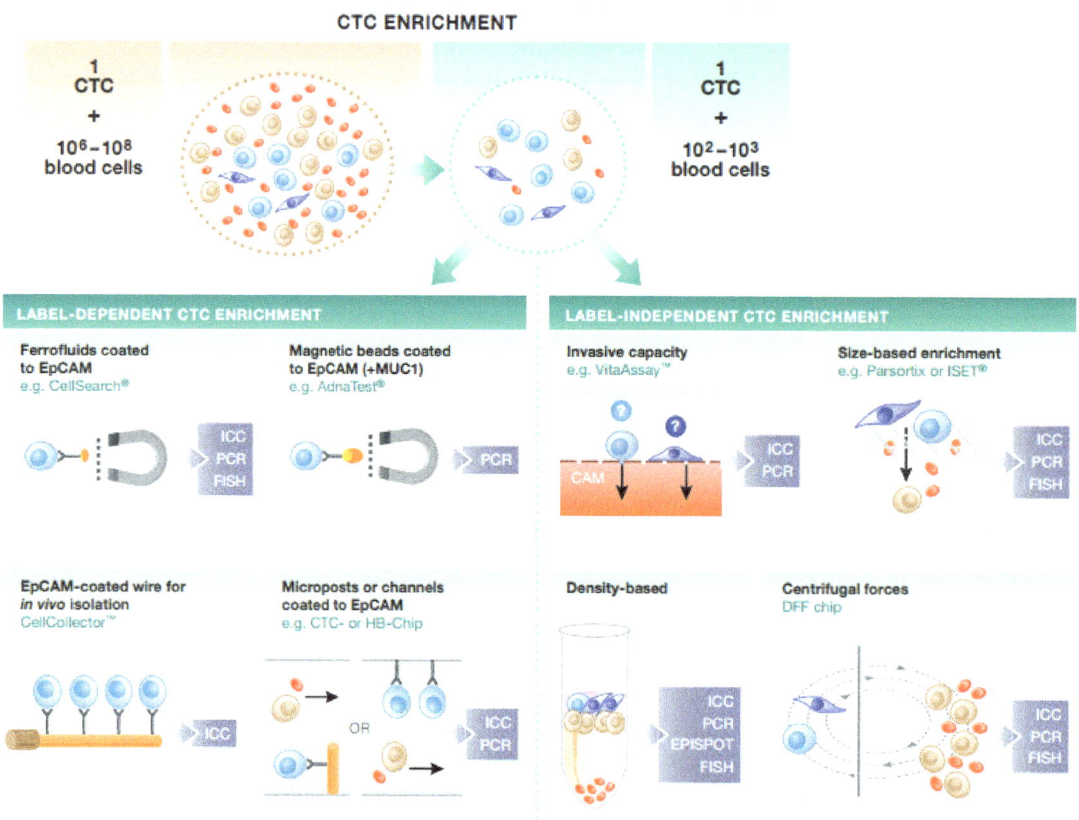

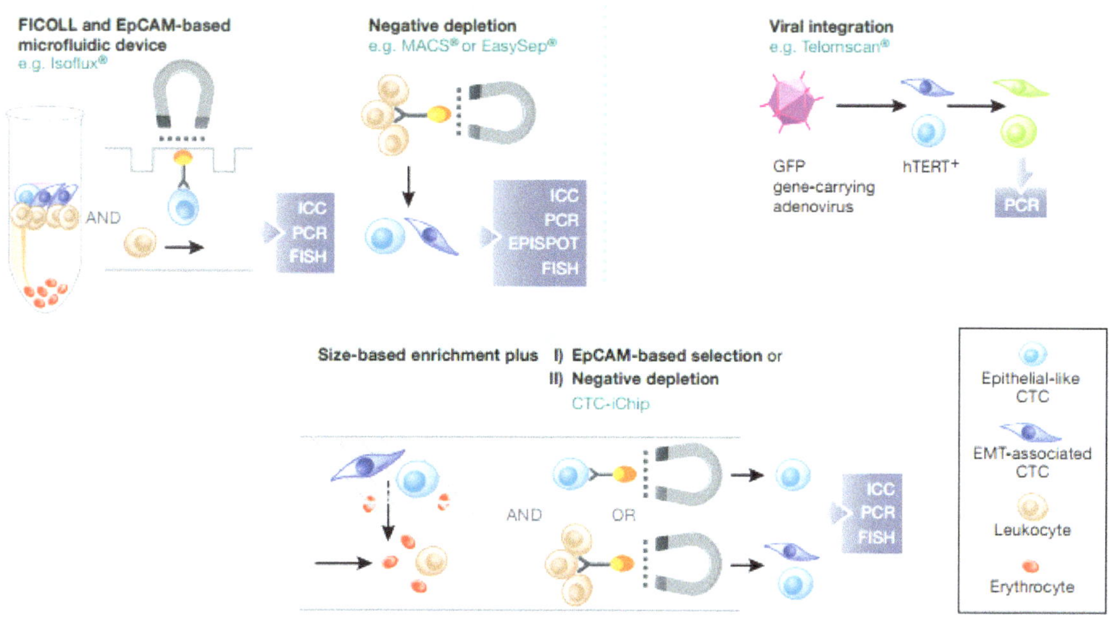

Figure 2: CTC enrichment and identification

Enrichment strategies for CTCs can be separated into label-dependent and label-independent techniques. Among label-dependent techniques, immunomagnetic-based assays targeting the EpCAM protein are the most commonly applied. Label-independent enrichment methods include size-based or density-based approaches. Additionally, negative depletion or the invasive capacity of tumor cells can be used. A combination of different enrichment strategies is also practicable. Captured tumor cells are ready for molecular characterization by immunocytochemistry (ICC) using antibodies for tumor-specific markers or by PCR approaches targeting tumor-specific mRNA or DNA sequences. Another possibility is to detect viable cells by protein secretion (EPISPOT). Additionally, fluorescence in situ hybridization (FISH) can be used for the detection of tumor-specific gene aberrations.

However, a major caveat is that metastasis formation might require EMT, which in turn might cause loss of CTC detection due to the down-regulation of EpCAM. EMT-related changes such as the down-regulation of EpCAM and Keratins or the up-regulation of EGFR and vimentin have already been described on CTCs and DTCs (Bartkowiak *et al*, 2009; Armstrong *et al*, 2011; Gorges *et al*, 2012; Joosse *et al*, 2012; Kasimir-Bauer *et al*, 2012; Gasch *et al*, 2013). Hence, EpCAM-based enrichment technologies should be re-worked for the comprehensive detection of epithelial (epithelial^{+}/mesenchymal^{-}), complete EMT (epithelial^{-}/mesenchymal^{+}), and intermediate EMT (epithelial^{+}/mesenchymal^{+}) CTC phenotypes.

Depletion of CD45-positive leukocytes (negative selection) is a preferred approach to capture CTCs lacking adequate expression of the EpCAM protein. Using this strategy, viable tumor cells have already been isolated, characterized, and cultured (Deneve *et al*, 2013). CD45 depletion can be combined with other label-independent methodologies (e.g., red blood cell lysis or density gradient centrifugation) to improve the yield and to displace unwanted erythrocytes (Liu *et al*, 2011). However, red blood cell lysis or density gradient centrifugation may lead to a loss of tumor cells resulting in false-negative results. The novel CTC-iChip can deplete hematopoietic cells combining this strategy with a size-based enrichment step (Ozkumur *et al*, 2013). In the antigen-independent mode, this system was able to isolate CTCs from several types of cancers that had lost or never had the epithelial marker characteristics (triple-negative breast cancer and melanoma). A combination of red blood cell lysis, depletion of CD45^{+} leukocytes, and size-based filtration has also been shown to be feasible in non-small cell lung, breast, colon, and esophageal cancer patients (Wu *et al*, 2014). Membranous filter devices like the ISET® system or the ScreenCell® approach appear to enrich CTCs with a high recovery rate as well. However, in particular, EMT-related CTCs may not be stiffer than leukocytes and CTCs are not always larger than leukocytes and might therefore not be retained by these filters. Moreover, even if tumor cells are captured on the membrane, they may be difficult to detach for further molecular examination. The Parsortix system is a novel separation device that captures viable tumor cells based on their physical properties. This system allows a size-based identification of cancer cells within the capture cassette device or harvesting of the enriched tumor cells outside the cassette. The harvested tumor cells are easily accessible and ready for downstream analysis, such as, next-generation sequencing (NGS), qPCR, FISH, ICC, or other cancer cell-related analytical methodologies. Another newly designed microfluidic technology called the JETTA™ microfluidic chip, uses a size- and deformability-based capture scheme to characterize CTCs (Riahi *et al*, 2014). The novelty of this platform is that it segregates CTCs into their own chambers enabling single cell analysis based on morphological, immunologic, or genetic criteria. The spiral biochip uses inherent centrifugal forces for continuous, size-based, and EpCAM-independent separation of tumor cells. This device can process 3 ml of whole blood within an hour and has low leukocyte contamination (441 ± 319.5 leukocytes/ml) (Hou *et al*, 2013).

In addition to the exploitation of specific physical properties, alternative approaches can be used to enrich CTCs. The CAM (cell adhesion matrix) methodology (Vita Assay) enriches invasive and viable CTCs. Patient-derived blood samples are added to CAM-coated tubes at which invasive tumor cells are captured based on their preferential adhesion to the CAM. CTCs attach to CAM proteins and digest them by exploiting the mechanism by which metastatic tumor cells invade into surrounding tissues. Using this technique, CTCs (mean 61 CTCs/ml) were already found in 28 of 54 (52%) stage I-III breast cancer patients (Lu *et al*, 2010).

In summary, isolation of CTCs can be attained with various different strategies. More and more approaches for the enrichment appear on the market or are under development in laboratories all over the world. However, independent and large clinical studies are now required to determine the robustness and clinical validity of these new assays beyond the initial proof of concept.

Identification and characterization of CTCs

Identification, enumeration, and molecular characterization of CTCs are likely to provide important insights into disease progression and might allow adaptation of therapeutic strategies. Differential CTC detection assays apply (or can be combined with) various strategies for the identification of tumor cells. Morphologic investigation together with fluorescence immunocytochemistry (ICC) is a common procedure for the identification and enumeration of CTCs after enrichment. The CellSearch® system classifies a CTC as positive event if the cell is $\geq$ 4 μm, DAPI$^+$(4,6-diamino-2-phenylindole), pan-keratin$^+$, and CD45$^-$. The additional 4th fluorescence channel is accessible for a user-defined detection of, for example, therapy-relevant markers such as the androgen receptor (AR), the prostate-specific antigen (PSA), the human epidermal growth factor receptor 2 (ERBB2), the epidermal growth factor receptor (EGFR) (Riethdorf *et al*, 2007; Gasch *et al*, 2013), or EMT-associated molecules (such as N-cadherin or vimentin) (Armstrong *et al*, 2011). Identification of CTCs by ICC is also applicable when using other enrichment strategies (microfluidic or size-based platforms, the Gilupi CellCollector™, red blood cell lysis, or density centrifugation-based procedures) (Nagrath *et al*,2007; Maheswaran *et al*, 2008; Saucedo-Zeni *et al*, 2012; Babayan *et al*, 2013). However, conventional microscopic examination by a researcher is cost- and time-consuming. The Ariol® system (Genetix USA) might be used for automated cell image capturing and analysis if the enriched cells are transferrable onto glass slides (Deng *et al*, 2008). Investigator microscopic filters set the limits for CTC enumeration and characterization by ICC; typically, the four different fluorescence channels are used for CTC characterization (two markers of interest), CD45 staining (to exclude false-positive findings), and nuclear counterstaining (DAPI or Hoechst33342).

Identification of CTCs by multiplex PCR targeting the numerous tumor-associated mRNA transcripts overcomes filter set limitations. CTC enrichment combined with a RT–PCR technology could already be used for the identification of tumor-related markers (*EpCAM*, *MUC1*, and*ERBB2*), EMT-associated transcripts (*PI3Kα* (*phosphatidylinositol 3-kinase alpha*), *Akt-2*, or *Twist1*), or stem cell markers such as *ALDH1*(*aldehyde dehydrogenase 1*) (Kasimir-Bauer *et al*, 2012). Additional transcripts such as *KRT-19*, *MAGE-A3*, and *PBGD* have also been used for the detection and molecular characterization of circulating tumor cells (Strati *et al*, 2013). Hence, quantitative RT–PCR confers high sensitivity to CTC identification, although uncertainties are foreseeable with respect to specific primer and target gene selection. Additionally, accurate quantification of CTCs is difficult because CTCs are heterogeneous and may express different levels of the genes considered for identification. Furthermore, precise cut-off levels between mRNA amounts of normal blood cells and tumor cells must be determined and validated since tumor-associated targets such as *KRT-19* mRNA have also been found in a small number of healthy individuals (Stathopoulou *et al*, 2003). Hence, a combination of different genes should be used to improve the specificity of RT–PCR for reliable CTC detection.

Another strategy for the identification of CTCs is to enumerate and analyze proteins specifically secreted by viable tumor cells (EPISPOT, EPithelial Immuno SPOT). CTCs are enriched by negative depletion and subsequently cultured on a membrane coated with antibodies that capture the secreted proteins. Afterward, the proteins are readily identifiable by immunofluorescence microscopy using fluorochrome-labeled secondary antibodies targeting the protein of interest. Using this strategy, viable CTCs and DTCs have been already identified by the detection of secreted KRT-19, MUC1, PSA, or FGF-2 (Alix-Panabieres *et al*, 2007; Alix-Panabieres, 2012).

Taken together, different strategies can be used for CTC identification and characterization. Frequently, protein- (ICC), RNA-, and/or DNA-based (PCR) approaches are implemented. Each strategy has advantages and drawbacks. A combination of different analytical methodologies is likely to be beneficial and may encompass the heterogeneity of CTCs and help to understand their role in metastasis formation.

CLINICAL IMPLICATIONS OF CIRCULATING TUMOR CELLS

DTCs in the bone marrow have been detected for all solid tumor types, suggesting that this compartment might be a preferred reservoir for blood-borne DTC from where they may re-disseminate to other distant organs such as liver or lungs where better growth conditions may exist. For example, DTCs are prognostic markers in colorectal cancer even though bone metastases are very rare in this cancer type (Lindemann *et al*,1992; Pantel *et al*, 1993). In breast cancer, where bone metastases are more frequent, DTCs in the bone marrow may predict metastatic relapse in other organs as well (Braun *et al*, 2000, 2005). However, it cannot be excluded that DTCs may home equally well to other organs such as lung or liver, where however they may not be so easily detected as in bone marrow. The observed correlation between DTCs in bone marrow and local relapse in breast cancer (Bidard *et al*, 2008) is consistent with the recent experimental findings in mice that these cells might even re-infiltrate their primary organs and promote tumor progression (Kim *et al*, 2009). If the bone marrow is a preferred reservoir for DTCs, drugs targeting the BM–tumor interactions (e.g., bisphosphonates or antibodies to the RANK ligand (Lewiecki, 2011; Gronich & Rennert, 2013)) might help to prevent metastatic or even local relapse.

Sequential peripheral blood drawings in particular for real-time monitoring of minimal residual disease in cancer patients undergoing systemic therapies are clearly more acceptable than repeated bone marrow aspirations. Indeed, many research groups are currently assessing the clinical value of CTC analyses, which so far has been proven to provide significant prognostic information in metastatic breast cancer (Zhang *et al*, 2012; Bidard *et al*, 2014), and other solid tumors such as of the prostate (de Bono *et al*, 2008), colorectal (Cohen *et al*, 2008), and lung (Krebs *et al*,2011; Hou *et al*, 2012) cancer, and seems to be superior over conventional imaging methods for response evaluation (Budd *et al*, 2006). However, the real challenge of CTC technologies is to monitor minimal residual disease in patients without signs of overt metastasis, as the CTC counts are very low in these patients. Promising results indicating that even such low CTC counts may have prognostic relevance have been recently published for several tumor entities such as breast cancer (Lucci *et al*, 2012; Rack *et al*, 2014), bladder cancer (Rink *et al*, 2011; Gazzaniga *et al*, 2014), and colorectal cancer (Deneve *et al*, 2013). Nevertheless, more sensitive methods and/or the analysis of larger amounts of blood (Saucedo-Zeni *et al*, 2012; Fischer *et al*, 2013) might be required to increase the robustness of CTC measurements in these patients. The implementation of leukapheresis or other innovative approaches for CTC enrichment (such as the CellCollector™) enables the screening for CTCs in blood volumes of 1.5-25 liters. The analysis of higher blood volumes may significantly improve the capturing rates of CTCs, which might help to predict therapy response and monitoring of the disease especially in early-stage disease.

Furthermore, molecular characterization of CTCs might be essential to identify therapeutic targets and contribute to more 'tailored' and personalized anti-metastatic therapies. In current clinical practice, the decision on targeted therapies is solely based on the analysis of the primary tumor although the therapy is directed against metastatic cells (Wan *et al*, 2013). However, metastatic relapse may occur many years after primary tumor diagnosis and surgical resection (Uhr & Pantel, 2011). Thus, selection pressure during the complex metastatic process might have originated a particular subclone of cancer cells that is underrepresented in the primary tumor and may therefore be easily missed using next-generation sequencing of single CTCs (Heitzer *et al*, 2013; Lohr *et al*, 2014). In particular, the microenvironment at the distant metastatic site can alter expression patterns as compared to their ascendants in the primary tumor (Korkaya & Wicha, 2013). All considered, the analysis of metastatic cells should provide additional information not otherwise obtainable from the primary tumor. However, biopsies of metastases are invasive procedures and certain locations (e.g., lung or brain) are especially difficult to access. Thus, the analysis of CTCs in the peripheral blood, a 'liquid biopsy', might become a much less invasive and cost-effective alternative (Alix-Panabieres & Pantel, 2013). Metastatic cells located at different organs may replenish the pool of CTCs and their analysis may provide useful information on actual targets and resistance mechanisms of systemic anti-cancer therapies (Wan *et al*, 2013), which may in turn lead to a better selection of cancer patients for targeted therapies. In breast cancer, ERBB2 is one of the most prominent targets for systemic therapy (Wan *et al*, 2013). Currently, all patients are stratified to trastuzumab (or other anti-ERBB2 therapies) by primary tumor tissue analysis only. Recent reports, however, have shown that ERBB2-positive CTCs can be detected also in patients with ERBB2-negative primary tumors (Riethdorf *et al*, 2010; Ignatiadis *et al*, 2011; Hartkopf *et al*, 2012), suggesting that there might exist additional patients that could benefit from ERBB2-directed therapies. Ongoing clinical studies (e.g., DETECT III (NCT01619111) trial in Germany and CTC-TREAT (NCT01548677) trial in Europe) will reveal whether the ERBB2 status of CTCs may predict response to ERBB2-directed therapies (Bidard *et al*, 2013a,b).

Another example for the use of CTCs as predictive biomarkers is the presence of ER-negative CTCs in breast cancer patients with ER-positive primary tumors (Babayan *et al*, 2013). ER is the most common therapeutic target in breast cancer, and 70–80% of patients have ER-positive primary tumors. However, ER-negative CTCs that may have escaped hormonal therapy to block ER-mediated growth frequently occur in these patients.

Genomic CTC analysis can also reveal gene mutations relevant for therapy resistance. For example, *KRAS* mutations are known to block the effect of therapeutic EGFR inhibition by antibodies or small inhibitors in colorectal cancer patients (Wan *et al*, 2013). The analysis of individual CTCs has shown a remarkable intra-patient *KRAS* mutation heterogeneity (i.e., *KRASWT* and *KRASMT* CTCs are present in the same patient) (Gasch *et al*, 2013). Indeed, the presence of *KRAS*-mutated CTCs in patients with *KRAS* wild-type primary colon carcinomas might be one explanation for failure of drug-mediated EGFR inhibition in these patients (Douillard *et al*, 2013; Peeters *et al*, 2013). At present, however, only an arbitrary section of the primary tumor is analyzed for *KRAS* mutations in colorectal cancer patients and the genomic heterogeneity of metastatic cells—the actual targets of systemic therapy–is not taken into consideration for therapy decisions.

In conclusion, the characterization of CTCs may have an important impact as companion diagnostics in future clinical trials testing new targeted therapies (Wan *et al*, 2013).

CONCLUSIONS AND OUTLOOK

Metastasis starts with cancer cell invasion from the primary tumor through the surrounding tissue into the bloodstream. Tumor cell motility may be acquired by different processes of which epithelial-to-mesenchymal transition (EMT) and amplification of centrosomes are currently considered to be the main ones. EMT leads to (partial) loss of the epithelial characteristics of cancer cells, whereas centrosome amplification-associated spread may retain the expression of epithelial markers. Because of these different dissemination processes, what the true features of the metastasis-initiating cells are and whether these cells can be found among the CTCs that can currently be detected in the blood of cancer patients remain to be fully established. CTC enrichment methods can broadly be divided into EpCAM-dependent cell selection that allows the identification of epithelial tumor cells and on the other hand the EpCAM-independent cell selection methods that may help to identify additional mesenchymal-like CTC subpopulations. Interestingly, the first report on grafting human-derived breast cancer CTCs in xenograft assays indicated that the putative metastasis-initiating cells expressed EpCAM, although at a reduced level (Baccelli *et al*, 2013). CTC detection approaches are usually based on the expression or secretion of specific proteins, DNA mutations, or RNA profiling. The development of new enrichment technologies are currently aimed at increased sensitivity and investigating increased blood volume using '*in vivo* strategies' to gain purity and enumeration of higher CTC counts.

Clinically, quantification of CTCs is of high value as these cancer cells generally represent the tumor (metastases) and facilitate real-time monitoring during systemic therapies by sequential peripheral blood sampling. Furthermore, molecular characterization of CTCs might enable the identification of therapeutic targets and contribute to personalized anti-metastatic therapies. Proof of the clinical relevance of the detection and characterization of CTCs has been substantially accumulating during the past decades. The use of xenograft models is a promising approach to gain further insights into the biology of tumor cell dissemination and may further help to test responses to newly designed therapies (Baccelli *et al*,2013; Hodgkinson *et al*, 2014; Yu *et al*, 2014).

In conclusion, analysis of CTCs in the peripheral blood (liquid biopsy) has a clear potential to further our understanding of the biology of tumor cell dissemination and to improve the management and possibly the prevention of metastatic disease in the near future.

REFERENCES

1. Aceto N, Bardia A, Miyamoto DT, Donaldson MC, Wittner BS, Spencer JA, Yu M, Pely A, Engstrom A, Zhu H, et al., editors. Circulating tumor cell clusters are oligoclonal precursors of breast cancer metastasis. Cell. 2014;158:1110–1122.
2. Alix-Panabieres C, Vendrell JP, Pelle O, Rebillard X, Riethdorf S, Muller V, Fabbro M, Pantel K. Detection and characterization of putative metastatic precursor cells in cancer patients. Clin Chem.2007;53:537–539.
3. Alix-Panabieres C. EPISPOT assay: detection of viable DTCs/CTCs in solid tumor patients. Recent Results Cancer Res. 2012;195:69–76.
4. Alix-Panabieres C, Schwarzenbach H, Pantel K. Circulating tumor cells and circulating tumor DNA. Annu Rev Med. 2012;63:199–215.
5. Alix-Panabieres C, Pantel K. Circulating tumor cells: liquid biopsy of cancer. Clin Chem.2013;59:110–118.
6. Alix-Panabieres C, Pantel K. Technologies for detection of circulating tumor cells: facts and vision.Lab Chip. 2014;14:57–62.
7. Armstrong AJ, Marengo MS, Oltean S, Kemeny G, Bitting RL, Turnbull JD, Herold CI, Marcom PK, George DJ, Garcia-Blanco MA. Circulating tumor cells from patients with advanced prostate and breast cancer display both epithelial and mesenchymal markers. Mol Cancer Res. 2011;9:997–1007.
8. Babayan A, Hannemann J, Spotter J, Muller V, Pantel K, Joosse SA. Heterogeneity of estrogen receptor expression in circulating tumor cells from metastatic breast cancer patients. PLoS One.2013;8:e75038.
9. Baccelli I, Schneeweiss A, Riethdorf S, Stenzinger A, Schillert A, Vogel V, Klein C, Saini M, Bauerle T, Wallwiener M, et al., editors. Identification of a population of blood circulating tumor cells from breast cancer patients that initiates metastasis in a xenograft assay. Nat Biotechnol. 2013;31:539–544.
10. Bartkowiak K, Wieczorek M, Buck F, Harder S, Moldenhauer J, Effenberger KE, Pantel K, Peter-Katalinic J, Brandt BH. Two-dimensional differential gel electrophoresis of a cell line derived from a breast cancer micrometastasis revealed a stem/progenitor cell protein profile. J Proteome Res.2009;8:2004–2014.
11. Bednarz-Knoll N, Alix-Panabieres C, Pantel K. Plasticity of disseminating cancer cells in patients with epithelial malignancies. Cancer Metastasis Rev. 2012;31:673–687.
12. Bidard FC, Vincent-Salomon A, Gomme S, Nos C, de RY, Thiery JP, Sigal-Zafrani B, Mignot L, Sastre-Garau X, Pierga JY. Disseminated tumor cells of breast cancer patients: a strong prognostic factor for distant and local relapse. Clin Cancer Res. 2008;14:3306–3311.
13. Bidard FC, Fehm T, Ignatiadis M, Smerage JB, Alix-Panabieres C, Janni W, Messina C, Paoletti C, Muller V, Hayes DF, et al., editors. Clinical application of circulating tumor cells in breast cancer: overview of the current interventional trials. Cancer Metastasis Rev. 2013a;32:179–188.
14. Bidard FC, Pierga JY, Soria JC, Thiery JP. Translating metastasis-related biomarkers to the clinic–progress and pitfalls. Nat Rev Clin Oncol. 2013b;10:169–179.
15. Bidard FC, Peeters DJ, Fehm T, Nole F, Gisbert-Criado R, Mavroudis D, Grisanti S, Generali D, Garcia-Saenz JA, Stebbing J, et al., editors. Clinical validity of circulating tumour cells in patients with metastatic breast cancer: a pooled analysis of individual patient data. Lancet Oncol. 2014;15:406–414.
16. de Bono JS, Scher HI, Montgomery RB, Parker C, Miller MC, Tissing H, Doyle GV, Terstappen LW, Pienta KJ, Raghavan D. Circulating tumor cells predict survival benefit from treatment in metastatic castration-resistant prostate cancer. Clin Cancer Res. 2008;14:6302–6309.
17. Brabletz T. To differentiate or not–routes towards metastasis. Nat Rev Cancer. 2012;12:425–436.
18. Bragado P, Estrada Y, Parikh F, Krause S, Capobianco C, Farina HG, Schewe DM, Aguirre-Ghiso JA. TGF-beta2 dictates disseminated tumour cell fate in target organs through TGF-beta-RIII and p38alpha/beta signalling. Nat Cell Biol. 2013;15:1351–1361.
19. Braun S, Pantel K, Muller P, Janni W, Hepp F, Kentenich CR, Gastroph S, Wischnik A, Dimpfl T, Kindermann G, et al., editors. Cytokeratin-positive cells in the bone marrow and survival of patients with stage I, II, or III breast cancer. N Engl J Med. 2000;342:525–533.
20. Braun S, Vogl FD, Naume B, Janni W, Osborne MP, Coombes RC, Schlimok G, Diel IJ, Gerber B, Gebauer G, et al., editors. A pooled analysis of bone marrow micrometastasis in breast cancer. N Engl J Med. 2005;353:793–802.
21. Budd GT, Cristofanilli M, Ellis MJ, Stopeck A, Borden E, Miller MC, Matera J, Repollet M, Doyle GV, Terstappen LW, et al., editors. Circulating tumor cells versus imaging–predicting overall survival in metastatic breast cancer. Clin Cancer Res. 2006;12:6403–6409.
22. Camara O, Kavallaris A, Noschel H, Rengsberger M,

Jorke C, Pachmann K. Seeding of epithelial cells into circulation during surgery for breast cancer: the fate of malignant and benign mobilized cells.World J Surg Oncol. 2006;4:67.

23. Celia-Terrassa T, Meca-Cortes O, Mateo F, de Paz AM, Rubio N, Arnal-Estape A, Ell BJ, Bermudo R, Diaz A, Guerra-Rebollo M, et al., editors. Epithelial-mesenchymal transition can suppress major attributes of human epithelial tumor-initiating cells. J Clin Invest. 2012;122:1849–1868.

24. Chambers AF, Groom AC, MacDonald IC. Dissemination and growth of cancer cells in metastatic sites. Nat Rev Cancer. 2002;2:563–572.

25. Cohen SJ, Punt CJ, Iannotti N, Saidman BH, Sabbath KD, Gabrail NY, Picus J, Morse M, Mitchell E, Miller MC, et al., editors. Relationship of circulating tumor cells to tumor response, progression-free survival, and overall survival in patients with metastatic colorectal cancer. J Clin Oncol.2008;26:3213–3221.

26. Cohen SJ, Punt CJ, Iannotti N, Saidman BH, Sabbath KD, Gabrail NY, Picus J, Morse MA, Mitchell E, Miller MC, et al., editors. Prognostic significance of circulating tumor cells in patients with metastatic colorectal cancer. Ann Oncol. 2009;20:1223–1229.

27. Cristofanilli M, Budd GT, Ellis MJ, Stopeck A, Matera J, Miller MC, Reuben JM, Doyle GV, Allard WJ, Terstappen LW, et al., editors. Circulating tumor cells, disease progression, and survival in metastatic breast cancer. N Engl J Med. 2004;351:781–791.

28. Deneve E, Riethdorf S, Ramos J, Nocca D, Coffy A, Daures JP, Maudelonde T, Fabre JM, Pantel K, Alix-Panabieres C. Capture of viable circulating tumor cells in the liver of colorectal cancer patients.Clin Chem. 2013;59:1384–1392.

29. Deng G, Herrler M, Burgess D, Manna E, Krag D, Burke JF. Enrichment with anti-cytokeratin alone or combined with anti-EpCAM antibodies significantly increases the sensitivity for circulating tumor cell detection in metastatic breast cancer patients. Breast Cancer Res. 2008;10:R69.

30. Douillard JY, Oliner KS, Siena S, Tabernero J, Burkes R, Barugel M, Humblet Y, Bodoky G, Cunningham D, Jassem J, et al., editors. Panitumumab-FOLFOX4 treatment and RAS mutations in colorectal cancer. N Engl J Med. 2013;369:1023–1034.

31. Douma S, Van LT, Zevenhoven J, Meuwissen R, Van GE, Peeper DS. Suppression of anoikis and induction of metastasis by the neurotrophic receptor TrkB. Nature. 2004;430:1034–1039.

32. El Touny LH, Vieira A, Mendoza A, Khanna C, Hoenerhoff MJ, Green JE. Combined SFK/MEK inhibition prevents metastatic outgrowth of dormant tumor cells. J Clin Invest. 2014;124:156–168.

33. Ell B, Mercatali L, Ibrahim T, Campbell N, Schwarzenbach H, Pantel K, Amadori D, Kang Y. Tumor-induced osteoclast miRNA changes as regulators and biomarkers of osteolytic bone metastasis. Cancer Cell. 2013;24:542–556.

34. Fischer JC, Niederacher D, Topp SA, Honisch E, Schumacher S, Schmitz N, Zacarias FL, Vay C, Hoffmann I, Kasprowicz NS, et al., editors. Diagnostic leukapheresis enables reliable detection of circulating tumor cells of nonmetastatic cancer patients. Proc Natl Acad Sci USA. 2013;110:16580–16585.

35. Fornvik D, Andersson I, Dustler M, Ehrnstrom R, Ryden L, Tingberg A, Zackrisson S, Aaltonen K. No evidence for shedding of circulating tumor cells to the peripheral venous blood as a result of mammographic breast compression. Breast Cancer Res Treat. 2013;141:187–195.

36. Friedl P, Gilmour D. Collective cell migration in morphogenesis, regeneration and cancer. Nat Rev Mol Cell Biol. 2009;10:445–457.

37. Friedl P, Wolf K. Proteolytic interstitial cell migration: a five-step process. Cancer Metastasis Rev.2009;28:129–135.

38. Friedl P, Alexander S. Cancer invasion and the microenvironment: plasticity and reciprocity. Cell.2011;147:992–1009.

39. Gasch C, Bauernhofer T, Pichler M, Langer-Freitag S, Reeh M, Seifert AM, Mauermann O, Izbicki JR, Pantel K, Riethdorf S. Heterogeneity of epidermal growth factor receptor status and mutations of KRAS/PIK3CA in circulating tumor cells of patients with colorectal cancer. Clin Chem. 2013;59:252–260.

40. Gazzaniga P, de BE, Raimondi C, Gradilone A, Busetto GM, De FE, Nicolazzo C, Giovannone R, Gentile V, Cortesi E, et al., editors. Circulating tumor cells detection has independent prognostic impact in high-risk non-muscle invasive bladder cancer. Int J Cancer. 2014;135:1978–1982.

41. Ghadimi BM, Sackett DL, Difilippantonio MJ, Schrock E, Neumann T, Jauho A, Auer G, Ried T. Centrosome amplification and instability occurs exclusively in aneuploid, but not in diploid colorectal cancer cell lines, and correlates with numerical chromosomal aberrations. Genes Chromosom Cancer.2000;27:183–190.

42. Godinho SA, Picone R, Burute M, Dagher R, Su Y, Leung CT, Polyak K, Brugge JS, Thery M, Pellman D. Oncogene-like induction of cellular invasion from centrosome amplification. Nature.2014;510:167–171.

43. Gorges TM, Tinhofer I, Drosch M, Rose L, Zollner TM, Krahn T, von AO. Circulating tumour cells escape from EpCAM-based detection due to epithelial-to-mesenchymal transition. BMC Cancer.2012;12:178.

44. Goss PE, Chambers AF. Does tumour dormancy offer a therapeutic target? Nat Rev Cancer.2010;10:871–877.

45. Gronich N, Rennert G. Beyond aspirin-cancer prevention with statins, metformin and bisphosphonates. Nat Rev Clin Oncol. 2013;10:625–642.

46. Hanahan D, Weinberg RA. Hallmarks of cancer: the next generation. Cell. 2011;144:646–674.

47. Harb W, Fan A, Tran T, Danila DC, Keys D, Schwartz M, Ionescu-Zanetti C. Mutational Analysis of Circulating Tumor Cells Using a Novel Microfluidic Collection Device and qPCR Assay. Transl Oncol. 2013;6:528–538.

48. Hartkopf AD, Banys M, Fehm T. HER2-positive DTCs/

CTCs in breast cancer. Recent Results Cancer Res. 2012;195:203–215.

49. Heitzer E, Auer M, Gasch C, Pichler M, Ulz P, Hoffmann EM, Lax S, Waldispuehl-Geigl J, Mauermann O, Lackner C, et al., editors. Complex tumor genomes inferred from single circulating tumor cells by array-CGH and next-generation sequencing. Cancer Res. 2013;73:2965–2975.

50. Hodgkinson CL, Morrow CJ, Li Y, Metcalf RL, Rothwell DG, Trapani F, Polanski R, Burt DJ, Simpson KL, Morris K, et al., editors. Tumorigenicity and genetic profiling of circulating tumor cells in small-cell lung cancer. Nat Med. 2014;20:897–903.

51. Hohensee I, Lamszus K, Riethdorf S, Meyer-Staeckling S, Glatzel M, Matschke J, Witzel I, Westphal M, Brandt B, Muller V, et al., editors. Frequent genetic alterations in EGFR-and HER2-driven pathways in breast cancer brain metastases. Am J Pathol. 2013;183:83–95.

52. Hou JM, Krebs MG, Lancashire L, Sloane R, Backen A, Swain RK, Priest LJ, Greystoke A, Zhou C, Morris K, et al., editors. Clinical significance and molecular characteristics of circulating tumor cells and circulating tumor microemboli in patients with small-cell lung cancer. J Clin Oncol. 2012;30:525–532.

53. Hou HW, Warkiani ME, Khoo BL, Li ZR, Soo RA, Tan DS, Lim WT, Han J, Bhagat AA, Lim CT. Isolation and retrieval of circulating tumor cells using centrifugal forces. Sci Rep. 2013;3:1259.

54. Ignatiadis M, Rothe F, Chaboteaux C, Durbecq V, Rouas G, Criscitiello C, Metallo J, Kheddoumi N, Singhal SK, Michiels S, et al., editors. HER2-positive circulating tumor cells in breast cancer. PLoS One. 2011;6:e15624.

55. Janni W, Vogl FD, Wiedswang G, Synnestvedt M, Fehm T, Juckstock J, Borgen E, Rack B, Braun S, Sommer H, et al., editors. Persistence of disseminated tumor cells in the bone marrow of breast cancer patients predicts increased risk for relapse–a European pooled analysis. Clin Cancer Res.2011;17:2967–2976.

56. Joosse SA, Hannemann J, Spotter J, Bauche A, Andreas A, Muller V, Pantel K. Changes in keratin expression during metastatic progression of breast cancer: impact on the detection of circulating tumor cells. Clin Cancer Res. 2012;18:993–1003.

57. Joosse SA, Pantel K. Biologic challenges in the detection of circulating tumor cells. Cancer Res.2013;73:8–11.

58. Kang Y, Pantel K. Tumor cell dissemination: emerging biological insights from animal models and cancer patients. Cancer Cell. 2013;23:573–581.

59. Kasimir-Bauer S, Hoffmann O, Wallwiener D, Kimmig R, Fehm T. Expression of stem cell and epithelial-mesenchymal transition markers in primary breast cancer patients with circulating tumor cells. Breast Cancer Res. 2012;14:R15.

60. Kim MY, Oskarsson T, Acharyya S, Nguyen DX, Zhang XH, Norton L, Massague J. Tumor self-seeding by circulating cancer cells. Cell. 2009;139:1315–1326.

61. Kishi S, Bayliss PE, Hanai JI. A prospective epigenetic paradigm between cellular senescence and epithelial-mesenchymal transition in organismal development and aging. Transl Res. 2014;165:241–249.

62. Korkaya H, Wicha MS. HER2 and breast cancer stem cells: more than meets the eye. Cancer Res.2013;73:3489–3493.

63. Krebs MG, Sloane R, Priest L, Lancashire L, Hou JM, Greystoke A, Ward TH, Ferraldeschi R, Hughes A, Clack G, et al., editors. Evaluation and prognostic significance of circulating tumor cells in patients with non-small-cell lung cancer. J Clin Oncol. 2011;29:1556–1563.

64. Lewiecki EM. New targets for intervention in the treatment of postmenopausal osteoporosis. Nat Rev Rheumatol. 2011;7:631–638.

65. Lindemann F, Schlimok G, Dirschedl P, Witte J, Riethmuller G. Prognostic significance of micrometastatic tumour cells in bone marrow of colorectal cancer patients. Lancet. 1992;340:685–689.

66. Lingle WL, Barrett SL, Negron VC, D'Assoro AB, Boeneman K, Liu W, Whitehead CM, Reynolds C, Salisbury JL. Centrosome amplification drives chromosomal instability in breast tumor development. Proc Natl Acad Sci USA. 2002;99:1978–1983.

67. Liu Z, Fusi A, Klopocki E, Schmittel A, Tinhofer I, Nonnenmacher A, Keilholz U. Negative enrichment by immunomagnetic nanobeads for unbiased characterization of circulating tumor cells from peripheral blood of cancer patients. J Transl Med. 2011;9:70.

68. Lohr JG, Adalsteinsson VA, Cibulskis K, Choudhury AD, Rosenberg M, Cruz-Gordillo P, Francis JM, Zhang CZ, Shalek AK, Satija R, et al., editors. Whole-exome sequencing of circulating tumor cells provides a window into metastatic prostate cancer. Nat Biotechnol. 2014;32:479–484.

69. Lu J, Fan T, Zhao Q, Zeng W, Zaslavsky E, Chen JJ, Frohman MA, Golightly MG, Madajewicz S, Chen WT. Isolation of circulating epithelial and tumor progenitor cells with an invasive phenotype from breast cancer patients. Int J Cancer. 2010;126:669–683.

70. Lu X, Mu E, Wei Y, Riethdorf S, Yang Q, Yuan M, Yan J, Hua Y, Tiede BJ, Lu X, et al., editors. VCAM-1 promotes osteolytic expansion of indolent bone micrometastasis of breast cancer by engaging alpha4beta1-positive osteoclast progenitors. Cancer Cell. 2011;20:701–714.

71. Lucci A, Hall CS, Lodhi AK, Bhattacharyya A, Anderson AE, Xiao L, Bedrosian I, Kuerer HM, Krishnamurthy S. Circulating tumour cells in non-metastatic breast cancer: a prospective study. Lancet Oncol. 2012;13:688–695.

72. Maheswaran S, Sequist LV, Nagrath S, Ulkus L, Brannigan B, Collura CV, Inserra E, Diederichs S, Iafrate AJ, Bell DW, et al., editors. Detection of mutations in EGFR in circulating lung-cancer cells. N Engl J Med. 2008;359:366–377.

73. McDonald DM, Baluk P. Significance of blood vessel leakiness in cancer. Cancer Res. 2002;62:5381–5385.

74. Meng S, Tripathy D, Frenkel EP, Shete S, Naftalis EZ, Huth JF, Beitsch PD, Leitch M, Hoover S, Euhus D, et al., editors. Circulating tumor cells in patients with breast cancer dormancy. Clin Cancer Res. 2004;10:8152–8162.

75. Mitchell MJ, King MR. Computational and experimental models of cancer cell response to fluid shear stress. Front Oncol. 2013;3:44.

76. Nagrath S, Sequist LV, Maheswaran S, Bell DW, Irimia D, Ulkus L, Smith MR, Kwak EL, Digumarthy S, Muzikansky A, et al., editors. Isolation of rare circulating tumour cells in cancer patients by microchip technology. Nature. 2007;450:1235–1239.

77. Nguyen DX, Massague J. Genetic determinants of cancer metastasis. Nat Rev Genet. 2007;8:341–352.

78. Nieto MA. Epithelial plasticity: a common theme in embryonic and cancer cells. Science.2013;342:1234850.

79. Ozkumur E, Shah AM, Ciciliano JC, Emmink BL, Miyamoto DT, Brachtel E, Yu M, Chen PI, Morgan B, Trautwein J, et al., editors. Inertial focusing for tumor antigen-dependent and-independent sorting of rare circulating tumor cells. Sci Transl Med. 2013;5:179ra47.

80. Pantel K, Schlimok G, Braun S, Kutter D, Lindemann F, Schaller G, Funke I, Izbicki JR, Riethmuller G. Differential expression of proliferation-associated molecules in individual micrometastatic carcinoma cells. J Natl Cancer Inst. 1993;85:1419–1424.

81. Pantel K, Passlick B, Vogt J, Stosiek P, Angstwurm M, Seen-Hibler R, Haussinger K, Thetter O, Izbicki JR, Riethmuller G. Reduced expression of plakoglobin indicates an unfavorable prognosis in subsets of patients with non-small-cell lung cancer. J Clin Oncol. 1998;16:1407–1413.

82. Pantel K, Brakenhoff RH. Dissecting the metastatic cascade. Nat Rev Cancer. 2004;4:448–456.

83. Parkinson DR, Dracopoli N, Petty BG, Compton C, Cristofanilli M, Deisseroth A, Hayes DF, Kapke G, Kumar P, Lee JS, et al., editors. Considerations in the development of circulating tumor cell technology for clinical use. J Transl Med. 2012;10:138.

84. Peeters M, Douillard JY, Van CE, Siena S, Zhang K, Williams R, Wiezorek J. Mutant KRAS codon 12 and 13 alleles in patients with metastatic colorectal cancer: assessment as prognostic and predictive biomarkers of response to panitumumab. J Clin Oncol. 2013;31:759–765.

85. Rack B, Schindlbeck C, Juckstock J, Andergassen U, Hepp P, Zwingers T, Friedl TW, Lorenz R, Tesch H, Fasching PA, et al., editors. Circulating tumor cells predict survival in early average-to-high risk breast cancer patients. J Natl Cancer Inst. 2014;106 : dju066.

86. Reymond N, d›Agua BB, Ridley AJ. Crossing the endothelial barrier during metastasis. Nat Rev Cancer. 2013;13:858–870.

87. Riahi R, Gogoi P, Sepehri S, Zhou Y, Handique K, Godsey J, Wang Y. A novel microchannel-based device to capture and analyze circulating tumor cells (CTCs) of breast cancer. Int J Oncol.2014;44:1870–1878.

88. Riethdorf S, Fritsche H, Muller V, Rau T, Schindlbeck C, Rack B, Janni W, Coith C, Beck K, Janicke F, et al., editors. Detection of circulating tumor cells in peripheral blood of patients with metastatic breast cancer: a validation study of the Cell Search system. Clin Cancer Res. 2007;13:920–928.

89. Riethdorf S, Wikman H, Pantel K. Review: biological relevance of disseminated tumor cells in cancer patients. Int J Cancer. 2008;123:1991–2006.

90. Riethdorf S, Muller V, Zhang L, Rau T, Loibl S, Komor M, Roller M, Huober J, Fehm T, Schrader I, et al., editors. Detection and HER2 expression of circulating tumor cells: prospective monitoring in breast cancer patients treated in the neoadjuvant GeparQuattro trial. Clin Cancer Res. 2010;16:2634–2645.

91. Rink M, Chun FK, Minner S, Friedrich M, Mauermann O, Heinzer H, Huland H, Fisch M, Pantel K, Riethdorf S. Detection of circulating tumour cells in peripheral blood of patients with advanced non-metastatic bladder cancer. BJU Int. 2011;107:1668–1675.

92. Rink M, Chun FK, Dahlem R, Soave A, Minner S, Hansen J, Stoupiec M, Coith C, Kluth LA, Ahyai SA, et al., editors. Prognostic role and HER2 expression of circulating tumor cells in peripheral blood of patients prior to radical cystectomy: a prospective study. Eur Urol. 2012;61:810–817.

93. Saucedo-Zeni N, Mewes S, Niestroj R, Gasiorowski L, Murawa D, Nowaczyk P, Tomasi T, Weber E, Dworacki G, Morgenthaler NG, et al., editors. A novel method for the in vivo isolation of circulating tumor cells from peripheral blood of cancer patients using a functionalized and structured medical wire.Int J Oncol. 2012;41:1241–1250.

94. Stathopoulou A, Gizi A, Perraki M, Apostolaki S, Malamos N, Mavroudis D, Georgoulias V, Lianidou ES. Real-time quantification of CK-19 mRNA-positive cells in peripheral blood of breast cancer patients using the lightcycler system. Clin Cancer Res. 2003;9:5145–5151.

95. Steinert G, Scholch S, Niemietz T, Iwata N, Garcia SA, Behrens B, Voigt A, Kloor M, Benner A, Bork U, et al., editors. Immune escape and survival mechanisms in circulating tumor cells of colorectal cancer. Cancer Res. 2014;74:1694–1704.

96. Strati A, Kasimir-Bauer S, Markou A, Parisi C, Lianidou ES. Comparison of three molecular assays for the detection and molecular characterization of circulating tumor cells in breast cancer. Breast Cancer Res. 2013;15:R20.

97. Tam WL, Weinberg RA. The epigenetics of epithelial-mesenchymal plasticity in cancer. Nat Med.2013;19:1438–1449.

98. Tewes M, Aktas B, Welt A, Mueller S, Hauch S, Kimmig R, Kasimir-Bauer S. Molecular profiling and predictive value of circulating tumor cells in patients with metastatic breast cancer: an option for monitoring response to breast cancer related therapies. Breast Cancer Res Treat. 2009;115:581–590.

99. Thege FI, Lannin TB, Saha TN, Tsai S, Kochman ML, Hollingsworth MA, Rhim AD, Kirby BJ. Microfluidic immunocapture of circulating pancreatic cells using parallel EpCAM and MUC1 capture: characterization, optimization and downstream analysis. Lab Chip. 2014;14:1775–1784.

100. Thiery JP. Epithelial-mesenchymal transitions in tumour

progression. Nat Rev Cancer. 2002;2:442–454.

101. Thiery JP, Acloque H, Huang RY, Nieto MA. Epithelial-mesenchymal transitions in development and disease. Cell. 2009;139:871–890.

102. Tsai JH, Donaher JL, Murphy DA, Chau S, Yang J. Spatiotemporal regulation of epithelial-mesenchymal transition is essential for squamous cell carcinoma metastasis. Cancer Cell.2012;22:725–736.

103. Uhr JW, Pantel K. Controversies in clinical cancer dormancy. Proc Natl Acad Sci USA.2011;108:12396–12400.

104. Uppal A, Wightman SC, Ganai S, Weichselbaum RR, An G. Investigation of the essential role of platelet-tumor cell interactions in metastasis progression using an agent-based model. Theor Biol Med Model. 2014;11:17.

105. Wan L, Pantel K, Kang Y. Tumor metastasis: moving new biological insights into the clinic. Nat Med.2013;19:1450–1464.

106. Wikman H, Vessella R, Pantel K. Cancer micrometastasis and tumour dormancy. APMIS.2008;116:754–770.

107. Wikman H, Lamszus K, Detels N, Uslar L, Wrage M, Benner C, Hohensee I, Ylstra B, Eylmann K, Zapatka M, et al., editors. Relevance of PTEN loss in brain metastasis formation in breast cancer patients. Breast Cancer Res. 2012;14:R49.

108. Wrage M, Ruosaari S, Eijk PP, Kaifi JT, Hollmen J, Yekebas EF, Izbicki JR, Brakenhoff RH, Streichert T, Riethdorf S, et al., editors. Genomic profiles associated with early micrometastasis in lung cancer: relevance of 4q deletion. Clin Cancer Res. 2009;15:1566–1574.

109. Wu S, Liu Z, Liu S, Lin L, Yang W, Xu J. Enrichment and enumeration of circulating tumor cells by efficient depletion of leukocyte fractions. Clin Chem Lab Med. 2014;52:243–251.

110. Yu M, Bardia A, Wittner BS, Stott SL, Smas ME, Ting DT, Isakoff SJ, Ciciliano JC, Wells MN, Shah AM, et al., editors. Circulating breast tumor cells exhibit dynamic changes in epithelial and mesenchymal composition. Science. 2013;339:580–584.

111. Yu M, Bardia A, Aceto N, Bersani F, Madden MW, Donaldson MC, Desai R, Zhu H, Comaills V, Zheng Z, et al., editors. Cancer therapy. Ex vivo culture of circulating breast tumor cells for individualized testing of drug susceptibility. Science. 2014;345:216–220.

112. Zhang L, Riethdorf S, Wu G, Wang T, Yang K, Peng G, Liu J, Pantel K. Meta-analysis of the prognostic value of circulating tumor cells in breast cancer. Clin Cancer Res. 2012;18:5701–5710.

CHAPTER 10

THE KYNURENINE PATHWAY IN STEM CELL BIOLOGY

Simon P. Jones,[1] Gilles J. Guillemin,[2] and Bruce J. Brew[1,3]

[1]St. Vincent's Centre for Applied Medical Research, The University of New South Wales, Sydney, Australia.

[2]Australian School of Advanced Medicine, Macquarie University, Sydney, Australia.

[3]Department of Neurology, St. Vincent's Hospital, Sydney, Australia.

ABSTRACT

The kynurenine pathway (KP) is the main catabolic pathway of the essential amino acid tryptophan. The KP has been identified to play a critical role in regulating immune responses in a variety of experimental settings. It is also known to be involved in several neuroinflammatory diseases including Huntington's disease, amyotrophic lateral sclerosis, and Alzheimer's disease. This review considers the current understanding of the role of the KP in stem cell biology. Both of these fundamental areas of cell biology have independently been the focus of a burgeoning research interest in recent years. A systematic review of how the two interact has not yet been conducted. Several inflammatory and infectious diseases in which the KP has been implicated include those for which stem cell therapies are being actively explored at a clinical level. Therefore, it is highly relevant to consider the evidence showing that the KP influences stem cell biology and impacts the functional behavior of progenitor cells.

INTRODUCTION

Stem Cells

A stem cell is defined by two fundamental properties: the capacity to self-renew and the ability to differentiate into mature cells. A hierarchy of stem cell potential exists, with pluripotent embryonic stem cells (ESCs) at the apex (Fig. 1). ESCs are derived from the inner cell mass of the developing blastocyst and can give rise to mature cells of all three germ layers. Due to a range of moral, bioethical, and technical issues, there are numerous hurdles to the clinical application of ESCs. Consequently, the last three to four decades have witnessed an increased interest in the use of adult stem cells. These cells can be isolated postnatally from a host of different organs and tissues. They typically have a more limited differentiation potential, often restricted to mature cells of one germ layer. Of the diverse range of adult stem cells the hematopoietic stem cell (HSC) represents perhaps the best-studied multipotent cell and can give rise to all cells of the blood. Another bone marrow resident stem cell is the mesenchymal stem cell (MSC). MSCs can differentiate into cells of mesodermal origin, typically osteoblasts, chondrocytes, and adipocytes. Neural stem cells

Citation: Simon P. Jones, Gilles J. Guillemin and Bruce J. Brew, The Kynurenine Pathway in Stem Cell Biology, doi: 10.4137/IJTR.S12626

(NSCs) represent a relatively recently identified organ-specific adult stem cell that can differentiate into neurons, astrocytes, and oligodendrocytes (Fig. 1).

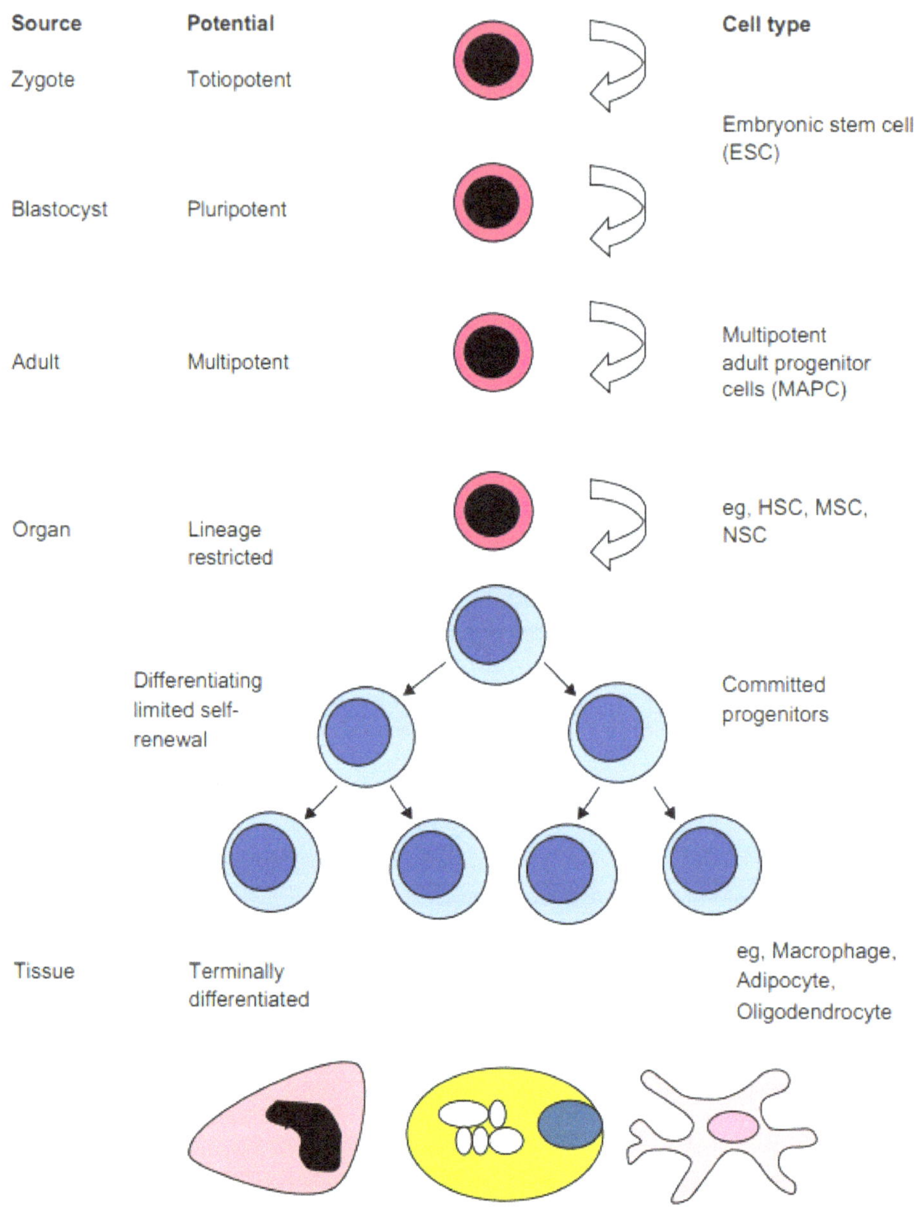

Figure 1: Hierarchy of stem cells.

The Kynurenine Pathway

Tryptophan is one of the 9 essential amino acids that the human body is unable to synthesize and thus must be provided through diet. Once absorbed by the body, tryptophan travels through the peripheral circulation. Tryptophan is the only amino acid that binds to albumin in the plasma with approximately 10%–15% of the total plasma tryptophan in the free form and 85%–90% transported bound to albumin, with these two states existing in equilibrium.[1] However, tryptophan can only be transported across the blood-brain barrier in its free form by the competitive and non-specific L-type amino acid transporter. Once in the central nervous system (CNS), tryptophan acts as a precursor to several metabolic pathways including general protein synthesis, serotonin/melatonin synthesis, and kynurenine production (Fig. 2).[1] In both the peripheral and central nervous systems, the kynurenine pathway (KP) represents the major route for the catabolism of L-tryptophan, resulting in the production of the essential co-factor pyridine nucleotide nicotinamide adenine dinucleotide (NAD^+) and other neuroactive intermediates (Fig. 2). Tryptophan is oxidized by cleavage of the indole ring, initiated either by tryptophan 2,3-dioxygenase (TDO2), indoleamine 2,3-dioxygenase 1 (IDO-1) or IDO-2.[2–5] TDO2 is primarily expressed in the liver,[6,7] but is also present in the CNS.[8] TDO2 is induced by its substrate tryptophan and by corticosteroids.[7] TDO2 can be inhibited by indoleamines and nicotinamide analogs as well as by some antidepressant drugs.[9] Extra-hepatically, IDO-1 is the predominant enzyme and can be found in most cell types, including macrophages, microglia, neurons, and astrocytes, but not in oligodendrocytes.[10–12] IDO-1 is up-regulated by several inflammatory molecules including lipopolysaccharides, amyloid peptides, and HIV proteins,[13–15] but its most potent activator is interferon gamma (IFN-γ).[5,16] IFN-γ induces both the gene expression and enzymatic activity of IDO-1.[17,18] While IDO-2 possesses similar structural and enzymatic activities to IDO-1, IDO-2 displays a different expression pattern and signaling pathway.[4,19] In the CNS, KP enzymes are variably expressed in most cell types,[11] including astrocytes,[20] neurons,[21] infiltrating macrophages and microglia,[22] oligodendrocytes,[12] and endothelial cells.[23] Infiltrating macrophages, activated microglia, and neurons express the full range of KP enzymes, whereas astrocytes and likely oligodendrocytes lack the crucial enzymes kynurenine 3-monooxygenase (KMO) and IDO-1, respectively.[12,20]

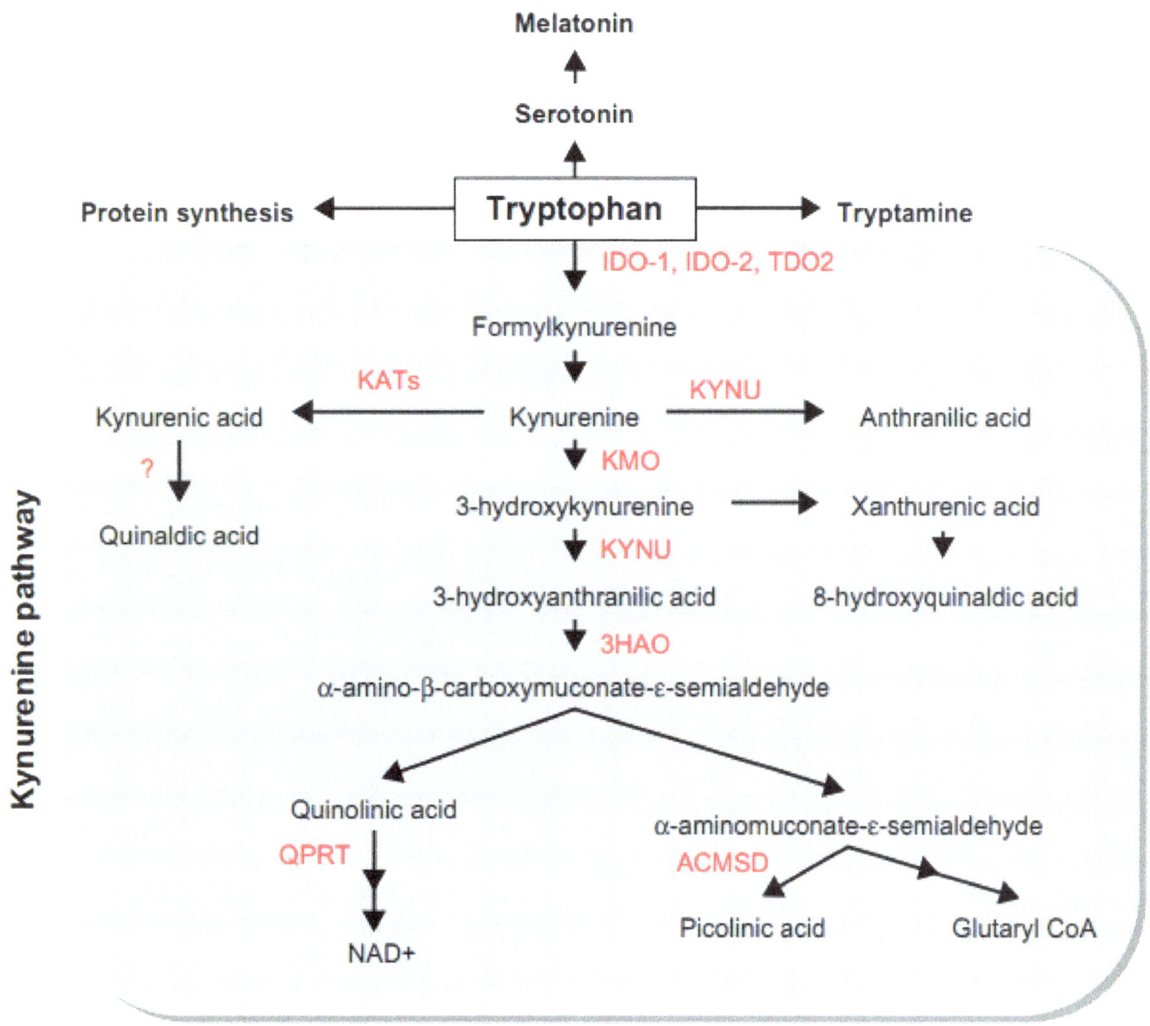

Figure 2: Overview of the kynurenine pathway of tryptophan metabolism.

In the production of KP metabolites, kynurenine (KYN) is the first stable intermediate formed. Subsequently, several other neuroactive intermediates are generated, including the free-radical generator, 3-hydroxyanthranilic acid (3HAA),[24] the excitotoxin and N-methyl-D-aspartate (NMDA) receptor agonist, quinolinic acid (QUIN),[25] the NMDA antagonist, kynurenic acid (KYNA),[26] and the neuroprotectant, picolinic acid (PIC).[27] Among KP metabolites, QUIN appears to be one of the most important in terms of biological activity.[28]

IDO-1 has recently been the focus of attention because of its potent immunosuppressive effects on T lymphocytes, resulting in part from tryptophan depletion and from direct effects of tryptophan catabolites.[29–32] Some kynurenines, such as QUIN and 3HAA, can selectively target immune cells undergoing activation, consequently suppressing T cell proliferation.[33,34] More recently, KYN is also involved in immuno-regulation through its ligand function for the Aryl hydrocarbon receptor (AhR).[35] KP metabolites can also act in concert to produce an additive effect.[36] IDO-1 up-regulation as well as accelerated and sustained degradation of tryptophan represent key indicators of inflammation. Indeed, inflammation and resulting immune activation lead to KP activation and the concomitant increased production of the excitotoxin QUIN.[37] To date, QUIN has been shown to be associated with the pathogenesis of a wide range of inflammatory diseases and disorders.[38–41]

EVIDENCE OF KP INVOLVEMENT IN STEM CELL BIOLOGY

Embryonic Stem Cells

Currently, very little is known regarding the role of the KP in ESC biology. As expected for such a critical metabolic pathway, the rate-limiting and key enzymes of the KP are expressed in human ESCs.[42] In a quest for biomarkers of developmental toxicity Cezar et al[42] identified that KP molecules, particularly TDO2, were up-regulated in ESCs when cells were treated with the anti-epileptic drug valproate. The authors proposed that the KP plays a role in the pathogenesis of neurodevelopmental disorders. They hypothesised that activation of the KP reduces the bioavailability of tryptophan, leading to reduced serotonin synthesis and subsequent neurodevelopmental defects. Aside from this study, there has been limited exploration of the KP in ESC biology.

Hematopoietic Stem Cells

HSCs represent a population of progenitor cells that, relative to other adult stem cells, have been well-characterized as precursors of cells of their respective tissue systems. The role of HSCs and their niche in hemopoiesis, transplantation biology, and treatment of malignancies has been a strong research focus in recent years. Cells of the hematopoietic lineage were among the first studied to show a link between the KP and immune regulation. IDO expression has been well-characterized in professional antigen-presenting cells.[43,44] In these studies, the activation of the KP in dendritic cells and macrophages was a potent mechanism for the regulation of T cell proliferation. In fact, IDO production from donor monocytes following hematopoietic stem cell transplantation is thought to be responsible for the depressed T cell function often observed in these patients.[45] Work on tryptophan catabolism in lineage committed hematopoietic cells has recently been extended to investigate activation of the KP in their progenitor precursor: the HSC.[46] In this study, the authors found that both acute and chronic stimulation with IFN-γ caused an increase in the production of kynurenine in $CD34^+$ HSC cultures. They hypothesized that IFN-γ mediated activation of the KP inhibited hematopoiesis. However, they did not observe any functional suppressive effect of IDO on erythropoiesis in their experiments. This may have occurred because tryptophan concentrations were not sufficiently depleted to starve HSCs. Kurz et al suggested that other bone marrow resident progenitor cells could therefore account for the inhibition of erythropoiesis observed in conditions such as anemia. These mesenchymal progenitors will be the focus of the next section.

Mesenchymal Stem Cells

As a purported population of adult stem cells, links between tryptophan metabolism and MSC biology have been widely investigated. MSCs are multipotent progenitors that were initially identified in the bone marrow,[47] where they are thought to play a physiological role in maintaining the hematopoietic stem cell niche.[48] Recently MSCs have also been isolated and expanded from a wide range of post-natal and fetal tissues,[49–51] with some investigators suggesting that these cells are also present in nearly all adult tissues.[52]The defining feature of MSCs is their ability to differentiate in various tissues of mesodermal origin, typically osteoblasts, chondrocytes and adipocytes.[53] Interestingly, it seems MSCs are not restricted to a mesodermal fate. They have also been shown to differentiate into endothelial cells[54] and neural cells.[55] This plasticity of MSCs, along with their ability to migrate to sites of inflammation, has attracted significant attention in the last decade for their potential use in transplant and regeneration studies.[56] Another aspect of MSC biology that makes them of particular interest is that they display potent immunomodulatory functions. MSCs from a variety of species inhibit the response of T cells to antigenic, mitogenic, and polyclonal stimuli.[57–59] This effect is targeted mainly at the level of T cell proliferation and has been further characterized as arrest of the cell cycle in the G_1 phase.[60] This has been shown to be a pro-survival property, with MSCs rescuing T cells from apoptosis.[61] The immunosuppressive effects of MSCs are not limited to T cells. B cell proliferation and differentiation,[62] dendritic cell maturation and antigen-presenting function,[63] and NK cell proliferation and cytotoxicity[64] are all targets of MSC-mediated suppression. There is also growing evidence that MSCs exert similar effects on the innate immune system (for a review of this subject[65]).

The mechanisms by which MSCs produce these compelling immunomodulatory functions remain unresolved. While cell-to-cell contact appears to be required for 'licensing' MSCs to become suppressive, the inhibitory effect is ultimately mediated by a soluble factor(s).[61] Several candidate molecules have been proposed, including transforming growth factor-β1,[58] hemeoxygenase-1,[66] prostaglandin E_2 (PGE_2),[67] and nitric oxide (NO).[68] Interestingly, activation of the KP has also been investigated as a potential mechanism for this immunosuppressive effect. Meisel et al showed that IDO could be induced in MSCs by exposure to IFN-γ in a dose-dependent manner. This finding was confirmed in co-cultures of MSCs and mixed lymphocyte reactions (MLR), in which significant IDO activity was detected compared to MSC or MLR cultures alone, suggesting that MSCs were the primary source of IDO activity. Importantly, in the MSC/MLR co-cultures, addition of tryptophan significantly restored T cell proliferation. The authors did not investigate the effect of IDO inhibitors in this system.[69] In support of these findings a more comprehensive study of the MSC immunomodulatory function found that the separate addition of two different competitive inhibitors of IDO, 1-methyl-tryptophan (1-MT) and norharmane, reduced the suppressive effect of MSCs on T cell proliferation.[70] A third study also confirmed a role for IDO in MSC-mediated suppression of allogeneic T cell proliferation. Treatment of MSC/MLR co-cultures with 1-MT significantly, but not completely, restored allo-driven proliferation of lymphocytes.[71] Interestingly, it has been documented that ESCs share similar immunosuppressive functions with MSCs.[72] However, it remains unclear whether the KP is involved in this inhibitory effect. In a recent study using murine ESCs and MSCs, the authors found that IDO production was not involved in stem cell-mediated immunosuppression in their system.[73]

Involvement of the KP in the immunosuppressive effects of MSCs has been supported by in vivo studies. Using experimental autoimmune encephalomyelitis (EAE), a murine model of multiple sclerosis, Matysiak et al showed that mice transplanted with MSCs showed significantly lower clinical scores and greater improvement than control mice with EAE.[74] The authors reported a higher than two-fold increase in the expression of IDO in the spleens of EAE mice treated with MSCs; upon closer analysis, it was found that $CD11c^+$ dendritic cells were the population predominantly expressing IDO. The use of 1-MT to treat MSC-transplanted EAE mice restored clinical scores to similar levels as those of control mice with

EAE and no transplanted MSCs. Therefore, blocking IDO activity led to a loss of the immunosuppressive function of MSCs in vivo and exacerbation of inflammatory disease.[74]

In addition to their effects on T cells, MSCs have been shown to inhibit NK cell proliferation and cytotoxicity, which is at least partially mediated by IDO.[75] Furthermore, in this study, addition of the PGE_2inhibitor NS-398 to MSC and NK cell co-cultures along with 1-MT (inhibitor of IDO) nearly fully recovered IL-2-stimulated NK cell proliferation. This suggests that mechanistic synergy exists between IDO and PGE_2in the immunosuppressive effect of MSCs. The authors speculate that the synthesis of IDO by MSCs is induced both directly by exposure to pro-inflammatory cytokines such as IFN-γ and indirectly by autocrine stimulation of cells by PGE_2. This hypothesis is supported by earlier work using dendritic cells in which PGE_2 exposure induced de novo expression of IDO mRNA.[76]

Despite the evidence outlined above, the hypothesis that the KP regulates MSC-mediated suppression of immune cells remains controversial. Gieseke and co-workers used MSCs cultured from the bone marrow of a boy with a frameshift mutation in the IFN-γ receptor 1 (R1). These $MSC^{IFN\gamma R1-}$ cells failed to respond to IFN-γ in vitro but displayed the same ability to inhibit PBMC proliferation as wild-type MSCs. Additionally, IDO expression was completely absent and non-inducible in $MSC^{IFN\gamma R1-}$ cells, yet they were still able to inhibit the proliferation of allogeneic PBMC.[77] These findings clearly contest the notion that IFN-γ-stimulated IDO expression is required for MSCs to exert their inhibitory effects on immune cells. However, this does not rule out the fact that other factors can upregulate IDO. More recently, Lanz et al demonstrated that murine MSCs suppress the activation of myelin specific T cells independently of IDO.[78] The pharmacological inhibition of IDO, using 1-MT, failed to rescue immunosuppression of MOG activated splenocytes. This can be explained by the observation that 1-MT is not a potent inhibitor of IDO.[79] Work from our laboratory and others has shown that 1-MT, along with other pharmacological inhibitors, at best only offers partial inhibition of IDO activity. Moreover, the authors showed that in an EAE model, IDO gene ablation ($IDO1^{-/-}$ MSC) did not attenuate the therapeutic effects of MSCs. Both wild-type MSCs and $IDO1^{-/-}$ MSCs reduced EAE clinical scores by a similar magnitude, potentially by inhibiting proinflammatory cytokines, notably IL-17.[78]

Interestingly, there appears to be more species differences between human and murine KP expression than what might be expected for an evolutionarily conserved pathway. Our group has found that in contrast to human MSCs, mouse MSCs do not experience significant IFN-γ inducible changes in the expression of several KP enzymes at the RNA level.[79] Such findings may explain the conflicting evidence outlined above. In the same paper, Croitoru-Lamoury et al showed that the inhibitory effects of KP activation are not limited to immune cells. In fact, IFN-γ inhibits the proliferation of both mouse and human MSCs through activation of the KP. The addition of excess tryptophan to cultures of MSCs stimulated with IFN-γ blocked the anti-proliferative activity of the cytokine on MSCs. Moreover, blocking IDO action by addition of 1-MT rescued cell viability of cytokine-stimulated MSCs. However, in long-term cultures of 50 days, addition of the IDO inhibitors norharmane and 1-MT did not recover the significant decrease in proliferation of IFN-γ-treated MSCs. The effect of IFN-γ-mediated activation of the KP on MSC proliferation was extended to investigate its effects on MSC differentiation. Based on the gene expression of recognized osteogenic and adipogenic differentiation markers (osteopontin, integrin-binding sialoprotein II, and adipsin, adipoQ, Fabp4 respectively), it was demonstrated that treatment with IFN-γ inhibited the gene expression of all these markers. Importantly, addition of norharmane to differentiation cultures partially recovered the gene expression of osteopontin, adipsin, and adipoQ. Similar findings were extended to the neural differentiation of MSCs.[79] Taken together, this evidence points to a role for the KP in modulating essential stem cell functions: proliferation and differentiation potential.

Neural Stem Cells

NSCs have been shown to express high levels of IDO mRNA and protein when activated by IFN-γ. This feature of NSCs was utilized in a cell-based assay to examine the effects of a number of naturally occurring anti-inflammatory phytochemicals on IDO expression. The authors suggested that the suppressive effect of some of these compounds on IDO may account for their observed anti-tumor and neuroprotective properties.[80] Our group previously established that much of the KP “machinery” is expressed in NSCs.[79]Using murine NSCs, transcripts encoding all the major KP enzymes, were detected under basal conditions. On-going experiments in our laboratory are examining the expression and function of the KP in neural progenitor cells. We hope to ascertain whether modulation of the KP can affect the survival, proliferation, and differentiation of neural progenitor cells and ultimately their functional ability to remyelinate damaged neurons. This has huge potential to offer novel therapeutic opportunities for the treatment of inflammatory and neurodegenerative disorders.

The KP has been implicated in the pathophysiology of depression and its associated deficits in neurogenesis. Zunsdain et al found that human hippocampal progenitor cells constitutively express IDO and TDO and that treatment with IL-1β significantly up-regulated IDO. IL-1β also induced a decrease in expression of the 3 major KAT enzyme isoforms, which catalyze the conversion of kynurenine (KYN) into the neuroprotectant kynurenic acid (KYNA). In contrast, transcripts for two enzymes that act along the neurotoxic branches of the KP, kynurenine 3-monooxygenase (KMO) and kynureninase (KYNU), were increased. Thus, the authors postulated that IL-1β inhibits neurogenesis in human hippocampal progenitor cells by increasing neurotoxicity and supressing neuroprotection, although they provided no direct evidence. Treatment with the KMO inhibitor Ro 61-8048 partially reversed the detrimental effects of IL-1β on neurogenesis, confirming that the observed effects were mediated via the KP.[81] Interestingly, the authors showed that IL-1β had contrasting effects on the differentiation and proliferation of neural progenitor cells; IL-1β inhibits neural differentiation while promoting proliferation of undifferentiated cells. It was postulated that this difference was facilitated through activation of the neurotoxic branch and inhibition of the neuroprotective branch of the KP during differentiation. In contrast, IL-1β activated both the neuroprotective and neurotoxic branches in proliferating cells. A major limitation of this study was the use of a progenitor cell line under in vitro experimental conditions. However, in vivo support for a role of the KP in NSC differentiation comes from a report exploring the function of the KP in neurogenesis and anxiety-related behavior using TDO$^{-/-}$ mice.[82] TDO$^{-/-}$ mice had significantly elevated plasma levels of tryptophan; however, concentrations of KYN and KYNA remained at physiological levels. IDO was suggested as the most likely candidate to compensate for the lack of TDO in maintaining downstream KP metabolite concentrations. In terms of the phenotype of neural progenitor cells, this study reported that there was a marked increase in the proliferation of NSCs in the subventricular zone of TDO$^{-/-}$ mice brains. This suggests that the KP is involved in regulating adult neurogenesis and possibly higher brain functions. Evidence to support the latter came from observations that TDO$^{-/-}$ mice displayed anxiety-related behavior in two classical behavior tests.

An investigation into the effect of alcohol exposure on neural development in the foetus, treated human ESCs, neural progenitors and neurons with ethanol in vitro and analysed the metabolome of human ESCs during these stages of neurogenesis.[83] Interestingly ethanol induced the significant alteration of tryptophan metabolism in human ESCs and NSCs. A number of KP metabolites were elevated upon alcohol exposure including 3-hydroxy-L-kynurenine, 5-hydroxy-L-kynurenine, L-kynurenine and indole-3-acetaldehyde. This may provide potential biomarkers to help with the diagnosis of foetal alcohol spectrum disorders (FASD) and also suggests aberrant tryptophan metabolism plays a mechanistic role in FASD.

Therapeutic Applications

This review has discussed the considerable body of evidence suggsting that the KP plays a significant role in the development and function of stem cells. The clinical application of this knowledge is in its infancy, but there is promising evidence that targeting stem cells through the manipulation of the KP is clinically beneficial. Several synthetic tryptophan catabolites (Tranilast, Teriflunomide, and Laquinimod) are in phase II and phase III clinical trials for the prevention of a number of autoimmune disorders, including multiple sclerosis.[84] Although it is thought the mechanistic action of these compounds is mediated largely through their immunosuppressive effect on T cells and NK cells, their influence on endogenous progenitor cells has not been investigated. Notably, in a recent double-blind, placebo controlled, phase III study investigating the efficacy of Teriflunomide in MS patients, treatment groups showed a significantly reduced risk of disability progression.[85] This suggests a regenerative action above and beyond a solely anti-inflammatory effect.

Two studies outlined in this review identify KP metabolites in stem cells as potential biomarkers for neurodevelopmental disorders.[42,83] In addition to their putative role as markers of disease, it is highly likely that the KP plays a mechanistic role in a number of cognitive and motor deficits associated with neurodegeneration. There is recent compelling evidence in both Drosophila[86] and mice[87] that the genetic and pharmacological inhibition of KMO reverses neurodegeneration in models of Alzheimer's disease and Huntington's disease. This effect was mediated by concomitant decrease in the neurotoxin 3-HK and increase in KYNA. Thus, use of KMO inhibitors should be explored at the clinical level. Interestingly, the mode of action of some antidepressant drugs has been linked to reducing KMO levels and stimulating KATs activity, which leads to the production of KYNA from KYN.[88] Ensuring the correct balance of neurotoxic and neuroprotective KP metabolites in future stem cell therapies must be of prime importance. This hypothesis is consistent with clinical data demonstrating a decrease in KYNA levels in depressed patients.[89]

CONCLUSION

In this review, we outlined the known links between the KP and stem cell biology. In ESCs and HSCs, there is relatively little known about the role of tryptophan metabolism in their function. However, in the case of MSCs and NSCs, there exists a rapidly growing body of evidence demonstrating a crucial link between the KP and the role of these progenitor cells in healthy and diseased tissue. The KP has significant potential for use in treating a wide range of inflammatory and degenerative diseases. The precise mechanistic role that progenitor cells play in this remains to be further elucidated. A better understanding of how KP metabolites impact stem cell populations may help to overcome some of the current barriers preventing their use in animal models and enhance future therapeutic transplantation. This review heralds an exciting time for the exploration of tryptophan metabolism as a modulator of stem cell behavior.

AUTHOR CONTRIBUTIONS

Wrote the first draft of the manuscript: SPJ. Contributed to the writing of the manuscript: GJG, BJB. Agree with manuscript results and conclusions: SPJ, GJG, BJB. Jointly developed the structure and arguments for the paper: SPJ, GJG, BJB. Made critical revisions and approved final version: SPJ, GJG, BJB. All authors reviewed and approved of the final manuscript.

DISCLOSURES AND ETHICS

As a requirement of publication the authors have provided signed confirmation of their compliance with ethical and legal obligations including but not limited to compliance with ICMJE authorship and competing interests guidelines, that the article is neither under consideration for publication nor published elsewhere, of their compliance with legal and ethical guidelines concerning human and animal research participants (if applicable), and that permission has been obtained for reproduction of any copyrighted material. This article was subject to blind, independent, expert peer review. The reviewers reported no competing interests.

REFERENCES

1. Ruddick JP, Evans AK, Nutt DJ, Lightman SL, Rook GA, Lowry CA. Tryptophan metabolism in the central nervous system: medical implications. Expert Rev Mol Med. 2006;8(20):1–27.
2. Tan L, Yu JT, Tan L. The kynurenine pathway in neurodegenerative diseases: Mechanistic and therapeutic considerations. J Neurol Sci. 2012;323(1–2):1–8.
3. Indoleamine 2,3-dioxygenase. A new vista in tryptophan metabolism. Acta Vitaminol Enzymol.1975;29(1–6):17–20.

4. Ball HJ, Sanchez-Perez A, Weiser S, et al. Characterization of an indoleamine 2,3-dioxygenase-like protein found in humans and mice. Gene. 2007;396(1):203–13.
5. Shimizu T, Nomiyama S, Hirata F, Hayaishi O. Indoleamine 2,3-dioxygenase. purification and some properties. J Biol Chem. 1978;253(13):4700–6.
6. Salter M, Pogson CI. The role of tryptophan 2,3-dioxygenase in the hormonal control of tryptophan metabolism in isolated rat liver cells. Effects of glucocorticoids and experimental diabetes. Biochem J.1985;229(2):499–504.
7. Miller CL, Llenos IC, Dulay JR, Barillo MM, Yolken RH, Weis S. Expression of the kynurenine pathway enzyme tryptophan 2,3-dioxygenase is increased in the frontal cortex of individuals with schizophrenia. Neurobiol Dis. 2004;15(3):618–29.
8. Ohira K, Hagihara H, Toyama K, et al. Expression of tryptophan 2,3-dioxygenase in mature granule cells of the adult mouse dentate gyrus. Mol Brain. 2010;3(1):26.
9. Davies NWS, Guillemin GJ, Brew BJ, Tryptophan Neurodegeneration. HIV-Associated Neurocognitive Disorder. Int J Tryptophan Res. 2010;3:1–20.
10. Guillemin GJ, Kerr SJ, Smythe GA, et al. Kynurenine pathway metabolism in human astrocytes: a paradox for neuronal protection. J Neurochem. 2001;78:1–13.
11. Guillemin GJ, Smythe G, Takikawa O, Brew BJ. Expression of indoleamine 2,3-dioxygenase and production of quinolinic acid by human microglia, astrocytes, and neurons. Glia. 2005;49(1):15–23.
12. Lim CK, Smythe GA, Stocker R, Brew BJ, Guillemin GJ. Characterization of the kynurenine pathway in human oligodendrocytes. International Congress Series. 2007;1304(1):213–7.
13. Fujigaki S, Saito K, Takemura M, et al. Species differences in L-tryptophan-kynurenine pathway metabolism: quantification of anthranilic acid and its related enzymes. Arch Biochem Biophys.1998;358(2):329–35.
14. Guillemin GJ, Smythe GA, Veas LA, Takikawa O, Brew BJ. A beta 1–42 induces production of quinolinic acid by human macrophages and microglia. Neuroreport. 2003;14(18):2311–5.
15. Takikawa O. Biochemical and medical aspects of the indoleamine 2,3-dioxygenase-initiated L-tryptophan metabolism. Biochem Biophys Res Commun. 2005;338(1):12–9.
16. Werner-Felmayer G, Werner ER, Fuchs D, Hausen A, Reibnegger G, Wachter H. Characteristics of interferon induced tryptophan metabolism in human cells in vitro. Biochim Biophys Acta.1989;1012(2):140–7.
17. Yasui H, Takai K, Yoshida R, Hayaishi O. Interferon enhances tryptophan metabolism by inducing pulmonary indoleamine 2,3-dioxygenase: its possible occurrence in cancer patients. Proc Natl Acad Sci U S A. 1986;83(17):6622–6.
18. Dai W, Gupta SL. Regulation of indoleamine 2,3-dioxygenase gene expression in human fibroblasts by interferon-gamma. Upstream control region discriminates between interferon-gamma and interferon-alpha. J Biol Chem. 1990;265(32):19871–7.
19. Metz R, Duhadaway JB, Kamasani U, Laury-Kleintop L, Muller AJ, Prendergast GC. Novel tryptophan catabolic enzyme IDO2 is the preferred biochemical target of the antitumor indoleamine 2,3-dioxygenase inhibitory compound D-1-methyl-tryptophan. Cancer Res. 2007;67(15):7082–7.
20. Guillemin GJ, Smith DG, Kerr SJ, et al. Characterisation of kynurenine pathway metabolism in human astrocytes and implications in neuropathogenesis. Redox Rep. 2000;5(2–3):108–11.
21. Guillemin GJ, Cullen KM, Lim CK, et al. Characterization of the kynurenine pathway in human neurons. J Neurosci. 2007;27(47):12884–92.
22. Guillemin GJ, Smith DG, Smythe GA, Armati PJ, Brew BJ. Expression of the kynurenine pathway enzymes in human microglia and macrophages. Adv Exp Med Biol. 2003;527:105–12.
23. Owe-Young R, Webster NL, Mukhtar M, et al. Kynurenine pathway metabolism in human blood-brain-barrier cells: implications for immune tolerance and neurotoxicity. J Neurochem. 2008;105(4):1346–57.
24. Goldstein LE, Leopold MC, Huang X, et al. 3-Hydroxykynurenine and 3-hydroxyanthranilic acid generate hydrogen peroxide and promote alpha-crystallin cross-linking by metal ion reduction. Biochemistry.2000;39(24):7266–75.
25. Stone TW, Perkins MN. Quinolinic acid: a potent endogenous excitant at amino acid receptors in CNS.Eur J Pharmacol. 1981;72(4):411–2.
26. Perkins MN, Stone TW. An iontophoretic investigation of the actions of convulsant kynurenines and their interaction with the endogenous excitant quinolinic acid. Brain Res. 1982;247(1):184–7.
27. Jhamandas K, Boegman RJ, Beninger RJ, Bialik M. Quinolinate-induced cortical cholinergic damage: modulation by tryptophan metabolites. Brain Res. 1990;529(1–2):185–91.
28. Guillemin GJ. Quinolinic acid, the inescapable neurotoxin. FEBS J. 2012;279(8):1356–65.
29. Munn DH, Shafizadeh E, Attwood JT, Bondarev I, Pashine A, Mellor AL. Inhibition of T cell proliferation by macrophage tryptophan catabolism. J Exp Med. 1999;189(9):1363–72.
30. Munn DH, Zhou M, Attwood JT, et al. Prevention of allogeneic fetal rejection by tryptophan catabolism. Science. 1998;281(5380):1191–3.
31. Terness P, Bauer TM, Rose L, et al. Inhibition of allogeneic T cell proliferation by indoleamine 2,3-dioxygenase-expressing dendritic cells: mediation of suppression by tryptophan metabolites. J Exp Med.2002;196(4):447–57.
32. Pilotte L, Larrieu P, Stroobant V, et al. Reversal of tumoral immune resistance by inhibition of tryptophan 2,3-dioxygenase. Proc Natl Acad Sci U S A. 2012;109(7):2497–502.
33. Frumento G, Rotondo R, Tonetti M, Damonte G, Benatti U, Ferrara GB. Tryptophan-derived catabolites are responsible for inhibition of T and natural

killer cell proliferation induced by indoleamine 2,3-dioxygenase. The J Exp Med. 2002;196(4):459–68.

34. Fallarino F, Grohmann U, Hwang KW, et al. Modulation of tryptophan catabolism by regulatory T cells.Nat Immunol. 2003;4(12):1206–12.

35. Opitz CA, Litzenburger UM, Sahm F, et al. An endogenous tumour-promoting ligand of the human aryl hydrocarbon receptor. Nature. 2011;478(7368):197–203.

36. Terness P, Bauer TM, Röse L, et al. Inhibition of allogeneic T cell proliferation by indoleamine 2,3-dioxygenase-expressing dendritic cells: mediation of suppression by tryptophan metabolites. J Exp Med.2002;196(4):447–57.

37. Moffett JR, Els T, Espey MG, Walter SA, Streit WJ, Namboodiri MA. Quinolinate immunoreactivity in experimental rat brain tumors is present in macrophages but not in astrocytes. Exp Neurol. 1997;144(2):287–301.

38. Vécsei L, Szalárdy L, Fülöp F, Toldi J. Kynurenines in the CNS: recent advances and new questions.Nat Rev Drug Discov. 2012;12(1):64–82.

39. Zádori D, Klivényi P, Vámos E, Fülöp F, Toldi J, Vécsei L. Kynurenines in chronic neurodegenerative disorders: future therapeutic strategies. J Neural Transm. 2009;116(11):1403–9.

40. Stone TW, Stoy N, Darlington LG. An expanding range of targets for kynurenine metabolites of tryptophan. Trends Pharmacol Sci. 2013;34(2):136–43.

41. Stone TW, Mackay GM, Forrest CM, Clark CJ, Darlington LG. Tryptophan metabolites and brain disorders. Clin Chem Lab Med. 2003;41(7):852–9.

42. Cezar GG, Quam JA, Smith AM, et al. Identification of small molecules from human embryonic stem cells using metabolomics. Stem Cells Dev. 2007;16(6):869–82.

43. Hwu P, Du MX, Lapointe R, Do M, Taylor MW, Young HA. Indoleamine 2,3-dioxygenase production by human dendritic cells results in the inhibition of T cell proliferation. J Immunol. 2000;164(7):3596–9.

44. Munn DH, Shafizadeh E, Attwood JT, Bondarev I, Pashine A, Mellor AL. Inhibition of T cell proliferation by macrophage tryptophan catabolism. J Exp Med. 1999;189(9):1363–72.

45. Hainz U, Obexer P, Winkler C, et al. Monocyte-mediated T-cell suppression and augmented monocyte tryptophan catabolism after human hematopoietic stem-cell transplantation. Blood. 2005;105(10):4127–34.

46. Kurz K, Gluhcheva Y, Zvetkova E, Konwalinka G, Fuchs D. Interferon-gamma-mediated pathways are induced in human CD34(+) haematopoietic stem cells. Immunobiology. 2010;215(6):452–7.

47. Pittenger MF, Mackay AM, Beck SC, et al. Multilineage potential of adult human mesenchymal stem cells. Science. 1999;284(5411):143–7.

48. Dazzi F, Ramasamy R, Glennie S, Jones SP, Roberts I. The role of mesenchymal stem cells in haemopoiesis. Blood Rev. 2006;20(3):161–71.

49. Campagnoli C, Roberts IA, Kumar S, Bennett PR, Bellantuono I, Fisk NM. Identification of mesenchymal stem/progenitor cells in human first-trimester fetal blood, liver, and bone marrow. Blood.2001;98(8):2396–402.

50. Sakaguchi Y, Sekiya I, Yagishita K, Muneta T. Comparison of human stem cells derived from various mesenchymal tissues: superiority of synovium as a cell source. Arthritis Rheum. 2005;52(8):2521–9.

51. Zvaifler NJ, Marinova-Mutafchieva L, Adams G, et al. Mesenchymal precursor cells in the blood of normal individuals. Arthritis Res. 2000;2(6):477–88.

52. da Silva Meirelles L, Chagastelles PC, Nardi NB. Mesenchymal stem cells reside in virtually all post-natal organs and tissues. J Cell Sci. 2006;119(Pt 11):2204–13.

53. Dominici M, Le Blanc K, Mueller I, et al. Minimal criteria for defining multipotent mesenchymal stromal cells. The International Society for Cellular Therapy position statement. Cytotherapy. 2006;8(4):315–7.

54. Reyes M, Lund T, Lenvik T, Aguiar D, Koodie L, Verfaillie CM. Purification and ex vivo expansion of postnatal human marrow mesodermal progenitor cells. Blood. 2001;98(9):2615–25.

55. Woodbury D, Schwarz EJ, Prockop DJ, Black IB. Adult rat and human bone marrow stromal cells differentiate into neurons. J Neurosci Res. 2000;61(4):364–70.

56. Ren G, Chen X, Dong F, et al. Concise review: mesenchymal stem cells and translational medicine: emerging issues. Stem Cells Transl Med. 2012;1(1):51–8.

57. Bartholomew A, Sturgeon C, Siatskas M, et al. Mesenchymal stem cells suppress lymphocyte proliferation in vitro and prolong skin graft survival in vivo. Exp Hematol. 2002;30(1):42–8.

58. Di Nicola M, Carlo-Stella C, Magni M, et al. Human bone marrow stromal cells suppress T-lymphocyte proliferation induced by cellular or nonspecific mitogenic stimuli. Blood. 2002;99(10):3838–43.

59. Krampera M, Glennie S, Dyson J, et al. Bone marrow mesenchymal stem cells inhibit the response of naive and memory antigen-specific T cells to their cognate peptide. Blood. 2003;101(9):3722–9.

60. Glennie S, Soeiro I, Dyson PJ, Lam EW, Dazzi F. Bone marrow mesenchymal stem cells induce division arrest anergy of activated T cells. Blood. 2005;105(7):2821–7.

61. Jones S, Horwood N, Cope A, Dazzi F. The antiproliferative effect of mesenchymal stem cells is a fundamental property shared by all stromal cells. J Immunol. 2007;179(5):2824–31.

62. Corcione A, Benvenuto F, Ferretti E, et al. Human mesenchymal stem cells modulate B-cell functions. Blood. 2006;107(1):367–72.

63. Ramasamy R, Fazekasova H, Lam EW, Soeiro I, Lombardi G, Dazzi F. Mesenchymal stem cells inhibit dendritic cell differentiation and function by preventing entry into the cell cycle. Transplantation.2007;83(1):71–6.

64. Sotiropoulou PA, Perez SA, Gritzapis AD, Baxevanis CN, Papamichail M. Interactions between human mesenchymal stem cells and natural killer cells. Stem Cells. 2006;24(1):74–85.

65. Le Blanc K, Mougiakakos D. Multipotent mesenchymal

stromal cells and the innate immune system. Nat Rev Immunol. 2012;12(5):383–96.

66. Chabannes D, Hill M, Merieau E, et al. A role for heme oxygenase-1 in the immunosuppressive effect of adult rat and human mesenchymal stem cells. Blood. 2007;110(10):3691–4.

67. Aggarwal S, Pittenger MF. Human mesenchymal stem cells modulate allogeneic immune cell responses. Blood. 2005;105(4):1815–22.

68. Sato K, Ozaki K, Oh I, et al. Nitric oxide plays a critical role in suppression of T-cell proliferation by mesenchymal stem cells. Blood. 2007;109(1):228–34.

69. Meisel R, Zibert A, Laryea M, Göbel U, Däubener W, Dilloo D. Human bone marrow stromal cells inhibit allogeneic T-cell responses by indoleamine 2,3-dioxygenase-mediated tryptophan degradation. Blood.2004;103(12):4619–21.

70. Krampera M, Cosmi L, Angeli R, et al. Role for interferon-gamma in the immunomodulatory activity of human bone marrow mesenchymal stem cells. Stem Cells. 2006;24(2):386–98.

71. Ryan JM, Barry F, Murphy JM, Mahon BP. Interferon-gamma does not break, but promotes the immunosuppressivecapacityofadulthumanmesenchymal stem cells. Clin Exp Immunol. 2007;149(2):353–63.

72. Koch CA, Geraldes P, Platt JL. Immunosuppression by embryonic stem cells. Stem Cells.2008;26(1):89–98.

73. Han KH, Ro H, Hong JH, et al. Immunosuppressive mechanisms of embryonic stem cells and mesenchymal stem cells in alloimmune response. Transpl Immunol. 2011;25(1):7–15.

74. Matysiak M, Stasiolek M, Orlowski W, et al. Stem cells ameliorate EAE via an indoleamine 2,3-dioxygenase (IDO) mechanism. J Neuroimmunol. 2008;193(1–2):12–23.

75. Spaggiari GM, Capobianco A, Abdelrazik H, Becchetti F, Mingari MC, Moretta L. Mesenchymal stem cells inhibit natural killer-cell proliferation, cytotoxicity, and cytokine production: role of indoleamine 2,3-dioxygenase and prostaglandin E2. Blood. 2008;111(3):1327–33.

76. Braun D, Longman RS, Albert ML. A two-step induction of indoleamine 2,3 dioxygenase (IDO) activity during dendritic-cell maturation. Blood. 2005;106(7):2375–81.

77. Gieseke F, Schutt B, Viebahn S, et al. Human multipotent mesenchymal stromal cells inhibit proliferation of PBMCs independently of IFN R1 signaling and IDO expression. Blood. 2007;110(6):2197–200.

78. Lanz TV, Opitz CA, Ho PP, et al. Mouse mesenchymal stem cells suppress antigen-specific TH cell immunity independent of indoleamine 2,3-dioxygenase 1 (IDO1) Stem Cells Dev. 2010;19(5):657–68.

79. Croitoru-Lamoury J, Lamoury FM, Caristo M, et al. Interferon-gamma regulates the proliferation and differentiation of mesenchymal stem cells via activation of indoleamine 2,3 dioxygenase (IDO) PloS ONE.2011;6(2):e14698.

80. Chen S, Corteling R, Stevanato L, Sinden J. Natural inhibitors of indoleamine 3,5-dioxygenase induced by interferon-gamma in human neural stem cells. Biochem Biophys Res Commun. 2012;429(1–2):117–23.

81. Zunszain PA, Anacker C, Cattaneo A, et al. Interleukin-1beta: a new regulator of the kynurenine pathway affecting human hippocampal neurogenesis. Neuro Psycho Pharmacology. 2012;37(4):939–49.

82. Kanai M, Funakoshi H, Takahashi H, et al. Tryptophan 2,3-dioxygenase is a key modulator of physiological neurogenesis and anxiety-related behavior in mice. Mol Brain. 2009;2:8.

83. Palmer JA, Poenitzsch AM, Smith SM, Conard KR, West PR, Cezar GG. Metabolic biomarkers of prenatal alcohol exposure in human embryonic stem cell-derived neural lineages. Alcohol Clin Exp Res.2012;36(8):1314–24.

84. Platten M, Ho PP, Steinman L. Anti-inflammatory strategies for the treatment of multiple sclerosis—tryptophan catabolites may hold the key. Drug Discovery Today: Therapeutic Strategies. 2006;3(3):401–8.

85. O'Connor P, Wolinsky JS, Confavreux C, et al. Randomized trial of oral teriflunomide for relapsing multiple sclerosis. N Engl J Med. 2011;365(14):1293–303.

86. Campesan S, Green EW, Breda C, et al. The kynurenine pathway modulates neurodegeneration in a Drosophila model of Huntington's disease. Curr Biol. 2011;21(11):961–6.

87. Zwilling D, Huang SY, Sathyasaikumar KV, et al. Kynurenine 3-monooxygenase inhibition in blood ameliorates neurodegeneration. Cell. 2011;145(6):863–74.

88. Kocki T, Wnuk S, Kloc R, Kocki J, Owe-Larsson B, Urbanska EM. New insight into the antidepressants action: modulation of kynurenine pathway by increasing the kynurenic acid/3-hydroxykynurenine ratio. J Neural Transm. 2012;119(2):235–43.

89. Myint AM, Kim YK, Verkerk R, Scharpe S, Steinbusch H, Leonard B. Kynurenine pathway in major depression: evidence of impaired neuroprotection. J Affect Disord. 2007;98(1–2):143–51.

CHAPTER 11 REPROGRAMMABLE CRISPR/CAS9-BASED SYSTEM FOR INDUCING SITE-SPECIFIC DNA METHYLATION

James I. McDonald[1], Hamza Celik[2], Lisa E. Rois[1], Gregory Fishberger[3], Tolison Fowler[1], Ryan Rees[3], Ashley Kramer[2], Andrew Martens[2], John R. Edwards[1], and Grant A. Challen[2,4]

[1] Center for Pharmacogenomics, Department of Medicine, Washington University in St. Louis School of Medicine, St. Louis, MO, USA

[2] Section of Stem Cell Biology, Division of Oncology, Department of Medicine, Washington University in St. Louis School of Medicine, St. Louis, MO 63110, USA

[3] College of Arts and Science, Washington University in St. Louis, St. Louis, MO 63130, USA

[4] Developmental, Regenerative and Stem Cell Biology Program, Division of Biology and Biomedical Sciences, Washington University in St. Louis School of Medicine, St. Louis, MO 63110, USA

ABSTRACT

Advances in sequencing technology allow researchers to map genome-wide changes in DNA methylation in development and disease. However, there is a lack of experimental tools to site-specifically manipulate DNA methylation to discern the functional consequences. We developed a CRISPR/Cas9 DNA methyltransferase 3A (DNMT3A) fusion to induce DNA methylation at specific loci in the genome. We induced DNA methylation at up to 50% of alleles for targeted CpG dinucleotides. DNA methylation levels peaked within 50 bp of the short guide RNA (sgRNA) binding site and between pairs of sgRNAs. We used our approach to target methylation across the entire CpG island at the *CDKN2A*promoter, three CpG dinucleotides at the *ARF* promoter, and the CpG island within the *Cdkn1a*promoter to decrease expression of the target gene. These tools permit mechanistic studies of DNA methylation and its role in guiding molecular processes that determine cellular fate.

INTRODUCTION

DNA methylation of CpG dinucleotides is a prominent epigenetic modification of the mammalian genome that can influence gene expression, and aberrant distribution of DNA methylation is associated with a spectrum of human disorders including cancers (Egger et al., 2004). Despite intensive study, it remains unclear which CpG dinucleotides must change methylation state in order to alter transcription. Genome-wide analyses have found associations between DNA methylation and reduced gene expression that occur both in the proximal promoter and downstream of the gene's transcription start site (TSS) (Bell et al., 2011; Bock et al., 2012; Lou et al., 2014; VanderKraats et al., 2013; Lund et al., 2014). However, evidence supports both that DNA methylation can cause a loss of expression, and that expression changes can alter DNA methylation patterns (Bestor et al., 2015; Busslinger et al., 1983). Here, we sought to develop tools for locus-specific epigenetic remodeling to directly address the role of DNA methylation in regulating gene expression.

Targeted DNA methylation approaches have been attempted by fusing DNA methyltransferase enzymes (DNMTs) to DNA-binding proteins such as zinc finger proteins (ZFPs) (Siddique et al., 2013), and transcriptional activator-like effector (TALE) (Bernstein et al., 2015). However, engineering

Citation: James I. McDonald, Hamza Celik, Lisa E. Rois, Gregory Fishberger, Tolison Fowler, Ryan Rees, Ashley Kramer,Andrew Martens, John R. Edwards, Grant A. Challen, Reprogrammable CRISPR/Cas9-based system for inducing site-specific DNA methylation, doi: 10.1242/bio.019067.

custom proteins for each targeted sequence is laborious and requires specialized expertise. Moreover, in these studies, induced DNA methylation of the targeted loci was relatively poor, with substantial off-target activity. An engineered form of the clustered, regularly interspaced, short palindromic repeat (CRISPR) system has emerged as an alternative for achieving site-specific DNA targeting (Jinek et al., 2012). Here, the Cas9 endonuclease is directed to genomic targets by engineered short guide RNAs (sgRNAs) (Jinek et al., 2012). Because the sgRNA is the DNA sequence-specific component of the system, it allows for efficient targeting of multiple regions due to the ease of design and synthesis of new sgRNAs (relative to engineering new custom proteins for each target site). A Cas9 mutant (D10A and H840A; henceforth referred to as dCas9) that lacks endonuclease activity but can still be recruited by sgRNA(s) (Jinek et al., 2012) has recently been used to target genes in mammalian cells for transcriptional activation (Perez-Pinera et al., 2013a,b; Maeder et al., 2013; Mali et al., 2013). Here, we demonstrate an easily reprogrammable CRISPR/dCas9 DNMT fusion capable of inducing site-specific DNA methylation.

RESULTS

To design a flexible system to target DNA methylation, we fused dCas9 to the catalytic domain of the *de novo* DNA methyltransferase DNMT3A (Fig. 1A). To test this system we targeted DNA methylation to the tumor suppressor gene *CDKN2A* (cyclin dependent kinase 2A), which inhibits progression through the cell cycle (Liggett and Sidransky, 1998). *CDKN2A* is one of the most frequently hypermethylated genes in The Cancer Genome Atlas (Ciriello et al., 2013), and numerous clinical studies show a negative correlation between *CDKN2A* methylation and expression in colorectal cancer (Shima et al., 2011). While it is generally assumed that *CDKN2A* methylation induces gene silencing, it has also been suggested that DNA methylation occurs after the loss of expression (Hinshelwood et al., 2009). From a literature search, we identified 17 publications that associate *CDKN2A* methylation with expression and/or cancer (Fig. S1.1). Overwhelmingly, these papers studied the differentially methylated region (cancer DMR, cDMR) on the 3′ end of the CpG island that overlapped the first exon of *CDKN2A* (Fig. 1B).

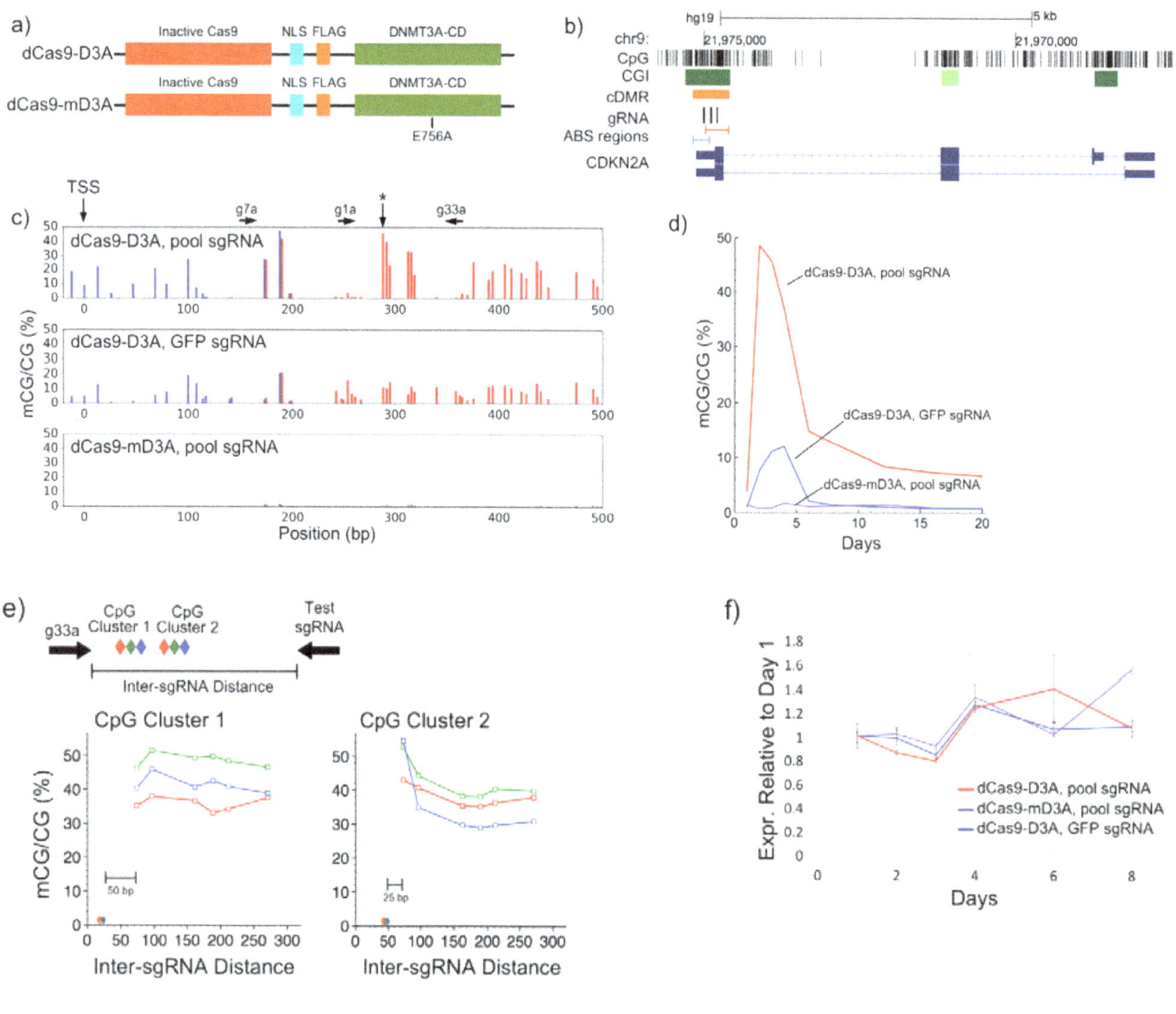

Figure 1: Site-specific induction of DNA methylation using a CRISPR-Cas9 DNMT fusion. (A) dCas9-DNMT3A CD fusion constructs. The E756A mutation inactivates the DNMT3A-CD. NLS, nuclear localization signal; FLAG, FLAG tag domain. (B) UCSC genome browser view showing the locations of the three *CDKN2A* sgRNA (g1a, g7a and g33a). The three sgRNA were validated to ensure they targeted this locus (Fig. S1.2). cDMR indicates the region from the literature where methylation changes are associated with expression changes (Fig. S1.1). (C) Induced DNA methylation at the *CDKN2A* promoter three days post-transfection. Colors correspond to the red and blue ABS regions in B.

Three CpGs were independently measured in both amplicons. sgRNA target sites are indicated above the graphs. Pool sgRNA indicates g1a, g7a, g33a were used simultaneously. Sanger sequencing validation is presented in Fig. S1.3, and non-CpG methylation data is presented inFig. S1.4. (D) Time course of the percent methylation data for the CpG marked with an asterisk in C. Additional CpGs are shown in Fig. S1.5. (E) Methylation induced by a pair of sgRNA decreases with increasing intervening distance. Distance is calculated relative to the 3′ end of the g33a sgRNA. Diamonds indicate the location each CpG monitored for methylation; whose color corresponds to appropriate line in the graph. Additional data from individual and paired sgRNA is presented inFig. S1.6. (F) *CDKN2A* expression for samples with induced methylation. Expression is normalized to day one for each respective sample. Error bars=mean±s.e.m. (n=1 performed in technical triplicate).

We computationally designed three sgRNAs (g1a, g7a and g33a) to target this region and test whether DNA methylation was sufficient to induce gene silencing. We validated the ability of each sgRNA to target the *CDKN2A* locus by transfecting them with active CRISPR/Cas9 and measuring the ability of Cas9 to cleave the locus (Fig. S1.2, Table S1). We then transfected HEK293T cells with the pool of three sgRNAs along with either a normal dCas9-DNMT3A catalytic domain (CD) fusion (dCas9-D3A) or one with a DNMT3A^{E756A} mutation (dCas9-mD3A, Fig. 1A), which abolishes DNA methyltransferase activity (Reither et al., 2003). Transfection efficiencies were >80-90% for all experiments as measured by co-transfection with GFP-containing plasmids. We analyzed DNA methylation levels for 20 days post-transfection using Illumina sequencing of two amplicon regions (amplicon bisulfite sequencing, ABS). ABS results were validated using Sanger bisulfite sequencing (R^2=0.83; Fig. S1.3), and DNA methylation levels at CpGs analyzed in both of two independent ABS amplicons showed strong correspondence (R^2=0.98; Fig. 1C). All CpGs had >100× sequencing coverage, with a median coverage of 15,200.

Over the 20-day time course we observed an increase in DNA methylation at the *CDKN2A* target locus that ranged from 20-43% at its peak on day three. Background methylation from transfection with dCas9-mD3A was consistently less than 1.5%, while background methylation from an off-target sgRNA, which controls for DNMT3A-CD overexpression, was less than 14% on day three. Induced DNA methylation levels were highest over a set of eight CpGs directly between the g33a and g7a sgRNA target sites (Fig. 1C). Increases in CHG and CHH methylation were minimal (Fig. S1.4). DNA methylation decreased rapidly after passaging the cells on day four, but stabilized 20 days post-transfection at 6-10% (Fig. 1D; Fig. S1.5). Despite the literature support for a negative correlation between expression and DNA methylation in this region, we did not observe a measurable effect on *CDKN2A* gene expression by RT-qPCR (Fig. 1F). This suggests that a limited increase in methylation in the region 100-400 bp downstream of the *CDKN2A* TSS is insufficient to trigger gene silencing.

Spatially, the induced DNA methylation spiked near the sgRNA target sites and dropped quickly toward background levels at surrounding CpG sites. Analysis of DNA methylation induced by single sgRNAs indicates that methylation occurs primarily within 50 bp of the sgRNA binding site (Fig. S1.6). Higher DNA methylation levels were often observed 3′ of the sgRNA binding site (Fig. S1.6). Our initial data from the three pooled sgRNAs suggested that CpG methylation was higher between pairs of sgRNAs. To investigate this effect, we transfected pairs of sgRNA with varying intervening distances and monitored methylation of six clustered CpGs between the sgRNA pairs using ABS (Fig. 1E). DNA methylation of the three CpGs (cluster 1) within 20 bp of the fixed sgRNA (g33a) did not change with addition of a second sgRNA 77 bp or further away. However, the methylation level increased from 30-40% to 42-53% when sgRNAs were paired within 80 bp and both sgRNAs were within 50 bp of CpG cluster 2 (Fig. 1E). This suggests the DNMT3A-CD activity at the target locus is additive.

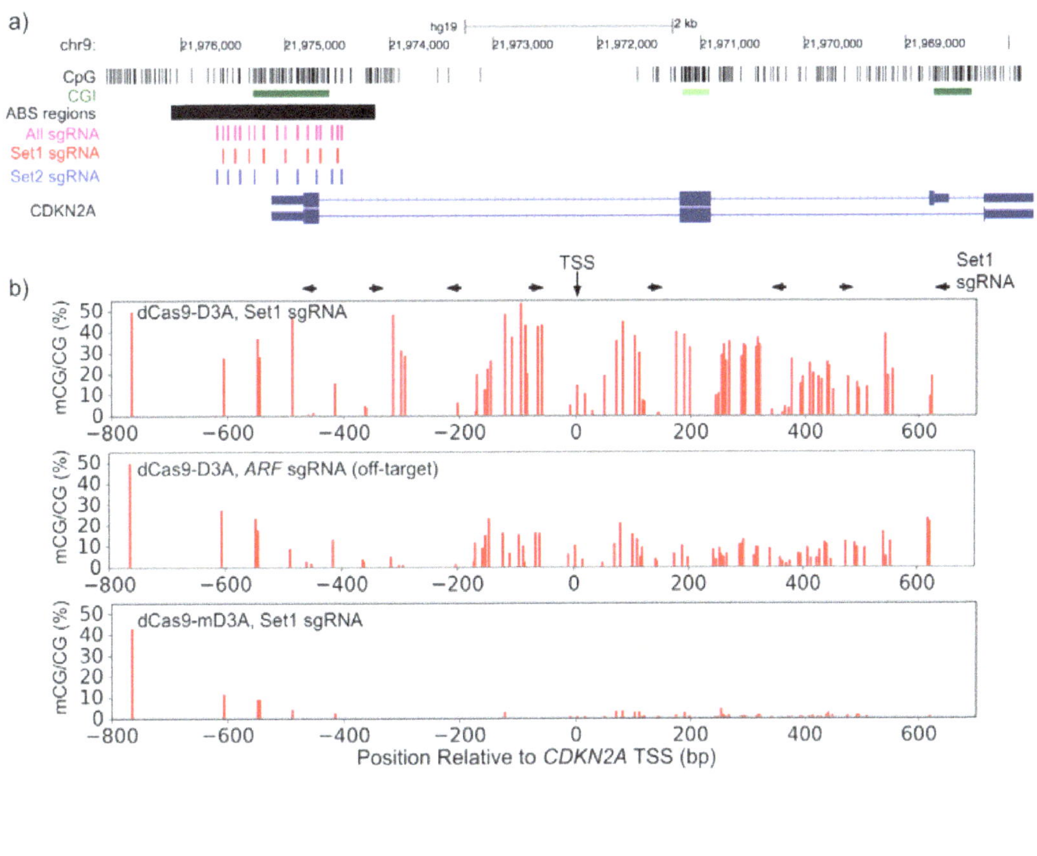

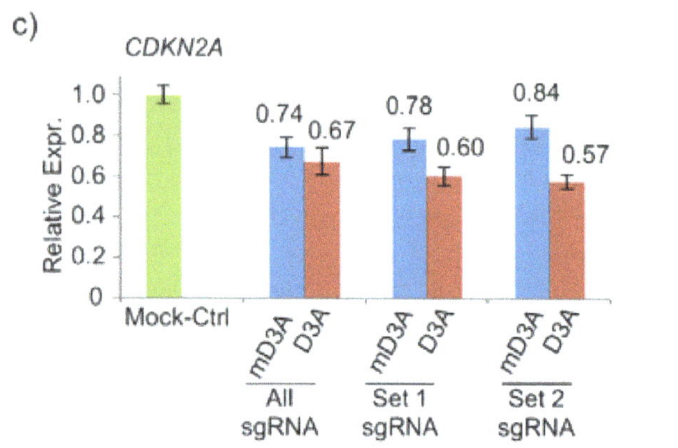

Figure 2: DNA methylation decreases gene expression at the *CDKN2A* **promoter in a context dependent manner.** (A) Locations of the 17 sgRNAs used in Sets 1 and 2 and regions sequenced in B. sgRNA coordinates are in Table S1. (B) Percent methylation is plotted for regions designated in A. Set 1 sgRNA target sites are indicated above the graphs. Data for All sgRNA, Set 2 sgRNA, and off-target SgRNA is presented in Fig. S2. (C) Methylation induced by *CDKN2A*-targeted sgRNA Set 1, Set 2, and All sgRNA, decreases gene expression. Relative expression of *CDKN2A* is normalized to a mock-treated control. Error bars=mean±s.e.m. (All, n=2; Set 1, n=1; Set 2, n=1; all performed in technical triplicate, paired t-test).

We next tested whether we could use our approach to methylate an entire CpG-island (CGI). We designed 17 sgRNAs (Fig. 2A; Table S1) to target DNA methylation across the *CDKN2A* CGI, which spans the TSS. We applied three combinations of sgRNAs (Set 1, 2, All) to test whether inducing DNA methylation of the entire CGI could decrease gene expression (Fig. 2A). ABS analysis of eight amplicons (minimum per CpG sequencing depth of 100) showed that the DNA methylation level increased to an average of 22% across the entire region with a peak of 54% (Fig. 2B). As an off-target negative control, we used three sgRNAs targeted to the *ARF* promoter located ~20 kb away. The average background methylation at *CDKN2A* after treatment with off-target *ARF* sgRNAs was 9% (Fig. 2B). The other two sgRNA sets (Set 2 and All) induced similar increases of methylation across the CGI overlapping the TSS of *CDKN2A* (Fig. S2).

Analysis of *CDKN2A* expression by RT-qPCR in all three sgRNA targeting experiments (Set 1, 2, All) indicated an average 39% decrease in *CDKN2A* mRNA expression after targeting with dCas9-D3A (Fig. 2C). Cells transfected with dCas9-mD3A showed a 16-26% reduction in *CDKN2A* expression, likely due to CRISPR inhibition (Fig. 2C). Across the three replicates (Set 1, 2, All) expression decreased by an average of 17% in dCas9-D3A relative to dCas9-mD3A (P<0.01 paired-t-test). This indicated that DNA methylation directly decreased *CDKN2A* expression, but targeting of the entire CGI was required to trigger this effect. Our results are consistent with other studies that find a similar reduction in gene expression after inducing methylation at the *CDKN2A* promoter using ZFP- and TALE-based systems (Bernstein et al., 2015; Cui et al., 2015).

To verify the effect of our system with a separate locus, we designed three inwardly-directed (5′ to 3′) sgRNAs that bracketed three CpG sites in the *ARF* promoter located 150-170 bp downstream of the TSS (Fig. 3A). *ARF*-targeted sgRNAs increased the DNA methylation level to 27-30% at these three CpG sites with less than 15% methylation induced in adjacent sites (Fig. 3B). Induced methylation of the*ARF* promoter was associated with a 19% decrease in its expression (Fig. 3C).

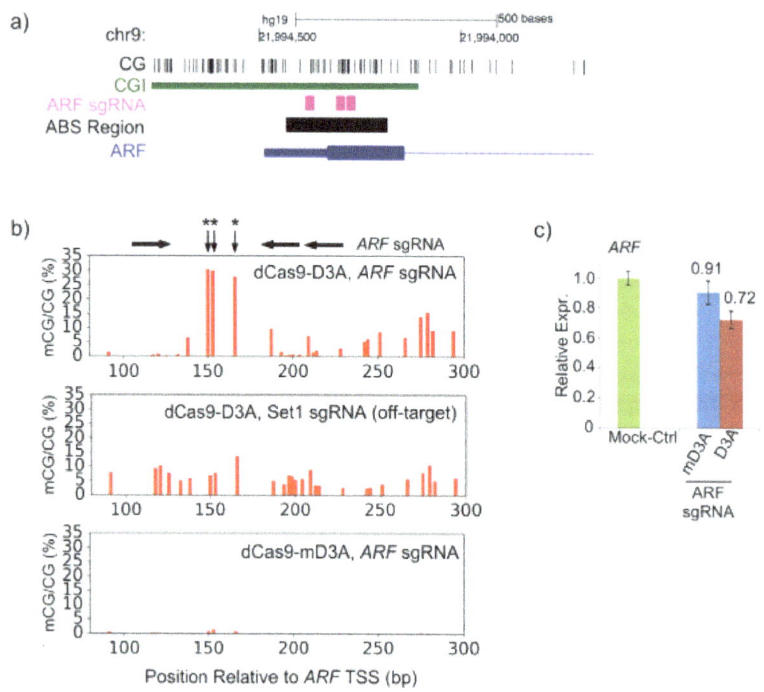

Figure 3: Induced methylation decreases *ARF* **expression.** (A) Locations of the three sgRNA used to induce DNA methylation at the*ARF* promoter. (B) Percent methylation is plotted for a region of exon 1 of the *ARF* locus. *ARF* sgRNA target sites and three targeted CpGs (asterisks) are indicated above. (C) Methylation induced by *ARF* targeted sgRNA decreases gene expression. Relative expression of *ARF* is normalized to a mock-treated control. Error bars=mean±s.e.m. (n=1 biological replicate performed in technical triplicate).

To further validate our design rules, we attempted to induce DNA methylation in a different system in a regulated fashion with a minimalistic combination of sgRNAs. dCas9-D3A and dCas9-mD3A were cloned into a doxycycline-inducible lentivirus where expression of the fusion protein is linked to mCherry via internal ribosome entry site (IRES), and targeted to the *Cdkn1a* locus with two rationally-designed sgRNAs. We designed two sgRNAs targeting the CpG island located in the *Cdkn1a* promoter **(Fig. 4A)** and cloned them into lentiviral vectors (in which sgRNAs and GFP are expressed constitutively from two different promoters) (Fig. S4.1) and transduced 32D cells (mouse IL3-dependent myeloid progenitor cell line with an unmethylated *Cdkn1a* promoter and active *Cdkn1a* expression). Additional negative controls were generated by transducing 32D cells with sgRNAs only, dCas9-D3A only, or sgRNAs plus dCas9-mD3A. Cells were then FACS-sorted by GFP and/or mCherry to establish stable cell lines. Analysis of the stable cell lines induced by doxycycline showed successful expression of dCas9-fusions and sgRNAs via mCherry and GFP signal (Fig. S4.2). In the absence of doxycycline, the expression of fusion constructs was silenced (Fig. S4.2).

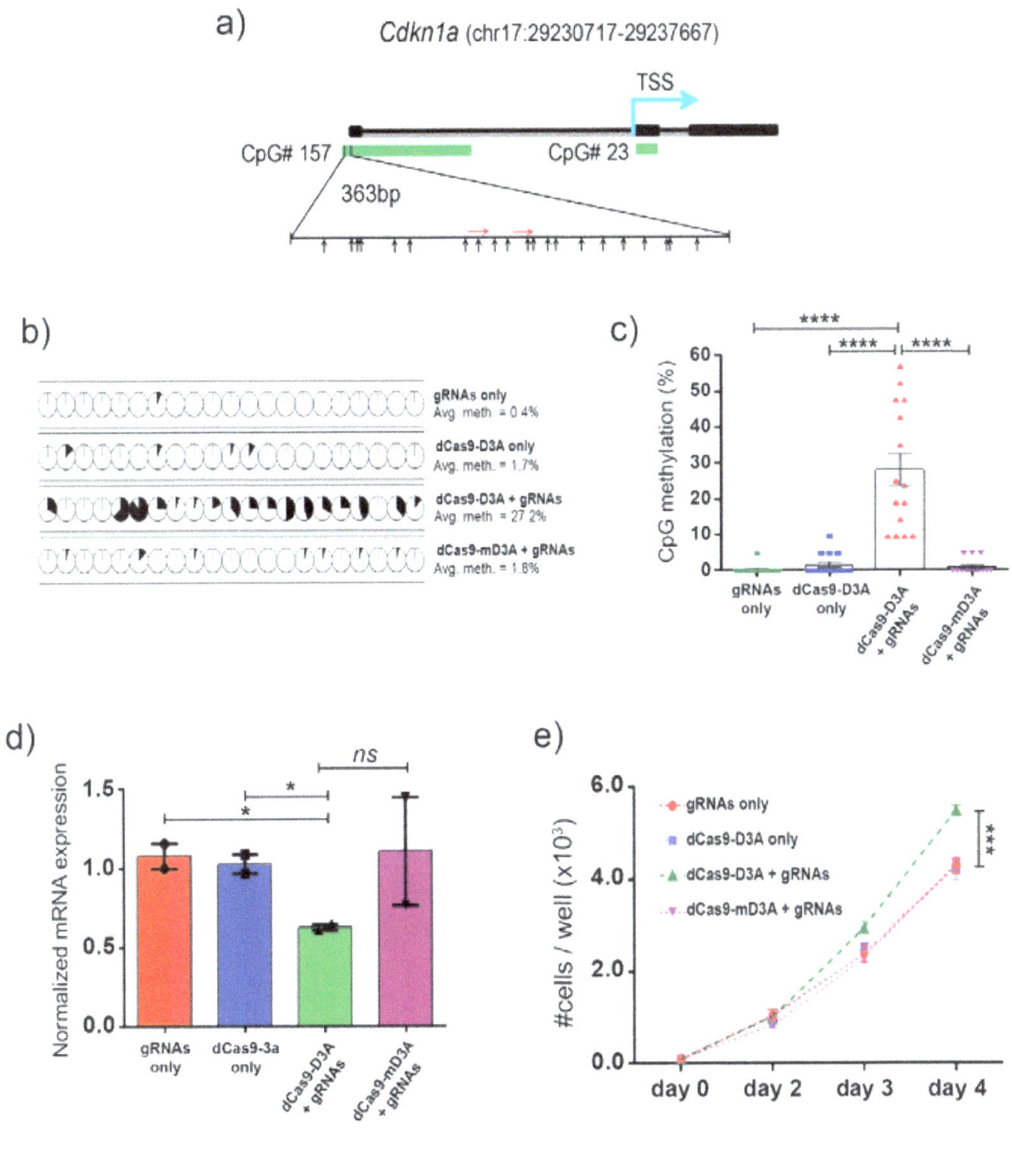

Figure 4: Targeted CpG methylation of the *Cdkn1a* **locus using dCas9-D3A fusion protein.** (A) Organization of the mouse *Cdkn1a*target locus with CpG islands indicated with green boxes. Blue arrow indicates the location of transcriptional start site (TSS). Red arrows indicate the relative position of sgRNA binding sites (5′ to 3′) within the 5′ CpG island. Black arrows indicate the relative positions of CpGs. (B) Bisulfite sequencing analysis of *Cdkn1a* promoter in 32D cells transduced with sgRNAs or dCas9-D3A only, and dCas9a-D3A or dCas9-mD3A with sgRNAs. Changes in methylation of specific CpGs are indicated by pie charts. Shading indicates the average amount of DNA methylation at each CpG (n=14-15 biological replicates representative of three independent experiments). (C) Average DNA methylation level of target locus within*Cdkn1a* promoter (unpaired t-test). (D) mRNA expression level of *Cdkn1a* (normalized to 18S) in transduced 32D cells (n=2 biological replicates performed in technical duplicate; unpaired t-test). (E) Growth kinetics of transduced 32D cells (n=2 biological replicates performed in technical quadruplicate; unpaired t-test). *$P<0.05$, ****$P<0.0001$, error bars=mean±s.e.m.

Stable cells were induced with doxycycline for eight days. *Cdkn1a* promoter DNA methylation and gene expression were then evaluated by bisulfite sequencing and RT-qPCR respectively. Cells transduced with dCas9-D3A together with sgRNAs showed an increase in *Cdkn1a* promoter DNA methylation by over 25% across the entire target region (and by as much as 40-85%

at individual CpGs around sgRNAs) compared to all control groups (Fig. 4B,C), and a reduction in *Cdkn1a* expression by approximately 40% (Fig. 4D). There was no difference in *Cdkn1a* promoter DNA methylation or expression between any control groups (Fig. 4B-D). Cells transduced with dCas9-D3A together with sgRNAs showed a growth advantage, consistent with an increase in proliferation due to repression of a cell cycle inhibitor (Fig. 4E). As with targeting of the human *CDKN2A* locus, these data suggested that induced DNA methylation pioneered by dCas9-D3A targeted with a minimal combination of sgRNAs can influence gene expression and generate subsequent downstream functional phenotypes.

We reverse cloned viral integrants to identify clones that contained either one of the two guides, or both sgRNAs together with dCas9-D3A. No induced promoter DNA methylation or reduction in gene expression was observed in clones expressing only a single sgRNA (Fig. 5A,B). In contrast, *Cdkn1a*promoter methylation was increased and its expression decreased in clones that expressed both sgRNAs (Fig. 5A,B). To confirm the presence of both sgRNAs were required to execute induced DNA methylation, clones that contained only a single sgRNA were transduced with their complementary sgRNA. Following this, after eight-days of induction, *Cdkn1a* DNA methylation increased across the promoter by up to 33% compared to the parental clone that contained only a single sgRNA (Fig. 5C).*Cdkn1a* expression was reduced by 40-50% in the complemented clones (Fig. 5D). These data indicate that targeting with multiple sgRNAs results in a synergism and induces robust DNA methylation. We also examined the stability of the induced DNA methylation by withdrawing doxycycline for eight days.*Cdkn1a* promoter DNA methylation and expression profiles remained largely unchanged (Fig. 5E,F) even though expression of dCas9-D3A was silenced (Fig. S4.2).

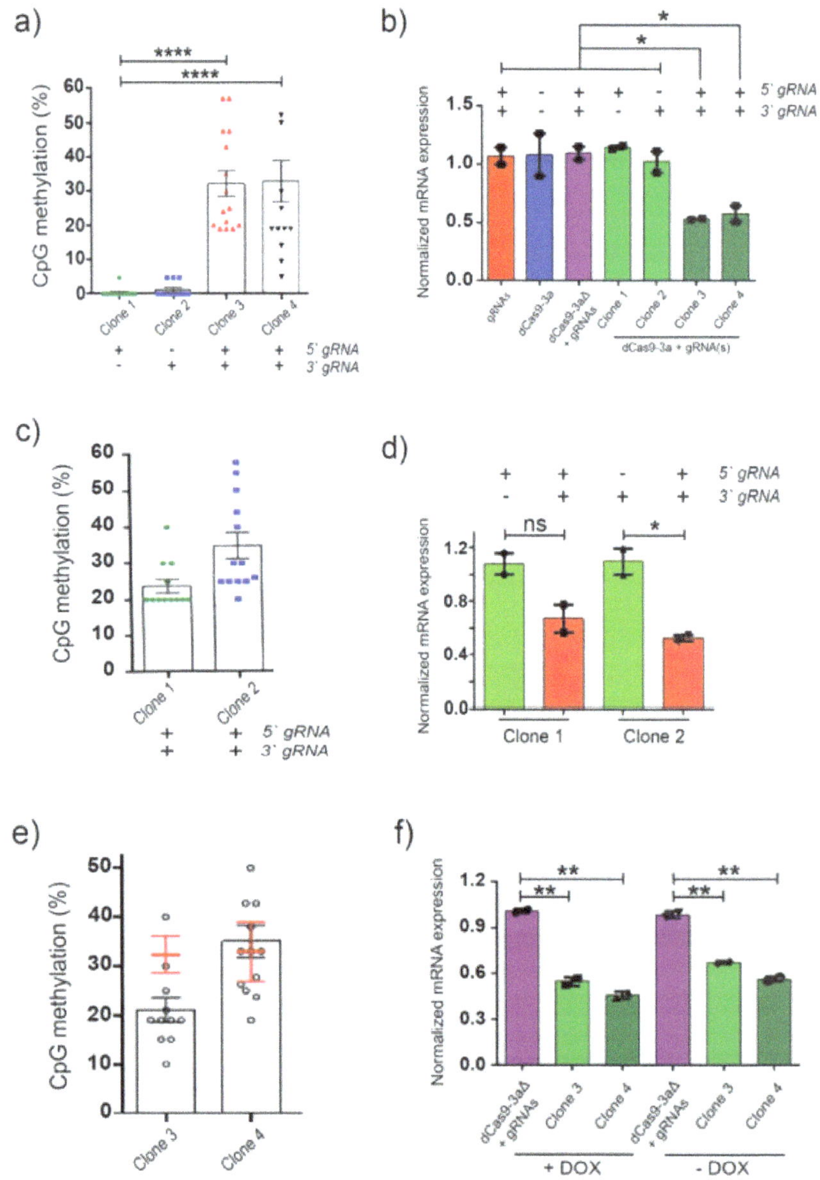

Figure 5: Multiple sgRNAs are required for the efficient establishment of targeted methylation. (A) Average DNA methylation level of target locus within *Cdkn1a* promoter (n=12-15 biological replicates representative of two independent experiments; unpaired t-test). Presence or absence of each sgRNA in each clone is shown (−; +) at the bottom of the graph. (B) Expression level of *Cdkn1a* mRNA in 32D cells transduced with sgRNAs, dCas9-D3A, dCas9-mD3A with sgRNAs or dCas9-D3A with either one or both sgRNAs (n=2 biological replicates performed in technical duplicate; unpaired t-test). Presence or absence of each sgRNA in each clone is shown (−; +). (C) Average promoter DNA meth-

ylation (n=12-15 biological replicates representative of two independent experiments) or (D) expression level of *Cdkn1a* in 32D cells in 32D cells with one sgRNA which were then transduced with their complementary sgRNA (n=2 biological replicates performed in technical duplicate; unpaired t-test). (E) Average DNA methylation of *Cdkn1a* target locus for clones following doxycycline withdrawal for eight days. Clones were previously induced for 12 days, with starting induced DNA methylation level shown by red error bars (n=12-16 biological replicates representative of two independent experiments). (F) Expression level of*Cdkn1a* in clones after eight days of de-induction (n=2 biological replicates performed in technical duplicate; unpaired t-test). *$P<0.05$; **$P<0.01$; ****$P<0.0001$, error bars=mean±s.e.m.

To assess non-specific activity in this system, we initially attempted to analyze off-target genomic areas with high homology to predicted binding regions of both *Cdkn1a* sgRNAs. However, no regions with high homology were found as these sgRNAs were designed based on their minimal off-target binding and the presence of a PAM-site. As an alternative, we measured the global DNA methylation levels between cells that contained fusion proteins (dCas9-D3A or dCas9-mD3A) in the presence of both sgRNAs (Fig. 6A), as well as the local DNA methylation profile of CpG islands located upstream and downstream of the *Cdkn1a* target locus (*Srsf3* and off-target *Cdkn1a* respectively), and the promoter CpG islands of the related cell cycle family members *Cdkn1b* and *Cdkn2d* (Fig. S6). *Srsf3* and off-target*Cdkn1a* are adjacent to *Cdkn1a* target locus on chromosome 17 (chr17), whereas *Cdkn1b* (chr6) and*Cdkn2d* (chr4) are distant loci. We found no statistically significant difference in global 5-methylcytosine (5-mC) levels between different groups or across the regions that were analyzed (Fig. 6A,B).

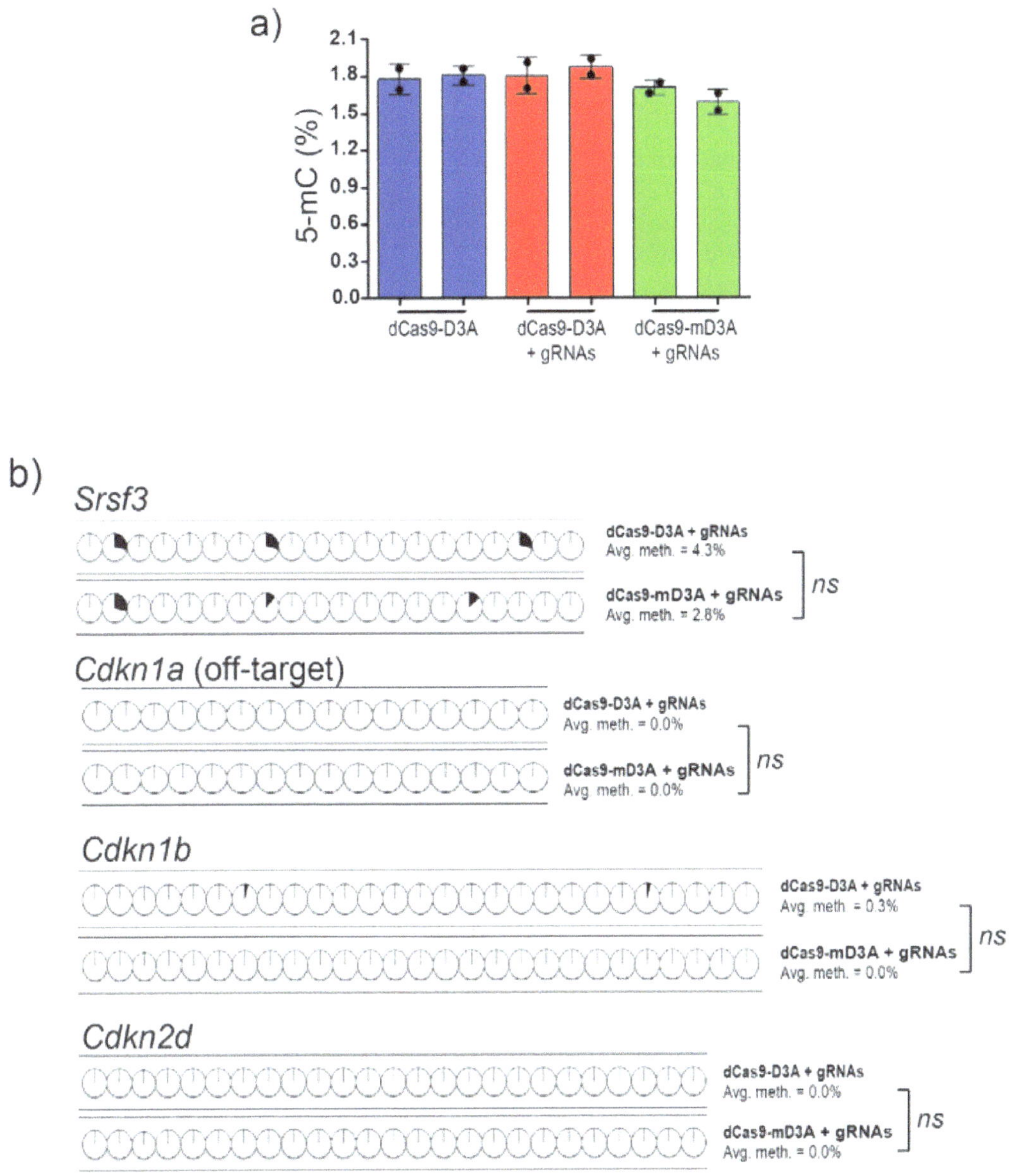

Figure 6: Limited off-target activity in dCas9-D3A driven targeted DNA methylation system. (A) Global 5-mC content levels in cells transduced with only dCas9-D3A, dCas9-D3A with sgRNAs, or dCas9-mD3A with sgRNAs targeting *Cdkn1a* locus (n=2 biological replicates performed in technical duplicate; unpaired t-test). (B) Bisulfite sequencing analysis of off-target CpG islands located upstream or downstream of *Cdkn1a* target locus (*Srsf3* and off-target *Cdkn1a*, respectively) or distant loci (*Cdkn1b* and *Cdkn2d*) in cells that are transduced with dCas9-3a or dCas9-3a with both sgRNAs (n=10-24). Source data inTable S5. Error bars=mean±s.e.m.

DISCUSSION

We provide an outline for using a modified CRISPR/dCas9 system to evaluate the functional relevance of DNA methylation at specific CpGs and described guidelines for its use. DNA methylation induction occurs within ~50 bp of a sgRNA target site and is strongest between two adjacent and inwardly directed sgRNA binding sites. Based on our design criteria, we designed sets of sgRNAs that induced methylation at the human *CDKN2A* and *ARF* promoters, and the mouse *Cdkn1a* promoters with similar efficiency. Induced methylation was sufficient to decrease expression of all three genes. Methylation increases and changes in expression were highly significant and reproducible either by using multiple distinct sgRNA combinations in the case of *CDKN2A* or at the clonal levels as observed for *Cdkn1a*. Moreover, the reduction in *Cdkn1a* expression clearly had functional consequences (increased proliferation) for the transduced cells. Though modest, the expression decreases caused by induced methylation are consistent with previously published results using ZFP and TALE fusions (Bernstein et al., 2015; Cui et al., 2015).

The effects of the induced methylation also appeared to be context dependent. While methylation of the entire CGI at the *CDKN2A* promoter repressed gene expression, inducing DNA methylation of a region 100-400 bp downstream of the *CDKN2A* TSS alone was insufficient to affect expression despite the frequent observation of a negative correlation between methylation and expression in this region. This indicates the importance of the flexibility to target multiple regions offered by our CRISPR/dCas9 DNMT fusion system.

Although approaches for targeted DNA methylation have been previously described, our method is advantageous for several reasons including; (1) the ease of designing new sgRNAs for targeting, (2) higher levels of induced DNA methylation, (3) little off-target activity. This approach can be used to interrogate the effects of DNA methylation on only a few CpG sites by bracketing them with sgRNAs, or can be used to test the effects of broader increases in DNA methylation by using many sgRNAs simultaneously. Further, the method described here provides a robust, reprogrammable approach to allow researchers to easily and thoroughly explore the functional roles of DNA methylation changes in development and disease.

MATERIALS AND METHODS

dCas9 fusion protein design and construction

The catalytic domain (CD) of human DNMT3A (amino acids 598 to 912 of NP_783328.1) both with and without the E756A mutation was cloned between the NheI and AgeI sites of pCMV_dCas9_VP64 (Addgene plasmid #49015, Cambridge, MA) with a NLS and FLAG tag linker. For lentivirus plasmids, TetO-FUW-OSKM (Addgene plasmid #20321), was digested with EcoRI and a multiple cloning site containing EcoRI, XbaI, NheI, AgeI, PspXI, AscI and EcoRI was cloned into this plasmid to generate TetO-FUW plasmid. An IRES-mCherry insert was amplified by PCR and cloned into AgeI- and AscI-digested TetO-FUW to generate TetO-FUW-IRES-mCherry. dCas9-D3A or dCas9-mD3A was cloned into this plasmid using XbaI and AgeI sites to generate TetO-dCas9-D3A or TetO-dCas9-mD3A. In the final plasmid, expression of both fusion proteins is controlled from the tetracycline operator (tetOP) and a minimal CMV promoter. The full amino acid sequence of the dCas9-3a or dCas9-3aΔ protein is shown inFig. S4.1. Plasmids sequences were validated by Sanger sequencing and prepared for transfection using a Qiagen Maxiprep kit. All plasmids are available in Addgene, and detailed protocol information is available at http://epigenomics.wustl.edu/epigenomeEditing and http://www.challenlab.com.

sgRNA design

Target sequences were entered into the MIT sgRNA design software (http://crispr.mit.edu/), the BROAD sgRNA design tool (http://www.broadinstitute.org/rnai/public/analysis-tools/sgrna-design-v1) (Doench et al., 2014), and the sgRNAcas9 tool (version 2.0.10) (Xie et al., 2014). The intersection of sgRNA target sites produced by all tools was taken for further analysis. sgRNA sequences that failed the BROAD test (score<0.2) were excluded. sgRNA were selected based on high BROAD scores and location relative to other sgRNAs. sgRNA coordinates and sequences are in Tables S1 and S2. Oligonucleotides corresponding to the target sites were annealed and cloned into MLM3636 (Addgene plasmid #43860). For lentivirus plasmids, U6-sgRNA coding sequences were amplified from pMLM3636 by PCR and cloned into an EcoRI and ClaI linearized pLVTHM lentiviral backbone (Addgene plasmid #12247) generating pL-sgRNA1 and pL-sgRNA2 in which GFP is expressed from EF1a promoter.

Cell culture, lentiviral production and doxycycline induction

HEK293T cells were acquired from ATCC (CRL-3216, Manassas, VA) and grown in DMEM supplemented with 10% FBS (Gibco, Waltham, MA), 1× Penicillin/Streptomycin (Gibco), and 2 mM GlutaMax (Gibco). For transfection experiments, 3×10^5 HEK293T cells were plated in a 60 mm dish. The next day, the cells were transfected with Lipofectamine LTX (Thermo Fisher Scientific, Waltham, MA). The Lipofectamine:DNA ratio was 3.5, with a total of 5.5 µg of plasmid DNA. The mass of Cas9-DNMT3A-CD fusion plasmid was equal to the total mass of the sgRNA plasmids. Since HEK293T cells incorporate either all plasmids or none, 0.5 or 0.7 µg of pMaxGFP was co-transfected in order to indicate the transfection efficiency. The plasmid DNA was first diluted in Gibco OptiMEM, then Lipofectamine LTX was added and mixed in by inversion. After 30 min, the transfection mixture was added dropwise to the cells and they were placed back in the incubator. 32D cells (ATCC) cells were maintained in RPMI supplemented with 10% FBS and 1% penicillin-streptomycin along with 5 ng/ml mouse interleukin-3 (IL-3). For lentiviral production, 293T cells were co-transfected with the packaging plasmids pMD2.G and psPAX2 along with the respective lentiviral plasmid (TetO- dCas9-3a, TetO-dCas9-3aΔ, pL-sgRNA1 and pL-sgRNA2). Viral supernatant was concentrated by centrifugation at 76,000 ***g*** for 1.5 h at 4°C. Cells were co-transduced with lentiviruses for the target gene expression along with an rtTA lentivirus (Addgene plasmid #20342) which is necessary for the activation of doxycycline controlled promoter in vectors containing the target gene. For transductions, a total of 1×10^5 32D cells were plated into 12-

well plates with DMEM containing 10% fetal bovine serum and 1× penicillin/streptomycin in the presence IL-3, spin-infected with lentivirus at MOI of 1:1 at 250 ***g*** for 2 h. 24 h post-transduction, cells were washed and re-plated in fresh media for expansion. For the induction of fusion proteins in transduced 32D cells, 0.25 μg/ml doxycycline (Sigma, D9891-100G) was added to the growth media. Every other day, a half-media change with (0.5 μg/ml) doxycycline was performed.

Sanger bisulfite sequencing

Genomic DNA (gDNA) was isolated using the PureLink Gemonic DNA Mini Kit (Invitrogen) or the Zymo Research (Irvine, CA) Quick gDNA MiniPrep kit and quantified with the Qubit dsDNA broad range assay (Thermo Fisher Scientific, Waltham, MA). gDNA was bisulfite converted with the Zymo Research EZ DNA methylation kit or the Epitech Bisulfite Kit (Qiagen) according to the manufacturer's instructions. All samples underwent bisulfite conversion with a high efficiency of at least 98% as determined by conversion of unmethylated, non-CpG cytosines. For *CDKN2A*, the target regions were amplified with the Qiagen PyroMark PCR kit and CDKN2A_B primers in Table S3. PCR products were cloned into the Promega pGEM-T Easy plasmid and transformed into NEB 10β competent cells. PCR products from individual colonies were sequenced by Sanger.

A nested PCR strategy was used to amplify *Cdkn1a* target genomic DNA site. In primary PCR reaction, 20 ng of total bisulfite converted DNA was used as a template with the primer pair F1 (GTTTAGATTTTTAGAGGGGAGGG) and R1 (CAAAAACTAAAAAAATAACTACCATCC). PCR was performed using Taq DNA polymerase (Invitrogen) using following cycling conditions; (1) 94°C for 3 min, (2) 95°C for 45 s, (3) 58°C for 58°C, (4) 72°C for 45 s and (5) 72°C for 10 min. Steps 2-4 were repeated for 40 cycles. For the second part of the nested PCR, 1.0 μl of the primary PCR reaction was used as a template and the same PCR conditions were used with primer pairs F1 and R2 (TCCTAAAATTCAAACTCTATATACC). The PCR amplicons were then gel extracted and cloned into pCR4-TOPO TA vector (Invitrogen) and resulting plasmids were transformed into TOP10 cells (Invitrogen). The target region was PCR amplified from different clones and the PCR products were sequenced by the Beckman Coulter Genomics (Danvers, MA, USA). Sanger bisulfite sequencing analysis was performed using BIQ Analyzer (Bock et al., 2005).

Amplicon bisulfite sequencing

gDNA was extracted, bisulfite converted and PCR amplified as above. A tenfold molar excess of Illumina sequencing Y-adapters was then annealed to 100 ng of PyroMark PCR product (total Y-adapter mass varies with PCR product length) with NEB Quick Ligase for 15 min. The ligations were purified over a 1.5% agarose II (ISC BioExpress, Kaysville, UT) gel in 1× TBE in order to remove incompletely ligated DNA. Custom 13 bp barcode index sequences were added via PCR using NEB Phusion (NEB, Ipswich, MA). The standard Phusion PCR protocol was followed but with the following primer concentrations (for a 50 μl reaction): 1 μl of Illumina Primer 1.0 (25 μM), 1 μl of Illumina PCR Primer 2.0 (0.5 μM), and 1 μl of the index primer (25 μM). The thermocycler protocol was the following: (1) 98°C, 30 s; (2) 98°C, 10 s; (3) 64°C, 30 s; (4) 72°C, 30 s; (5) Return to step two 11 times; (6) 72°C 5 min; (7) 4°C hold. The PCR products were again purified over a 1.5% agarose II gel in 1× TBE, and their concentration was measured with the Qubit dsDNA high sensitivity kit. Individually indexed samples were pooled and submitted for sequencing.

Amplicon bisulfite sequencing data were checked for quality using fastQC, adaptor and poor quality sequence (quality less than 20) was trimmed using fqtrim, and the trimmed sequences were mapped to the target sequences using Bismark (Krueger and Andrews, 2011).

ELISA methylation analysis

For quantification of total 5-mC level, the ELISA-based MethylFlash Methylated DNA Quantification Kit (Colorimetric; all from Epigentek, distributed by BioCat GmbH, Heidelberg, Germany) were utilized to quantify the amount of 5-mC in the DNA of transduced 32D cell lines. Quantification was performed in technical and biological duplicates according to the manufacturer's instructions. 100 ng of total DNA isolated from each transfected cells was used for 5-mC quantification. For absolute 5-mC quantifications, a standard curve was generated by plotting the various concentrations of the positive control provided with the kit against the optical density (OD) at 450 nm.

Expression analysis

For human cells, RNA was extracted with the Zymo Research Quick mRNA Miniprep kit. RNA concentration was measured with the Qubit RNA BR kit. RNA integrity was determined by visualizing rRNA bands using agarose gel electrophoresis. Reverse transcription was performed using the Bio-Rad iScript Reverse Transcriptase kit. Quantitative reverse-transcription polymerase chain reaction (RT-qPCR) was performed with the Bio-Rad iTaq Universal with SYBR Green reagent on an Applied Biosystems Viia7 instrument. The thermocycler protocol was the following: (1) 95°C, 20 s; (2) 95°C, 3 s; (3) 60°C, 20 s; for 40 cycles. qPCR primers are listed in Table S4. A melt curve was performed to indicate there was not off-target amplification. Data was analyzed as described by Hellemans et al. (2007) using the geometric mean of *ACTB*, *GAPDH*, and *RPL0* as an internal control. The All sgRNA sample represents data from two independent transfection experiments. The data for all remaining samples derives from technical replication using the same RNA sample. *P*-values were calculated with paired sample *t*-tests on the normalized levels of gene expression.

For mouse cells, total RNA was isolated from 5×10^3 FACS-sorted cells using NucleoSpin RNA XS (Macherey-Nagel) and reverse transcribed with the SuperScript VILO kit (Life Technologies). cDNA input was standardized and real-time polymerase chain reaction (PCR) was performed with TaqMan master Mix (Applied Biosystems), 18 s-rRNA probe (VIC-MGB; Applied Biosystems), and *Cdkn1a* gene probe (FAM-MGB; Mm04205640_g1, Applied Biosystems) on a StepOnePlus Real-Time PCR System (Life Technologies). Samples were normalized to expression of 18S and fold change determined by the ΔΔCt method.

Western blot

32D cells transduced with dCas9-D3A or dCas9-mD3A with or without sgRNAs were grown in the presence of doxycycline for 12 days, collected and washed with ice cold PBS twice. Cells were then lysed in complete RIPA buffer containing protease inhibitors (Santa Cruz Biotechnology). 20 µg of protein lysates were separated on 10% SDS-PAGE gels and transferred to nitrocellulose membranes (Millipore). Membranes were subsequently probed to detect fusion proteins using primary antibodies recognizing Cas9 (Active Motif) or β-actin (Santa Cruz) and detection was performed using horseradish-peroxidase-conjugated secondary mouse antibody (Santa Cruz) and chemiluminescence (Millipore).

ACKNOWLEDGEMENTS

The authors thank the Genome Engineering and iPSC Center at Washington University School of Medicine for production and validation of sgRNAs. We thank Jessica Hoisington-Lopez for sequencing assistance, and we thank C. Scholsberg for critical reading of this manuscript. We thank the Alvin J. Siteman Cancer Center at Washington University School of Medicine and Barnes-Jewish Hospital for the use of the Siteman Flow Cytometry Core, which provided cell sorting and analysis services. The Siteman Cancer Center is supported in part by NCI Cancer Center Support Grant #P30 CA91842.

FOOTNOTES

Author contributions

J.I.M and H.C were equally responsible for cell culture, methylation and expression experiments, data analysis, and writing the manuscript. L.E.R. performed methylation analyses and assisted with writing the manuscript. T.F. performed methylation analyses. G.F. and R.R. performed cloning of target loci for sequencing analysis. A.K. helped with cell culture and off-target analysis. A.M. designed the cloning of dCas9-D3A and dCas9-mD3A fusion proteins. J.R.E and G.A.C supervised the work, edited and approved the manuscript.

Funding

We acknowledge the following funding sources: National Institutes of Health [NIDDK R01DK102428], the Edward Mallinckrodt Jr. Foundation, the American Society of Hematology, Alex's Lemonade Stand Foundation for Childhood Cancer, the Children's Discovery Institute, the Sidney Kimmel Foundation for Cancer Research and the V Foundation for Cancer Research (to G.A.C.), the Siteman Cancer Center, U.S. Department of Defense Congressionally Directed Medical Research Program for Breast Cancer [W81XWH-11-1-0401], and the National Institute of General Medicine Sciences [NIGMS 5R01GM108811, NLM R21LM011199] (to J.R.E.), and National Institutes of Health T32 CMB Training Grant [2T32GM007067-37] for pre-doctoral support to J.I.M.

REFERENCES

1. Bell, J. T., Pai, A. A., Pickrell, J. K., Gaffney, D. J., Pique-Regi, R., Degner, J. F., Gilad, Y. and Pritchard, J. K. (2011). DNA methylation patterns associate with genetic and gene expression variation in HapMap cell lines. Genome Biol. 12, R10. doi:10.1186/gb-2011-12-1-r10
2. Bernstein, D. L., Le Lay, J. E., Ruano, E. G. and Kaestner, K. H. (2015). TALE-mediated epigenetic suppression of CDKN2A increases replication in human fibroblasts. J. Clin. Invest. 125, 1998-2006. doi:10.1172/JCI77321
3. Bestor, T. H., Edwards, J. R. and Boulard, M. (2015). Notes on the role of dynamic DNA methylation in mammalian development. Proc. Natl. Acad. Sci. USA 112, 6796-6799. doi:10.1073/pnas.1415301111
4. Bock, C., Reither, S., Mikeska, T., Paulsen, M., Walter, J. and Lengauer, T. (2005). BiQ analyzer: visualization and quality control for DNA methylation data from bisulfite sequencing. Bioinformatics 21, 4067-4068. doi:10.1093/bioinformatics/bti652
5. Bock, C., Beerman, I., Lien, W.-H., Smith, Z. D., Gu, H., Boyle, P., Gnirke, A., Fuchs, E., Rossi, D. J. and Meissner, A. (2012). DNA methylation dynamics during in vivo differentiation of blood and skin stem cells. Mol. Cell 47, 633-647. doi:10.1016/j.molcel.2012.06.019
6. Busslinger, M., Hurst, J. and Flavell, R. A. (1983). DNA methylation and the regulation of globin gene expression. Cell 34, 197-206. doi:10.1016/0092-8674(83)90150-2
7. Ciriello, G., Miller, M. L., Aksoy, B. A., Senbabaoglu, Y., Schultz, N. and Sander, C. (2013). Emerging landscape of oncogenic signatures across human cancers. Nat. Genet. 45, 1127-1133. doi:10.1038/ng.2762
8. Cui, C., Gan, Y., Gu, L., Wilson, J., Liu, Z., Zhang, B. and Deng, D. (2015). P16-specific DNA methylation by engineered zinc finger methyltransferase inactivates gene transcription and promotes cancer metastasis. Genome Biol. 16, 252. doi:10.1186/s13059-015-0819-6
9. Doench, J. G., Hartenian, E., Graham, D. B., Tothova, Z., Hegde, M., Smith, I., Sullender, M., Ebert, B. L., Xavier, R. J. and Root, D. E. (2014). Rational design of highly active sgRNAs for CRISPR-Cas9-mediated gene inactivation. Nat. Biotechnol. 32, 1262-1267. doi:10.1038/nbt.3026
10. Egger, G., Liang, G., Aparicio, A. and Jones, P. A. (2004). Epigenetics in human disease and prospects for epigenetic therapy. Nature 429, 457-463. doi:10.1038/nature02625
11. Hellemans, J., Mortier, G., De Paepe, A., Speleman, F. and Vandesompele, J. (2007). qBase relative quantification framework and software for management and automated analysis of real-time quantitative PCR data. Genome Biol. 8, R19. doi:10.1186/gb-2007-8-2-r19
12. Hinshelwood, R. A., Melki, J. R., Huschtscha, L. I., Paul, C., Song, J. Z., Stirzaker, C., Reddel, R. R. and Clark, S. J. (2009). Aberrant de novo methylation of the p16INK4A CpG island is initiated post gene silencing in association with chromatin remodelling and mimics nucleosome positioning. Hum. Mol. Genet. 18, 3098-3109. doi:10.1093/hmg/ddp251
13. Jinek, M., Chylinski, K., Fonfara, I., Hauer, M., Doudna, J. A. and Charpentier, E. (2012). A programmable dual-RNA–guided DNA endonuclease in adaptive bacterial immunity. Science 337, 816-821. doi:10.1126/

science.1225829

14. Krueger, F. and Andrews, S. R. (2011). Bismark: a flexible aligner and methylation caller for Bisulfite-Seq applications. Bioinformatics 27, 1571-1572. doi:10.1093/bioinformatics/btr167Abstract/FREE Full Text

15. Liggett, W. H. and Sidransky, D. (1998). Role of the p16 tumor suppressor gene in cancer. J. Clin. Oncol. 16, 1197-1206.

16. Lou, S., Lee, H.-M., Qin, H., Li, J.-W., Gao, Z., Liu, X., Chan, L. L., Kl Lam, V., So, W.-Y., Wang, Y. et al. (2014). Whole-genome bisulfite sequencing of multiple individuals reveals complementary roles of promoter and gene body methylation in transcriptional regulation. Genome Biol. 15, 6. doi:10.1186/s13059-014-0408-0

17. Lund, K., Cole, J. J., VanderKraats, N. D., McBryan, T., Pchelintsev, N. A., Clark, W., Copland, M., Edwards, J. R. and Adams, P. D. (2014). DNMT inhibitors reverse a specific signature of aberrant promoter DNA methylation and associated gene silencing in AML. Genome Biol. 15, 406. doi:10.1186/s13059-014-0406-2

18. Maeder, M. L., Linder, S. J., Cascio, V. M., Fu, Y., Ho, Q. H. and Joung, J. K. (2013). CRISPR RNA-guided activation of endogenous human genes. Nat. Methods 10, 977-979. doi:10.1038/nmeth.2598Web of Science

19. Mali, P., Aach, J., Stranges, P. B., Esvelt, K. M., Moosburner, M., Kosuri, S., Yang, L. and Church, G. M. (2013). CAS9 transcriptional activators for target specificity screening and paired nickases for cooperative genome engineering. Nat. Biotechnol. 31, 833-838. doi:10.1038/nbt.2675

20. Perez-Pinera, P., Kocak, D. D., Vockley, C. M., Adler, A. F., Kabadi, A. M., Polstein, L. R., Thakore, P. I., Glass, K. A., Ousterout, D. G., Leong, K. W. et al. (2013a). RNA-guided gene activation by CRISPR-Cas9-based transcription factors. Nat. Methods 10, 973-976. doi:10.1038/nmeth.2600

21. Perez-Pinera, P., Ousterout, D. G., Brunger, J. M., Farin, A. M., Glass, K. A., Guilak, F., Crawford, G. E., Hartemink, A. J. and Gersbach, C. A. (2013b). Synergistic and tunable human gene activation by combinations of synthetic transcription factors. Nat. Methods 10, 239-242. doi:10.1038/nmeth.2361

22. Reither, S., Li, F., Gowher, H. and Jeltsch, A. (2003). Catalytic mechanism of DNA-(cytosine-C5)-methyltransferases revisited: covalent intermediate formation is not essential for methyl group transfer by the murine Dnmt3a enzyme. J. Mol. Biol. 329, 675-684. doi:10.1016/S0022-2836(03)00509-6

23. Shima, K., Nosho, K., Baba, Y., Cantor, M., Meyerhardt, J. A., Giovannucci, E. L., Fuchs, C. S. and Ogino, S. (2011). Prognostic significance of CDKN2A (p16) promoter methylation and loss of expression in 902 colorectal cancers: cohort study and literature review. Int. J. Cancer 128, 1080-1094. doi:10.1002/ijc.25432

24. Siddique, A. N., Nunna, S., Rajavelu, A., Zhang, Y., Jurkowska, R. Z., Reinhardt, R., Rots, M. G., Ragozin, S., Jurkowski, T. P. and Jeltsch, A. (2013). Targeted methylation and gene silencing of VEGF-A in human cells by using a designed Dnmt3a–Dnmt3L single-chain fusion protein with increased DNA methylation activity. J. Mol. Biol. 425, 479-491. doi:10.1016/j.jmb.2012.11.038

25. VanderKraats, N. D., Hiken, J. F., Decker, K. F. and Edwards, J. R. (2013). Discovering high-resolution patterns of differential DNA methylation that correlate with gene expression changes. Nucleic Acids Res. 41, 6816-6827. doi:10.1093/nar/gkt482

26. Xie, S., Shen, B., Zhang, C., Huang, X. and Zhang, Y. (2014). sgRNAcas9: a software package for designing CRISPR sgRNA and evaluating potential off-target cleavage sites. PLoS ONE 9, e100448. doi:10.1371/journal.pone.0100448

CHAPTER 12

TECHNICAL ADVANCE: TRANSCRIPTION FACTOR, PROMOTER, AND ENHANCER UTILIZATION IN HUMAN MYELOID CELLS

Anagha Joshi[1] Christopher Pooley[1] Tom C. Freeman[1] Andreas Lennartsson[4] Magda Babina[5] Christian Schmidl[6] Teunis Geijtenbeek[7] the FANTOM Consortium[2] Tom Michoe[l] Jessica Severin[3,8] Masayoshi Itoh[2,3,8] Timo Lassmann[3,8] Hideya Kawaji[2,3,8] Yoshihide Hayashizaki[2,8] Piero Carninci[3,8] Alistair R. R. Forrest[3,8] Michael Rehli[6] and David A. Hume[1]

[1]The Roslin Institute and Royal (Dick) School of Veterinary Studies, The University of Edinburgh, Scotland, United Kingdom;

[2] RIKEN Preventive Medicine and Diagnosis Innovation Program, Tsurumi-ku, Yokohama, Japan;

[3] RIKEN Center for Life Science Technologies, Division of Genomic Technologies, Tsurumi-ku, Yokohama, Japan;

[4] Department of Biosciences and Nutrition, Karolinska Institute, Huddinge, Sweden;

[5] Department of Dermatology and Allergy, Charit´e Universit¨atsmedizin Berlin, Germany;

[6] Department of Internal Medicine III, University Hospital, University of Regensburg, Germany;

[7] Department of Experimental Immunology, Academic Medical Center, Amsterdam, the Netherlands; and

[8]RIKEN Omics Science Center, Tsurumi-ku, Yokohama, Japan

ABSTRACT

The generation of myeloid cells from their progenitors is regulated at the level of transcription by combinatorial control of key transcription factors influencing cell-fate choice. To unravel the global dynamics of this process at the transcript level, we generated transcription profiles for 91 human cell types of myeloid origin by use of CAGE profiling. The CAGE sequencing of these samples has allowed us to investigate diverse aspects of transcription control during myelopoiesis, such as identification of novel transcription factors, miRNAs, and noncoding RNAs specific to the myeloid lineage. We further reconstructed a transcription regulatory network by clustering coexpressed transcripts and associating them with enriched cis-regulatory motifs. With the use of the bidirectional expression as a proxy for enhancers, we predicted over 2000 novel enhancers, including an enhancer 38 kb downstream of *IRF8* and an intronic enhancer in the *KIT* gene locus. Finally, we highlighted relevance of these data to dissect transcription dynamics during progressive maturation of granulocyte precursors. A multifaceted analysis of the myeloid transcriptome is made available (www.myeloidome.roslin.ed.ac.uk). This high-quality dataset provides a powerful resource to study transcriptional regulation during myelopoiesis and to infer the likely functions of unannotated genes in human innate immunity.

Citation: Anagha Joshi, Christopher Pooley,Tom C. Freeman, Andreas Lennartsson, Magda Babina, Christian Schmidl, Teunis Geijtenbeek¶, the FANTOM Consortium, Tom Michoel, Jessica Severin, Masayoshi Itoh, Timo Lassmann, Hideya Kawaji, Yoshihide Hayashizaki, Piero Carninci, Alistair R. R. Forrest, Michael Rehli§ and David A. Hume, Technical Advance: Transcription factor, promoter, and enhancer utilization in human myeloid cells

INTRODUCTION

Myeloid cells are a family of innate immune cells derived from pluripotent stem cells via a common, committed progenitor. They differentiate to become innate immune effector cells in response to various lineage-restricted growth factors, including CSF-1, GM-CSF, IL-3, IL-34, FLT3, and stem cell factor [1]. Each myeloid cell type has a distinct function and a set of gene products facilitating that function. Many genome-scale expression datasets have been generated in mouse and human systems to define the genes that distinguish different myeloid cell types [2, 3] and the underlying transcriptional regulatory networks [4, 5]. For example, the gene-expression profiles of human hematopoietic lineages, including progenitor populations, highlighted transcription modules that distinguish the broad erythroid, myeloid, and lymphoid classes [4]. We have generated related data based on meta analysis of human microarray data in the public domain [6].

Microarray-based data enable expression characterization at the gene level and not at the transcript level. To overcome this limitation, the FANTOM5 consortium used CAGE to assess promoter use in numerous primary cells and tissues, including progenitor and mature myeloid cell populations. CAGE tag sequencing produces a transcript-specific expression proxy enabling characterization of alternate promoter use [7]. Accordingly, transcription factors with multiple distinct promoters and distinct functional isoforms can be associated with their likely targets. Many widely expressed genes may still have specific functions in myeloid lineage divergence or cell-specific function. For example, differences in expression of metabolic enzymes distinguish subsets of human monocytes [8]. CAGE sequencing also detects products of the bidirectional promoter activity of active enhancers [9]. In this paper, a subset of 91 samples was selected from the FANTOM5 dataset (975 samples) [10], and we have dissected the myeloid cell populations separately to identify coregulated sets of promoters, novel enhancers, miRNAs, and noncoding RNAs specific to myeloid lineages. We highlight the use of these data to dissect transcript dynamics during progressive maturation of granulocyte precursors. The data reveal a much greater level of transcriptional complexity underlying myeloid lineage divergence than the classic binary view [11]. This work is part of the FANTOM5 project. Data download, genomic tools, details of the CAGE library generation and clustering, and copublished manuscripts are available at http://fantom.gsc.riken.jp/5/.

MATERIALS AND METHODS

Ninety-one myeloid cell types were selected from the entire FANTOM5 dataset submitted to the DNA Data Bank of Japan database. They contain 18 distinct cell types in the myeloid lineage with at least 2 biologic replicates for each cell type and 22 AML samples. A description of sample preparation, quality control, and data processing is provided in Supplemental Methods. Supplemental Table 1provides the details of each sample, as well as sequencing details, including number of TSSs identified in each sample. In brief, CAGE tags for each library were aligned to the genome (hg19), neighboring tags were grouped into clusters, and further individual TSSs were identified by use of decomposition-based peak identifications [10]. The TSSs were annotated based on known transcript 5′ ends within 500 bases and summarized into regions. The TSSs with at least 5 tags/million were selected for further analysis (details in Supplemental materials). Coexpression visualization and cluster analysis were performed by use of BioLayout *Express*^{3D} [12]. Initially, a sample-to-sample Pearson correlation matrix was calculated, graphs were visualized at a Pearson correlation cut-off value of 0.58 (see Fig. 2), and nodes were colored according to cell-type grouping to describe the similarity relations between samples. To identify loose relationships between samples, a heatmap of sample-to-sample Pearson's correlation coefficients was also generated (Supplemental Fig. 1). Following this, a TSS level analysis was performed. A Pearson correlation matrix was calculated, comparing the expression profiles of the 106,709 TSSs, and a correlation cutoff, $r = 0.9$, was used to construct a graph. The resultant graph (nodes = 34,868) was then clustered by use of Markov cluster algorithm [13] with an inflation value of 2.2. Clusters with at least 20 tags and with 5 or more genes were selected for an in-depth analysis. To gain insights into the transcriptional mechanisms behind the control of these clusters, we performed cis-regulatory motif enrichment analysis by use of JASPAR [14, 15] motifs and HOMER [16]. We also calculated enrichment of each cluster with respect to the genome-wide binding patterns of transcription factors from ENCODE. Bidirectional expression was used as a proxy for predicting enhancer elements (details in Supplemental Methods). The functional enrichment was computed by use of a hypergeometric test with false discovery rate correction.

To facilitate analysis of the myeloid systems described in this manuscript, we have made the dataset available at www.myeloidome.roslin.ed.ac.uk. This web portal consists of 7 analysis tabs, namely, samples, clusters, genome browser, enhancers, miRNAs, lincRNAs, and granulopoiesis (explained in detail in the sections to follow). Each analysis tab highlights 1 aspect of use of these data. For example, the «clusters» tab lists the TSSs and corresponding genes in each cluster along with cis-regulatory control, whereas the «enhancers» tab provides cell type-specific enhancer locations and associated gene predictions. The «genome browser» tab provides a visualization of the myeloid sample data in the ZENBU genome browser [17], enabling users to view the expression of a peak across the 91 myeloid samples. The ZENBU genome browser system was developed for the FANTOM5 project to enable the interactive exploration of the FANTOM5 data sets. ZENBU can perform data-processing manipulations to allow the CAGE data to be visualized in many different ways.

RESULTS

Transcriptional profiling during normal and malignant myelopoiesis

To analyze the events in myeloid differentiation, we identified a total of 91 progenitor and mature myeloid primary cell populations and AML cell-line samples from the FANTOM5 dataset. Eighteen distinct cell types in myeloid lineage with at least 2 biologic replicates for each cell type and 22 AML samples were then analyzed by use of CAGE to assess promoter use during myelopoiesis. On average, 4 million CAGE tags mapped to hg19 assembly in each sample (Supplemental Table 1), and ~7000 TSSs were identified for each sample.

The sample identity was consistent with expression of known lineage-specific growth regulatory receptors elaborated further with examples. *CSF1R* encodes the receptor for the macrophage growth and differentiation factor, M-CSF (CSF1).*CSF1R* gene transcription in myeloid cells initiated from a broad, purine-rich proximal promoter region with low but detectable expression in purified HPCs, and in CMPs. *CSF1R* transcription was induced in the GMPs, and increased in the transition from classic (CD14$^+$, CD16$^-$) to nonclassic (CD14$^-$, CD16$^+$) monocytes.*CSF1R* was highly expressed in immature LCs isolated from skin {validated by their strong expression of *CD207* (langerin) and *CD1a* [18]} and the monocyte-derived DCs. In contrast, mature LCs isolated from afferent lymph (CD207$^-$ and CD1a$^-$) and PDCs lacked detectable *CSF1R* and instead, expressed *FLT3*. The absence of*CSF1R* from the PDC in humans contrasts with its functional role in PDC in mice [19]. The FANTOM5 data confirm the existence of an alternative *CSF1R* TSS within the 3′ UTR of the upstream platelet-derived growth factor type -type locus, used specifically in trophoblast and placental libraries [20] (data not shown). Across the dataset, *CSF1R* expression varies inversely with expression of *FLT3*, consistent with proposed roles of the 2 factors in lineage divergence between monocyte-macrophages and classic DCs [21]. In addition to the mature LCs and the PDCs,*FLT3* mRNA was expressed by HPCs and committed GMPs and in several AML lines, consistent with its involvement in myeloid leukemia [22] (Supplemental Fig. 1). The G-CSFR (*CSF3R*) was highly expressed by circulating neutrophils and eosinophils. Neutrophil PMN samples were purified from bone marrow and expressed much lower levels of *CSF3R* mRNA, implying a large increase in receptor expression as mature cells enter the circulation. Interestingly, *CSF3R* was also expressed by the classic CD14$^+$, CD16$^-$ monocytes but almost ablated in the CD16$^+$ nonclassic subset (Supplemental Fig. 2). Finally, *CD117* (KIT), a surface marker of HPCs, was expressed in HPCs, CMPs, and GMPs. It was massively expressed in mast cells, and contrary to reports in the mouse [23], there was minimal expression in DCs or LCs (Supplemental Fig. 3). As expected, given the known role of IL-5 in allergy [24], the mast cells also specifically expressed high levels of the IL-5R. The *EMR1* gene product, recognized by mAb F4/80, is widely used as a macrophage differentiation marker in mice [25]. Recent reports have claimed that its expression is restricted to eosinophils in humans [26]. However, in the FANTOM5 dataset, *EMR1*transcription was detected in monocytes but not in eosinophils. It was elevated in the CD16$^+$ monocyte subset, but by contrast to the mouse, where it is CSF1 inducible [25], it was ablated in all of the culture-derived macrophage/DC populations.

In summary, expression profiling of known key surface markers and regulators provides confidence on the identity of the cells (**Fig. 1A**) and the biologic relevance of downstream analysis.

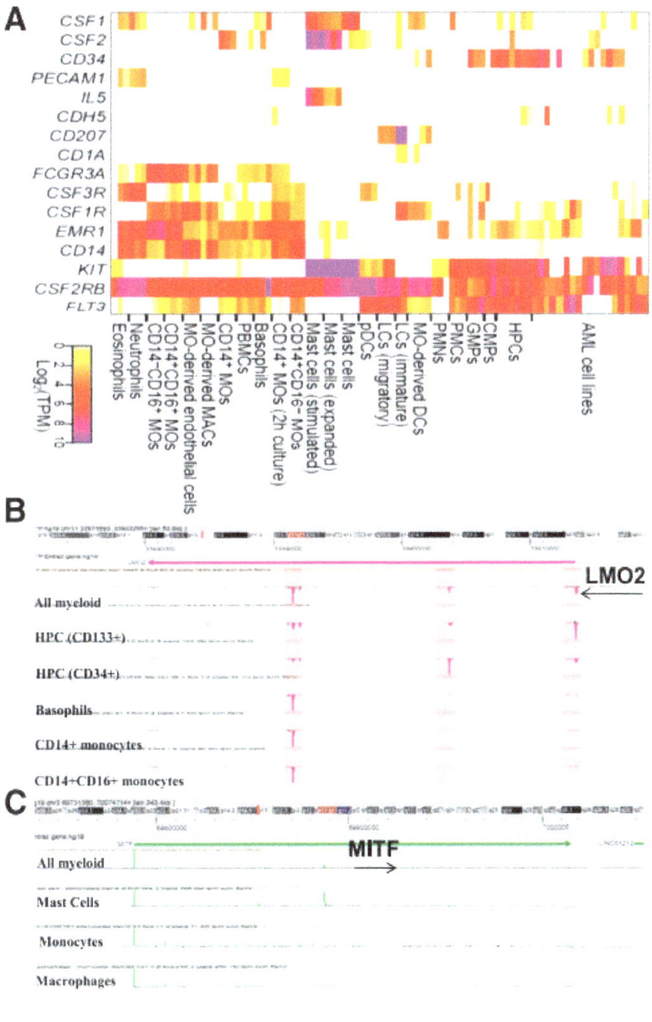

Figure 1: Expression of key surface markers and transcription factors across myeloid lineage.

- Expression [log2 transcripts per million (TPM)] of key surface markers and transcription factors across 91 myeloid samples. (B) Normalized CAGE tags in the LMO2 gene locus in the ZENBU genome browser, highlighting differential expression of the 3 alternate transcripts in myeloid lineage. (C) Normalized CAGE tags in the MITF gene locus in the ZENBU genome browser with 4 alternate TSSs.

Alternate promoter use is a common feature of transcription regulators

CAGE-based expression profiling measures expression at multiple promoters in the same gene locus. We selected TSSs with more than 5 tags/million in at least 1 sample, resulting in a total of 106,709 TSSs (from a total of 3.5 million TSSs across 975 samples in the FANTOM5 dataset). Of these, 39,913 (40%) mapped to promoter regions identified by GENCODE 19, and 14,755 (15%) were associated with intragenic regions. The TSSs in the promoter or the gene body were ascribed to that gene name. The distribution of alternate TSSs to number of genes shows an approximate power-law relationship with an average of 5 TSSs per gene. The 1500 genes with 10 or more TSSs are enriched for "transcription regulation" (*P* value: 4.0e-19), regulation of apoptosis, and cell death (*P* value: 4.2e-20). They include transcription factors, such as *FOSB*, *FOXP1*, and *KLF6*, and cytokines, such as *IL-8*and *IL-1B*.

In individual samples, on average, 8000 TSSs map to ~6000 unique genes, with an average of 1.2 TSSs per gene. This fraction is significantly lower than the 5 TSSs per gene observed across multiple cell types ($P < 2.2e\text{-}16$, Fisher test), indicating that alternate TSSs are used to control cell type-specific functions. Eosinophils and neutrophils have the highest number of expressed TSSs (~11,000) with a ratio of 1.8 TSSs/gene.

Many of the known transcription regulators of myeloid lineages, including *LMO2,RARA*, *CEBPE*, *FOXP1*, *MITF*, mothers against decapentaplegic homolog 2, *KLF6*, and ETS translocation-variant gene 6,, have several alternate TSSs, some of which alter the N terminus of the protein. For example, LMO2 protein is a T cell oncogene, reportedly expressed in blood and endothelial progenitors in mice [27]. The human *LMO2* gene was reported to generate 3 alternate transcripts: LMO2-a, -b, and -c (Fig. 1B). LMO2-a and -b encode for the same 158 aa protein but differ in the 5′ UTR length [28]. The 2 distal TSSs (LMO-a/-b) were detected in HPCs and the AML samples but differed in relative expression (Supplemental Fig. 4). They were undetectable in committed progenitors, CMP or GMP. The same promoters were used in several CML lines, including K562, included in the primary FANTOM5 dataset. The proximal *LMO2* promoter (LMO-2c), on the other hand, highlights a function in monocyte differentiation and lineage divergence. LMO-2c transcript was barely detectable in progenitor populations (CMP, GMP) and highly expressed in monocytes but not at all in granulocytes or LCs.

RARA is implicated in myeloid differentiation through its involvement in the well-characterized t(15;17) translocation, which produces a promyelocytic leukemia-RARA fusion protein in AML [29]. At least 5 distinct transcripts with differing 5′ ends have been identified so far, and all are supported by the CAGE data. The major TSS is induced in all mature myeloid cells compared with the low levels in progenitors. Three more proximal TSSs, as well as antisense transcription associated with intronic enhancers, are restricted to mature myeloid cells and induced in these cells compared with progenitors (CMP and GMP). The transcription factor CEBPE has 2 reported promoters, forming multiple isoforms, including an inhibitory one, regulated by alternate TSS and splicing [30]. The data herein identified at least 5 TSSs. However, they were detected only in the HL60 cell line and were not detected in any primary cell population. Interestingly, a previous report on the association of CEBPE mutations in humans with specific granule deficiency in neutrophils seems to involve ectopic overexpression and does not imply a physiologic function [31].

Monocytes and macrophages in the mouse express all 4 members of the MITF family (*MITF*, *TFEC*, *TFEB*, and *TFE3*) [32], and these gene products have been implicated in regulating phagocyte-specific promoters associated with lysosomal hydrolases. All 4 members of the MITF family were differentially regulated in human myeloid cells from alternative promoters. MITF itself was expressed in differentiated macrophages and mast cells from the most distal of the 4 promoters (Fig. 1C). As reported in the mouse [21], TFEC is a monocyte-macrophage differentiation marker and absent from progenitors or granulocytes. These data confirm the existence of a 100 kb upstream TSS within the TFEC locus, also restricted to monocyte-macrophages. Finally, TFEB has several promoters, all of which were detected in myeloid cells. The expression was absent from stem cells and progenitors and induced in monocytes and granulocytes but not in mast cells. The MITF proteins can form heterodimers and homodimers with other MITF family members [33] and can interact with other lineage-specific factors, such as PU.1, so their complex expression pattern provides considerable scope for combinatorial function.

Sample level analysis identifies signature gene sets in myeloid cells

To establish relatedness among the 91 samples, we created a sample-to-sample correlation matrix. The visual representation of this matrix at Pearson correlation coefficient cutoff of 0.58 by use of BioLayout *Express*^{3D} [12] shows that different samples of the same or related cell type, such as neutrophils and eosinophils, cluster together (**Fig. 2A**). Monocyte-derived macrophages (grown in CSF-1), DCs (grown in GM-CSF), and endothelial progenitors (grown in vascular endothelial growth factor) form 1 tight cluster. The monocyte samples cluster together regardless of the expression of CD16 or partial activation. The progenitor cells cluster together and lie closest to the AMLs, with which they share expression of cell cycle-related genes, such as proliferating cell nuclear antigen (data not shown). The dataset includes 3 mast cell populations, primary cells, and cells expanded in vitro and stimulated via the Fc*e*R1. Although they exhibit quite distinct gene-expression profiles [34], the cluster analysis reveals their relatedness to each other and distinguishes them from other myeloid populations.

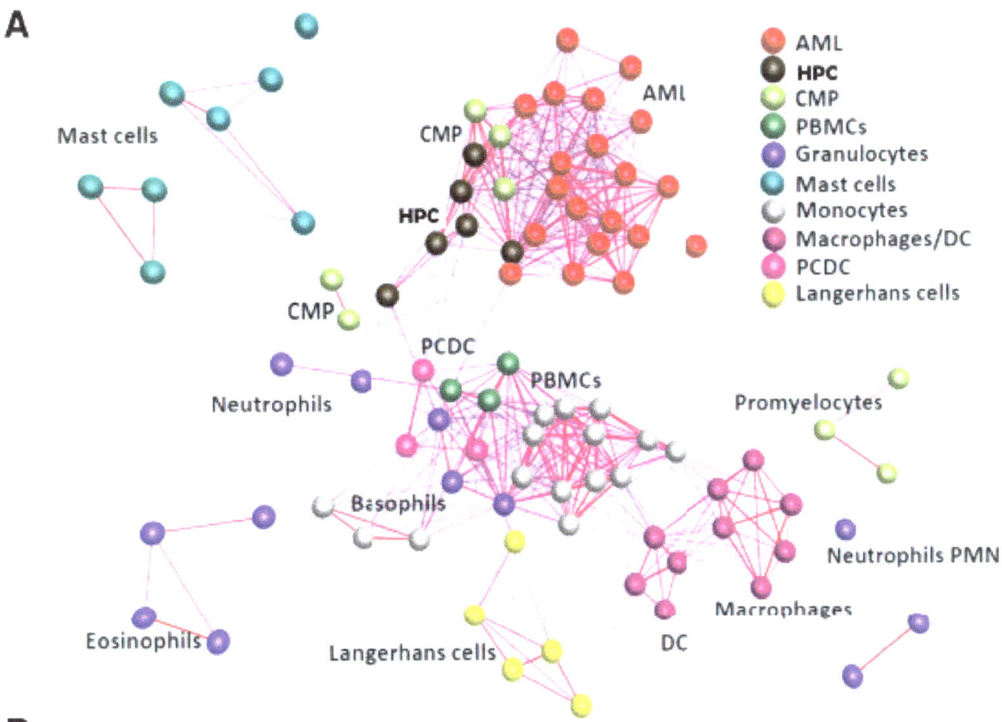

B

Cell Type	TF family	Consensus sequence motif
AML HPC GMP	E2F	
Basophil, CD14+ endothelial progenitors CD14+ monocytes	BZIP	
CD14+CD16+monocytes	IRF	
Eosinophils	NFkB-p65	
Macrophages	HLH	
Mast cells	GATA	

Figure 2: Sample clustering and cis-regulatory sequence motif enrichment.

(A). Clustering and visualization of 91 myeloid samples with the Pearson correlation cutoff of 0.58 by use of BioLayout*Express*[3D], highlighting the replicates as well as related sample clusters together. PCDC, plasmocytoid dentritic cells. (B) Cis-regulatory sequence motif enrichments along with sequence logos enriched only in specific cell type(s) calculated by use of HOMER software. TF, Transcription factor; BZIP, basic region-leucine zipper motif; HLH, helix-loop-helix.

Three sequence motifs (PU.1, GABPA, and TATA-binding protein) were enriched in all samples. PU.1 (SPI1) is an essential regulator controlling diverse aspects of early hematopoiesis and myeloid cell differentiation. The PU.1-expressing CMPs were found to have myeloid potential, whereas PU.1-deficient CMPs had erythroid potential, suggesting that PU.1 expression determines the developmental fate of a CMP [35]. Likewise, GABPA is also a key transcription factor required for myeloid differentiation [36]. Apart from these globally enriched transcription factor motifs, 6 motif families were restricted to 1 or a few cell types (Fig. 2B). For example, the E2F motif is enriched in progenitor populations and AML, whereas the GATA motif is enriched only in mast cells. GATA-1 and -2 have long been known for their importance in mast cell development (mastopoiesis) and mast cell function (such as cytokine responses), where a multitude of mast cell marker genes (FCER1A, MS4A2, KIT, and IL-1RL1/ST2) depends on GATA-1 and/or -2 for their transcription. The "samples" analysis tab in the web browser provides the genes and TSSs in each of 19 cell types together with genes specifically up- and down-regulated in that cell type, along with the predicted transcription regulators in each sample.

Clustering of Coexpressed TSSs

To identify sets of promoters that share transcriptional regulation [37], we used the clustering tool BioLayout *Express*^{3D} [12] to identify 162 clusters of coexpressed TSSs (**Fig. 3A**). Over 100 clusters contained TSSs highly expressed only in 1 or a few cell types. For example, cluster 1 was $CD14^{+}$ monocyte specific, whereas cluster 2 was neutrophil and eosinophil specific. Cluster 68 distinguished LCs from other myeloid cell types, whereas clusters 13 and 20 further distinguished between the 2 LC populations (described in detail below). Several clusters contain genes highly expressed in a single AML sample, demonstrating that each AML sample has its unique signature genes. The clusters tab of the web browser provides the TSS positions, corresponding genes, predicted transcription regulators, as well as a graphical representation of the average expression dynamics for all clusters. We noted the presence of well-known transcription factors in multiple clusters, demonstrating the prevalence of use of alternate promoters to achieve cell type-specific gene regulation (Fig. 3B). For example, different promoter use by *GFI1* places it into 2 clusters: cluster 11 and 86 (Fig. 3C). Cluster 11 consists of genes highly expressed only in PDCs, whereas cluster 86 consists of genes highly expressed in neutrophils. With the use of the ENCODE chromatin immunoprecipitation sequencing data and cis-regulatory motif enrichment, we predict TCF family members to regulate cluster 11. TCF family members control developmental processes by forming homo- or heterodimers with other E proteins, such as the development of T cells, B cells, and PDCs at a transcriptional level [38]. TCF4 belongs to cluster 11.

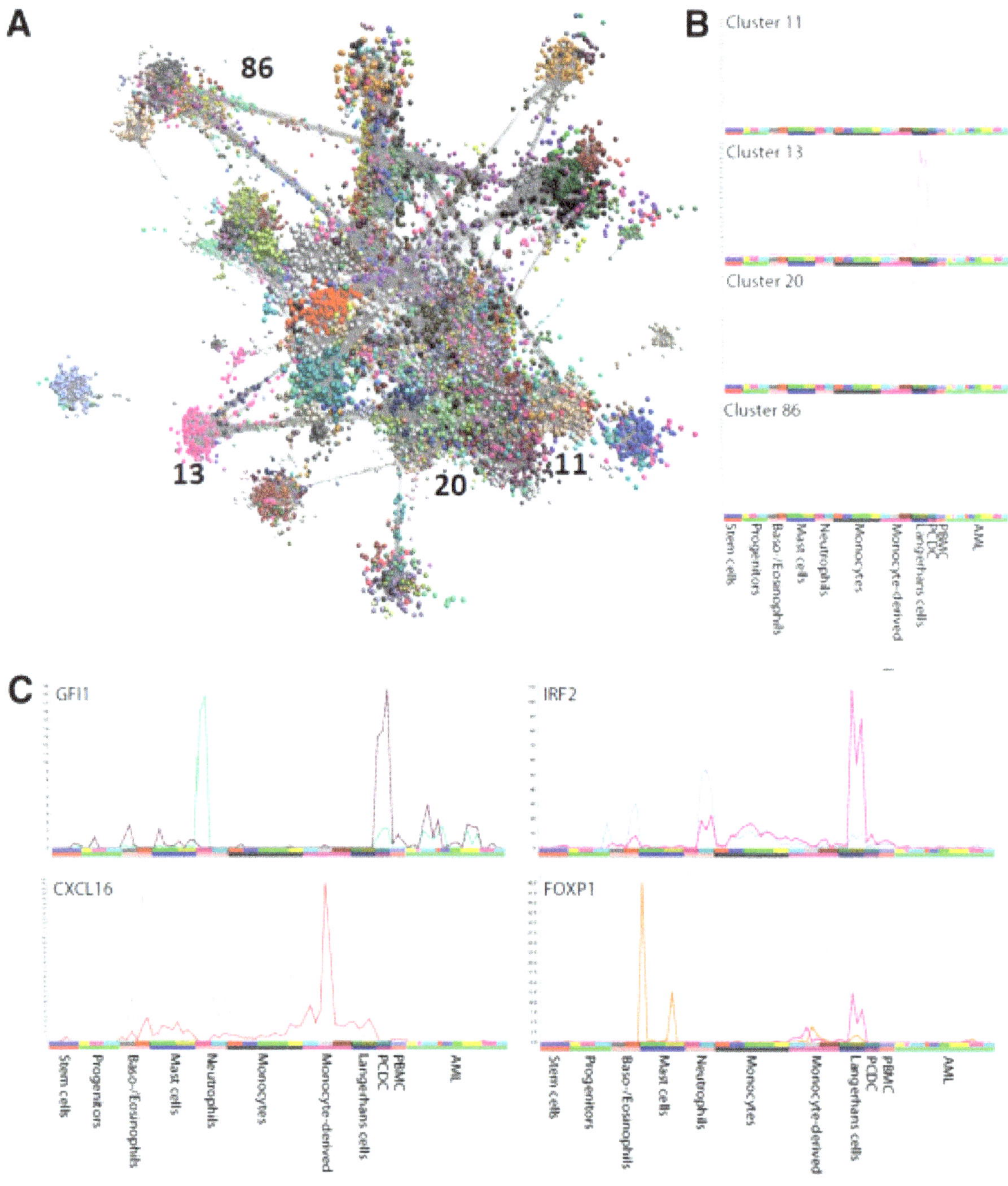

Figure 3: Gene clustering and alternate transcription regulation.

(A) The clustering and visualization of all TSSs by use of BioLayout *Express*^{3D}, with each of the TSSs belonging to each of 162 clusters indicated in a different color. (B) Expression profiles of 4 clusters (clusters 11, 13, 20, and 86) across 91 myeloid samples, indicating that many clusters are expressed in a specific cell type, providing a list of signature genes for that cell type, including distinction between immature and mature LCs (Clusters 13 and 20). (C) Four transcription factors with cell type-specific alternate transcripts across 91 cell types grouped into 11 subgroups labeled on the *x*-axis. The 2 (colored) profiles in each panel indicate 2 alternate transcripts expressed in distinct cell types. For example, Gfi1 places it into clusters 11 (expression profile in brown color) and 86 (expression profile in green color). Cluster 11 consists of genes highly expressed only in PDCs, whereas cluster 86 consists of genes highly expressed in neutrophils.

As noted above, the LC populations are distinguished from each other and from monocyte-derived DCs by their relative expression of *CSF1R* and *FLT3*. The LCs contribute to 2 large clusters—13 and 20—that distinguish the LC from monocyte-derived DC and other cell populations (Fig. 3A). Both clusters contain *CD207*(langerin), expressed from distinct promoters. Cluster 13 contains a set of promoters highly enriched in the mature LC, lower in the "immature" LC, and largely absent from monocyte-derived DCs. The cluster includes *CCR7*, the central regulator of DC migration [39, 40]; the transcription factor *RELB*, also associated with DC function in humans [41]; as well as many class II MHC genes. Cluster 20 is enriched in the immature LC relative to migratory DC and interestingly, contains a distinct *CSF1R* promoter, as well as *CD1a*, many other class II MHC genes, and the transcription factor class II, MHC, transactivator.

Enhancer analysis in the myeloid lineage predicts cell type-specific enhancers downstream of 2 key regulators: IRF8 and KIT

CAGE tag profiling detects candidate enhancers based on bidirectional transcription [9]. From the large FANTOM5 dataset, 20,301 enhancers were active in myeloid lineages (the complete list is available from the web resource). For the small number of genes studied in detail previously, the results confirm prior knowledge. For example, the data confirm the presence of a major myeloid-specific enhancer ~16–17 kb upstream of the major TSS of PU.1 (SPI1), the location of a SNP that is more frequent in AML patients [42].A further 10 of our enhancer regions that overlapped with a set of 2000 high-confidence enhancers identified by ENCODE [43]. Not surprisingly, as the enhancer predictions from ENCODE come from diverse sets of cell types, these 10 regions were active across multiple cell types and were associated with housekeeping genes and cell-cycle machinery, such as cell division protein kinase 6. Among the set of enhancers detected in myeloid cells, 196 were clearly restricted to a specific cell type. One-half of these was restricted to eosinophils and neutrophils. The genomic location, associated gene, as well as the expression profile across myeloid lineages of these putative enhancers are available from the enhancers tab of the web resource.

One example of these lineage-specific enhancers is associated with *IRF8*. The *IRF8*mRNA was expressed in GMPs and maintained in monocytes. *IRF8* mutations in humans are associated with monocyte deficiency [44], including a SNP associated with variation in the monocyte count in humans [45]. It was induced further in PDCs but was switched completely off in granulocytes. With the reinforcement of the complex interactions of enhancers and promoters, at least 5 candidate enhancers are evident upstream, downstream, and within introns of the *IRF8* locus. We focused on a candidate enhancer region (chr16:85969764-85970393; hg19), specifically active in GMPs (**Fig. 4**). This enhancer is located 38 kb downstream of the *IRF8* gene TSS and was associated with *IRF8*, based on coexpression [9]. In the FANTOM4 study of THP1 monocytic differentiation to macrophages in response to PMA, this enhancer was identified as a site of H3K9 acetylation based on whole genome tiling arrays [38]. The peak of H3K9 acetylation was reduced as IRF8 expression was extinguished in response to PMA. This region was the only one marked by H3K27Ac (a chromatin mark associated with active enhancers) in the otherwise repressed locus (as indicated by H3K27me3) in HPCs. It was strongly associated with H3K27Ac, DHS (indicating transcription factor binding), in *IRF8*-expressing monocytes and B cells and bound by PU.1 in monocytes (Fig. 4), a factor recently implicated in the remodeling of the *IRF8* locus [46].

The candidate enhancer is also highly conserved between human and mouse (ecrbrowser.dcode.org) and also marked as a PU.1-bound enhancer in mouse macrophages. The data outlined above indicate that this bidirectionally transcribed region acts as an enhancer that is most likely associated with transcriptional activation of the *IRF8* gene during hematopoietic differentiation. Another example is of an enhancer region chr4:55527478-55527674 (hg19), specific to mast cells, located 3.5 kb downstream of the main promoter of *KIT* (in the first intron) that shows an expression correlation with the *KIT* main promoter. This region is highly conserved across 100 vertebrates, including 2 highly conserved transcription factor-binding sites, supporting its role as a regulatory region. Moreover, this region is bound by *FOS*, *STAT3*, and *FOXP2* in the MCF10A-Er-Src cell line and contains a strongly enriched DNase-hypersensitive site from ENCODE data. This developmental gene locus has been characterized in great detail for cell type-specific chromatin looping [47].

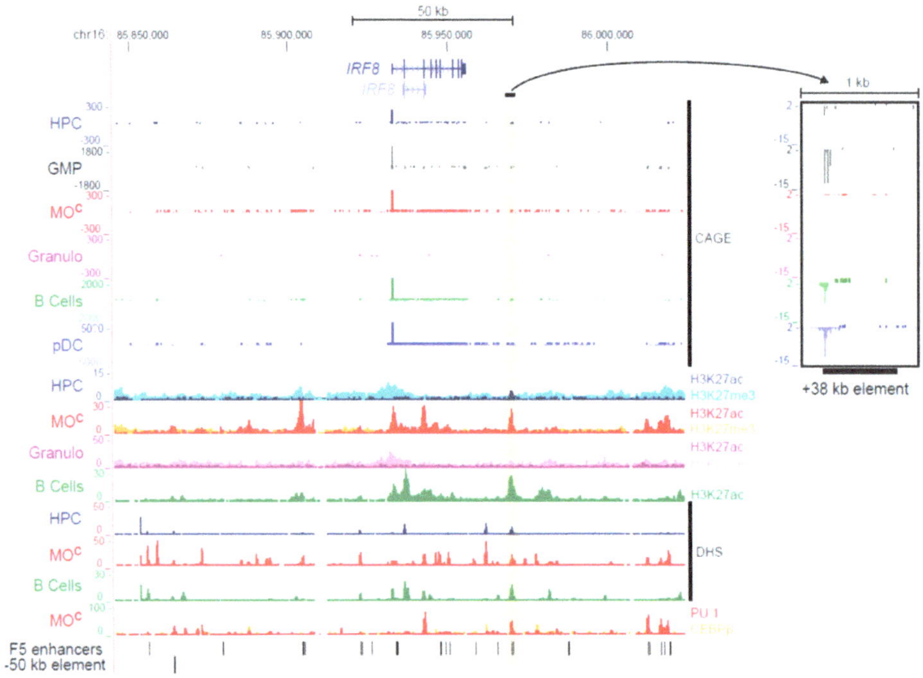

Figure 4: The +38 kb enhancer downstream of the IRF8 gene locus.

Normalized CAGE tag counts in the IRF8 gene locus across selected myeloid samples with a zoom-in view of the putative +38 kb enhancer element associated with the IRF8 gene. The view also indicates peaks in this enhancer region for DHS, as well as H3K27ac (enhancer mark) in HPC and mature cell types. Moreover, there is PU.1 binding at the +38 kb enhancer region (the last track). MO^c, monocytes.

Transcription Dynamics of miRNAs and Non-coding RNAs in Myeloid Lineages

Many short, noncoding RNAs, in particular, miRNAs, act at the post-transcriptional level to control gene expression. Two hundred eleven miRNAs are expressed in myeloid lineages, including known regulators, during myelopoiesis. For example, miR-223 is preferentially expressed in myeloid cells and is involved in granulopoietic regulation [48]. lincRNAs also function in transcriptional regulation through several mechanisms. To identify robust, myeloid-specific lincRNAs, we first identified a set of CAGE-defined TSSs that do not lie within a gene body or promoter region of protein-coding genes, and 6286 of these were found to overlap with lincRNA locations from ENSEMBL. The list of miRNAs and lincRNAs expressed in myeloid lineages, along with their expression profile, is available from the "miRNAs" and "lincRNAs" tabs within the web resource, respectively.

To validate these datasets against current literature, we examined the HOX gene clusters, known for the prevalence of its intergenic, noncoding transcription. miRNAs and lincRNAs have been identified as regulators of HOX genes [49]. miRNA-196 was expressed specifically only in AML and progenitor cell populations (Supplemental Fig. 8), in keeping with evidence that miR-196a and miR-196b contribute to hematopoietic differentiation, proliferation, and AML [50]. On the 5′ and 3′ edges of the *HOXA* cluster, there are 2 lincRNA loci: the 3′ element called HOTAIRM1 [51] and the 5′ element called HOTTIP [52]. HOTAIRM1 is expressed specifically in myeloid lineage of the hematopoietic tree [51], and was detectably transcribed in all 91 samples (Supplemental Fig. 9). HOTTIP is a 3764 nucleotide, spliced, polyadenylated RNA, transcribed just upstream of *HOXA13*, and was expressed specifically in AML samples (Supplemental Fig. 10).

Transcriptional Dynamics During Granulopoiesis

The dataset provides a development of the granulocyte series, from HPCs through CMPs to promyelocytes and segmented neutrophils in the marrow and thence, to mature neutrophils in the circulation (**Fig. 5A**). We clustered 11,091 genes, dynamically expressed in the granulopoiesis series, to identify 38 sets of coexpressed gene sets. The expression clusters detected with CAGE are entirely consistent with the literature on progressive maturation of granulocytes [53] (Fig. 5B). We highlight this case by use of exemplar clusters that exhibit a clear, stage-specific expression profile (Fig. 5C). The first cluster (cluster 3) is HPC specific and proximal to cluster 4, which contains RUNX1 and T cell acute lymphocytic leukemia 1, the 2 transcription factors indispensable for generating HPCs [35]. The down-regulation of these stem cell-specific genes during granulopoiesis is consistent with the ability of RUNX1 to suppress myelopoiesis [35]. The second cluster (cluster 5), enriched in CD34$^+$ progenitor cells, showed a functional enrichment for positive regulation of cell proliferation (*P* value: 1.4e-3) and contains transcription factors, such as JUN and NMYC. The third cluster (cluster 18) was enriched for mitochondrial matrix and lumen (*P* value: 7.7e-7) and contains the primary granule genes, such as myeloperoxidase, that are known to be highly expressed in GM progenitors, whereas the fourth cluster (cluster 38), enriched for response to microbicidal activity (*P* value: 1.7e-3), contains secondary granule proteins, such as lactoferrin, expressed in promyelocytes and marrow neutrophils and down-regulated in circulating neutrophils. The tertiary granule protein matrix metalloproteinase 9 and the components of the phagocyte oxidase, such as cytochrome B-245, are induced only at the marrow neutrophil stage and belong to the fifth cluster (cluster 11). Finally, chemotactic factor receptors, such as the formyl peptide receptor 1 and the complement receptor 1, are induced on blood neutrophils (as well as eosinophils, basophils, and CD14^{+} monocytes) and belong to the sixth cluster (cluster 1), enriched for inflammation response (*P* value: 1.4e-8) and myeloid leukocyte activation (*P* value: 1.2e-6). On the principle of guilt by association, other genes in these clusters, including those with little informative annotation, are likely to have functions in granulocyte maturation (for example, potential partners of the key transcription factors). Therefore, we have provided the lists of genes, including transcription factors and miRNAs, along with the graphical representation of the expression profile, for each cluster in the «granulopoiesis» tab of the web resource.

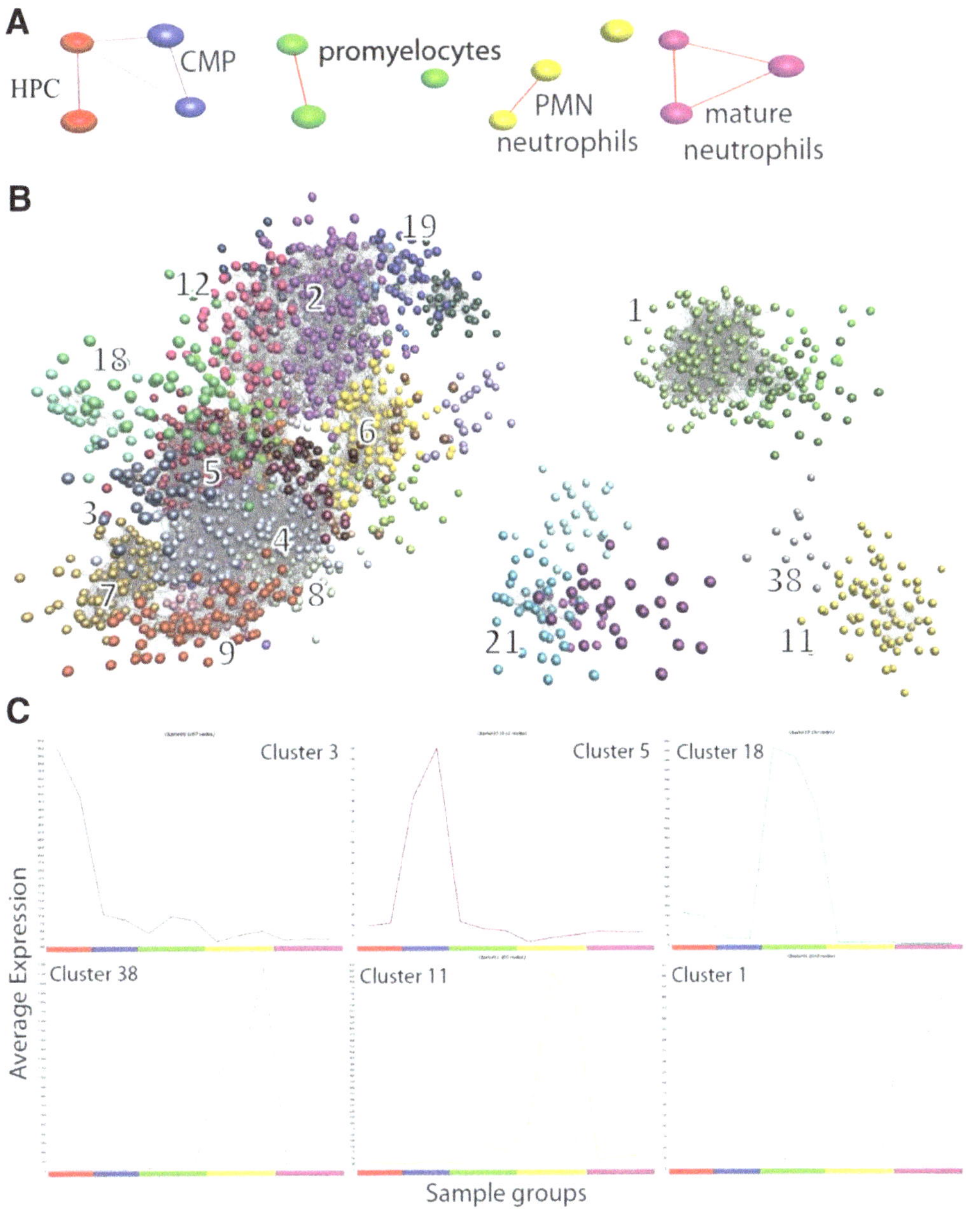

Figure 5: Temporal expression dynamics during granulopoiesis.

(A) The clustering of samples and visualization of granulopoiesis by use of BioLayout *Express*^{3D} , highlighting the replicates that cluster together (HPC, red; CMP, blue; promyelocytes, green; PMN neutrophils, yellow; mature neutrophils, purple), and the samples follow the maturation/ developmental order (from left to right) during granulopoiesis. (B) The clustering of genes and visualization of granulopoiesis by use of BioLayout *Express*^{3D}, with a few exemplar clusters numbered. (C) The expression profile for 6 clusters: cluster 3, 5, 18, 38, 11, and 1, serially expressed during the granulopoiesis series.

DISCUSSION

Myeloid cells are essential players in innate immunity, derived from a common, committed progenitor. The FANTOM5 dataset includes hematopoietic progenitors and many distinct mature, primary human cell populations of myeloid origin, as well as diverse AML lines. We further validated sample identity by expression profiles of key surface markers, such as CSF1R, and CSF1R expression was largely excluded from various granulocyte lineages and mast cells. This finding contrasts with mouse granulocytes, where CSF1R mRNA (but not CSF1R protein) is highly expressed [54]. Nevertheless, many genes expressed outside of the myeloid lineages distinguish myeloid cell types and may use distinct promoters and enhancers. We have carried out a more-detailed analysis of this subset of FANTOM5 data, permitting the identification of novel transcripts, transcription factors, miRNAs, and noncoding RNAs specific to individual myeloid cell types that can be used as markers of differentiation. With the use of the bidirectional expression as a proxy for enhancers, we predicted over 2000 novel enhancers, including a +38 kb enhancer in the IRF8 and an intronic enhancer in the KIT gene locus. Finally, we highlighted relevance of these data to dissect transcription dynamics during progressive maturation of granulocyte precursors. We have made all analysis available to the research community through a web resource.

SNPs associated with human disease susceptibility in genome-wide association studies are associated with promoters and enhancers [9, 10]. Perhaps the greatest use of the FANTOM5 myeloid dataset is for a detailed (re)analysis of individual loci associated with human inflammatory diseases for hypothesis generation. For example, variation at the *NOD2* (caspase recruitment domain-containing protein 15) has been associated with susceptibility to Crohn's disease [55]. The FANTOM5 data reveals that *NOD2* is monocyte specific, and expression is ablated in monocyte-derived macrophages. Interestingly, *NOD2* is located head to head with*SNX20*. A single report in the literature implicated SNX20 in P selectin-mediated myeloid cell adhesion [56]. There are 4 myeloid-specific enhancers between the 2 genes, and they are perfectly coexpressed. Hence, *SNX20* may contribute to Crohn's disease susceptibility associated with variation at this locus.

Taken together, the high-quality dataset provides a powerful resource to study transcriptional regulation during myelopoiesis and to infer the likely functions of unannotated genes in human innate immunity.

AUTHORSHIP

A.J. and D.A.H. designed the study, performed analysis, interpreted data, and wrote the manuscript. C.P. and J.S. developed the web resource. T.C.F. performed the analysis. T.C.F., M.R., and T.M. generated figures. The labs of M.B., A.L., T.G., M.R., and C.S. generated samples for CAGE profiling, M.I. produced the data. T.L. and H.K. performed CAGE data processing and management. A.R.R.F., P.C., and Y.H. were involved in the FANTOM5 concept, sample recruitment, and consortium management.

ACKNOWLEDGMENTS

A.J. is supported by a chancellor's fellowship from the University of Edinburgh. C.P. is funded by the Scottish government through the Strategic Partnership for Animal Science Excellence (SPASE). T.C.F. and D.A.H. are funded by an Institute Strategic Grant from the Biotechnology and Biological Sciences Research Council (Grant Number BB/JO1446X/1) and are part of the FANTOM5 project. FANTOM5 is supported through a research grant from the Japanese Ministry of Education, Culture, Sports, Science and Technology (MEXT) (to Y.H.), and a grant of the Innovative Cell Biology by Innovative Technology (Cell Innovation Program) from MEXT (to Y.H.). This work was supported by a research grant from the MEXT to the RIKEN Center for Life Science Technologies, and RIKEN Preventive Medicine and Diagnosis Innovation Program (to Y.H.). The authors thank all members of the FANTOM5 consortium for contributing to generation of samples and analysis of the dataset and thank GeNAS for data production. RIKEN Omics Science Center ceased to exist as of April 1, 2013, due to RIKEN reorganization.

REFERENCES

1. Clark, S. C., Kamen, R. (1987) The human hematopoietic colony-stimulating factors. Science 236, 1229–1237.
2. Coppe, A., Ferrari, F., Bisognin, A., Danieli, G. A., Ferrari, S., Bicciato, S., Bortoluzzi, S. (2009) Motif discovery in promoters of genes co-localized and co-expressed during myeloid cells differentiation. Nucleic Acids Res. 37, 533–549.
3. Li, Z., Herold, T., He, C., Valk, P. J., Chen, P., Jurinovic, V., Mansmann, U., Radmacher, M. D., Maharry, K. S., Sun, M., Yang, X., Huang, H., Jiang, X., Sauerland, M. C., Büchner, T., Hiddemann, W., Elkahloun, A., Neilly, M. B., Zhang, Y., Larson, R. A., Le Beau, M. M., Caligiuri, M. A., Döhner, K., Bullinger, L., Liu, P. P., Delwel, R., Marcucci, G., Lowenberg, B., Bloomfield, C. D., Rowley, J. D., Bohlander, S. K., Chen, J. (2013) Identification of a 24-gene prognostic signature that improves the European LeukemiaNet risk classification of acute myeloid leukemia: an international collaborative study. J. Clin. Oncol. 31, 1172–1181.
4. Novershtern, N., Subramanian, A., Lawton, L. N., Mak, R. H., Haining, W. N., McConkey, M. E., Habib, N., Yosef, N., Chang, C. Y., Shay, T., Frampton, G. M., Drake, A. C., Leskov, I., Nilsson, B., Preffer, F., Dombkowski, D., Evans, J. W., Liefeld, T., Smutko, J. S., Chen, J., Friedman, N., Young, R. A., Golub, T. R., Regev, A., Ebert, B. L. (2011) Densely interconnected transcriptional circuits control cell states in human hematopoiesis. Cell 144, 296–309.
5. Miller, J. C., Brown, B. D., Shay, T., Gautier, E. L., Jojic, V., Cohain, A., Pandey, G., Leboeuf, M., Elpek, K. G., Helft, J., Hashimoto, D., Chow, A., Price, J., Greter, M., Bogunovic, M., Bellemare-Pelletier, A., Frenette, P. S., Randolph, G. J., Turley, S. J., Merad, M.; Immunological Genome Consortium. (2012) Deciphering the transcriptional network of the dendritic cell lineage. Nat. Immunol. 13, 888–899.
6. Mabbott, N. A., Baillie, J. K., Brown, H., Freeman, T. C., Hume, D. A. (2013) An expression atlas of human primary cells: inference of gene function from coexpression networks. BMC Genomics 14, 632.
7. Kanamori-Katayama, M., Itoh, M., Kawaji, H., Lassmann, T., Katayama, S., Kojima, M., Bertin, N., Kaiho, A., Ninomiya, N., Daub, C. O., Carninci, P., Forrest, A. R., Hayashizaki, Y. (2011) Unamplified cap analysis of gene expression on a single-molecule sequencer. Genome Res. 21, 1150–1159.
8. Schmidl, C., Renner, K., Peter, K., Eder, R., Lassmann, T., Balwierz, P. J., Itoh, M., Nagao-Sato, S., Kawaji, H., Carninci, P., Suzuki, H., Hayashizaki, Y., Andreesen, R., Hume, D. A., Hoffmann, P., Forrest, A. R., Kreutz, M. P., Edinger, M., Rehli, M.; FANTOM consortium. (2014)

Transcription and enhancer profiling in human monocyte subsets. Blood 123, e90–e99.

9. Andersson, R., Gebhard, C., Miguel-Escalada, I., Hoof, I., Bornholdt, J., Boyd, M., Chen, Y., Zhao, X., Schmidl, C., Suzuki, T., Ntini, E., Arner, E., Valen, E., Li, K., Schwarzfischer, L., Glatz, D., Raithel, J., Lilje, B., Rapin, N., Bagger, F. O., Jørgensen, M., Andersen, P. R., Bertin, N., Rackham, O., Burroughs, A. M., Baillie, J. K., Ishizu, Y., Shimizu, Y., Furuhata, E., Maeda, S., Negishi, Y., Mungall, C. J., Meehan, T. F., Lassmann, T., Itoh, M., Kawaji, H., Kondo, N., Kawai, J., Lennartsson, A., Daub, C. O., Heutink, P., Hume, D. A., Jensen, T. H., Suzuki, H., Hayashizaki, Y., Müller, F., Forrest, A. R., Carninci, P., Rehli, M., Sandelin, A.; FANTOM Consortium. (2014) An atlas of active enhancers across human cell types and tissues. Nature 507, 455–461.

10. Forrest, A. R., Kawaji, H., Rehli, M., Baillie, J. K., de Hoon, M. J., Haberle, V., Lassman, T., Kulakovskiy, I. V., Lizio, M., Itoh, M., Andersson, R., Mungall, C. J., Meehan, T. F., Schmeier, S., Bertin, N., Jørgensen, M., Dimont, E., Arner, E., Schmidl, C., Schaefer, U., Medvedeva, Y. A., Plessy, C., Vitezic, M., Severin, J., Semple, C., Ishizu, Y., Young, R. S., Francescatto, M., Alam, I., Albanese, D., Altschuler, G. M., Arakawa, T., Archer, J. A., Arner, P., Babina, M., Rennie, S., Balwierz, P. J., Beckhouse, A. G., Pradhan-Bhatt, S., Blake, J. A., Blumenthal, A., Bodega, B., Bonetti, A., Briggs, J., Brombacher, F., Burroughs, A. M., Califano, A., Cannistraci, C. V., Carbajo, D., Chen, Y., Chierici, M., Ciani, Y., Clevers, H. C., Dalla, E., Davis, C. A., Detmar, M., Diehl, A. D., Dohi, T., Drabløs, F., Edge, A. S., Edinger, M., Ekwall, K., Endoh, M., Enomoto, H., Fagiolini, M., Fairbairn, L., Fang, H., Farach-Carson, M. C., Faulkner, G. J., Favorov, A. V., Fisher, M. E., Frith, M. C., Fujita, R., Fukuda, S., Furlanello, C., Furino, M., Furusawa, J., Geijtenbeek, T. B., Gibson, A. P., Gingeras, T., Goldowitz, D., Gough, J., Guhl, S., Guler, R., Gustincich, S., Ha, T. J., Hamaguchi, M., Hara, M., Harbers, M., Harshbarger, J., Hasegawa, A., Hasegawa, Y., Hashimoto, T., Herlyn, M., Hitchens, K. J., Ho Sui, S. J., Hofmann, O. M., Hoof, I., Hori, F., Huminiecki, L., Iida, K., Ikawa, T., Jankovic, B. R., Jia, H., Joshi, A., Jurman, G., Kaczkowski, B., Kai, C., Kaida, K., Kaiho, A., Kajiyama, K., Kanamori-Katayama, M., Kasianov, A. S., Kasukawa, T., Katayama, S., Kato, S., Kawaguchi, S., Kawamoto, H., Kawamura, Y. I., Kawashima, T., Kempfle, J. S., Kenna, T. J., Kere, J., Khachigian, L. M., Kitamura, T., Klinken, S. P., Knox, A. J., Kojima, M., Kojima, S., Kondo, N., Koseki, H., Koyasu, S., Krampitz, S., Kubosaki, A., Kwon, A. T., Laros, J. F., Lee, W., Lennartsson, A., Li, K., Lilje, B., Lipovich, L., Mackay-Sim, A., Manabe, R., Mar, J. C., Marchand, B., Mathelier, A., Mejhert, N., Meynert, A., Mizuno, Y., de Lima Morais, D. A., Morikawa, H., Morimoto, M., Moro, K., Motakis, E., Motohashi, H., Mummery, C. L., Murata, M., Nagao-Sato, S., Nakachi, Y., Nakahara, F., Nakamura, T., Nakamura, Y., Nakazato, K., van Nimwegen, E., Ninomiya, N., Nishiyori, H., Noma, S., Noma, S., Noazaki, T., Ogishima, S., Ohkura, N., Ohimiya, H., Ohno, H., Ohshima, M., Okada-Hatakeyama, M., Okazaki, Y., Orlando, V., Ovchinnikov, D. A., Pain, A., Passier, R., Patrikakis, M., Persson, H., Piazza, S., Prendergast, J. G., Rackham, O. J., Ramilowski, J. A., Rashid, M., Ravasi, T., Rizzu, P., Roncador, M., Roy, S., Rye, M. B., Saijyo, E., Sajantila, A., Saka, A., Sakaguchi, S., Sakai, M., Sato, H., Savvi, S., Saxena, A., Schneider, C., Schultes, E. A., Schulze-Tanzil, G. G., Schwegmann, A., Sengstag, T., Sheng, G., Shimoji, H., Shimoni, Y., Shin, J. W., Simon, C., Sugiyama, D., Sugiyama, T., Suzuki, M., Suzuki, N., Swoboda, R. K., 't Hoen, P. A., Tagami, M., Takahashi, N., Takai, J., Tanaka, H., Tatsukawa, H., Tatum, Z., Thompson, M., Toyodo, H., Toyoda, T., Valen, E., van de Wetering, M., van den Berg, L. M., Verado, R., Vijayan, D., Vorontsov, I. E., Wasserman, W. W., Watanabe, S., Wells, C. A., Winteringham, L. N., Wolvetang, E., Wood, E. J., Yamaguchi, Y., Yamamoto, M., Yoneda, M., Yonekura, Y., Yoshida, S., Zabierowski, S. E., Zhang, P. G., Zhao, X., Zucchelli, S., Summers, K. M., Suzuki, H., Daub, C. O., Kawai, J., Heutink, P., Hide, W., Freeman, T. C., Lenhard, B., Bajic, V. B., Taylor, M. S., Makeev, V. J., Sandelin, A., Hume, D. A., Carninci, P., Hayashizaki, Y.; FANTOM Consortium and the RIKEN PMI and CLST (DGT). (2014) A promoter-level mammalian expression atlas. Nature 507, 462–470.

11. Krause, D. S. (2002) Regulation of hematopoietic stem cell fate. Oncogene 21, 3262–3269.

12. Freeman, T. C., Goldovsky, L., Brosch, M., van Dongen, S., Mazière, P., Grocock, R. J., Freilich, S., Thornton, J., Enright, A. J. (2007) Construction, visualisation, and clustering of transcription networks from microarray expression data. PLOS Comput. Biol. 3, 2032–2042. MedlineGoogle Scholar

13. Enright, A. J., Van Dongen, S., Ouzounis, C. A. (2002) An efficient algorithm for large-scale detection of protein families. Nucleic Acids Res. 30, 1575–1584.

14. Bryne, J. C., Valen, E., Tang, M.-H. E., Marstrand, T., Winther, O., da Piedade, I., Krogh, A., Lenhard, B., Sandelin, A. (2008) JASPAR, the open access database of transcription factor-binding profiles: new content and tools in the 2008 update. Nucleic Acids Res. 36, D102–D106.

15. Jolma, A., Yan, J., Whitington, T., Toivonen, J., Nitta, K. R., Rastas, P., Morgunova, E., Enge, M., Taipale, M., Wei, G., Palin, K., Vaquerizas, J. M., Vincentelli, R., Luscombe, N. M., Hughes, T. R., Lemaire, P., Ukkonen, E., Kivioja, T., Taipale, J. (2013) DNA-binding specificities of human transcription factors. Cell 152, 327–339.

16. Heinz, S., Benner, C., Spann, N., Bertolino, E., Lin, Y. C., Laslo, P., Cheng, J. X., Murre, C., Singh, H., Glass, C. K. (2010) Simple combinations of lineage-determining transcription factors prime cis-regulatory elements required for macrophage and B cell identities. Mol. Cell 38, 576–589.

17. Severin, J., Lizio, M., Harshbarger, J., Kawaji, H., Daub, C. O., Hayashizaki, Y., Bertin, N., Forrest, A. R.; FANTOM Consortium. (2014) Interactive visualization and analysis of large-scale sequencing datasets using ZENBU. Nat. Biotechnol. 32, 217–219.

18. De Jong, M. A. W. P., de Witte, L., Geijtenbeek, T. B. H.

(2010) Isolation of immature primary Langerhans cells from human epidermal skin. Methods Mol. Biol. 595, 55–65.

19. MacDonald, K. P. A., Rowe, V., Bofinger, H. M., Thomas, R., Sasmono, T., Hume, D. A., Hill, G. R. (2005) The colony-stimulating factor 1 receptor is expressed on dendritic cells during differentiation and regulates their expansion. J. Immunol. 175, 1399–1405.
20. Visvader, J., Verma, I. M. (1989) Differential transcription of exon 1 of the human c-fms gene in placental trophoblasts and monocytes. Mol. Cell. Biol. 9, 1336–1341.
21. Hettinger, J., Richards, D. M., Hansson, J., Barra, M. M., Joschko, A. C., Krijgsveld, J., Feuerer, M. (2013) Origin of monocytes and macrophages in a committed progenitor. Nat. Immunol. 14, 821–830.
22. Zeisig, B. B., Kulasekararaj, A. G., Mufti, G. J., So, C. W. E. (2012) SnapShot: acute myeloid leukemia. Cancer Cell 22, 698–698.e1.
23. Zhang, Y., Harada, A., Wang, J. B., Zhang, Y. Y., Hashimoto, S., Naito, M., Matsushima, K. (1998) Bifurcated dendritic cell differentiation in vitro from murine lineage phenotype-negative c-kit+ bone marrow hematopoietic progenitor cells. Blood 92, 118–128.
24. Molfino, N. A., Gossage, D., Kolbeck, R., Parker, J. M., Geba, G. P. (2012) Molecular and clinical rationale for therapeutic targeting of interleukin-5 and its receptor. Clin. Exp. Allergy 42, 712–737.
25. Gordon, S., Hamann, J., Lin, H.-H., Stacey, M. (2011) F4/80 and the related adhesion-GPCRs. Eur. J. Immunol. 41, 2472–2476.
26. Legrand, F., Tomasevic, N., Simakova, O., Lee, C. C., Wang, Z., Raffeld, M., Makiya, M. A., Palath, V., Leung, J., Baer, M., Yarranton, G., Maric, I., Bebbington, C., Klion, A. D. (2014) The eosinophil surface receptor epidermal growth factor-like module containing mucin-like hormone receptor 1 (EMR1): a novel therapeutic target for eosinophilic disorders. J. Allergy Clin. Immunol. 133, 1439–1447.e1–8, CrossRefGoogle Scholar
27. Yamada, Y., Warren, A. J., Dobson, C., Forster, A., Pannell, R., Rabbitts, T. H. (1998) The T cell leukemia LIM protein Lmo2 is necessary for adult mouse hematopoiesis. Proc. Natl. Acad. Sci. USA 95, 3890–3895.
28. Wang, Q., Zhang, M., Wang, X., Yuan, W., Chen, D., Royer-Pokora, B., Zhu, T. (2007) A novel transcript of the LMO2 gene, LMO2-c, is regulated by GATA-1 and PU.1 and encodes an antagonist of LMO2. Leukemia 21, 1015–1025. MedlineGoogle Scholar
29. Takatsuki, H., Umemura, T., Sadamura, S., Yamashita, S., Goto, T., Abe, Y., Yufu, Y., Inaba, S., Nishimura, J., Nawata, H. (1995) Detection of minimal residual disease by reverse transcriptase polymerase chain reaction for the PML/RAR alpha fusion mRNA: a study in patients with acute promyelocytic leukemia following peripheral stem cell transplantation. Leukemia 9, 889–892. MedlineGoogle Scholar
30. Yamanaka, R., Kim, G. D., Radomska, H. S., Lekstrom-Himes, J., Smith, L. T., Antonson, P., Tenen, D. G., Xanthopoulos, K. G. (1997) CCAAT/enhancer binding protein epsilon is preferentially up-regulated during granulocytic differentiation and its functional versatility is determined by alternative use of promoters and differential splicing. Proc. Natl. Acad. Sci. USA 94, 6462–6467.
31. Khanna-Gupta, A., Sun, H., Zibello, T., Lee, H. M., Dahl, R., Boxer, L. A., Berliner, N. (2007) Growth factor independence-1 (Gfi-1) plays a role in mediating specific granule deficiency (SGD) in a patient lacking a gene-inactivating mutation in the C/EBPepsilon gene. Blood 109, 4181–4190.
32. Rohan, P. J., Stechschulte, D. J., Li, Y., Dileepan, K. N. (1997) Macrophage function in mice with a mutation at the microphthalmia (mi) locus. Proc. Soc. Exp. Biol. Med. 215, 269–274.
33. Hemesath, T. J., Steingrímsson, E., McGill, G., Hansen, M. J., Vaught, J., Hodgkinson, C. A., Arnheiter, H., Copeland, N. G., Jenkins, N. A., Fisher, D. E. (1994) Microphthalmia, a critical factor in melanocyte development, defines a discrete transcription factor family. Genes Dev. 8, 2770–2780.
34. Motakis, E., Guhl, S., Ishizu, Y., Itoh, M., Kawaji, H., de Hoon, M., Lassmann, T., Carninci, P., Hayashizaki, Y., Zuberbier, T., Forrest, A. R., Babina, M.; FANTOM consortium. (2014) Redefinition of the human mast cell transcriptome by deep-CAGE sequencing. Blood 123, e58–e67.
35. Rosenbauer, F., Tenen, D. G. (2007) Transcription factors in myeloid development: balancing differentiation with transformation. Nat. Rev. Immunol. 7, 105–117.
36. Yang, Z.-F., Drumea, K., Cormier, J., Wang, J., Zhu, X., Rosmarin, A. G. (2011) GABP transcription factor is required for myeloid differentiation, in part, through its control of Gfi-1 expression. Blood 118, 2243–2253.
37. Joshi, A., Van de Peer, Y., Michoel, T. (2008) Analysis of a Gibbs sampler method for model-based clustering of gene expression data. Bioinformatics 24, 176–183.
38. Suzuki, H., Forrest, A. R. R., van Nimwegen, E., Daub, C. O., Balwierz, P. J., Irvine, K. M., Lassmann, T., Ravasi, T., Hasegawa, Y., de Hoon, M. J., Katayama, S., Schroder, K., Carninci, P., Tomaru, Y., Kanamori-Katayama, M., Kubosaki, A., Akalin, A., Ando, Y., Arner, E., Asada, M., Asahara, H., Bailey, T., Bajic, V. B., Bauer, D., Beckhouse, A. G., Bertin, N., Björkegren, J., Brombacher, F., Bulger, E., Chalk, A. M., Chiba, J., Cloonan, N., Dawe, A., Dostie, J., Engström, P. G., Essack, M., Faulkner, G. J., Fink, J. L., Fredman, D., Fujimori, K., Furuno, M., Gojobori, T., Gough, J., Grimmond, S. M., Gustafsson, M., Hashimoto, M., Hashimoto, T., Hatakeyama, M., Heinzel, S., Hide, W., Hofmann, O., Hörnquist, M., Huminiecki, L., Ikeo, K., Imamoto, N., Inoue, S., Inoue, Y., Ishihara, R., Iwayanagi, T., Jacobsen, A., Kaur, M., Kawaji, H., Kerr, M. C., Kimura, R., Kimura, S., Kimura, Y., Kitano, H., Koga, H., Kojima, T., Kondo, S., Konno, T., Krogh, A., Kruger, A., Kumar, A., Lenhard, B., Lennartsson, A., Lindow, M., Lizio, M., Macpherson, C., Maeda, N., Maher, C. A., Maqungo, M., Mar, J., Matigian, N. A., Matsuda, H.,

Mattick, J. S., Meier, S., Miyamoto, S., Miyamoto-Sato, E., Nakabayashi, K., Nakachi, Y., Nakano, M., Nygaard, S., Okayama, T., Okazaki, Y., Okuda-Yabukami, H., Orlando, V., Otomo, J., Pachkov, M., Petrovsky, N., Plessy, C., Quackenbush, J., Radovanovic, A., Rehli, M., Saito, R., Sandelin, A., Schmeier, S., Schönbach, C., Schwartz, A. S., Semple, C. A., Sera, M., Severin, J., Shirahige, K., Simons, C., St Laurent, G., Suzuki, M., Suzuki, T., Sweet, M. J., Taft, R. J., Takeda, S., Takenaka, Y., Tan, K., Taylor, M. S., Teasdale, R. D., Tegnér, J., Teichmann, S., Valen, E., Wahlestedt, C., Waki, K., Waterhouse, A., Wells, C. A., Winther, O., Wu, L., Yamaguchi, K., Yanagawa, H., Yasuda, J., Zavolan, M., Hume, D. A., Arakawa, T., Fukuda, S., Imamura, K., Kai, C., Kaiho, A., Kawashima, T., Kawazu, C., Kitazume, Y., Kojima, M., Miura, H., Murakami, K., Murata, M., Ninomiya, N., Nishiyori, H., Noma, S., Ogawa, C., Sano, T., Simon, C., Tagami, M., Takahashi, Y., Kawai, J., Hayashizaki, Y.; FANTOM Consortium, ; Riken Omics Science Center. (2009) The transcriptional network that controls growth arrest and differentiation in a human myeloid leukemia cell line. Nat. Genet. 41, 553–562.

39. Comerford, I., Harata-Lee, Y., Bunting, M. D., Gregor, C., Kara, E. E., McColl, S. R. (2013) A myriad of functions and complex regulation of the CCR7/CCL19/CCL21 chemokine axis in the adaptive immune system. Cytokine Growth Factor Rev. 24, 269–283.

40. Platt, A. M., Randolph, G. J. (2013) Dendritic cell migration through the lymphatic vasculature to lymph nodes. Adv. Immunol. 120, 51–68.

41. O'Sullivan, B. J., MacDonald, K. P., Pettit, A. R., Thomas, R. (2000) RelB nuclear translocation regulates B cell MHC molecule, CD40 expression, and antigen-presenting cell function. Proc. Natl. Acad. Sci. USA 97, 11421–11426.

42. Steidl, U., Steidl, C., Ebralidze, A., Chapuy, B., Han, H. J., Will, B., Rosenbauer, F., Becker, A., Wagner, K., Koschmieder, S., Kobayashi, S., Costa, D. B., Schulz, T., O'Brien, K. B., Verhaak, R. G., Delwel, R., Haase, D., Trümper, L., Krauter, J., Kohwi-Shigematsu, T., Griesinger, F., Tenen, D. G. (2007) A distal single nucleotide polymorphism alters long-range regulation of the PU.1 gene in acute myeloid leukemia. J. Clin. Invest. 117, 2611–2620.

43. Kheradpour, P., Ernst, J., Melnikov, A., Rogov, P., Wang, L., Zhang, X., Alston, J., Mikkelsen, T. S., Kellis, M. (2013) Systematic dissection of regulatory motifs in 2000 predicted human enhancers using a massively parallel reporter assay. Genome Res. 23, 800–811.

44. Hambleton, S., Salem, S., Bustamante, J., Bigley, V., Boisson-Dupuis, S., Azevedo, J., Fortin, A., Haniffa, M., Ceron-Gutierrez, L., Bacon, C. M., Menon, G., Trouillet, C., McDonald, D., Carey, P., Ginhoux, F., Alsina, L., Zumwalt, T. J., Kong, X. F., Kumararatne, D., Butler, K., Hubeau, M., Feinberg, J., Al-Muhsen, S., Cant, A., Abel, L., Chaussabel, D., Doffinger, R., Talesnik, E., Grumach, A., Duarte, A., Abarca, K., Moraes-Vasconcelos, D., Burk, D., Berghuis, A., Geissmann, F., Collin, M., Casanova, J. L., Gros, P. (2011) IRF8 mutations and human dendritic-cell immunodeficiency. N. Engl. J. Med. 365, 127–138.

45. Crosslin, D. R., McDavid, A., Weston, N., Zheng, X., Hart, E., de Andrade, M., Kullo, I. J., McCarty, C. A., Doheny, K. F., Pugh, E., Kho, A., Hayes, M. G., Ritchie, M. D., Saip, A., Crawford, D. C., Crane, P. K., Newton, K., Carrell, D. S., Gallego, C. J., Nalls, M. A., Li, R., Mirel, D. B., Crenshaw, A., Couper, D. J., Tanaka, T., van Rooij, F. J., Chen, M. H., Smith, A. V., Zakai, N. A., Yango, Q., Garcia, M., Liu, Y., Lumley, T., Folsom, A. R., Reiner, A. P., Felix, J. F., Dehghan, A., Wilson, J. G., Bis, J. C., Fox, C. S., Glazer, N. L., Cupples, L. A., Coresh, J., Eiriksdottir, G., Gudnason, V., Bandinelli, S., Frayling, T. M., Chakravarti, A., van Duijn, C. M., Melzer, D., Levy, D., Boerwinkle, E., Singleton, A. B., Hernandez, D. G., Longo, D. L., Witteman, J. C., Psaty, B. M., Ferrucci, L., Harris, T. B., O'Donnell, C. J., Ganesh, S. K., Larson, E. B., Carlson, C. S., Jarvik, G. P.; CHARGE Hematology Working Group, ; Electronic Medical Records and Genomics (eMERGE) Network. (2013) Genetic variation associated with circulating monocyte count in the eMERGE Network. Hum. Mol. Genet. 22, 2119–2127.

46. Schönheit, J., Kuhl, C., Gebhardt, M. L., Klett, F. F., Riemke, P., Scheller, M., Huang, G., Naumann, R., Leutz, A., Stocking, C., Priller, J., Andrade-Navarro, M. A., Rosenbauer, F. (2013) PU.1 level-directed chromatin structure remodeling at the Irf8 gene drives dendritic cell commitment. Cell Reports 3, 1617–1628.

47. Jing, H., Vakoc, C. R., Ying, L., Mandat, S., Wang, H., Zheng, X., Blobel, G. A. (2008) Exchange of GATA factors mediates transitions in looped chromatin organization at a developmentally regulated gene locus. Mol. Cell 29, 232–242.

48. Johnnidis, J. B., Harris, M. H., Wheeler, R. T., Stehling-Sun, S., Lam, M. H., Kirak, O., Brummelkamp, T. R., Fleming, M. D., Camargo, F. D. (2008) Regulation of progenitor cell proliferation and granulocyte function by microRNA-223. Nature 451, 1125–1129.

49. Hornstein, E., Mansfield, J. H., Yekta, S., Hu, J. K., Harfe, B. D., McManus, M. T., Baskerville, S., Bartel, D. P., Tabin, C. J. (2005) The microRNA miR-196 acts upstream of Hoxb8 and Shh in limb development. Nature 438, 671–674.

50. Chen, C., Zhang, Y., Zhang, L., Weakley, S. M., Yao, Q. (2011) MicroRNA-196: critical roles and clinical applications in development and cancer. J. Cell. Mol. Med. 15, 14–23.

51. Zhang, X., Lian, Z., Padden, C., Gerstein, M. B., Rozowsky, J., Snyder, M., Gingeras, T. R., Kapranov, P., Weissman, S. M., Newburger, P. E. (2009) A myelopoiesis-associated regulatory intergenic noncoding RNA transcript within the human HOXA cluster. Blood 113, 2526–2534.

52. Wang, K. C., Yang, Y. W., Liu, B., Sanyal, A., Corces-Zimmerman, R., Chen, Y., Lajoie, B. R., Protacio, A., Flynn, R. A., Gupta, R. A., Wysocka, J., Lei, M., Dekker, J., Helms, J. A., Chang, H. Y. (2011) A long noncoding RNA maintains active chromatin to coordinate homeotic gene expression. Nature 472, 120–124.

53. Fiedler, K., Brunner, C. (2012) The role of transcription factors in the guidance of granulopoiesis. Am. J. Blood Res. 2, 57–65. MedlineGoogle Scholar

54. Sasmono, R. T., Ehrnsperger, A., Cronau, S. L., Ravasi, T., Kandane, R., Hickey, M. J., Cook, A. D., Himes, S. R., Hamilton, J. A., Hume, D. A. (2007) Mouse neutrophilic granulocytes express mRNA encoding the macrophage colony-stimulating factor receptor (CSF-1R) as well as many other macrophage-specific transcripts and can transdifferentiate into macrophages in vitro in response to CSF-1. J. Leukoc. Biol. 82, 111–123.

55. Van Limbergen, J., Radford-Smith, G., Satsangi, J. (2014) Advances in IBD genetics. Nat. Rev. Gastroenterol. Hepatol. 11, 372–385.

56. Schaff, U. Y., Shih, H. H., Lorenz, M., Sako, D., Kriz, R., Milarski, K., Bates, B., Tchernychev, B., Shaw, G. D., Simon, S. I. (2008) SLIC-1/sorting nexin 20: a novel sorting nexin that directs subcellular distribution of PSGL-1. Eur. J. Immunol. 38, 550–564.

CHAPTER 13

APPLICATION OF CHROMOGENIC IN SITU HYBRIDIZATION (CISH) FOR HUMAN PAPILLOMAVIRUS (HPV) GENOTYPING

Ekaterina Kldiashvili[1] and Sophio Bojgua[2]

[1]Georgian Telemedicine Unio (Association), Tbilisi, Georgia

[2]New Vision University, Tbilisi, Georgia

ABSTRACT

Background: The research aimed evaluation of the effectiveness of CISH for HPV genotyping on cervical smears and compare the results of genotyping with cytopathology findings of routine screening.

Methods and Findings: Gynecological cytology cases (n=1000) were taken from the clinical laboratory. Cases were diagnosed routinely by application of liquid based cytology method. Patients with diagnosed atypia were recalled for obtaining of material for HPV genotyping. This has been performed by usage of CISH method.

Conclusion: CISH is effective and easy for implementation method for HPV genotyping on cervical smears. There has been revealed 92.2% concordance in average between genotyping results and cytopathology findings of routine screening.

INTRODUCTION

Obvious decreasing in cervical cancer mortality has been observed in countries with organized cervical cancer screening programs [1]. Liquid based cytology is a technique that enables cells to be suspended in liquid medium and spread in monolayer for better morphological assessment. It includes the preparation and evaluation of cells collected in liquid fixative. It is being introduced to improve the sensitivity of the Pap test [2]. Liquid based preparations are increasingly being used both for gynecologic and non-gynecologic cytology, including fine needle aspirations [3]. Two technologies, ThinPrep (Hologic, Marlborough, MA, USA) and BD Sure Path (BD Diagnostics – TriPath, Burlington, NC, USA) have been more widely used [2]. The advantages of liquid based cytology include improved sensitivity and specificity since fixation is better and nuclear details are well-preserved. Abnormal cells are not obscured or diluted by other epithelial or inflammatory cells. There is, therefore, a lower rate of unsatisfactory cervical cytology samples. The residual cell suspension can be used to make further cytological preparations or used for other tests like detection of human papillomavirus DNA. Other ancillary techniques like immunocytochemistry can also be performed on the residual sample [4]. The more widely used technologies for liquid based cytology require expensive equipment [2].

Nearly all sexually active men and women are exposed to human papillomavirus (HPV) at some stage in their lives; it does not cause health problems. HPVs that infect the anogenital tract fall into two broad groups: those that cause warts (low-

Citation: Kldiashvili E, Bojgua S. Application of Chromogenic In Situ Hybridization (CISH) for Human Papillomavirus (HPV) Genotyping. Arch Cancer Res. 2016, 4:2.

risk) and those associated with cancer (high-risk). Persistent infection with high-risk HPV types causes all cervical cancer, most vulvar, vaginal and anal cancers; approximately half of penile cancers, as well as an increasing subset of oropharyngeal cancers, and HPV is also implicated on cancer precursor conditions in the cervix, anus, vulva and vagina. In some instances, HPV status will determine the approach to cancer treatment. The rising number of HPVrelated cancers is a major public health issue. The concept of a virus causing cancer is frightening. The association of cervical precancer with HPV has clear psychosocial adverse effects. Health professionals must be prepared to discuss HPV status because affected patients may want to know the cause of their condition and may question the implications for their sexual partners. Although discussions about HPV between patients and health professionals are becoming more common in cervical disease, patients express concern about the stigma attached to sexual transmission. There is little rigorous research into how clinicians communicate with patients for the other cancers associated with HPV. Patients and the public know very little about HPV. Despite the introduction of HPV vaccination in schools, and HPV testing within cervical screening programmes in the UK and several other countries, systematic reviews demonstrate consistently poor knowledge and lack of awareness that HPV is a sexually transmitted causative factor for cervical cancer. Furthermore, women who are found to be HPV positive during cervical screening experience distress, anxiety and a notable lack of understanding. Health care professionals do not know enough about HPV-associated cancers, a developing area of research where there are still many uncertainties and good quality patient information is lacking [5].

More than 200 types of HPV have been recognized on the basis of DNA sequence data showing genomic differences. Based on their association with cervical cancer and precursor lesions, HPV is grouped to high-risk and low-risk HPV types. Low-risk HPV types include types 6, 11, 42, 43, and 44. High-risk HPV types include types 16, 18, 31, 33, 34, 35, 39, 45, 51, 52, 56, 58, 59, 66, 68, and 70 [6]. It is recognized that persistent infection of the HPV is required for the development of invasive cervical cancer [7]. While infection with HPV is common, especially in sexually active young women, most infections are transient and are characterized by self-recurrence without clinical consequences. However, some women develop persistent HPV infections and are at risk for cervical cancer and its precursors. Cervical cancer is heralded as being the third most common cancer in women followed by breast and colorectal cancer. World Health Organization (WHO) has reported that approximately 530,000 women worldwide are diagnosed with cervical cancer, and the mortality of incidence ratio changed from 52 to 88% in developing countries [8]. Although cervical cancer screening programs by cytology tests (e.g., conventional Pap-smears, LBS) has decreased the incidence and death rate in many countries in the past few decades, cervical cancer still remains a leading cause of death in women due to the high rate of false positive results is limited reproducibility of cytology tests [9,10]. Besides, the sensitivity and specificity of the test have been questioned. The low sensitivity of cytology tests would put women at risk of developing invasive cervical cancer. Thus, much concern has arisen recently to develop a better screening test and/or design for disease prevention, especially the role of HPV testing.

Our research aimed evaluation of the effectiveness of CISH for high-risk HPV DNA (16/18/31/33/35; 16/18; and 31/33) detection and genotyping on cervical smears and compare the results of genotyping with cytopathology findings of routine screening.

METHODS

One thousand patients attending gynecologist at outpatient clinic for routine visit during 2015 (from February 2015 until November 2015) were selected for study. These were 18-65 years old nonvaccinated for HPV gynecologically asymptomatic females. The median age of screened group was 37 years. Specific inclusion criteria as not been used for patients recruitment. Informed consent from all cases was taken.

All cases were taken by usage of the ThinPrep reagents (Hologic). The cervical smear was obtained by rover cervical brushes and washed in the sampling solution ThinPrep (Hologic). One package of sampling materials (cervical brush and vial with sampling solution ThinPrep) has been used per patient. After obtaining and before laboratory processing the samples were stored at room temperature. The delay time between obtaining of samples and their laboratory processing did not exceed 2 hours. The smears have been prepared on glass slides by the application of the ThinPrep 2000 Processor (Hologic) accordingly with the provided for gynecology sample instructions, the program #4 of the processor has been used. One glass slide has been prepared for each screened patient. Prepared wet smears have been fixed in absolute alcohol during 30 min and stained accordingly with Papanicolau staining protocol (http://www.nottingham.ac.uk/ pathology/protocols/ papcytol.html). The Bethesda 2001 System terminology (http://nih.techriver.net/bethesdaTable.php) has been used for reporting of cervical smears. The average time required for processing and reporting of the sample was 4 hours (92.7% of cases). The Papanicolau stained smears were evaluated by light microscopy (Konus, 5601-Biorex-2) under x4, x10, x40 and x100 objective lens. The stained smears have been archived accordingly with Georgian requirements to medical data storage and documentation.

The re-usage of the rest of samples containing ThinPrep solution containing samples for additional smear preparation was impossible because of insufficient for smear preparation amount of liquid. Therefore, patients with diagnosed atypia were recalled within 5 working days (to avoid self-recurrence and/or new infection cases) for obtaining of material for HPV genotyping. The short time period has been elaborated to exclude and avoid situation of viral clearance and/or possible new infections. The HPV genotyping has been performed by CISH method on conventional smears. The smears have been fixed in absolute alcohol during 30 min. HPV specific DNA detection in the smear were performed by ZytoFast HPV type 16/18/31/33/35 Probe Kit (ZytoVision) and HPV genotyping by ZytoFast HPV type 16/18 and type 31/33 Probe Kits (ZytoVision) according to the manufacturer's instructions. Positive and negative controls were used as reference for the color appearance and reaction quality assurance.

Briefly, for enzyme digestion the smears were treated by pepsin solution on 37ºC during 5 minutes and then washed by distilled water at room temperature. For fixation smears were incubated in 1% formaldehyde solution (5 min, room temperature). After heat pretreatment step (incubation in EDTA for 15 min, 98°C; wash distilled water for 1 min, room temperature) air dried smears were hybridized with dig-labeled probes (5 μL, 5 min at 75°C and 60 min at 37°C). The detection of HPV DNA in smears was performed by rabbit-anti-DIG (30 min at 37°C) and anti-rabbit-AP-polymer (30 min, 37°C). For counterstain Mayer's hematoxylin was used (1 min, room temperature). The stained smears were evaluated by light microscopy (Konus, 5601-Biorex-2) under x40 and x100 objective lens. Smears with stained red cells (AP-Permanent Red) on the blue background were considered positive for HPV DNA (16/18/31/33/35) and HPV concrete genotype (16/18, 31/33).

RESULTS

Sampling Quality

In our study one thousand cases were studied. The cellularity was low in liquid based cytology as compared with conventional smears. The nuclei overlap to the lesser extent, so better cell morphology can be studied in the liquid based cytology technique.

Screening Results

894 cases (89.40%) were interpreted as the negative for intraepithelial lesion (NILM), epithelial cell abnormalities were diagnosed in 103 (10.3%) cases, and glandular cell abnormalities were diagnosed in 3 (0.3%) cases. Cases with abnormal cervical cytology included 79 cases (76.7%) with atypical squamous cells of undetermined significance (ASC-US); 16 cases (15.53%) with a diagnosis of atypical squamous cells of undetermined significance, suspicious but not diagnostic for high grade squamous intraepithelial neoplasia (ASC-H); 6 cases (5.83%) with a diagnosis of low grade squamous intraepithelial lesion (LSIL); 2 case (1.94%) with a diagnosis of high grade squamous intraepithelial lesion (HSIL). The results of screening are summarized in Table 1.

Table 1: The screening results of intraepithelial lesion, epithelial cell abnormalities, glandular cell abnormalities

Category **Number of cases**	**NILM**	**ASCUS**	**ASC-H**	**LSIL**	**HSIL**	**AGC**
Totally Ã¢ÂÂ 1000 cases (%)	894 (89.4%)	79 (76.7%)	16 (15.53%)	6 (5.83%)	2 (1.94%)	3 (0.3%)

HPV

HPV DNA has been revealed in 95 cases (92.2%) from 103 cases with diagnosed epithelial cell abnormalities. These cases were positive for high-risk HPV of 16/18/31/33/35 type. 16/18 types of HPV were revealed in 78 cases (82.11%), 31/33 types of HPV were revealed in 17 cases (17.89%). 8 cases (7.8%) were negative for HPV DNA. All these cases were diagnosed as ASCUS during LBC screening. The correlation between HPV DNA detection and genotyping by application of 16/18 and 31/33 HPV types probes with LBC screening diagnosis are summarized in Table 2.

Table 2: The correlation between HPV DNA detection and genotyping

CISH HPV **LBC screening diagnosis**	**HPV DNA (16/18/31/33/35)**	**HPV DNA (16/18)**	**HPV DNA (31/33)**
ASCUS	71 (74.73%)	51 (53.68%)	16 (16.84%)
ASC-H	16 (16.85%)	8 (8.42%)	6 (6.31%)
LSIL	6 (6.32%)	5 (5.26%)	1 (1.05%)
HSIL	2 (2.10%)	2 (2.10%)	--

DISCUSSION

The article presents a small pilot study of 1000 women in Georgia. As a result of the review and analysis of the data it can be concluded, that the liquid based cytology technique gives results comparable to the conventional technique with better morphology [3]. Furthermore, among the specific objectives of a pilot study was standardization of the classification system for cervical lesions to improve communication and clinical care. For this purpose the trainings for healthcare professionals have been performed. These trainings present the data on the comprehensive organization of screening, standards of cervical smear obtaining, sample processing tasks and etc.Cancer is an emerging public health issue in Georgia [11]. While the data available are known to be underestimates of true cancer incidence and cancer-related mortality, the statistics portend a troubling outlook for women in particular. The leading female cancers in Georgia are breast, cervical and ovarian – all in the reproductive system. Unfortunately, about half of all cancers are diagnosed in the late stages, obviating the opportunity for effective treatment.

Screening for cervical cancer is now being widely discussed as part of the larger context of a general effort to improve women's health through education, disease prevention and early intervention [12]. These discussions have established the following: 1) most women have never had a Pap test, 2) different classification (e.g., the Bethesda System, Papanicolau, CIN) are used to communicate results of cytology tests, 3) the primary risk factor for cervical cancer, human papillomavirus, is having a partner with previous relationships.

The Pap test is the most successful screening tests utilized in medicine. However, for this test to be effective, three things must occur. First, an adequate specimen must be obtained from the patient and submitted to the laboratory. Second, the specimen must be properly prepared, screened, and interpreted. Third, it is imperative that when communicating the diagnosis, the laboratory uses terminology that is clearly understood by the clinician. To this end, several classification schemes and sets of terminology to provide more meaningful diagnostic information have been developed. The most informative and adequate is the Bethesda 2001 System (TBS) [8]. It is not only a simple listing of diagnostic terms, but is a comprehensive way to report cervical cytology that incorporates a descriptive diagnosis and evaluation of specimen adequacy with a heavy emphasis on clinical relevance [9,10].

The benefits of screening and early intervention are clear. Early intervention is available and can be performed in Georgia at minimal cost. There is no standardized approach for obtaining smears or interpreting results. The resulting ambiguity makes it difficult for clinicians to compare results of Pap-test, negatively affecting patient care. By the implementation of the present pilot the introduction of standardized and LBC approaches for cervical cancer screening has been performed.

It has been revealed, that LBC improved sensitivity and specificity of cervical cancer screening since fixation is better and nuclear details are well-preserved, the amount of unsatisfactory samples is decreased. It has been also revealed, that CISH is effective and easy for implementation method for HPV DNA detection and genotyping on cervical smears. The efficacy of the high-risk HPV type (16/18/31/33/35) probe for detection of HPV infection has been confirmed [12]. Furthermore, the prevalence of 16/18 type of HPV in atypical cases has been confirmed by the present pilot study. The 92.2% concordance in average between genotyping results and cytopathology findings of routine screening has been revealed. 7.8% cases of atypical cases were HPV negative; all these cases were diagnosed as ASCUS during LBC screening. It has been concluded, that severe inflammation as well as specific inflammation (i.e., induced by *Candida* spp.) can be misdiagnosed. Taking into account the atypical cases follow up guidelines, it has been concluded, that shift towards HPV DNA based cervical cancer screening will be effective and appropriate. The increased costs of cervical cancer screening will compensated by the higher specificity and prolonged screening intervals.

FUNDING

The research has been performed during implementation of the project 'Cervical Cancer Screening Using Telemedicine Resources in Georgia' (#007G). This project was supported by 700 for Science and Hologic.

REFERENCES

1. Arbyn M, Raifu AO, Weiderpass E, Bray F, Anttila A (2009). Trends of cervical cancer mortality in the member states of the European Union. Eur J Cancer 45: 2640Ã¢ÂÂ2648.
2. Kavatkar AN, Nagwanshi CA, Dabak SM (2008). Study of manual method of liquid based cervical cytology. Indian J Pathol Microbiol 51: 190-194.
3. Hoda RS (2007) Non-gynecologic cytology on liquid-based preparations: A morphologic review of facts and artifacts. Diagn Cytopathol 35: 621-634.
4. Ronco G, Cuzick J, Pierotti P, Cariaggi MP, Dalla P, et al. (2007) Accuracy of liquid based versus conventional cytology: Overall results of new technologies for cervical cancer screening: Randomized controlled trial BMJ 335:28.
5. Hendry M, Pasterfield D, Adams R, Evans M, Fiander A, et al. (2016) Talking about human papillomavirus and cancer: protocol for a patient-centred study to develop scripted consultations. BMJ 6:e011205.
6. Burd EM (2003) Human papillomavirus and cervical cancer. Clinical Microbiology Reviews1-17.
7. Walboomers JM, Jacobs MV, Manos MM, Bosch FX, Kummer JA, et al. (1999) Human papillomavirus is a necessary cause of invasive cervical cancer worldwide. J Pathol189:12Ã¢ÂÂ19.
8. World Health Organization (2013) Comprehensive cervical cancer control: A guide to essential practice.
9. Katz IT, Wright AA (2006) Preventing cancer in developing world. N Engl J Med 354: 1110.
10. Debby L, Sankaranarayanan R (2006) Secondary prevention of cervical cancer. Int J Gynecology & Obstetrics 94: S65-S70.
11. Gvamichava R, Lomtadze N, Alibegashvili T, Charkviani T, Beruchashvili T, et al. (2012) Cancer screening program in Georgia (results of 2011). Georgian Med News 7-15.
12. Smith MA, Gertig D, Hall M, Simms K, Lew JB, et al. (2016) Transitioning from cytology-based screening to HPV-based screening at longer intervals: implications for resource use. MBC Health Service Research 16:147.

CHAPTER 14

MOLECULAR CLASSIFICATION OF CANCER WITH THE 92-GENE ASSAY IN CYTOLOGY AND LIMITED TISSUE SAMPLES

Elena F. Brachtel[1], Theresa N. Operaña[2], Peggy S. Sullivan[3], Sarah E. Kerr[4], Karen A. Cherkis[2], Brock E. Schroeder[2], Sarah M. Dry[3] and Catherine A. Schnabel[2]

[1] Department of Pathology, Massachusetts General Hospital, Boston, Massachusetts, USA

[2] Biotheranostics, Inc., San Diego, California, USA

[3] Department of Pathology and Laboratory Medicine, David Geffen School of Medicine, University of California Los Angeles, Los Angeles, California, USA

[4] Department of Laboratory Medicine and Pathology, Mayo Clinic, Rochester, Minnesota, USA

ABSTRACT

Background: Detailed molecular evaluation of cytology and limited tissue samples is increasingly becoming the standard for cancer care. Reproducible and accurate diagnostic approaches with reduced demands on cellularity are an ongoing unmet need. This study evaluated the performance of a 92-gene assay for molecular diagnosis of tumor type/subtype in cytology and limited tissue samples.

Methods: Clinical validation of accuracy for the 92-gene assay in limited tissue samples such as cytology cell blocks, core biopsies and small excisions was conducted in a blinded multi-institutional study (N = 109, 48% metastatic, 53% grade II and III). Analytical success rate and diagnostic utility were evaluated in a consecutive series of 644 cytology cases submitted for clinical testing.

Results: The 92-gene assay demonstrated 91% sensitivity (95% CI [0.84, 0.95]) for tumor classification, with high accuracy maintained irrespective of specimen type (100%, 92%, and 86% in FNA/cytology cell blocks, core biopsies, and small excisions, respectively; $p = 0.26$). The assay performed equally well for metastatic versus primary tumors (90% vs 93%, $p = 0.73$), and across histologic grades (100%, 90%, 89%, in grades I, II, and III, respectively; $p = 0.75$). In the clinical case series, a molecular diagnosis was reported in 87% of the 644 samples, identifying 23 different tumor types and allowing for additional mutational analysis in selected cases.

Conclusions: These findings demonstrate high accuracy and analytical success rate of the 92-gene assay, supporting its utility in the molecular diagnosis of cancer for specimens with limited tissue.

Citation: Elena F. Brachtel, Theresa N. Operaña, Peggy S. Sullivan, Sarah E. Kerr, Karen A. Cherkis, Brock E. Schroeder, Sarah M. Dry and Catherine A. Schnabel, Molecular classification of cancer with the 92-gene assay in cytology and limited tissue samples, DOI: 10.18632/oncotarget.8449.

INTRODUCTION

While molecular testing in oncology continues to develop and validate new methodologies employing reduced cellularity, such as circulating tumor cells (CTCs) and cell-free tumor material, the current standard of care depends on appropriate management of tumor tissue biopsy specimens for diagnosis and ancillary studies [1, 2]. Often, cytology samples from fine needle aspirations (FNAs), fluids or core biopsies are the only tissue samples available for diagnosis and advanced molecular interrogation (Figure 1). As the tissue samples obtained by minimally invasive diagnostic procedures continue to decrease in size, the demand increases for detailed diagnostic evaluation and molecular testing using these limited tissue samples [3]. In addition to new sampling procedures such as endoscopically guided fine needle aspirations, new genomic technologies such as next generation sequencing (NGS) are being used by clinicians to simultaneously evaluate multiple genes/signaling pathways to identify actionable mutations for treatment planning [4]. Use of tissue conserving technologies will ensure that specimens are available for diagnostic and molecular analyses as oncology disease management evolves.

Limited tissue samples are small and often exhausted during routine histology and immunohistochemical studies. In response to continuing developments and in an effort to conserve tissue samples, molecular tests for diagnosis and targeted therapy approaches are being incorporated into clinical practice and molecular testing guidelines for various tumor types and subtypes [5, 6]. Several diagnostic assays for gene expression-based classification of tumors requiring minimal tissue are already in use [7, 8]. These molecular tools provide independent quantitative data that is standardized and complementary to routine morphology and immunohistochemistry (IHC) approaches [8-10]. Integrating gene expression profiling into diagnostic algorithms for samples that have limited cellular content or lack definitive morphological and immunohistochemical features allows the remaining tissue to be used for reflex biomarker testing [11]. The goal of these various approaches is to ensure accurate diagnosis and allocation of tissue from small samples for the appropriate ancillary studies.

The 92-gene assay is a molecular classifier that predicts tumor type and subtype through an algorithmic-based comparison of a tumor's gene expression profile to a reference database of known tumor types. The assay distinguishes between 28 main tumor types and 50 subtypes, which represents >95% of all solid tumors based on incidence [8, 12]. The assay employs a real-time RT-PCR, tissue-sparing platform, requiring approximately 300-500 non-necrotic tumor cells, that interrogates expression of transcription factors, plasma membrane proteins, and tissue specific tumor markers not commonly evaluated during routine IHC analyses to assess over 350 unique biological functions [13]. In a previously published validation study, the 92-gene assay demonstrated 87% (95% CI, 84-89%) overall accuracy, and notably, statistically similar performance in metastatic and high grade tumors [8]. In a head-to-head comparison vs IHC, the 92-gene assay demonstrated a statistically significant improvement in diagnostic accuracy in poorly-to-undifferentiated metastatic tumors (79% vs 69%; P = 0.019) [14]. Additionally, when the molecular diagnosis provided by the 92-gene assay was used to direct therapy in treatment-naïve patients with carcinoma of unknown primary, a 3.4 month increase in overall survival was observed when compared to a pre-specified historical cohort [15]. In the current study, the specific performance and analytical success rate of the 92-gene assay in tumors from limited tissue and cytology specimens were further characterized to assess its diagnostic utility within the current landscape of precision oncology.

RESULTS

Specimen and perfomance characteristics of clinical validation cohort

Specimen and patient characteristics for clinical validation cohort

Case characteristics for the clinical validation cohort (N = 109) are summarized in Table 1. Approximately half (48%) of all tumor specimens were from metastatic lesions. Tumors were primarily intermediate or high grade (I = 5%, II = 18%, III = 35%). Forty-two percent of cases were ungraded including metastatic lesions and primary cases in which tumor grading was not standard procedure during the diagnostic evaluation or not traditionally conducted in certain tumors such as pheochromocytomas and gastrointestinal stromal tumors. Specimens with 50% or less tumor content, greater than 20% fibrosis and 5-30% necrosis comprised 29%, 25%, and 36% of the cases analyzed, respectively. The biopsy types evaluated for this data set included FNA/ cell blocks (N = 20, 18%), core / other biopsies (N = 52, 48%), and small excisions (N = 37, 34%) (Table 1 and Figure 1). The cases were distributed among 17 biopsy sites and included 24 different tumor types (Figure 2A and 2B).

Table 1: Tumor characteristics for the clinical validation cohort (*N* = 109)

Characteristic	N (%)
Gender	
Male	54 (49)
Female	55 (51)
Primary	57 (52)
Metastatic	52 (48)
Histologic Grade	
Grade I	5 (5)
Grade II	20 (18)
Grade III	38 (35)
Not graded	46 (42)
Tumor content	
≤ 50%	32 (29)
> 50%	77 (71)
Fibrosis	
< 20%	82 (75)
≥ 20%	27 (25)
Necrosis	
Non-necrotic	70 (64)
5-30%	39 (36)
Inflammation	
< 20%	95 (87)
≥ 20%	14 (13)
Biopsy Type	
Fine needle aspirate (FNA)/ cell block	20 (18)
Core and other biopsies	52 (48)
Small excisions	37 (34)

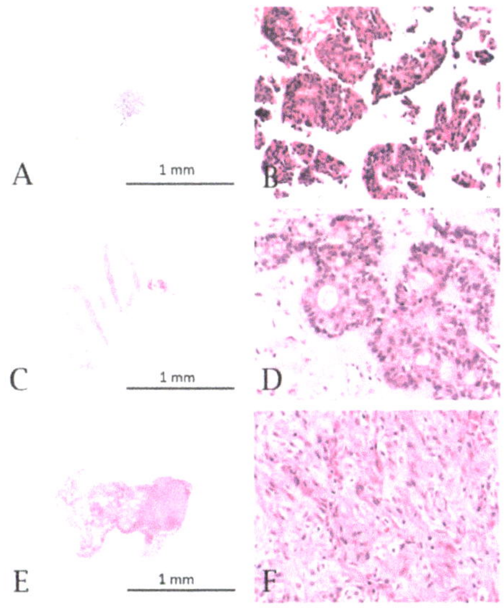

Figure 1: Hematoxylin and eosin stained specimens from limited tissue sample types: A. & B. Fine needle aspirate (FNA)/cell block showing breast adenocarcinoma (10X and 400X). C. & D. Core biopsy showing colorectal adenocarcinoma metastatic to the liver (10X and 400X). E.& F. Small excisional biopsy showing renal cell carcinoma metastatic to the lung (10X and 400X).

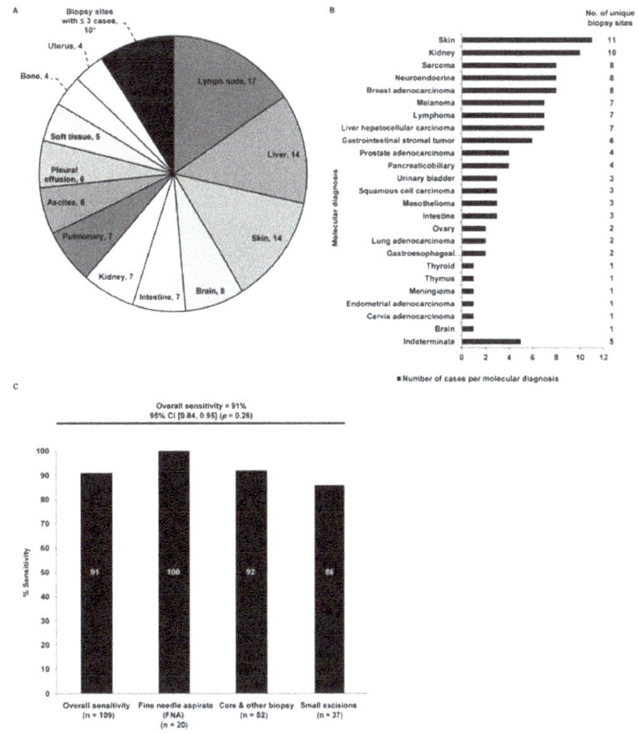

Figure 2: A. Distribution of biopsy sites from clinical validation cohort. *Biopsy sites with three or fewer cases were combined into a single category, which includes: Breast (3), Adrenal (2), Mediastinum (2), Prostate (2), Retroperitoneum (1). B. Distribution of molecular diagnoses as determined by the 92-gene assay in the limited tissue data set. Molecular diagnoses labeled on the y-axis. Number of unique biopsy sites for each molecular diagnosis is labeled to the right of the graph. C. Sensitivity of the 92-gene assay for each specimen type in the limited tissue data set (p = 0.26). Specimen type labeled along the x-axis. Sensitivity expressed as a percentage is shown by each column. Exact values per specimen type are shown in white. Overall sensitivity for the 92-gene assay was 91%.

92-gene assay clinical validation

The 92-gene assay predicted a molecular diagnosis in 95% (N = 104) of cases (the remaining 5 cases were indeterminate) when compared to the gold standard of adjudicated histopathological diagnoses with integration of clinical findings and ancillary testing. The overall sensitivity in these cases was 91% [95% confidence interval (CI) 0.84 - 0.95; Figure 2C]. There was no difference in sensitivities between primary and metastatic cases (p = 0.73; Table 2). It was previously demonstrated in a large blinded validation study that the assay maintained high accuracy irrespective of specimen type (limited tissue vs non-limited tissue) [8]. In the current study, the performance of the 92-gene assay in the limited tissue subset was further analyzed and stratified based on biopsy types. Performance of the assay was consistent across biopsy types (FNA/cell blocks = 100%, core and other biopsies = 92%, small excisions = 86%; p = 0.26; Figure 2C) and across histologic grades (I = 100%, II = 90%, III = 89%, not graded = 93%; p = 0.84; Table 2). No statistical difference in performance was observed across graded tumors (histologic grades I, II and III; p = 0.75).

Table 2: 92-gene assay performance by clinical subset in the clinical validation cohort (N = 109)

Clinical variables	N (%)	Sensitivity	*P*
Disease Type			
Primary	57 (52)	93%	0.73
Metastatic	52 (48)	90%	
Histologic Grade			
Grade I	5 (5)	100%	0.84
Grade II	20 (18)	90%	
Grade III	38 (35)	89%	
Not graded	46 (42)	93%	

Analysis of cytology clinical case specimens

Distribution of biopsy sites and molecular diagnoses

Of 644 cytology clinical cases submitted for testing with the 92-gene assay as part of routine clinical care, the 92-gene assay had an analytical success rate of 87%, while 13% of cases failed analytical QC due to insufficient RNA or poor RNA quality. Submitted cases were distributed among 18 biopsy sites with liver (n = 125), pleural effusion (n = 111), ascites (n = 71) and bone (n = 61) contributing to over 50% of the cases (Figure 3A). The 92-gene assay predicted 23 different tumor types; the four most common molecular diagnoses were pancreaticobiliary (19%; n = 105), lung adenocarcinoma (11%; n = 60), ovary (9%; n = 49) and urinary bladder (8%; n = 47) (Figure 3B). These four most common molecular diagnoses were detected across a large number of biopsy sites (pancreaticobiliary, n = 14; lung adenocarcinoma, n = 10; ovary, n = 9; and urinary bladder, n = 9; Figure 3B). Biopsies from the liver, pleural effusions, bone, lymph node, peritoneal effusions, lung, and soft tissue all had more than 10 different tumor types predicted by the 92-gene assay. Thirty-five cases (6%) were indeterminate, in which the assay did not report a probability of at least 70% for a single site diagnosis; however, the 92-gene assay test report for indeterminate cases does provide additional information that can help to establish a diagnosis, including a list of tumor types with some degree of gene expression overlap with known tumors within the reference database, as well as tumor types and subtypes that can be ruled out with a 95% confidence [8].

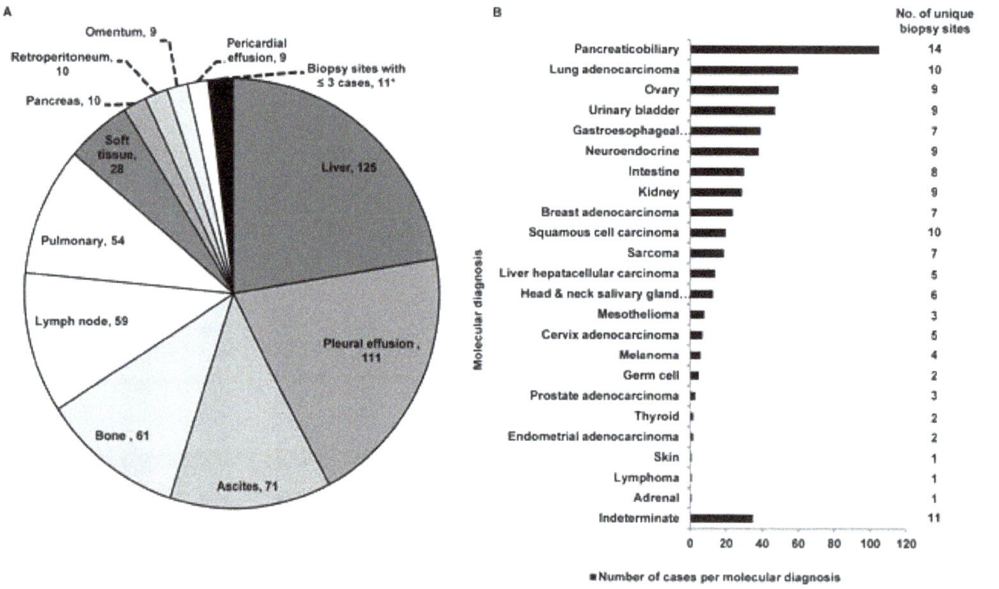

Figure 3: A. Distribution of biopsy sites within the clinical cytology case series that passed analytical QC ($N = 558$). *Biopsy sites with three or fewer cases were combined into a single category, which includes: Mediastinum (3), Bowel (2), Breast (2), Kidney (2), Adrenal (1), Peri-aortic (1), Thyroid (1). B. Distribution of molecular diagnoses as determined by the 92-gene assay in the clinical cytology data set that passed analytical QC ($N= 558$). Molecular diagnoses labeled on the y-axis. Number of unique biopsy sites for each molecular diagnosis is labeled to the right of the graph.

Site-specific Biomarker Testing

Thirty-seven cases with a 92-gene assay molecular diagnosis of lung (n = 22), colorectal (n = 6), gastric (n = 4), breast (n = 2), and melanoma (n = 3) from the clinical case dataset had ancillary biomarker testing performed from the same laboratory (Biotheranostics, Inc.) (Supplementary Table 1). Of these, 36 cases had successful biomarker testing. The most common biomarkers analyzed were *EGFR* mutations (n = 20), *KRAS* mutations (n = 12), *BRAF* mutation (n = 10), *PIK3CA* (n = 8) and *ALK* rearrangements (n = 8) (Supplementary Table 1). Within this group, 53% (n = 19) of cases had more than one predictive biomarker assessed. Of the cases predicted by the 92-gene assay to be lung and colorectal (n=28), *EGFR* and *KRAS* testing were performed in 18 and 5 cases, respectively. *EGFR* mutations were detected in 4 of the lung predictions and 4 of the colorectal predictions were identified to be wild type for *KRAS*.

Table 3: Summary of the clinical characteristics, molecular diagnosis from 92-gene assay and results from predictive biomarker testing for an exemplar patient case from the clinical case cohort

Clinicopathologic characteristics	
Age	64 y
Gender	Female
Biopsy site	Pleural effusion
Pathology diagnosis	Adenocarcinoma
Immunohistochemical analysis	
Cam5.2	Positive
CK7	Positive
MOC-31	Positive
CA 19-9	Positive
CA125	Positive
TTF-1	Positive (weak)
CK20	Negative
Calretinin	Negative
WT-1	Negative
D2-40	Negative
SP-A	Negative
Napsin A	Negative
CDX-2	Negative
ER	Negative
PR	Negative
BRST-2	Negative
Mammoglobin	Negative
92-gene assay molecular diagnosis	Lung adenocarcinoma
Biomarker analysis	
EGFR mutation	Exon 19 deletion
ALK rearrangement	Not detected
ROS-1 rearrangement	Not detected

H & E stain

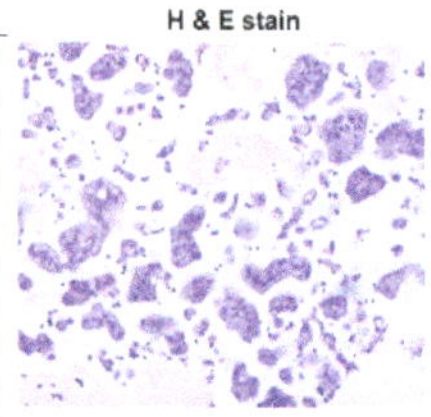

20X

In one case, a 64 year old female presented with malignant cells in a pleural effusion (Table 3). The pathology report indicated a diagnosis of adenocarcinoma with possible primary sites including the lung, pancreaticobiliary tract, and genital tract based on immunohistochemical stains. The 92-gene assay provided a molecular diagnosis of lung adenocarcinoma (probability = 96%), with all other suspected primary sites ruled out with 95% confidence. Predictive biomarker analysis detected a deletion of exon 19 in *EGFR*, while *ALK* rearrangement and *ROS-1* rearrangement were not detected (Table 3). These data indicate potential benefit from treatment with targeted molecular therapies. In another case, an 81 year old female presented with malignant cells in a pericardial effusion and was initially suspected to have a pancreaticobiliary or upper gastrointestinal malignancy (Table 4). The 92-gene assay resulted in a diagnosis of lung adenocarcinoma (probability = 90%) with an *EGFR* L858R mutation, enabling eligibility for EGFR targeted therapy options.

Table 4: Summary of the clinical characteristics, molecular diagnosis from 92-gene assay and results from predictive biomarker testing from a clinical patient case utilizing the 92-gene assay and NGS testing

Clinicopathologic characteristics	
Age	81 y
Gender	Female
Biopsy site	Pericardial fluid
Pathology diagnosis	Pancreaticobiliary vs Upper GI
Immunohistochemical analysis	
CEA	Positive
CK7	Positive
MUC4	Positive (weak)
CK5/6	Positive (weak)
TTF-1	Negative
Napsin A	Negative
CDX-2	Negative
CK20	Negative
ER	Negative
PAX-8	Negative
GATA-3	Negative
Calretinin	Negative
92-gene assay molecular diagnosis	Lung adenocarcinoma
Biomarker analysis	
EGFR mutation	Exon 21 L858R
KIT mutation	Exon 10 M541L
TP53 mutation	R273S
ALK rearrangement	Not detected
ROS-1 rearrangement	Not detected
RET rearrangement	Not detected
PD-L1 expression	Not detected

H & E stain

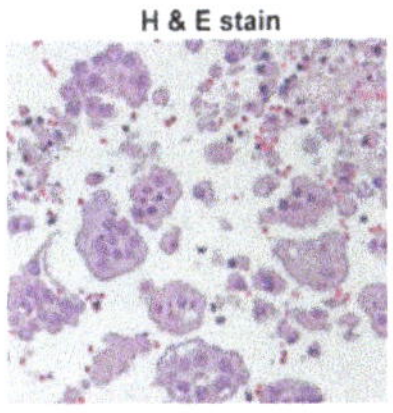

20X

DISCUSSION

Advances in genomics and cancer-targeting technologies have provided new diagnostic and treatment options for an increasing range of metastatic malignancies [16, 17]. In cases where small amounts of tissue are obtained from minimally invasive biopsies, the extent of diagnostic work-up for accurate diagnosis and the technical success of genomic assays that now are often recommended to define targeted treatment options may be limited [18]. Thoughtful use of complementary techniques such as IHC, fluorescence in situ hybridization (FISH), mutational and gene expression-based analyses is critical for optimizing disease management in oncology. Implementation of a tissue-sparing platform, such as the 92-gene assay, provides a standardized approach for tumor classification in cases wherein limited biopsy specimen is available. In addition to the 92-gene assay, several validated molecular cancer classifiers that analyze either messenger RNA or microRNA are commercially available [8, 9, 19]. Prior studies of microarray-based classifiers have previously assessed feasibility in a small cohort of cytologic specimens (n = 27), however, blinded validation of performance was not conducted [20]. Findings from this study show that the 92-gene assay demonstrated high sensitivity (91%) in the blinded validation of the limited tissue cohort (Figure 2C). Importantly, the accuracy of the assay was maintained across specimen types (Figure 2C), clinically relevant subsets (histologic grade and disease type) (Table 2), and is comparable to the accuracy observed in non-limited specimens (excisions and resections) from the previously published 790-sample blinded validation of the 92-gene assay [8].

Examination of the analytical success rate of the 92-gene assay in the consecutive clinical cytology cases illustrated high success rate (87%), irrespective of lower tumor content and cellularity than required by the assay based on histopathological examination. The 92-gene assay distinguished multiple tumor types from a range of biopsy sites (Figures 2A, 2B, and 3). The observed spectrum of molecular diagnoses in the cytology clinical case series (23 distinct tumor types) highlights the difficulty in establishing a diagnosis and treatment plan when analyzing metastatic and poorly differentiated tumors from a limited tissue specimen. A key contributing factor to the analytical success of the 92-gene assay is its ability to interrogate multiple biologic pathways while consuming a minimal amount of tissue from specimens with a range of cellularity [8, 12]. Furthermore, the 92-gene assay has been optimized to manage additional sample complexity arising from tissue preservation methodologies like the use of FFPE samples, which can cause degradation of the nucleic acid content within the specimen.

Utilization of tissue sparing technologies can impact downstream analyses on patient tissue samples. While not the objective of this study, the 92-gene assay in the clinical case series allowed for additional biomarker or mutation analysis of actionable targets within a tumor type. Thirty-six of the 37 cytology cases with a diagnosis of lung-, colorectal-, gastric-, breast adenocarcinoma, and melanoma had additional biomarker testing performed at the same lab (Supplementary Table 1). These additional analyses can provide useful information with respect to treatment planning and individualized response to therapy (Table 3 and Table 4). These data illustrate that molecular techniques such as the 92-gene assay may allow combined diagnostic and biomarker testing for limited tissue specimens to maximize their diagnostic yield (Supplementary Table 1).

New techniques such as NGS provide an increasing amount of information while decreasing tissue consumption, however, knowledge of histology and tumor type continue to be critical for individualizing treatment. Growing understanding of cancer biology has led to an emerging form of clinical trial design, termed "Basket Trials," based on the hypothesis that a molecular alteration predicts response to a targeted therapy independent of tumor histology. Initial results from these studies have shown that integration of tumor type/cellular context continues to be fundamental in the era of precision medicine and one cannot extrapolate the effectiveness of a targeted therapy uniformly to all cancer types with a matching molecular alteration [21, 22]. These data underscore the importance of having an accurate diagnosis for metastatic cancer patients to enable responsible precision medicine. For example, in both case studies presented, the molecular diagnosis from the 92-gene assay in conjunction with downstream biomarker analyses provided actionable information including eligibility and efficacy for specific targeted therapies and clinical trial protocols that can be used to tailor each patient's treatment regimen.

In addition to providing a diagnosis, pathologists are responsible for the strategic use of tissue specimens in molecular studies [23-25]. Use of a tissue-based, diagnostic algorithm which incorporates molecular testing may help preserve the tissue specimen and allow for further evaluation (Figure 4). Diagnostic algorithms should be flexible according to individual clinical and pathological characteristics (*e.g.* working diagnosis, histologic appearance, presence of metastases, and tumor location), as well as the needs and case distributions at a clinical practice (Figure 4). Clinicians and pathologists, as members of an oncology care team may implement tissue-sparing diagnostic algorithms based on consensus decisions custom-tailored to their patient population. For example, rapid interpretation of FNA slides is a means to ensure adequate tissue sampling while the diagnostic procedure is in progress; this requires trained personnel and can be done on-site in radiology or endoscopy suites or remotely with the help of digital imaging technology [26, 27].

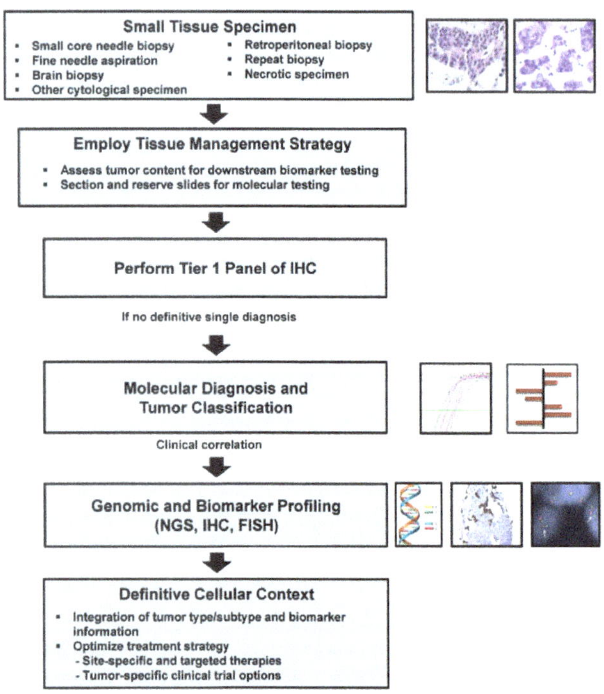

Figure 4: Proposed tissue-based diagnostic algorithm. Adapted from (Schnabel & Erlander, Expert Opin. Med. Diag. 2012). Abbreviations: IHC = Immunohistochemistry, FISH = Fluorescence in-situ hybridization, NGS = Next generation sequencing.

Quality control mechanisms and appropriate handling are needed to ensure tissue sample preservation and avoid exhaustion of blocks [28]. Additionally, effective reporting mechanisms have to be in place to communicate not only the diagnosis but also quality and quantity of tumor cells in any given tissue sample [5]. The goal of combining these various sampling and analytic approaches is to ensure accurate diagnosis and allocation of tissue from small samples for the appropriate ancillary studies at the point of care.

Key findings of this study demonstrate the high accuracy and analytical success rate of the 92-gene assay in limited tissue samples, providing further support of the diagnostic utility of the 92-gene assay for cytology and limited tissue samples. A limitation of the study was that although a high analytical success rate was observed in clinical cytology specimens, clinical validation by definitive follow up could not be performed at this point. Further studies to include clinical follow up and patient consent are needed to define these parameters in more detail. Additionally, further studies assessing the economic impact, effect on time to treatment, and patient outcomes after molecular diagnosis are merited. Incorporation of molecular classifiers such as the 92-gene assay that have been designed as a tissue sparing technology may serve as a safeguard for tissue preservation, especially in those cases where predictive biomarker analysis is warranted, and mitigate the risks associated with repeated biopsy procedures.

MATERIALS AND METHODS

Patient cases that were examined in this report included 1) samples from a prospective blinded clinical validation and 2) a case series from clinical testing.

Blinded clinical validation

A validation cohort consisting of N=109 limited tissue samples that were part of a larger 790-sample blinded validation were evaluated. Details of case selection, adjudication, sample processing and analysis of the samples for the validation cohort (N = 109) were previously described [8]. In brief, formalin-fixed paraffin-embedded (FFPE) tissue samples of known specimen type and histopathologic diagnoses that met study inclusion criteria were selected with waived patient consent by the respective IRBs. Representative slides were scanned, uploaded to a digital pathology platform (Aperio®), and diagnostically adjudicated. Definitive histopathological diagnoses (gold standard) were established utilizing ancillary tests and integration of clinical, radiographic and follow-up information, and where consensus was obtained during adjudication by at least two board certified study pathologists. Samples were categorized as "limited" in the parent study based on specimen type and confirmed by microscopic review by the adjudicating pathologists. Limited tissue samples were defined as cell blocks from fine needle aspiration (FNA) samples (N = 20), core biopsies and other small biopsy specimens such as skin punch or shave biopsies and curettings (N = 52), and small excisions with tumor spanning less than 1 cm (N = 37), while non-limited tissue samples were from surgical procedures where entire tumors were available for analysis. Inclusion criteria were specified in the protocol of the parent study; (i) tumors processed less than 6 years prior to testing, (ii) diagnosis contained within the assay panel, (iii) at least 40% tumor available in a markable area on the hematoxylin and eosin (H&E) slide, (iv) minimal necrosis, and (v) result within the analytical QC requirement of the assay.

Clinical case series

The 92-gene assay was performed on 644 consecutive cytology samples submitted as part of routine clinical evaluation. The study was approved by Western Institutional Review Board (WIRB) with waived individual patient consent. Selection criteria for clinical cytology cases included any FNA biopsy or cell block containing cytologic body fluid specimens (pleural effusions, ascites, and pericardial effusions). In cytology cases where the number of tumor cells was lower than the suggested minimum requirements for the 92-gene assay, the assay was performed with fewer tumor cells at the discretion of the reviewing pathologist, and results reported for cases that did not exceed the PCR analytical cutoff value for internal controls (cycling threshold > 29). Additionally, biomarker analysis (*EGFR*, *KRAS*, *BRAF*, *PIK3CA*, and *ALK* rearrangements) was performed for cases in which the clinician requested biomarker testing subsequent to the 92-gene assay results of lung-, colorectal-, gastric-, and breast carcinoma and melanoma.

92-Gene assay

Cases were analyzed and molecular diagnosis predicted using the 92-gene assay (CancerTYPE ID®, Biotheranostics, Inc.) as described previously [8, 12]. Tumor cells were captured and enriched via laser microdissection (LMD) and RNA extracted from the specimen. Real-time RT-PCR analysis yields a collective gene expression profile from 87 tumor-related genes and 5 internal reference genes. An algorithm compares the patient's gene expression profile to expression profiles from a reference database composed of known tumor types thereby generating a molecular diagnosis reported as a percent probability [8, 12]. Input requirements for the 92-gene assay include approximately 300 to 500 non-necrotic tumor cells. Sample input range is generally between 0.2 nanograms to 200 nanograms of RNA. Cases exceeding the PCR analytical cut-off for internal controls (cycling threshold (C_t) >29) were considered quality control failures.

Data analyses

Overall sensitivity (i.e., diagnostic accuracy) of the 92-gene assay was calculated by comparing the molecular diagnosis to the definitive histopathological diagnosis obtained by consensus adjudication by two board certified pathologists who integrated morphology, ancillary testing, clinical and radiologic findings during the validation study (gold standard). Number of concordant cases was then divided by the total number of cases classifiable by the assay. Cases were categorized as unclassifiable or indeterminate if the algorithm did not report a probability of at least 70% for a single site diagnosis. Specificity was calculated as the number of true negative results divided by the sum of true negatives and false positives. Both sensitivity and specificity calculations employed a two-sided 95% confidence interval. Fisher's exact test was used to compare diagnostic accuracy between clinical subsets. The performance of the 92-gene assay in the limited tissue subset (N = 109) was further examined according to biopsy type (FNA/ cell block, core biopsy, other small biopsies and small excisions). Analysis of the performance of the 92-gene assay with respect to disease type and histologic grade was also conducted. Analytical success rate of the 92-gene assay for the clinical samples was defined as the proportion of cases passing pathology review, analytical quality control, and completion of 92-gene assay testing.

ACKNOWLEDGMENTS

The authors thank Yi Zhang for providing statistical analysis support for the manuscript.

CONFLICTS OF INTEREST

The following co-authors KAS, TNO, BES, and CAS are employees of Biotheranostics, Inc. The following authors PSS, SMD, SEK, and EFB received commercial research grants from Biotheranostics, Inc.

REFERENCES

1. Heitzer E, Ulz P, and Geigl JB. Circulating tumor DNA as a liquid biopsy for cancer. Clin Chem. 2015; 61:112-23.
2. Ilie M, Hofman V, Long E, Bordone O, Selva E, Washetine K, Marquette CH, and Hofman P. Current challenges for detection of circulating tumor cells and cell-free circulating nucleic acids, and their characterization in non-small cell lung carcinoma patients. What is the best blood substrate for personalized medicine? Ann Transl Med. 2014; 2:107.
3. Dumur CI, Idowu MO, and Powers CN. Targeting tyrosine kinases in cancer: the converging roles of cytopathology and molecular pathology in the era of genomic medicine. Cancer Cytopathol. 2013; 121:61-71.
4. Koboldt DC, Steinberg KM, Larson DE, Wilson RK, and Mardis ER. The next-generation sequencing revolution and its impact on genomics. Cell. 2013; 155:27-38.
5. Cagle PT, Sholl LM, Lindeman NI, Alsabeh R, Divaris DX, Foulis P, Lee G, Neal JW, Nowak JA, Yu PP, and Members of the Cancer Biomarker Reporting Workgroup CoAP. Template for reporting results of biomarker testing of specimens from patients with non-small cell carcinoma of the lung. Arch Pathol Lab Med. 2014; 138:171-4.
6. Rafael OC, Aziz M, Raftopoulos H, Vele OE, Xu W, and Sugrue C. Molecular testing in lung cancer: fine-needle aspiration specimen adequacy and test prioritization prior to the CAP/IASLC/AMP Molecular Testing Guideline publication. Cancer Cytopathol. 2014; 122:454-8.
7. Hur J, Lee HJ, Nam JE, Kim YJ, Hong YJ, Kim HY, Kim SK, Chang J, Kim JH, Chung KY, Lee HS, and Choi BW. Additional diagnostic value of tumor markers in cytological fluid for diagnosis of non-small-cell lung cancer. BMC Cancer. 2012; 12:392.
8. Kerr SE, Schnabel CA, Sullivan PS, Zhang Y, Singh V, Carey B, Erlander MG, Highsmith WE, Dry SM, and Brachtel EF. Multisite validation study to determine performance characteristics of a 92-gene molecular cancer classifier. Clin Cancer Res. 2012; 18:3952-60.
9. Pillai R, Deeter R, Rigl CT, Nystrom JS, Miller MH, Buturovic L, and Henner WD. Validation and reproducibility of a microarray-based gene expression test for tumor identification in formalin-fixed, paraffin-embedded specimens. J Mol Diagn. 2011; 13:48-56.
10. Sokilde R, Vincent M, Moller AK, Hansen A, Hoiby PE, Blondal T, Nielsen BS, Daugaard G, Moller S, and Litman T. Efficient identification of miRNAs for classification of tumor origin. J Mol Diagn. 2014; 16:106-15.
11. Varadhachary GR and Raber MN. Carcinoma of unknown primary site. N Engl J Med. 2014; 371:2040.
12. Erlander MG, Ma XJ, Kesty NC, Bao L, Salunga R, and

Schnabel CA. Performance and clinical evaluation of the 92-gene real-time PCR assay for tumor classification. J Mol Diagn. 2011; 13:493-503.

13. Ma XJ, Patel R, Wang X, Salunga R, Murage J, Desai R, Tuggle JT, Wang W, Chu S, Stecker K, Raja R, Robin H, Moore M, et al. Molecular classification of human cancers using a 92-gene real-time quantitative polymerase chain reaction assay. Arch Pathol Lab Med. 2006; 130:465-73.
14. Weiss LM, Chu P, Schroeder BE, Singh V, Zhang Y, Erlander MG, and Schnabel CA. Blinded comparator study of immunohistochemical analysis versus a 92-gene cancer classifier in the diagnosis of the primary site in metastatic tumors. J Mol Diagn. 2013; 15:263-9.
15. Hainsworth JD, Rubin MS, Spigel DR, Boccia RV, Raby S, Quinn R, and Greco FA. Molecular gene expression profiling to predict the tissue of origin and direct site-specific therapy in patients with carcinoma of unknown primary site: a prospective trial of the Sarah Cannon research institute. J Clin Oncol. 2013; 31:217-23.
16. Larkin J, Ascierto PA, Dreno B, Atkinson V, Liszkay G, Maio M, Mandala M, Demidov L, Stroyakovskiy D, Thomas L, de la Cruz-Merino L, Dutriaux C, Garbe C, et al. Combined vemurafenib and cobimetinib in BRAF-mutated melanoma. N Engl J Med. 2014; 371:1867-76.
17. Shaw AT and Engelman JA. Ceritinib in ALK-rearranged non-small-cell lung cancer. N Engl J Med. 2014; 370:2537-9.
18. Karnes HE, Duncavage EJ, and Bernadt CT. Targeted next-generation sequencing using fine-needle aspirates from adenocarcinomas of the lung. Cancer Cytopathol. 2014; 122:104-13.
19. Rosenwald S, Gilad S, Benjamin S, Lebanony D, Dromi N, Faerman A, Benjamin H, Tamir R, Ezagouri M, Goren E, Barshack I, Nass D, Tobar A, et al. Validation of a microRNA-based qRT-PCR test for accurate identification of tumor tissue origin. Mod Pathol. 2010; 23:814-23.
20. Stancel GA, Coffey D, Alvarez K, Halks-Miller M, Lal A, Mody D, Koen T, Fairley T, and Monzon FA. Identification of tissue of origin in body fluid specimens using a gene expression microarray assay. Cancer Cytopathol. 2012; 120:62-70.
21. Hyman DM, Puzanov I, Subbiah V, Faris JE, Chau I, Blay JY, Wolf J, Raje NS, Diamond EL, Hollebecque A, Gervais R, Elez-Fernandez ME, Italiano A, et al. Vemurafenib in Multiple Nonmelanoma Cancers with BRAF V600 Mutations. N Engl J Med. 2015; 373:726-36.
22. Le Tourneau C, Delord JP, Goncalves A, Gavoille C, Dubot C, Isambert N, Campone M, Tredan O, Massiani MA, Mauborgne C, Armanet S, Servant N, Bieche I, et al. Molecularly targeted therapy based on tumour molecular profiling versus conventional therapy for advanced cancer (SHIVA): a multicentre, open-label, proof-of-concept, randomised, controlled phase 2 trial. Lancet Oncol. 2015; 16:1324-34.
23. Cagle PT, Allen TC, and Olsen RJ. Lung cancer biomarkers: present status and future developments. Arch Pathol Lab Med. 2013; 137:1191-8.
24. Schnabel CA and Erlander MG. Gene expression-based diagnostics for molecular cancer classification of difficult to diagnose tumors. Expert Opin Med Diagn. 2012; 6:407-19.
25. Travis WD, Brambilla E, Noguchi M, Nicholson AG, Geisinger K, Yatabe Y, Ishikawa Y, Wistuba I, Flieder DB, Franklin W, Gazdar A, Hasleton PS, Henderson DW, et al. Diagnosis of lung cancer in small biopsies and cytology: implications of the 2011 International Association for the Study of Lung Cancer/American Thoracic Society/European Respiratory Society classification. Arch Pathol Lab Med. 2013; 137:668-84.
26. da Cunha Santos G, Boerner SL, and Geddie WR. Maximizing the yield of lymph node cytology: Lessons learned from rapid onsite evaluation of image- and endoscopic-guided biopsies of hilar and mediastinal lymph nodes. Cancer Cytopathol. 2011; 119:361-6.
27. Schmidt RL, Witt BL, Lopez-Calderon LE, and Layfield LJ. The influence of rapid onsite evaluation on the adequacy rate of fine-needle aspiration cytology: a systematic review and meta-analysis. Am J Clin Pathol. 2013; 139:300-8.
28. Kapp JR, Diss T, Spicer J, Gandy M, Schrijver I, Jennings LJ, Li MM, Tsongalis GJ, de Castro DG, Bridge JA, Wallace A, Deignan JL, Hing S, et al. Variation in pre-PCR processing of FFPE samples leads to discrepancies in BRAF and EGFR mutation detection: a diagnostic RING trial. J Clin Pathol. 2015; 68:111-8.

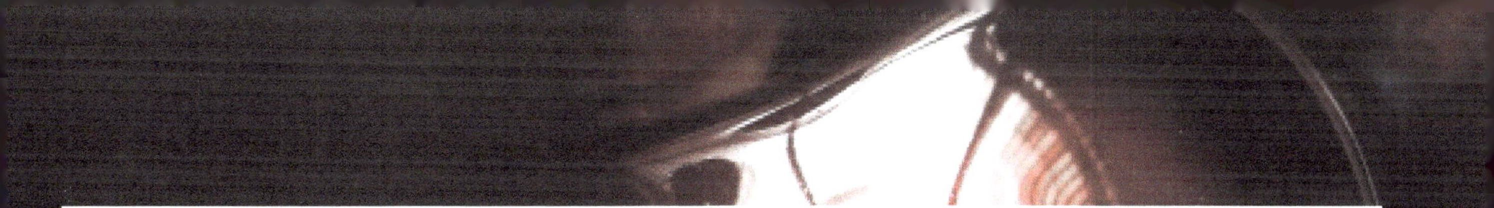

CHAPTER 15

CHANGING PERSPECTIVE ON ONCOMETABOLITES: FROM METABOLIC SIGNATURE OF CANCER TO TUMORIGENIC AND IMMUNOSUPPRESSIVE AGENTS

Mauro Corrado[1,2], Luca Scorrano[1,3] and Silvia Campello[2,4]

[1] Dulbecco-Telethon Institute, Venetian Institute of Molecular Medicine, Padova, Italy

[2] IRCCS Fondazione Santa Lucia, Roma, Italy

[3] Department of Biology, University of Padova, Padova, Italy

[4] Department of Biology, University of Roma Tor Vergata, Roma, Italy

ABSTRACT

During tumorigenesis, the shift from oxidative phosphorylation to glycolysis in ATP production accounts for the dramatic change in the cellular metabolism and represents one of the major steps leading to tumour formation. The so-called Warburg effect is currently considered something more than a mere modification in the cellular metabolism. The paradox that during cancer cell proliferation the increase in energy need is supplied by glycolysis can be only explained by taking into account the many roles that intermediates of glycolysis or TCA cycle play in cellular physiology, besides energy production. Recent studies have shown that metabolic intermediates induce changes in chromatin structure or drive neo-angiogenesis. In this review, we present some of the latest findings in the study of cancer metabolism with particular attention to how tumour metabolism and its microenvironment can favour tumour growth and aggressiveness, by hijacking and dampening the anti-tumoral immune response.

INTRODUCTION

Cancer formation and progression pass through a multistep process in which cancer cells acquire new properties and deregulate many homeostatic pathways. This, to enable them to sustain cell proliferation, escape cell death, disrupt the original niche of development by inducing neovascularization, increase the dimension of the tumoral mass and eventually metastasize. All these properties have been considered as hallmarks of cancer, as described by Hanahan and Weinberg in their seminal review in 2000 [1]. Nonetheless, years of advances in oncology have rendered that picture rather incomplete. As a result, researchers and oncologists have now turned their attention to characteristics which were previously under-valued, including the metabolic reprogramming of cancer cells and the role of tumour-invading inflammatory cells and inflammation [2].

In cancer cells, the tight balance between cell proliferation and cell dismissal is altered, the balancing point being tipped toward the former [3]. This uncontrolled cell proliferation needs energetic support, and is accompanied by changes in the metabolism, shifting from oxidative phosphorylation to glycolysis [4]. When cells are in a non-proliferative state and mitochondria are functional, pyruvate - the final product of

Citation: Mauro Corrado, Luca Scorrano and Silvia Campello, Changing perspective on oncometabolites: from metabolic signature of cancer to tumorigenic and immunosuppressive agents, DOI: 10.18632/oncotarget.8727.

glycolysis - is imported into the organelles and is completely oxidized in the TCA cycle to produce reducing equivalents (NADH); these molecules then fuel the mitochondria respiratory chain to produce ATP through oxidative phosphorylation [5]. Conversely, Otto Warburg observed that, even in the presence of oxygen, cancer cells rely instead on glycolysis for their metabolic needs, hence his coining the term "aerobic glycolysis" [6],[7]. However, for decades the "Warburg effect" was considered only as a metabolic signature of cancer, or an adaptation to an environment with low oxygen concentration within the tumoral mass. Moreover, the discovery of oncogenes and onco-suppressors shed light on the genetic basis of cancer, diverting attention away from cancer metabolism [8],[9],[10]. The discovery of the mitochondrial localization of the onco-protein Bcl-2 extended the role of these organelles from metabolism to the control of the apoptotic cascade, therefore presenting a novel scenario for mitochondria and metabolism in cancer development [11],[12],[13],[14]. Interestingly, the metabolic switch towards glycolysis was shown to be not limited to cancer cells, but shared with many other highly proliferative cell types: cells during embryogenesis, stem cells upon growth factors stimulation, T cells after antigen activation are a few examples [15],[16],[17]. Moreover, a connection between cancer and altered metabolism was clearly established when mutations in genes encoding for the two TCA cycle enzymes succinate dehydrogenase (SDH) [18] and fumarate hydratase (FH) [19] were identified in human tumours.

Despite these observations, a paradox emerges: why would active proliferating cells use a less efficient method to produce ATP, compared to oxidative phosphorylation? One explanation could be that, when resources are scarce, cells turn to an inefficient method to produce at least a minimal amount of ATP; however, this is not the case for cancer cells that are supplied with glucose and nutrients by the neo-angiogenic blood vessels. A more convincing explanation suggests that cancer cells have metabolic requirements that go beyond ATP production: the intermediates derived from glycolysis and TCA cycle could be diverted towards biosynthetic pathways or could activate signalling cascades, thus enabling the characteristic hallmarks of cancer to occur. As a consequence, changes in metabolism can sustain cell proliferation by feeding anabolic pathways; moreover, they can also modify the tumour microenvironment by altering the complex interaction and crosstalk between cancer cells and normal cells surrounding, or infiltrating, the tumoral mass [4],[20]. Indeed, we can speculate that the newly depicted hallmarks of cancer, glycolytic metabolism and pro-inflammatory microenvironment, are mutually regulated and influence each other so as to determine a tumour's overall capacity for development.

Aerobic glycolysis can be triggered by mutations in oncogenes, whose broad effect on cellular reprogramming involves changes in the regulation of metabolism. Alternatively, and even more interestingly, it can arise directly from mutations in genes involved in glycolysis or TCA cycle [21]. In the latter case, a direct link between altered metabolism and cancer is provided (Figure 1). In a second step, the oncogenic properties acquired by metabolic intermediates, accumulating as a consequence of the Warburg effect, will influence the tumour growth. Consequently, by this mean they will amplify pro-oncogenic signals, while -in some cases, they will exert a paracrine effect on cancer-infiltrating inflammatory cells.

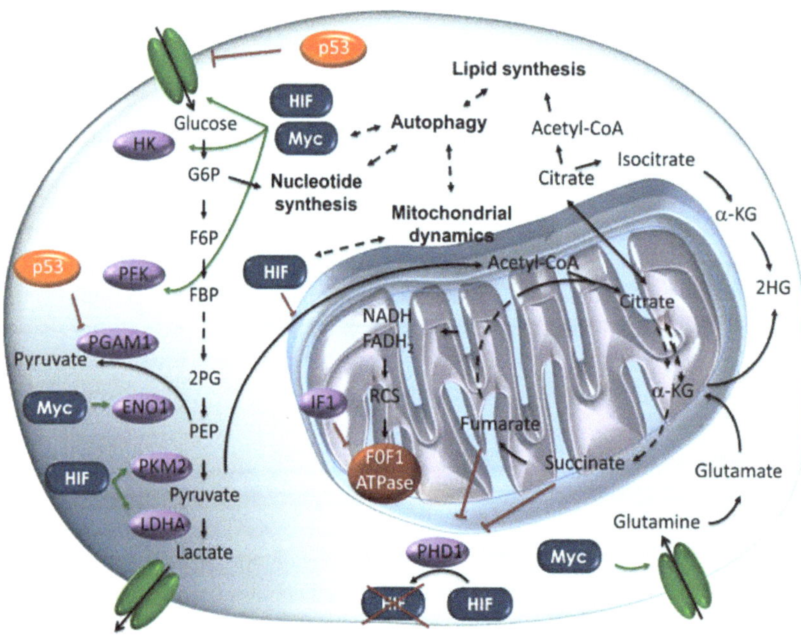

Figure 1: Schematic representation of the metabolic pathways altered in cancer cells. Metabolic pathways in cancer cells are directly controlled by the main oncogenes and oncosuppressors. This schematic picture depicts our current knowledge about how glycolysis and oxidative phosphorylation are inter-regulated through the synthesis of nucleotides and lipids. Key steps promoting the "Warburg effect" in cancer cells are shown. In *orange* onco-suppressor proteins, while in *blue* onco-proteins are shown. In *violet*, onco-proteins' direct targets in the metabolic cascade are depicted. *Dashed lines* represent biunique interactions. Abbreviations: α-KG, α-Ketoglutarate; ENO1, Enolase1; FBP, Fructose-1,6-biphosphate; F6P, Fructose-6-phosphate; G6P, Glucose-6-phosphate, HK, hexokinase; HIF, hypoxia inducible factor; IF1, ATPase inhibitory factor 1; LDHA, Lactate dehydrogenase A; PGAM1, phosphoglycerate mutase 1; PFK, phosphofructokinase; PEP, Phosphoenolpyruvate; PHD1, prolyl hydroxylase domain 1; PKM2, pyruvate kinase isoform 2; 2HG, 2-hydroxyglutarate; 2PG, 2-phosphoglycerate; RCS, Respiratory Chain Supercomplexes.

CLASSIC ONCOGENIC SIGNALS DIVERT METABOLISM FROM OXIDATIVE PHOSPHORYLATION TO GLYCOLYSIS

Many tumours are driven by mutations in oncogenes or oncosuppressor genes such as c-Myc, RAS and p53. Interestingly, it has been shown that alterations in their function are responsible for the metabolic reprogramming observed in many cancer cells (Figure 1).

The multifaceted oncogene c-Myc is a master regulator of cellular growth and metabolism in cancer cells [22]. Although in some cancer types c-Myc has a primary oncogenic role, its DNA sequence being translocated downstream of promoters of either the light or the heavy immunoglobulin chain [23], its activity is usually up-regulated post-transcriptionally by other oncogenic signals [24]. The first hint that c-Myc has a direct role in up-regulating glycolysis in cancer came from the observation that LDH-A, the enzyme converting pyruvate to lactate, is a putative target of c-Myc [25]. Since then, different enzymes involved in glycolysis have been found over-expressed in a c-Myc-dependent manner (glucose transporter - GLUT1, hexokinase 2 - HK2, phosphofructokinase - PFKM and Enolase 1 - ENO1) [26],[27],[28]. Interestingly, not only glucose metabolism, but also glutaminolysis is increased in a c-Myc dependent manner, thus causing cancer cells to become addicted to glutamine and sensitive to its withdrawal [29].

Other oncogenes also play a role in regulating metabolism. The MAP kinase pathway (RAS-RAF-MEK-ERK pathway) is often altered in cancer. In particular, single-point mutations in RAS lead to its constitutively active signalling in many solid cancers such as pancreas, lung and colorectal cancer [30]. Recently, an unexpected link between the RAS pathway and mitochondria has been described. Indeed, activated RAS (H-RAS) mediates the translocation of the signal transducer and activator of transcription 3 (STAT3) to mitochondria, where it regulates mtDNA transcription, so altering electron transport and increasing lactate production [31],[32]. Also RAF, which acts downstream RAS, is able to inhibit oxidative phosphorylation and down-regulate the master regulator of mitochondrial biogenesis PGC1α [33]. These observations show that, despite its complex role during cancer formation and progression, the MAP kinase pathway also impinges on cancer metabolism.

To note, c-Myc and RAS have also a prominent role in regulating the catabolic process of autophagy [34],[35], of which the metabolic regulator mTOR is the major inhibitor, most likely responding to the new metabolic demands during cancer progression [36]. To give some insights, c-Myc shows an intricate relationship with the autophagic signalling machinery component AMBRA1. This factor, indeed, is able to facilitate PP2A-dependent dephosphorylation and degradation of c-Myc, in conditions in which autophagy is active whilst mTOR and cell proliferation, hence cancer development, are inhibited [37].

More in general, the role of autophagy and organelle quality control (especially of mitochondria) during cancer development has long been debated. Briefly, the common believe today is that autophagy plays different roles in different stages of cancer development. In healthy tissues, or in early stages of cancer, autophagy represents a pivotal anti-tumoral defence. Instead, when cancer is established by autophagy-unrelated mutations, up-regulation of autophagy facilitates cell survival and metabolic adaptation of cancer cells. Indeed, the intriguing crosstalk between autophagy and cancer goes beyond the goal of this review and for this reason we direct the interested reader to other outstanding reviews [36],[38],[39].

Regarding onco-suppressors, the observations that the mutated forms of p53 have oncogenic properties led to the initial misclassification of p53 as an oncogene. Further studies showed that p53 mutations are gain-of-function mutations and that the wild-type protein is instead an onco-suppressor [40]. The p53 onco-suppressor activity is manifested when it is stabilized and its transcriptional activity is up-regulated in response to a variety of stress stimuli. This is in order to mediate apoptosis, DNA repair, cell cycle arrest or senescence, to maintain genome integrity and finally to limit cancer development [41]. Interestingly, recent studies also revealed a role for p53 in the regulation of metabolism [42]. In particular, wild-type p53 negatively regulates glycolysis, through negative regulation or mis-localization of glucose transporters (GLUT1 and GLUT4) [43] and other glycolytic enzymes (phosphoglycerate mutase - PGM; pyruvate dehydrogenase kinase 2 - PDK2) [44],[45]. Moreover, p53 also shunts glucose to the pentose phosphate pathway and to NADPH production [46]. NADPH production has a double role both in anabolic pathways and in restoring the reduced form of the anti-oxidant glutathione (GSH) [47], a molecule in the first line of defence *versus* the oxidative stress responsible for increased mutational rate. Hence, in the case of mutated p53, the electron transport chain (ETC) is compromised and cells switch to glycolysis to overcome the block in ATP production. In addition to directly regulating the expression of glycolytic enzymes and components of the ETC, p53 can repress glycolysis by inhibiting the AKT/mTOR and NF-kB signalling pathways whose activity is strongly up-regulated in cancer cells [48],[49]. Thus, the different "negative" gain-of-function properties of p53 include promotion of cell proliferation, angiogenesis, migration, invasion, metastatization, chemoresistence [40] and, importantly for the purposes of this review, the capacity to switch metabolism towards glycolysis [50],[51] (Figure 1).

WHEN MITOCHONDRIA ARE UPSTREAM OF CANCER DEVELOPMENT

Although of great interest, only recently the metabolic effects of tumour-associated mutations have been investigated; as such, a complete picture of the role of mutations in oncogenes and oncosuppressors in cancer metabolism is still lacking. Besides the mutations so far described, alterations in mitochondrial metabolism have also been directly linked to cancer development. Interestingly, inhibition of oxidative phosphorylation leads to a loss or inactivation of p53 due to the generation of reactive oxygen species [52]. Moreover, a number of mutations in genes directly implicated in glycolysis or TCA cycle have been shown to promote tumour formation [53].

Indeed, mutations in either fumarate hydratase (FH) or succinate dehydrogenase (SDH B, C and D), both enzymes of the TCA cycle, are known to promote different cancer types, from leiomyoma, leiomyosarcoma and renal cell carcinoma (FH mutations) to paraganglioma and pheochromocytoma

(SDH mutations) [19],[18]. Interestingly, these mutations give rise to pseudo-hypoxia, a condition in which the complex metabolic and transcriptomic changes, usually taking place under low oxygen concentration, occur under normoxic conditions. Pseudo-hypoxia enhances cancer formation and aggressiveness [54],[55]. Indeed, all solid cancers experience cycling changes in oxygen pressure, varying from normal oxygen concentration to hypoxia or even anoxia in the innermost part of the tumoral mass [56]. During acute hypoxia, hypoxia inducible factor (HIF)1α - otherwise degraded - is stabilized and induces the transcription of its target genes including specific angiogenic and pro-metastatic factors [57]. The oncometabolites succinate and fumarate, which both accumulate due to mutations inactivating SDH or FH, inhibit HIF1α degradation, so causing pseudo-hypoxia and favouring tumour progression [58],[59],[60],[61]. Moreover, fumarate can affect and alter many different pathways, thus contributing to complex metabolic dysfunctions involved in tumour formation. For example, it can alter metabolism of urea [62] or the anti-oxidant response [63],[64],[65],[66].

Of note, SDH is the Complex II of the mitochondrial respiratory chain, feeding the OXPHOS with $FADH_2$ reduced equivalents. In fact, in addition to mutations in the nuclear DNA-encoded SDH, mutations in mtDNA-encoded subunits of the Complex I, III and IV have been associated with tumour pre-conditioning or development [67],[68].

Further, inhibition of OXPHOS can also be achieved in cancer cells by direct inhibition of F_1F_0-ATPase, mediated by over-expression of its specific inhibitor IF_1 [69]. The latter controls mitochondrial function and cell survival [70],[71] by directly acting on the mitochondrial structure and on F(1)F(o)-ATP synthase activity [69], thus representing a bad-prognosis predictor in a plethora of cancer types (ranging from liver to bladder and gastric cancer) [72],[73],[74].

In addition to mutations, alterations of the catalytic activity of metabolic enzymes can also be a secondary effect of microenvironmental signals. Hypoxia itself reduces the activity of SDH and components of the mitochondrial respiratory chain, so contributing to the metabolic shift toward glycolysis [75],[76],[77].

In addiction to hypoxic and pesudo-hypoxic signalling, also metabolic intermediates can "sensitize" back to the nucleus and control gene expression, thus unveiling an unexpected connection between oncometabolites and epigenetics. For instance, 2-hydroxyglutarate, whose accumulation is driven by mutations in the enzyme isocitrate dehydrogenase (IDH), was described to varying DNA methylation [78]. In similar studies, it was shown that increased cytoplasmic levels of acetyl-CoA and acidic pH, both due to the inhibition of TCA cycle and to the accumulation of lactic acid in the cytoplasm, modulate DNA acetylation [79],[80]. It would be of great interest to analyse whether these epigenetic alterations are stochastically distributed on the DNA or, still more intriguingly, if they are targeted to specific transcription factors or possible hot spots in promoters of cancer- or metabolism- related genes.

As a final remark of this paragraph, we would like to mention that dysfunctional mitochondria are able to signal back to the autophagy machinery to selectively remove them, through a quality control process called mitophagy. Failure on modulating mitophagy upon mitochondrial oncogenic stimuli is an intriguing cancer-related phenotype, representing a field of research that still remains to be comprehensively approached [81],[82].

CANCER-RELATED INFLAMMATION IS DRIVEN BY ONCOMETABOLITES

Pathologic analysis of human cancers' biopsies often shows an inflammatory microenvironment with recruitment to the tumoral mass of cells from both the innate and adaptive branch of the immune response [83]. Inflammation's role in the context of cancer development has long been debated. Historically, according to the immune surveillance model, inflammation was thought to be part of the attempt of the immune system to eradicate cancer cells [84]. To support this idea, there are indeed several pieces of evidence of increased tumour formation rates in mouse models where components of the immune system are genetically ablated [85],[86]. Moreover, human immunocompromised individuals show a higher probability of developing virus-derived cancer types [87]. In particular, the roster of anti-tumoral immune cells includes CD8+ cytotoxic T (CTLs), CD4+ Th1 helper lymphocytes and natural killer (NK) cells. However, more recent evidence has challenged the classic immune surveillance model by showing that some specific immune cells have, instead, the role of promoting cancer [88],[89],[90] (Figure 2). Indeed, in particular macrophages (polarized to an M2 subtype), mast cells, neutrophils and some subtype of T and B lymphocytes are able to favour cancer. To this end, these cells supply growth factors to sustain cell proliferation, survival factors to escape apoptosis, molecules or enzymes to modify extracellular matrix and facilitate invasion and metastatization, or to inhibit the anti-tumoral counterparts of the immune response [91],[92],[93]. Noteworthy, apart from its role in the cellular quality control, autophagy has been related to an unconventional secretion pathway of inflammatory mediators, such as Interleukin-1β and Interleukin-18 [94],[95].

Of great interest, some publications have started to link glycolytic metabolism of cancer cells to the inflammatory state of the tumour microenvironment. Both processes would be engaged in a positive feedback in which oncometabolites act as paracrine molecules that modulate the tumour-infiltrating cell polarity and activity (Figure 2). For example, succinate accumulation has been shown to mediate HIF1α-dependent expression of the pro-inflammatory chemokine IL-1-β by bone-marrow-derived macrophages (BMDMs) [96]. Although this has been described as the result of lipopolysaccharide-mediated glycolytic shift in the context of microbial infection, it is reasonable to speculate that a similar mechanism could be involved in cancer. Two observations support this hypothesis. First, the concentration of circulating succinate (and HIF1α activity) in patients with tumours harbouring mutations in SDH is higher than in the healthy population [54],[97]; second, succinate can signal through its receptor GPR91 on dendritic cells in order to modulate the immune response [98].

Thus, macrophages seem to have a pivotal role in the crosstalk between cancer signalling and the inflammatory response. This is supported by the role of chemokines produced by the majority of the solid tumours, such as the colony-stimulating factor 1 (CSF1), which is a potent chemoattractant for macrophages [99]. Two main subpopulations of these immune cells are reported to invade the tumour: the classically activated macrophages (M1 population) and the alternatively activated ones (M2). M1 macrophages usually promote antigen presentation and immune activation, while M2 show pro-tumoral properties [100]. Remarkably, tumour-derived TGF-β1 shifts the population of macrophages towards a M2 phenotype [101]. On a feedback loop, squamous cell carcinoma (SCC) cells are differently sensitive to monocyte-derived TGF-β [102]. TGF-β responding SCC-stem cells cycle more slowly than TGF-β insensitive counterparts, responsible for tumour growth. TGF-β responding cells show increased invasiveness and resistance to anti-cancer therapies due to an enhanced antioxidant power mediated by a TGF-β-dependent change in metabolism, which ensures an adequate NADPH pool to reduce the oxidized form of glutathione [102]. The increase in the invasive properties associated with metabolic changes is further supported by the metabolic alterations during epithelial to mesenchymal transition (EMT) and metastasis formation. During EMT, it has been shown that TGF-β signaling mediates a transient increase in mitochondrial respiration to promote migration [103]. Moreover, super invasive properties of cancer cells have been linked to increased mitochondrial activity [104], together with an increased fragmentation of the organelles facilitating cell movement [105],[106],[107]. This property correlates with the observation that mitochondria fragment and localize at the posterior area of T lymphocytes during chemotaxis, where they provide the energy required for the myosin machineries to drive migration [108],[109].

The functional versatility of mitochondria is indeed paralleled by the dynamic nature of the organelles, which can fuse and divide through the action of the so-called mitochondria-shaping proteins (OPA1, MFN1, MFN2, DRP1, FIS1, among others) [110]. The mitochondrial shape is involved in many mechanisms contributing to cancer development and progression. For instance, mitochondria fragment during cell division in a CDK1/DRP1-dependent manner to achieve their stochastic distribution between daughter cells in highly proliferative tissues [111]. In addition, highly glycolytic cells show fragmented mitochondria when compared to cells relying on OXPHOS [112]. Moreover, DRP1 and OPA1 are both involved in the mitochondrial shape changes occurring upon apoptotic stimulation [113]. In particular, OPA1 is a multifaceted protein: it is able of keeping the mitochondria cristae shape under control, so confining the pro-apoptotic cytochrome *c* inside the cristae (and thus controlling apoptosis) [114],[115]; however, it also organizes the quaternary structure of the respiratory chain complexes in super-complexes boosting OXPHOS [116]. Future studies will be necessary to dissect the anti-apoptotic *versus* the pro-OXPHOS roles of this protein in cancer development. In addition to metabolism and apoptosis, as discussed above, the shape of mitochondria also controls cell migration during metastatization. Indeed, fragmented mitochondria re-localize within specific areas of migrating cells to generate the ATP necessary to "fuel" the cell motor myosin, so promoting migration and invasion [108],[105],[106],[107].

A complete analysis of the role of different mitochondria-shaping proteins in cancer development is still missing, although it represents a promising area for new ground-breaking discoveries in cancer biology.

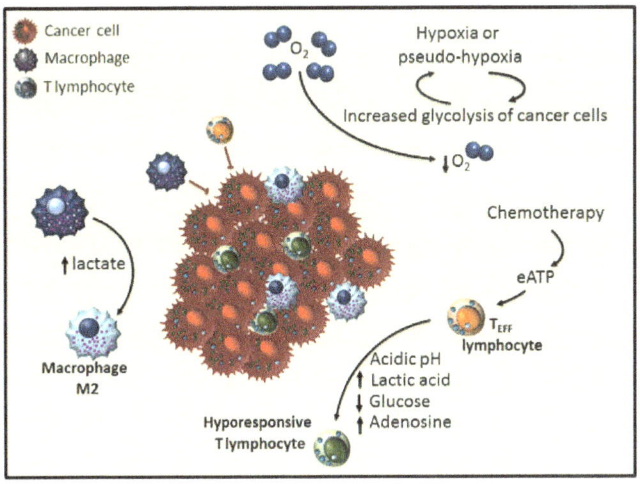

Figure 2: Tumour microenvironment hijacks anti-tumoral immune response. Cancer cells alter the immune response through metabolic competition establishing an inflammatory microenvironment. Increased glycolysis in tumoral cells depletes the tumour microenvironment of glucose and amino acids, so making T_{EFF} cells hyporesponsive to tumour antigens. Moreover, high lactic acid levels generated by tumoral cells engage macrophages in a differentiation pathway towards a pro-inflammatory and tumour-promoting M2 subtype. Chemotherapy induces immunogenic cancer cell death, which through increased extracellular ATP (eATP) is able to partially re-activate T_{EFF} cells.

A CLOSER LOOK AT TUMOUR-DERIVED LACTIC ACID: HOW TUMOUR-ASSOCIATED MACROPHAGES ARE FUNCTIONALLY POLARIZED

To enhance this intricate relationship between cancer metabolism and tumour-invading immune cells a seminal paper from Medzhitov's group shed light on a novel mechanism of communication between cancer cells and tumour-associated macrophages, based on tumour-derived lactic acid [20]. In this paper, the authors show that tumour-associated macrophages (TAM) - when compared to the peritoneal macrophages or to the tumoral cells - are characterized by a higher expression of the vascular endothelial growth factor (VEGF) and Arginase 1 (ARG1), both of which support tumour growth by neovascularization and by providing metabolic substrates respectively [20]. Remarkably, ARG1 can induce depletion of arginine from the microenvironment, thus leading to inhibition of cytotoxic CD8+ T lymphocyte function and their immune surveillance role [117]. Tumour conditioned medium shows the same capacity to induce VEGF and ARG1 expression in an HIF1α dependent manner [20], as indicated by the fact that their expression is prevented in an HIF1α null background. This suggests that some tumour-derived and secreted molecules can mediate VEGF and ARG1 expression in macrophages through a mechanism involving HIF1α stabilization under normoxic conditions. Strikingly, lactic acid - specifically produced by tumoral cells in many cancer types, in which mitochondrial respiration is inhibited and pyruvate is converted into lactic acid by the pyruvate kinase isoform M2 (PKM2) - mediates this communication loop between cancer and tumour invading cells [20]. Interestingly, the higher concentration of lactic acid, the larger and more aggressive are the tumoral masses observed. The uptake of lactic acid by TAM from the stroma is facilitated by the acidic pH also produced by the glycolytic metabolism of cancer cells. Eventually, *in vivo* co-injection of tumoral cells with lactate medium-conditioned macrophages results in tumours bigger than in the case of control-medium cultured macrophages, so confirming the pro-tumoral phenotype acquired by these immune cells [20]. In addition, the interaction between PKM2 and Transglutaminase type 2 (TG2), observed in a model of human fibrosarcoma, has been shown to significantly modulate autophagy, this facilitating the metabolic shift towards aerobic glycolysis [118],[119].

Tumour cell-derived lactic acid is also able to inhibit monocyte maturation to dendritic cells [120],[121]. This complex response to lactic acid could be an ancestral response reminiscent of the immune recognition of bacterial infection, in line with the observation that bacteria are able to inhibit oxidative phosphorylation and increase glycolysis in infected cells through LPS [96]. Above all, the discovery of this pro-tumoral paracrine effect of lactic acid could lead to the investigation of previously unexplored mechanisms endowing cancer cells with resistance to anti-cancer therapies. There is already evidence that accumulation of succinate and lactate facilitates epithelial to mesenchymal transition [122],[123]. One could also speculate, for example, that oncometabolites also exert a similar role in tropism of metastasis. Although current opinion favor the passive homing of metastasis (that is, metastasis forms in the first place where invasive or circulating cancer cells are entrapped), it is intriguing to speculate that homing of cancer cells could occur in eligible tissues according to the metabolic microenvironment, in a mechanism similar to that described for chemokines [124]. From this viewpoint, the tissue tropism of metastasis could be mediated by the favorable conditions in which to adapt and survive in the new micro-environmental niche. In line with this, as computational studies already suggested, normalizing the function of some metabolic targets could also inhibit cancer cell migration [125].

METABOLIC COMPETITION CONTROLS ANTI-TUMORAL FUNCTION OF T LYMPHOCYTES

Tumour microenvironment and imbalance of metabolites have also an important role in shaping and dampening the adaptive immune response (Figure 2) *versus* the tumoral cells, so resulting in a more aggressive tumour and poorer prognosis for the patients [126],[100]. Many publications in recent years showed that nutrient availability controls immune response by reducing the number and function of tumour-invading lymphocytes (TILs) while promoting an immunosuppressive environment [120],[20],[127],[128] (Figure 2).

T lymphocytes have characteristic metabolic features. In particular, naïve and memory T cells rely on mitochondrial respiration to fulfil their energetic demands [129],[130]. When engaged by an antigen and proper co-stimulation, T cells become effectors of the immune response, shifting their metabolism to a highly glycolytic one, this being more suitable for intense proliferation and production of cytokines [131],[132]. Importantly, cancer cells and the opponents fighting them share the same metabolic phenotype, this generating a strong competition for nutrients in the tumoral area. Pearce's group showed that glucose consumption by the tumoral cells restricts T lymphocytes, reducing mTOR activity, glycolytic capacity and IFN-γ production, thus favouring tumour progression [127]. By decreasing the immune response in a coordinate fashion, lactate generated by the glycolytic tumoral mass is also able to dampen T effector lymphocytes glycolysis in a negative feedback loop through the inhibition of lactic acid release from T cells [120]. Moreover, although in a model of rheumatoid arthritis, lactate accumulation has been shown to inhibit motility of CD4 and CD8 T lymphocytes [133].

Regulatory T cells play an important role in shaping an immunosuppressive environment in the tumour [2] and they have been shown to metabolically rely on fatty acid oxidation [134],[135], as well as on glycolysis, in conditions different than cancer [136]. Nonetheless, the interplay between cancer cell metabolism and the actively immunosuppressive population of T_{reg} cells has to be further investigated.

Tumour microenvironment is a complex milieu, where some molecules are depleted while others abound. *In vivo* studies showed that extracellular ATP levels are higher at tumour sites, when compared to tumour-free tissues [137]. Indeed, cancer cell demise is usually accompanied by release of ATP in an immunogenic fashion, especially in response to chemotherapic agents [138]. Extracellular ATP (eATP) is a potent alert signal for the immune system and differently targets various tumour-infiltrating cells. ATP acts, indeed, as an autocrine or paracrine co-stimulator for IL-2 production by activated T lymphocytes, thus regulating intracellular calcium waves and T cell activation and motility [139],[140],[141],[142]. Moreover, eATP attracts dendritic cell (DC) precursors into the tumour bed, so facilitating their stabilization in the proximal area of dying cells, and their development in mature cells with the capacity of presenting tumour-associated antigens [143].

To counteract the pro-immunogenic role of ATP, its levels can be reduced by the ecto-enzymes CD39 and CD73 (able to progressively hydrolyse ATP to adenosine), which are highly expressed on the immunosuppressive CD4+ FoxP3+ regulatory T cells [144],[145], on the intra-tumoral CD8+ FoxP3+ regulatory T cells [146] and on other immune cells subjected to TGFβ stimulation in the tumour microenvironment [147].

Reduced rates of glycolysis in T cells lead to the inhibition of T cell function by different molecular mechanisms. On the one hand, low levels of phosphoenolpyruvate (PEP, a glycolytic intermediate) result in unopposed activity of SERCA pumps, which reduce cytoplasmic calcium and, thus, the NFAT signalling necessary for translating a proper activation of the TCR [128]. On the other hand, low glycolytic activity leads to the interaction between GAPDH (not engaged in metabolic activity) and IFN-γ mRNA, suppressing its translation [148],[132].

Given the promising results already observed in clinics, special focus has now being put on immune checkpoint blockade exploitation, so as to treat a range of tumours more efficiently [149],[150]. The expression of inhibitory checkpoint receptors PD-1, Lag3, CTLA-4 increases in tumour infiltrating T lymphocytes, contributing to the dampening of the immune response [151],[152]. Interestingly, it has been shown that these receptors also have a role in the metabolic modification of TILs. For example, PD-1-expressing CD8+ T cells fail to fully activate glycolysis upon TCR engagement [152]. Moreover, treatment with checkpoint blockade antibodies against CTLA-4, PD-1, PD-L1 restores glucose availability in tumour microenvironment, re-establishing the condition for proper T effector cell function [127]. In particular, the action of PD-L1 directly on tumours reduces glycolysis by impinging on mTOR activation and on the expression of glycolytic enzymes, this being sufficient to restore T cell activity [127].

Nevertheless, nutrient availability issues do not stop with glucose. Tumour microenvironment is depleted of amino acids as well. Some of these are necessary for immune function of T cells [153]. High expression of Arginase-1 in myeloid-derived suppressor cells generates low levels of arginine, which in turn are responsible for reduced expression of TCR components limiting its proper activation [154]. Depletion of tryptophan and accumulation of immunosuppressive tryptophan metabolites are instead mediated by Indoleamine 2,3-dioxygenase (IDO) and are associated with poor prognosis in different cancer types, including endometrial and ovarian cancer [155],[156]. A similar mechanism involves the depletion of cysteine from the tumour microenvironment mediated by myeloid cells [157].

Of note, culturing tumour-specific T cells *in vitro* for a short time (6-24hours), in nutrient deplete condition, allows the recovery of effector function [158]. It comes as no surprise that some frontier anti-cancer treatments are trying to use the modulation of metabolism of cancer cells and/or T cells to improve their anti-cancer properties. Whatever the result, certainly more attention should be given to the therapeutic use of drugs impinging on tumour cell glycolysis since the same therapies would also modulate the function of the anti-tumoral TILs [159]. Fascinatingly, a recent paper by Restifo's group proposes a simple and clinically feasible method to identify - by cytofluorimetric analysis of mitochondrial membrane potential - the T cell population with the best metabolic fitness to survive and accomplish long-term effector functions [160]. The authors show that isolating T cells based only on their lower mitochondrial membrane potential - which likely parallels the fact that these cells have active OXPHOS - allows the selection of cells with superior antitumour activity [160]. This method may be applied in different clinical settings that rely on adoptive transfer strategies; coupled with approaches aiming at metabolic re-activation of T cells, it could provide a promising strategy for cancer treatment.

CONCLUSIONS

Alterations in metabolism have long been considered only as a consequence (rather than as a pivotal factor) of tumour formation and progression. Despite this oversight, in the recent decades mitochondria and metabolism have been returned to the scientific crime scene to be further investigated for their role in cancer. We know now that the metabolic shift from mitochondrial respiration to glycolysis is not a mere modification of cellular metabolism. The metabolic intermediates, which accumulate due to the Warburg effect, acquire new functions impinging on a plethora of mechanisms ranging from pro-angiogenic to epigenetic alterations, from pro-inflammatory to immune-evasive effects (Figure 3). Additionally, an outstanding crosstalk has been discovered between cancer cells and tumour-infiltrating immune cells where the oncometabolites modify, or inhibit, the function of the latter to favour cancer cells' proliferation, tumour expansion, and metastasis formation (Figure 3). New therapeutic approaches could be drawn up to restore tumoral cells to mitochondrial respiration; this would limit the pro-tumoral advantages of glycolysis on tumour aggressiveness itself, and would boost the anti-tumoral immune response in order to synergize the efforts in the battle against cancer.

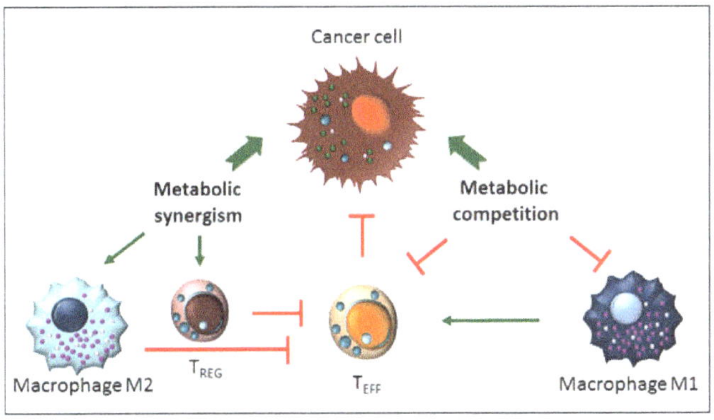

Figure 3: Metabolic competition and metabolic synergism in the tumour microenvironment. Metabolism influences the function of tumour-invading immune cells. Metabolic competition dampens the anti tumoral properties of M1 macrophages and T_{EFF} cells. Metabolic synergism, instead, favors the development of pro-tumoral immune T_{REG} cells and M2 macrophages.

ACKNOWLEDGMENTS

We are extremely grateful to C Frezza for helpful suggestions and for critically reading the manuscript.

FUNDING

Our lab is supported by the Italian Ministry of Health (GR-2011-02351643).

REFERENCES

1. Hanahan D, Weinberg R A. The hallmarks of cancer. Cell. 2000; 100, 57-70.
2. Hanahan D, Weinberg R A. Hallmarks of cancer: the next generation. Cell. 2011; 144, 646-674.
3. Evan G I, Vousden K H. Proliferation, cell cycle and apoptosis in cancer. Nature. 2001; 411, 342-348.
4. Vander Heiden M G, Cantley L C, Thompson C B. Understanding the Warburg effect: the metabolic requirements of cell proliferation. Science. 2009; 324, 1029-1033.
5. Slater E C. Mechanism of oxidative phosphorylation. Annu Rev Biochem. 1977; 46, 1015-1026.
6. Warburg O, Wind F, Negelein E. The metabolism of tumors in the body. J Gen Physiol. 1927; 8, 519-530.
7. Warburg O. On respiratory impairment in cancer cells. Science. 1956; 124, 269-270.
8. McAllister R M. Search for oncogenes in human rhabdomyosarcoma cells. Prog Immunobiol Stand. 1971; 5, 237-242.
9. Newmark P. Tyrosine phosphorylation and oncogenes. Nature. 1981; 292, 15-16.

10. Meier H, Huebner R J. Host-gene control of C-type tumor virus-expression and tumorigenesis: relevance of studies in inbred mice to cancer in man and other species. Proc Natl Acad Sci U S A. 1971; 68, 2664-2668.

11. McDonnell T J, Deane N, Platt F M, Nunez G, Jaeger U, McKearn J P, Korsmeyer S J. bcl-2-immunoglobulin transgenic mice demonstrate extended B cell survival and follicular lymphoproliferation. Cell. 1989; 57, 79-88.

12. Vaux D L, Cory S, Adams J M. Bcl-2 gene promotes haemopoietic cell survival and cooperates with c-myc to immortalize pre-B cells. Nature. 1988; 335, 440-442.

13. Hockenbery D, Nunez G, Milliman C, Schreiber R D, Korsmeyer S J. Bcl-2 is an inner mitochondrial membrane protein that blocks programmed cell death. Nature. 1990; 348, 334-336.

14. Liu X, Kim C N, Yang J, Jemmerson R, Wang X. Induction of apoptotic program in cell-free extracts: requirement for dATP and cytochrome c. Cell. 1996; 86, 147-157.

15. Harvey A J, Kind K L, Thompson J G. REDOX regulation of early embryo development. Reproduction. 2002; 123, 479-486.

16. Rafalski V A, Mancini E, Brunet A. Energy metabolism and energy-sensing pathways in mammalian embryonic and adult stem cell fate. J Cell Sci. 2012; 125, 5597-5608.

17. Wang R, Green D R. Metabolic checkpoints in activated T cells. Nat Immunol. 2012; 13, 907-915.

18. Baysal B E, Ferrell R E, Willett-Brozick J E, Lawrence E C, Myssiorek D, Bosch A, van der Mey A, Taschner P E, Rubinstein W S, Myers E N, Richard C W r, Cornelisse C J, Devilee P et al. Mutations in SDHD, a mitochondrial complex II gene, in hereditary paraganglioma. Science. 2000; 287, 848-851.

19. Tomlinson I P, Alam N A, Rowan A J, Barclay E, Jaeger E E, Kelsell D, Leigh I, Gorman P, Lamlum H, Rahman S, Roylance R R, Olpin S, Bevan S et al. Germline mutations in FH predispose to dominantly inherited uterine fibroids, skin leiomyomata and papillary renal cell cancer. Nat Genet. 2002; 30, 406-410.

20. Colegio O R, Chu N Q, Szabo A L, Chu T, Rhebergen A M, Jairam V, Cyrus N, Brokowski C E, Eisenbarth S C, Phillips G M, Cline G W, Phillips A J, Medzhitov R. Functional polarization of tumour-associated macrophages by tumour-derived lactic acid. Nature. 2014; 513, 559-563.

21. Frezza C. The role of mitochondria in the oncogenic signal transduction. Int J Biochem Cell Biol. 2014; 48, 11-17.

22. Miller D M, Thomas S D, Islam A, Muench D, Sedoris K. c-Myc and cancer metabolism. Clin Cancer Res. 2012; 18, 5546-5553.

23. Lombardi L, Newcomb E W, Dalla-Favera R. Pathogenesis of Burkitt lymphoma: expression of an activated c-myc oncogene causes the tumorigenic conversion of EBV-infected human B lymphoblasts. Cell. 1987; 49, 161-170.

24. Sears R, Nuckolls F, Haura E, Taya Y, Tamai K, Nevins J R. Multiple Ras-dependent phosphorylation pathways regulate Myc protein stability. Genes Dev. 2000; 14, 2501-2514.

25. Shim H, Dolde C, Lewis B C, Wu C S, Dang G, Jungmann R A, Dalla-Favera R, Dang C V. c-Myc transactivation of LDH-A: implications for tumor metabolism and growth. Proc Natl Acad Sci U S A. 1997; 94, 6658-6663.

26. Dang C V, Le A, Gao P. MYC-induced cancer cell energy metabolism and therapeutic opportunities. Clin Cancer Res. 2009; 15, 6479-6483.

27. Kim J W, Gao P, Liu Y C, Semenza G L, Dang C V. Hypoxia-inducible factor 1 and dysregulated c-Myc cooperatively induce vascular endothelial growth factor and metabolic switches hexokinase 2 and pyruvate dehydrogenase kinase 1. Mol Cell Biol. 2007; 27, 7381-7393.

28. Osthus R C, Shim H, Kim S, Li Q, Reddy R, Mukherjee M, Xu Y, Wonsey D, Lee L A, Dang C V. Deregulation of glucose transporter 1 and glycolytic gene expression by c-Myc. J Biol Chem. 2000; 275, 21797-21800.

29. Wise D R, DeBerardinis R J, Mancuso A, Sayed N, Zhang X Y, Pfeiffer H K, Nissim I, Daikhin E, Yudkoff M, McMahon S B, Thompson C B. Myc regulates a transcriptional program that stimulates mitochondrial glutaminolysis and leads to glutamine addiction. Proc Natl Acad Sci U S A. 2008; 105, 18782-18787.

30. Pylayeva-Gupta Y, Grabocka E, Bar-Sagi D. RAS oncogenes: weaving a tumorigenic web. Nat Rev Cancer. 2011; 11, 761-774.

31. Gough D J, Corlett A, Schlessinger K, Wegrzyn J, Larner A C, Levy D E. Mitochondrial STAT3 supports Ras-dependent oncogenic transformation. Science. 2009; 324, 1713-1716.

32. Hu Y, Lu W, Chen G, Wang P, Chen Z, Zhou Y, Ogasawara M, Trachootham D, Feng L, Pelicano H, Chiao P J, Keating M J, Garcia-Manero G et al. K-ras(G12V) transformation leads to mitochondrial dysfunction and a metabolic switch from oxidative phosphorylation to glycolysis. Cell Res. 2012; 22, 399-412.

33. Lee M H, Lee S E, Kim D W, Ryu M J, Kim S J, Kim S J, Kim Y K, Park J H, Kweon G R, Kim J M, Lee J U, De Falco V, Jo Y S et al. Mitochondrial localization and regulation of BRAFV600E in thyroid cancer: a clinically used RAF inhibitor is unable to block the mitochondrial activities of BRAFV600E. J Clin Endocrinol Metab. 2011; 96, E19-30.

34. Guo J Y, Chen H Y, Mathew R, Fan J, Strohecker A M, Karsli-Uzunbas G, Kamphorst J J, Chen G, Lemons J M, Karantza V, Coller H A, Dipaola R S, Gelinas C et al. Activated Ras requires autophagy to maintain oxidative metabolism and tumorigenesis. Genes Dev. 2011; 25, 460-470.

35. Tsuneoka M, Umata T, Kimura H, Koda Y, Nakajima M, Kosai K, Takahashi T, Takahashi Y, Yamamoto A. c-myc induces autophagy in rat 3Y1 fibroblast cells. Cell Struct Funct. 2003; 28, 195-204.

36. Galluzzi L, Pietrocola F, Bravo-San Pedro J M, Amaravadi R K, Baehrecke E H, Cecconi F, Codogno

P, Debnath J, Gewirtz D A, Karantza V, Kimmelman A, Kumar S, Levine B et al. Autophagy in malignant transformation and cancer progression. EMBO J. 2015; 34, 856-880.

37. Cianfanelli V, Fuoco C, Lorente M, Salazar M, Quondamatteo F, Gherardini P F, De Zio D, Nazio F, Antonioli M, D'Orazio M, Skobo T, Bordi M, Rohde M et al. AMBRA1 links autophagy to cell proliferation and tumorigenesis by promoting c-Myc dephosphorylation and degradation. Nat Cell Biol. 2015; 17, 706.

38. Goldsmith J, Levine B, Debnath J. Autophagy and cancer metabolism. Methods Enzymol. 2014; 542, 25-57.

39. White E, Mehnert J M, Chan C S. Autophagy, Metabolism, and Cancer. Clin Cancer Res. 2015; 21, 5037-5046.

40. Muller P A, Vousden K H. p53 mutations in cancer. Nat Cell Biol. 2013; 15, 2-8.

41. Vousden K H, Prives C. Blinded by the Light: The Growing Complexity of p53. Cell. 2009; 137, 413-431.

42. Liu J, Zhang C, Hu W, Feng Z. Tumor suppressor p53 and its mutants in cancer metabolism. Cancer Lett. 2015; 356, 197-203.

43. Zhang C, Liu J, Wu R, Liang Y, Lin M, Liu J, Chan C S, Hu W, Feng Z. Tumor suppressor p53 negatively regulates glycolysis stimulated by hypoxia through its target RRAD. Oncotarget. 2014; 5, 5535-5546. doi: 10.18632/oncotarget.2137.

44. Kondoh H, Lleonart M E, Gil J, Wang J, Degan P, Peters G, Martinez D, Carnero A, Beach D. Glycolytic enzymes can modulate cellular life span. Cancer Res. 2005; 65, 177-185.

45. Contractor T, Harris C R. p53 negatively regulates transcription of the pyruvate dehydrogenase kinase Pdk2. Cancer Res. 2012; 72, 560-567.

46. Bensaad K, Tsuruta A, Selak M A, Vidal M N, Nakano K, Bartrons R, Gottlieb E, Vousden K H. TIGAR, a p53-inducible regulator of glycolysis and apoptosis. Cell. 2006; 126, 107-120.

47. Pai E F, Schulz G E. The catalytic mechanism of glutathione reductase as derived from x-ray diffraction analyses of reaction intermediates. J Biol Chem. 1983; 258, 1752-1757.

48. Feng Z, Levine A J. The regulation of energy metabolism and the IGF-1/mTOR pathways by the p53 protein. Trends Cell Biol. 2010; 20, 427-434.

49. Kawauchi K, Araki K, Tobiume K, Tanaka N. p53 regulates glucose metabolism through an IKK-NF-kappaB pathway and inhibits cell transformation. Nat Cell Biol. 2008; 10, 611-618.

50. Freed-Pastor W A, Mizuno H, Zhao X, Langerod A, Moon S H, Rodriguez-Barrueco R, Barsotti A, Chicas A, Li W, Polotskaia A, Bissell M J, Osborne T F, Tian B et al. Mutant p53 disrupts mammary tissue architecture *via* the mevalonate pathway. Cell. 2012; 148, 244-258.

51. Zhang C, Liu J, Liang Y, Wu R, Zhao Y, Hong X, Lin M, Yu H, Liu L, Levine A J, Hu W, Feng Z. Tumour-associated mutant p53 drives the Warburg effect. Nat Commun. 2013; 4, 2935.

52. Bartesaghi S, Graziano V, Galavotti S, Henriquez N V, Betts J, Saxena J, A D, Karlsson A, Martins L M, Capasso M, Nicotera P, Brandner S, De Laurenzi V et al. Inhibition of oxidative metabolism leads to p53 genetic inactivation and transformation in neural stem cells. Proc Natl Acad Sci U S A. 2015; 112, 1059-1064.

53. Gaude E, Frezza C. Defects in mitochondrial metabolism and cancer. Cancer Metab. 2014; 2, 10.

54. Gimenez-Roqueplo A P, Favier J, Rustin P, Mourad J J, Plouin P F, Corvol P, Rotig A, Jeunemaitre X. The R22X mutation of the SDHD gene in hereditary paraganglioma abolishes the enzymatic activity of complex II in the mitochondrial respiratory chain and activates the hypoxia pathway. Am J Hum Genet. 2001; 69, 1186-1197.

55. Pollard P J, Briere J J, Alam N A, Barwell J, Barclay E, Wortham N C, Hunt T, Mitchell M, Olpin S, Moat S J, Hargreaves I P, Heales S J, Chung Y L et al. Accumulation of Krebs cycle intermediates and over-expression of HIF1alpha in tumours which result from germline FH and SDH mutations. Hum Mol Genet. 2005; 14, 2231-2239.

56. Rofstad E K, Galappathi K, Mathiesen B, Ruud E B. Fluctuating and diffusion-limited hypoxia in hypoxia-induced metastasis. Clin Cancer Res. 2007; 13, 1971-1978.

57. Wang G L, Jiang B H, Rue E A, Semenza G L. Hypoxia-inducible factor 1 is a basic-helix-loop-helix-PAS heterodimer regulated by cellular O2 tension. Proc Natl Acad Sci U S A. 1995; 92, 5510-5514.

58. Selak M A, Armour S M, MacKenzie E D, Boulahbel H, Watson D G, Mansfield K D, Pan Y, Simon M C, Thompson C B, Gottlieb E. Succinate links TCA cycle dysfunction to oncogenesis by inhibiting HIF-alpha prolyl hydroxylase. Cancer Cell. 2005; 7, 77-85.

59. Bruick R K, McKnight S L. A conserved family of prolyl-4-hydroxylases that modify HIF. Science. 2001; 294, 1337-1340.

60. Hirsila M, Koivunen P, Gunzler V, Kivirikko K I, Myllyharju J. Characterization of the human prolyl 4-hydroxylases that modify the hypoxia-inducible factor. J Biol Chem. 2003; 278, 30772-30780.

61. Isaacs J S, Jung Y J, Mole D R, Lee S, Torres-Cabala C, Chung Y L, Merino M, Trepel J, Zbar B, Toro J, Ratcliffe P J, Linehan W M, Neckers L. HIF overexpression correlates with biallelic loss of fumarate hydratase in renal cancer: novel role of fumarate in regulation of HIF stability. Cancer Cell. 2005; 8, 143-153.

62. Zheng L, Mackenzie E D, Karim S A, Hedley A, Blyth K, Kalna G, Watson D G, Szlosarek P, Frezza C, Gottlieb E. Reversed argininosuccinate lyase activity in fumarate hydratase-deficient cancer cells. Cancer Metab. 2013; 1, 12.

63. Adam J, Hatipoglu E, O'Flaherty L, Ternette N, Sahgal N, Lockstone H, Baban D, Nye E, Stamp G W, Wolhuter K, Stevens M, Fischer R, Carmeliet P et al. Renal cyst formation in Fh1-deficient mice is independent of the Hif/Phd pathway: roles for fumarate in KEAP1 succination and Nrf2 signaling. Cancer Cell. 2011; 20, 524-537.

64. Ooi A, Wong J C, Petillo D, Roossien D, Perrier-Trudova

V, Whitten D, Min B W, Tan M H, Zhang Z, Yang X J, Zhou M, Gardie B, Molinie V et al. An antioxidant response phenotype shared between hereditary and sporadic type 2 papillary renal cell carcinoma. Cancer Cell. 2011; 20, 511-523.

65. Sullivan L B, Martinez-Garcia E, Nguyen H, Mullen A R, Dufour E, Sudarshan S, Licht J D, Deberardinis R J, Chandel N S. The proto-oncometabolite fumarate binds glutathione to amplify ROS-dependent signaling. Mol Cell. 2013; 51, 236-248.

66. Ternette N, Yang M, Laroyia M, Kitagawa M, O'Flaherty L, Wolhulter K, Igarashi K, Saito K, Kato K, Fischer R, Berquand A, Kessler B M, Lappin T et al. Inhibition of mitochondrial aconitase by succination in fumarate hydratase deficiency. Cell Rep. 2013; 3, 689-700.

67. Brandon M, Baldi P, Wallace D C. Mitochondrial mutations in cancer. Oncogene. 2006; 25, 4647-4662.

68. Lu J, Sharma L K, Bai Y. Implications of mitochondrial DNA mutations and mitochondrial dysfunction in tumorigenesis. Cell Res. 2009; 19, 802-815.

69. Campanella M, Casswell E, Chong S, Farah Z, Wieckowski M R, Abramov A Y, Tinker A, Duchen M R. Regulation of mitochondrial structure and function by the F1Fo-ATPase inhibitor protein, IF1. Cell Metab. 2008; 8, 13-25.

70. Faccenda D, Tan C H, Seraphim A, Duchen M R, Campanella M. IF1 limits the apoptotic-signalling cascade by preventing mitochondrial remodelling. Cell Death Differ. 2013; 20, 686-697.

71. Santacatterina F, Sanchez-Cenizo L, Formentini L, Mobasher M A, Casas E, Rueda C B, Martinez-Reyes I, Nunez de Arenas C, Garcia-Bermudez J, Zapata J M, Sanchez-Arago M, Satrustegui J, Valverde A M et al. Down-regulation of oxidative phosphorylation in the liver by expression of the ATPase inhibitory factor 1 induces a tumor-promoter metabolic state. Oncotarget. 2016; 7, 490-508. doi: 10.18632/oncotarget.6357.

72. Faccenda D, Campanella M. Molecular Regulation of the Mitochondrial F(1)F(o)-ATPsynthase: Physiological and Pathological Significance of the Inhibitory Factor 1 (IF(1)). Int J Cell Biol. 2012; 2012, 367934.

73. Yin T, Lu L, Xiong Z, Wei S, Cui D. ATPase inhibitory factor 1 is a prognostic marker and contributes to proliferation and invasion of human gastric cancer cells. Biomed Pharmacother. 2015; 70, 90-96.

74. Wu J, Shan Q, Li P, Wu Y, Xie J, Wang X. ATPase inhibitory factor 1 is a potential prognostic marker for the migration and invasion of glioma. Oncol Lett. 2015; 10, 2075-2080.

75. Frezza C, Zheng L, Tennant D A, Papkovsky D B, Hedley B A, Kalna G, Watson D G, Gottlieb E. Metabolic profiling of hypoxic cells revealed a catabolic signature required for cell survival. PLoS One. 2011; 6, e24411.

76. Merlo A, de Quiros S B, Secades P, Zambrano I, Balbin M, Astudillo A, Scola B, Aristegui M, Suarez C, Chiara M D. Identification of a signaling axis HIF-1alpha/microRNA-210/ISCU independent of SDH mutation that defines a subgroup of head and neck paragangliomas. J Clin Endocrinol Metab. 2012; 97, E2194-200.

77. Puissegur M P, Mazure N M, Bertero T, Pradelli L, Grosso S, Robbe-Sermesant K, Maurin T, Lebrigand K, Cardinaud B, Hofman V, Fourre S, Magnone V, Ricci J E et al. miR-210 is overexpressed in late stages of lung cancer and mediates mitochondrial alterations associated with modulation of HIF-1 activity. Cell Death Differ. 2011; 18, 465-478.

78. Figueroa M E, Abdel-Wahab O, Lu C, Ward P S, Patel J, Shih A, Li Y, Bhagwat N, Vasanthakumar A, Fernandez H F, Tallman M S, Sun Z, Wolniak K et al. Leukemic IDH1 and IDH2 mutations result in a hypermethylation phenotype, disrupt TET2 function, and impair hematopoietic differentiation. Cancer Cell. 2010; 18, 553-567.

79. Cai L, Sutter B M, Li B, Tu B P. Acetyl-CoA induces cell growth and proliferation by promoting the acetylation of histones at growth genes. Mol Cell. 2011; 42, 426-437.

80. McBrian M A, Behbahan I S, Ferrari R, Su T, Huang T W, Li K, Hong C S, Christofk H R, Vogelauer M, Seligson D B, Kurdistani S K. Histone acetylation regulates intracellular pH. Mol Cell. 2013; 49, 310-321.

81. Lu H, Li G, Liu L, Feng L, Wang X, Jin H. Regulation and function of mitophagy in development and cancer. Autophagy. 2013; 9, 1720-1736.

82. Chourasia A H, Boland M L, Macleod K F. Mitophagy and cancer. Cancer Metab. 2015; 3, 4.

83. Dvorak H F. Tumors: wounds that do not heal. Similarities between tumor stroma generation and wound healing. N Engl J Med. 1986; 315, 1650-1659.

84. Swann J B, Smyth M J. Immune surveillance of tumors. J Clin Invest. 2007; 117, 1137-1146.

85. Teng M W, Swann J B, Koebel C M, Schreiber R D, Smyth M J. Immune-mediated dormancy: an equilibrium with cancer. J Leukoc Biol. 2008; 84, 988-993.

86. Kim R, Emi M, Tanabe K. Cancer immunoediting from immune surveillance to immune escape. Immunology. 2007; 121, 1-14.

87. Vajdic C M, van Leeuwen M T. Cancer incidence and risk factors after solid organ transplantation. Int J Cancer. 2009; 125, 1747-1754.

88. Coffelt S B, Lewis C E, Naldini L, Brown J M, Ferrara N, De Palma M. Elusive identities and overlapping phenotypes of proangiogenic myeloid cells in tumors. Am J Pathol. 2010; 176, 1564-1576.

89. DeNardo D G, Andreu P, Coussens L M. Interactions between lymphocytes and myeloid cells regulate pro- *versus* anti-tumor immunity. Cancer Metastasis Rev. 2010; 29, 309-316.

90. Johansson M, Denardo D G, Coussens L M. Polarized immune responses differentially regulate cancer development. Immunol Rev. 2008; 222, 145-154.

91. Murdoch C, Muthana M, Coffelt S B, Lewis C E. The role of myeloid cells in the promotion of tumour angiogenesis. Nat Rev Cancer. 2008; 8, 618-631.

92. De Palma M, Murdoch C, Venneri M A, Naldini L, Lewis C E. Tie2-expressing monocytes: regulation of tumor angiogenesis and therapeutic implications. Trends Immunol. 2007; 28, 519-524.

93. Biswas S K, Mantovani A. Macrophage plasticity and interaction with lymphocyte subsets: cancer as a paradigm. Nat Immunol. 2010; 11, 889-896.

94. Rubartelli A, Cozzolino F, Talio M, Sitia R. A novel secretory pathway for interleukin-1 beta, a protein lacking a signal sequence. EMBO J. 1990; 9, 1503-1510.

95. Dupont N, Jiang S, Pilli M, Ornatowski W, Bhattacharya D, Deretic V. Autophagy-based unconventional secretory pathway for extracellular delivery of IL-1beta. EMBO J. 2011; 30, 4701-4711.

96. Tannahill G M, Curtis A M, Adamik J, Palsson-McDermott E M, McGettrick A F, Goel G, Frezza C, Bernard N J, Kelly B, Foley N H, Zheng L, Gardet A, Tong Z et al. Succinate is an inflammatory signal that induces IL-1beta through HIF-1alpha. Nature. 2013; 496, 238-242.

97. Dahia P L, Ross K N, Wright M E, Hayashida C Y, Santagata S, Barontini M, Kung A L, Sanso G, Powers J F, Tischler A S, Hodin R, Heitritter S, Moore F et al. A HIF1alpha regulatory loop links hypoxia and mitochondrial signals in pheochromocytomas. PLoS Genet. 2005; 1, 72-80.

98. Rubic T, Lametschwandtner G, Jost S, Hinteregger S, Kund J, Carballido-Perrig N, Schwarzler C, Junt T, Voshol H, Meingassner J G, Mao X, Werner G, Rot A et al. Triggering the succinate receptor GPR91 on dendritic cells enhances immunity. Nat Immunol. 2008; 9, 1261-1269.

99. Wyckoff J, Wang W, Lin E Y, Wang Y, Pixley F, Stanley E R, Graf T, Pollard J W, Segall J, Condeelis J. A paracrine loop between tumor cells and macrophages is required for tumor cell migration in mammary tumors. Cancer Res. 2004; 64, 7022-7029.

100. Mantovani A, Allavena P, Sica A, Balkwill F. Cancer-related inflammation. Nature. 2008; 454, 436-444.

101. Martinez F O, Sica A, Mantovani A, Locati M. Macrophage activation and polarization. Front Biosci. 2008; 13, 453-461.

102. Oshimori N, Oristian D, Fuchs E. TGF-beta Promotes Heterogeneity and Drug Resistance in Squamous Cell Carcinoma. Cell. 2015; 160, 963-976.

103. Jiang L, Xiao L, Sugiura H, Huang X, Ali A, Kuro-O M, Deberardinis R J, Boothman D A. Metabolic reprogramming during TGFbeta1-induced epithelial-to-mesenchymal transition. Oncogene. 2014; 0,

104. Porporato P E, Payen V L, Perez-Escuredo J, De Saedeleer C J, Danhier P, Copetti T, Dhup S, Tardy M, Vazeille T, Bouzin C, Feron O, Michiels C, Gallez B et al. A mitochondrial switch promotes tumor metastasis. Cell Rep. 2014; 8, 754-766.

105. Ferreira-da-Silva A, Valacca C, Rios E, Populo H, Soares P, Sobrinho-Simoes M, Scorrano L, Maximo V, Campello S. Mitochondrial dynamics protein Drp1 is overexpressed in oncocytic thyroid tumors and regulates cancer cell migration. PLoS One. 2015; 10, e0122308.

106. Zhao J, Zhang J, Yu M, Xie Y, Huang Y, Wolff D W, Abel P W, Tu Y. Mitochondrial dynamics regulates migration and invasion of breast cancer cells. Oncogene. 2013; 32, 4814-4824.

107. Wan Y Y, Zhang J F, Yang Z J, Jiang L P, Wei Y F, Lai Q N, Wang J B, Xin H B, Han X J. Involvement of Drp1 in hypoxia-induced migration of human glioblastoma U251 cells. Oncol Rep. 2014; 32, 619-626.

108. Campello S, Lacalle R A, Bettella M, Manes S, Scorrano L, Viola A. Orchestration of lymphocyte chemotaxis by mitochondrial dynamics. J Exp Med. 2006; 203, 2879-2886.

109. da Silva A F, Mariotti F R, Maximo V, Campello S. Mitochondria dynamism: of shape, transport and cell migration. Cell Mol Life Sci. 2014; 71, 2313-2324.

110. Corrado M, Scorrano L, Campello S. Mitochondrial dynamics in cancer and neurodegenerative and neuroinflammatory diseases. Int J Cell Biol. 2012; 2012, 729290.

111. Taguchi N, Ishihara N, Jofuku A, Oka T, Mihara K. Mitotic phosphorylation of dynamin-related GTPase Drp1 participates in mitochondrial fission. J Biol Chem. 2007; 282, 11521-11529.

112. Hackenbrock C R. Ultrastructural bases for metabolically linked mechanical activity in mitochondria. I. Reversible ultrastructural changes with change in metabolic steady state in isolated liver mitochondria. J Cell Biol. 1966; 30, 269-297.

113. Suen D F, Norris K L, Youle R J. Mitochondrial dynamics and apoptosis. Genes Dev. 2008; 22, 1577-1590.

114. Frezza C, Cipolat S, Martins de Brito O, Micaroni M, Beznoussenko G V, Rudka T, Bartoli D, Polishuck R S, Danial N N, De Strooper B, Scorrano L. OPA1 controls apoptotic cristae remodeling independently from mitochondrial fusion. Cell. 2006; 126, 177-189.

115. Cipolat S, Rudka T, Hartmann D, Costa V, Serneels L, Craessaerts K, Metzger K, Frezza C, Annaert W, D'Adamio L, Derks C, Dejaegere T, Pellegrini L et al. Mitochondrial rhomboid PARL regulates cytochrome c release during apoptosis *via* OPA1-dependent cristae remodeling. Cell. 2006; 126, 163-175.

116. Cogliati S, Frezza C, Soriano M E, Varanita T, Quintana-Cabrera R, Corrado M, Cipolat S, Costa V, Casarin A, Gomes L C, Perales-Clemente E, Salviati L, Fernandez-Silva P et al. Mitochondrial cristae shape determines respiratory chain supercomplexes assembly and respiratory efficiency. Cell. 2013; 155, 160-171.

117. Bronte V, Zanovello P. Regulation of immune responses by L-arginine metabolism. Nat Rev Immunol. 2005; 5, 641-654.

118. Altuntas S, Rossin F, Marsella C, D'Eletto M, Diaz-Hidalgo L, Farrace M G, Campanella M, Antonioli M, Fimia G M, Piacentini M. The transglutaminase type 2 and pyruvate kinase isoenzyme M2 interplay in autophagy regulation. Oncotarget. 2015; 6, 44941-44954. doi: 10.18632/oncotarget.6759

119. Rossin F, D'Eletto M, Falasca L, Sepe S, Cocco S, Fimia G M, Campanella M, Mastroberardino P G, Farrace M G, Piacentini M. Transglutaminase 2 ablation leads to mitophagy impairment associated with a metabolic shift towards aerobic glycolysis. Cell Death Differ. 2015; 22,

408-418.

120. Fischer K, Hoffmann P, Voelkl S, Meidenbauer N, Ammer J, Edinger M, Gottfried E, Schwarz S, Rothe G, Hoves S, Renner K, Timischl B, Mackensen A et al. Inhibitory effect of tumor cell-derived lactic acid on human T cells. Blood. 2007; 109, 3812-3819.

121. Dietl K, Renner K, Dettmer K, Timischl B, Eberhart K, Dorn C, Hellerbrand C, Kastenberger M, Kunz-Schughart L A, Oefner P J, Andreesen R, Gottfried E, Kreutz M P. Lactic acid and acidification inhibit TNF secretion and glycolysis of human monocytes. J Immunol. 2010; 184, 1200-1209.

122. Aspuria P J, Lunt S Y, Varemo L, Vergnes L, Gozo M, Beach J A, Salumbides B, Reue K, Wiedemeyer W R, Nielsen J, Karlan B Y, Orsulic S. Succinate dehydrogenase inhibition leads to epithelial-mesenchymal transition and reprogrammed carbon metabolism. Cancer Metab. 2014; 2, 21.

123. Baumann F, Leukel P, Doerfelt A, Beier C P, Dettmer K, Oefner P J, Kastenberger M, Kreutz M, Nickl-Jockschat T, Bogdahn U, Bosserhoff A K, Hau P. Lactate promotes glioma migration by TGF-beta2-dependent regulation of matrix metalloproteinase-2. Neuro Oncol. 2009; 11, 368-380.

124. Joyce J A, Pollard J W. Microenvironmental regulation of metastasis. Nat Rev Cancer. 2009; 9, 239-252.

125. Yizhak K, Le Devedec S E, Rogkoti V M, Baenke F, de Boer V C, Frezza C, Schulze A, van de Water B, Ruppin E. A computational study of the Warburg effect identifies metabolic targets inhibiting cancer migration. Mol Syst Biol. 2014; 10, 744.

126. Mockler M B, Conroy M J, Lysaght J. Targeting T cell immunometabolism for cancer immunotherapy; understanding the impact of the tumor microenvironment. Front Oncol. 2014; 4, 107.

127. Chang C H, Qiu J, O'Sullivan D, Buck M D, Noguchi T, Curtis J D, Chen Q, Gindin M, Gubin M M, van der Windt G J, Tonc E, Schreiber R D, Pearce E J et al. Metabolic Competition in the Tumor Microenvironment Is a Driver of Cancer Progression. Cell. 2015; 162, 1229-1241.

128. Ho P C, Bihuniak J D, Macintyre A N, Staron M, Liu X, Amezquita R, Tsui Y C, Cui G, Micevic G, Perales J C, Kleinstein S H, Abel E D, Insogna K L et al. Phosphoenolpyruvate Is a Metabolic Checkpoint of Anti-tumor T Cell Responses. Cell. 2015; 162, 1217-1228.

129. van der Windt G J, O'Sullivan D, Everts B, Huang S C, Buck M D, Curtis J D, Chang C H, Smith A M, Ai T, Faubert B, Jones R G, Pearce E J, Pearce E L. CD8 memory T cells have a bioenergetic advantage that underlies their rapid recall ability. Proc Natl Acad Sci U S A. 2013; 110, 14336-14341.

130. van der Windt G J, Everts B, Chang C H, Curtis J D, Freitas T C, Amiel E, Pearce E J, Pearce E L. Mitochondrial respiratory capacity is a critical regulator of CD8+ T cell memory development. Immunity. 2012; 36, 68-78.

131. Verbist K C, Wang R, Green D R. T cell metabolism and the immune response. Semin Immunol. 2012; 24, 399-404.

132. Gubser P M, Bantug G R, Razik L, Fischer M, Dimeloe S, Hoenger G, Durovic B, Jauch A, Hess C. Rapid effector function of memory CD8+ T cells requires an immediate-early glycolytic switch. Nat Immunol. 2013; 14, 1064-1072.

133. Haas R, Smith J, Rocher-Ros V, Nadkarni S, Montero-Melendez T, D'Acquisto F, Bland E J, Bombardieri M, Pitzalis C, Perretti M, Marelli-Berg F M, Mauro C. Lactate Regulates Metabolic and Pro-inflammatory Circuits in Control of T Cell Migration and Effector Functions. PLoS Biol. 2015; 13, e1002202.

134. Michalek R D, Gerriets V A, Jacobs S R, Macintyre A N, MacIver N J, Mason E F, Sullivan S A, Nichols A G, Rathmell J C. Cutting edge: distinct glycolytic and lipid oxidative metabolic programs are essential for effector and regulatory CD4+ T cell subsets. J Immunol. 2011; 186, 3299-3303.

135. Lochner M, Berod L, Sparwasser T. Fatty acid metabolism in the regulation of T cell function. Trends Immunol. 2015; 36, 81-91.

136. De Rosa V, Galgani M, Porcellini A, Colamatteo A, Santopaolo M, Zuchegna C, Romano A, De Simone S, Procaccini C, La Rocca C, Carrieri P B, Maniscalco G T, Salvetti M et al. Glycolysis controls the induction of human regulatory T cells by modulating the expression of FOXP3 exon 2 splicing variants. Nat Immunol. 2015; 16, 1174-1184.

137. Pellegatti P, Raffaghello L, Bianchi G, Piccardi F, Pistoia V, Di Virgilio F. Increased level of extracellular ATP at tumor sites: *in vivo* imaging with plasma membrane luciferase. PLoS One. 2008; 3, e2599.

138. Martins I, Wang Y, Michaud M, Ma Y, Sukkurwala A Q, Shen S, Kepp O, Metivier D, Galluzzi L, Perfettini J L, Zitvogel L, Kroemer G. Molecular mechanisms of ATP secretion during immunogenic cell death. Cell Death Differ. 2014; 21, 79-91.

139. Schenk U, Westendorf A M, Radaelli E, Casati A, Ferro M, Fumagalli M, Verderio C, Buer J, Scanziani E, Grassi F. Purinergic control of T cell activation by ATP released through pannexin-1 hemichannels. Sci Signal. 2008; 1, ra6.

140. Yip L, Woehrle T, Corriden R, Hirsh M, Chen Y, Inoue Y, Ferrari V, Insel P A, Junger W G. Autocrine regulation of T-cell activation by ATP release and P2X7 receptors. FASEB J. 2009; 23, 1685-1693.

141. Wang C M, Ploia C, Anselmi F, Sarukhan A, Viola A. Adenosine triphosphate acts as a paracrine signaling molecule to reduce the motility of T cells. EMBO J. 2014; 33, 1354-1364.

142. Goere D, Flament C, Rusakiewicz S, Poirier-Colame V, Kepp O, Martins I, Pesquet J, Eggermont A, Elias D, Chaput N, Zitvogel L. Potent immunomodulatory effects of the trifunctional antibody catumaxomab. Cancer Res. 2013; 73, 4663-4673.

143. Ma Y, Adjemian S, Yang H, Catani J P, Hannani D, Martins I, Michaud M, Kepp O, Sukkurwala A Q, Vacchelli E, Galluzzi L, Zitvogel L, Kroemer G. ATP-dependent recruitment, survival and differentiation of

dendritic cell precursors in the tumor bed after anticancer chemotherapy. Oncoimmunology. 2013; 2, e24568.

144. Borsellino G, Kleinewietfeld M, Di Mitri D, Sternjak A, Diamantini A, Giometto R, Hopner S, Centonze D, Bernardi G, Dell'Acqua M L, Rossini P M, Battistini L, Rotzschke O et al. Expression of ectonucleotidase CD39 by Foxp3+ Treg cells: hydrolysis of extracellular ATP and immune suppression. Blood. 2007; 110, 1225-1232.

145. Deaglio S, Dwyer K M, Gao W, Friedman D, Usheva A, Erat A, Chen J F, Enjyoji K, Linden J, Oukka M, Kuchroo V K, Strom T B, Robson S C. Adenosine generation catalyzed by CD39 and CD73 expressed on regulatory T cells mediates immune suppression. J Exp Med. 2007; 204, 1257-1265.

146. Parodi A, Battaglia F, Kalli F, Ferrera F, Conteduca G, Tardito S, Stringara S, Ivaldi F, Negrini S, Borgonovo G, Simonato A, Traverso P, Carmignani G et al. CD39 is highly involved in mediating the suppression activity of tumor-infiltrating CD8+ T regulatory lymphocytes. Cancer Immunol Immunother. 2013; 62, 851-862.

147. Regateiro F S, Howie D, Nolan K F, Agorogiannis E I, Greaves D R, Cobbold S P, Waldmann H. Generation of anti-inflammatory adenosine by leukocytes is regulated by TGF-beta. Eur J Immunol. 2011; 41, 2955-2965.

148. Chang C H, Curtis J D, Maggi L B J, Faubert B, Villarino A V, O'Sullivan D, Huang S C, van der Windt G J, Blagih J, Qiu J, Weber J D, Pearce E J, Jones R G et al. Posttranscriptional control of T cell effector function by aerobic glycolysis. Cell. 2013; 153, 1239-1251.

149. Gubin M M, Zhang X, Schuster H, Caron E, Ward J P, Noguchi T, Ivanova Y, Hundal J, Arthur C D, Krebber W J, Mulder G E, Toebes M, Vesely M D et al. Checkpoint blockade cancer immunotherapy targets tumour-specific mutant antigens. Nature. 2014; 515, 577-581.

150. Kreamer K M. Immune Checkpoint Blockade: A New Paradigm in Treating Advanced Cancer. J Adv Pract Oncol. 2014; 5, 418-431.

151. Baitsch L, Legat A, Barba L, Fuertes Marraco S A, Rivals J P, Baumgaertner P, Christiansen-Jucht C, Bouzourene H, Rimoldi D, Pircher H, Rufer N, Matter M, Michielin O et al. Extended co-expression of inhibitory receptors by human CD8 T-cells depending on differentiation, antigen-specificity and anatomical localization. PLoS One. 2012; 7, e30852.

152. Parry R V, Chemnitz J M, Frauwirth K A, Lanfranco A R, Braunstein I, Kobayashi S V, Linsley P S, Thompson C B, Riley J L. CTLA-4 and PD-1 receptors inhibit T-cell activation by distinct mechanisms. Mol Cell Biol. 2005; 25, 9543-9553.

153. Mockler M B, Conroy M J, Lysaght J. Targeting T cell immunometabolism for cancer immunotherapy; understanding the impact of the tumor microenvironment. Front Oncol. 2014; 4, 107.

154. Rodriguez P C, Quiceno D G, Zabaleta J, Ortiz B, Zea A H, Piazuelo M B, Delgado A, Correa P, Brayer J, Sotomayor E M, Antonia S, Ochoa J B, Ochoa A C. Arginase I production in the tumor microenvironment by mature myeloid cells inhibits T-cell receptor expression and antigen-specific T-cell responses. Cancer Res. 2004; 64, 5839-5849.

155. Uyttenhove C, Pilotte L, Theate I, Stroobant V, Colau D, Parmentier N, Boon T, Van den Eynde B J. Evidence for a tumoral immune resistance mechanism based on tryptophan degradation by indoleamine 2,3-dioxygenase. Nat Med. 2003; 9, 1269-1274.

156. Ino K, Yoshida N, Kajiyama H, Shibata K, Yamamoto E, Kidokoro K, Takahashi N, Terauchi M, Nawa A, Nomura S, Nagasaka T, Takikawa O, Kikkawa F. Indoleamine 2,3-dioxygenase is a novel prognostic indicator for endometrial cancer. Br J Cancer. 2006; 95, 1555-1561.

157. Srivastava M K, Sinha P, Clements V K, Rodriguez P, Ostrand-Rosenberg S. Myeloid-derived suppressor cells inhibit T-cell activation by depleting cystine and cysteine. Cancer Res. 2010; 70, 68-77.

158. Wang S F, Fouquet S, Chapon M, Salmon H, Regnier F, Labroquere K, Badoual C, Damotte D, Validire P, Maubec E, Delongchamps N B, Cazes A, Gibault L et al. Early T cell signalling is reversibly altered in PD-1+ T lymphocytes infiltrating human tumors. PLoS One. 2011; 6, e17621.

159. Munn D H, Bronte V. Immune suppressive mechanisms in the tumor microenvironment. Curr Opin Immunol. 2015; 39, 1-6.

160. Sukumar M, Liu J, Mehta G U, Patel S J, Roychoudhuri R, Crompton J G, Klebanoff C A, Ji Y, Li P, Yu Z, Whitehill G D, Clever D, Eil R L et al. Mitochondrial Membrane Potential Identifies Cells with Enhanced Stemness for Cellular Therapy. Cell Metab. 2015.

CHAPTER 16

INSIGHTS IN CELL SCIENCE – TOWARDS A QUANTITATIVE APPROACH OF FLUORESCENCE MICROSCOPY THAT UNRAVELS CELLULAR FUNCTION

Ben Loos

Department of Physiological Sciences, Stellenbosch University, South Africa

MINI REVIEW

It is undoubtedly so, the time for fluorescence microscopy in cellular sciences has come, allowing the unraveling of the 'inside' of cells and their molecular itinererary in a novel and most precise manner. This is supported by the in 2014 awarded Noble price in chemistry, received for the development of the superresolution fluorescence microscopy techniques photo-activation localization microscopy (PALM), stochastic optical reconstruction microscopy (STORM) and stimulated emission depletion (STED). These techniques provide for the first time a platform for subdiffraction molecular imaging complementing the growing arsenal of fluorescence based techniques such as live cell imaging, photo-activation, fluorescence resonance energy transfer (FRET) or fluorescence recovery after photo-bleaching (FRAP). The challenge however remains, not to be drowning in a sea of imagedata, that may be building up to terabytes of data, but to be able to extract meaningful and statistically sound data points, that are suitable to inform on cellular function and dysfunction [1]. More so, it is envisaged to expand our fluorescent toolbox in a manner that not only allows the utilization of fluorescence based image data in a most powerful manner, but also that may contribute towards enabling predictive and preventative medicine [2]. Here we comment on some of the recently developed and employed applications and trends, that enable to powerfully provide "Insights in cell science", both visually and quantitatively. In doing so, it is hoped to offer a perspective on current applications, trends and challenges that define our present research landscape of cellular imaging. This may assist in better and more carefully designed experimental settings that are most informative and well aligned with the powers of fluorescence microscopy technology. Cellular decision making centers around maintaining homeostasis and cell viability. When the cell is unable to meet its metabolic demands, or is exposed to some form of cellular injury, cell death will be induced [3]. Although it is highly important to assess the overall contribution of cell death modalities, life cell imaging in combination with FRET-based caspase-sensors [4] allow for the quantification of the exact onset of cell death induction (Figure 1A) [5].

This enables to inform dynamically about cellular decision making, to trace the fate of individual cells and to assess the time span particular cells were able to maintain viability. Such an approach, based on long series of time lapses may reveal

Citation: Ben Loos, Insights in Cell Science – Towards a Quantitative Approach of Fluorescence Microscopy that Unravels Cellular Function, http://stemcells.imedpub.com/insights-in-cell-science--towards-a-quantitative-approach-of-fluorescence-microscopy-that-unravels-cellular-function.pdf a more accurate understanding of cellular pathology and cell disease pathogenesis [6] compared to a static approach that merely quantifies parameters at few selected time points. Integral part in this context is the availability of ATP, and its presence in regions of ATP consuming processes. The tubulin network (Figure 1B) and ATP dependent molecular motors such as kinesin and dynein [7] ensure the transport of mitochondria and autophagosomes to regions where either ATP is required (such as synaptic processes) or where fusion processes with other organelles (such as lysosomes in the perinuclear region) take place [8]. In the past, it has been a major challenge to visualize ATP, and usually luciferase-based methods have been employed to quantify overall ATP availability [9]. However, recently, the development of sensitive sensors, that are based on fluorescence resonance energy transfer (FRET) have been developed [10] and successfully employed to visualize and localize intracellular ATP hotspots (Figure 2a). Such an approach may not only allow to assess where intracellular ATP demand is highest or lowest and how such localization may change in disease or intervention, but it may also, through live cell imaging, enable the quantification of the duration a cell is able to maintain favorable ATP availability. This enhances greatly our understanding of metabolic processes and cellular decision making during metabolic adaptations or perturbations and cellular injury, such as in cancer or neurodegeneration. The precise labeling of selective proteins through either a transfection- or immuno-fluorescence based approach enables to visualize specific organelles of interest. The employment of super-resolution techniques, such as superresolution structured illumination microscopy (SR-SIM) allows to do implement such approach at enhanced resolving power, in a three-dimensional space. This in turn enables to not only visualize organelles of interest, but, through optical z-stacking, to acquire complete cellular organelle pool sizes with their subsequent quantification.

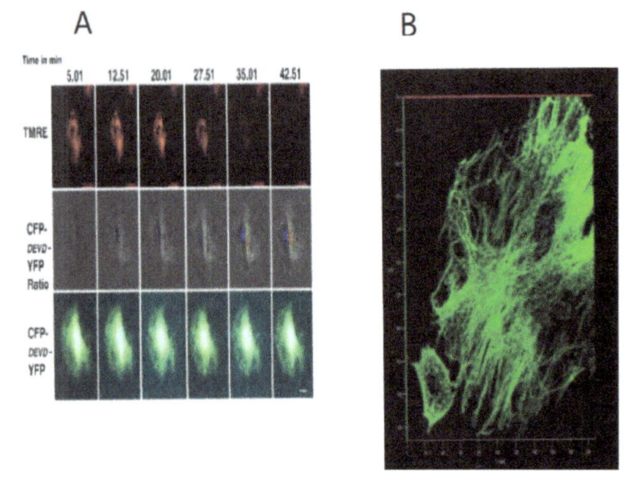

Figure 1: Apoptosis onset (1A) and the microtubule network (1B). Cells were transfected with a CFP-DEVD-YFP construct and loaded with tetramethylrhodamineethylester (TMRE) to monitor the onset of caspase-3 activation and mitochondrial depolarization. B: Acetylated tubulin allows for the assessment of tubulin network structural integrity. Image was acquired from z-planes and projected in maximum intensity. (From Loos et al., 2011)

This is powerful, since, although higher resolution may be achieved through transmission electron microscopy (TEM), complete cell volume and organelle pool size quantification is enabled using a subdiffraction limit resolving power. In that manner, for example lysosomes (Figure 2B) can be visualized and their distribution expressed as complete lysosomal pool size nA [11]. SR-SIM also enables to assess the mitochondrial network at much greater detail (Figure 2C), allowing its quantification more precisely than using confocal-based images, by employing form factor analyses or aspect ratio approaches [12]. Together with techniques such as photoactivation-based mitochondrial fission and fusion dynamics [13], network properties can be assessed in their dynamic nature, which is more sensitive and by far more rapid in response compared to a change in protein expression levels. Here, fluorescence microscopy techniques compile a powerful tool box to understand mitochondria at their nexus in cell death and survival. If combined with pool size analyses of associated organelles (such as autophagosomes, (Figure 2D), the dynamic behavior of interactions such as mitophagy or protein degradation through macroautophagy can be better understood and quantitatively described.

Advances in organic chemistry and molecular biology as well as material sciences have enabled the research community to choose from a wide range of classes of fluorescent probes for imaging in cell biology. Here, a combination of inherently binding fluorochromes, fluorescent proteins and an immunofluorescence approach allows the concomitant assessment of a number of parameters of interest (Figure 3). Such multi-colour imaging in turn enables the assessment of organelle and/or protein-protein interactions through colocalization analysis in a powerful manner, reporting on cell function and dysfunction. Multi-parameter imaging enables a better understanding of the relationship between the parts of interest in the cellular system. If performed in combination with super-resolution techniques, such intermolecular relationships can report on function and localization in a most precise manner [14].

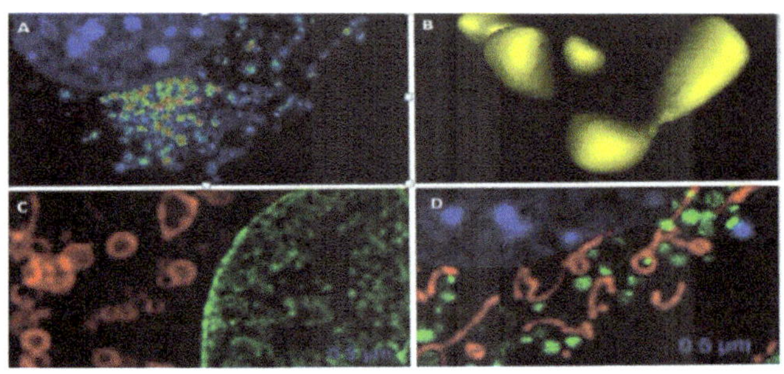

Figure 2: GT 1-7 neuronal cells indicating ATP (A) targeted to the mitochondria, using a FRET construct. Superresolution structured illumination microscopy (SRSIM) enables the visualization of labeled organelles down to 80 nm spatial x-y resolution. Fluorescent micrographs indicate lysosomes labeled for the lysosomal associated membrane protein-1 (B), mitochondria (C) and the interaction between GFPLC3 labeled autophagosomes and mitochondria (D).

The large amount of data that is being generated through fluorescence microscopy based techniques demands an integrative approach, in order to interpret distribution and intensities of fluorescent signal in a meaningful manner. This will become of even greater importance as we venture into PALM- 3D and super-resolution time lapse acquisition techniques. Here, systems biology and computational modeling may become an even more crucial part in the future to integrate the vast amount of biological information, as quantification of molecular elements of a biological system is at the heart of this discipline [2]. This becomes particularly clear in current trends using fluorescence-based image data at a multi-scale level [15], i.e. from molecular density mapping based on PALM [16] to in vivo whole organ imaging based on light sheet microscopy [17]. With no doubt such an approach will enable us to map proteins in cells and tissues spatiotemporally like never before, building dynamic and most comprehensive organ systems based on real fluorescence microscopy derived data. When doing so, and when doing so quantitatively, we are likely one step closer towards enabling predictive or preventative medicine. Taken together, we can look forward to an exciting future, where insights in cell science will matter greatly and where the use of fluorescence microscopy techniques will play a fundamental and integral part.

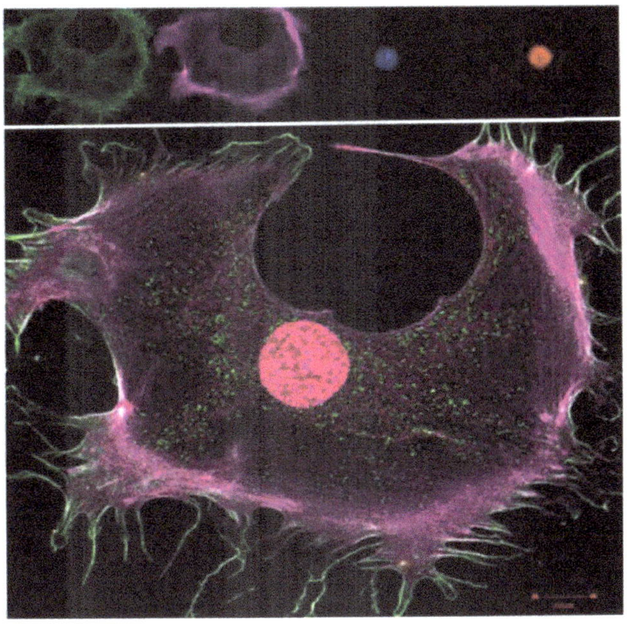

Figure 3: A combination of targeting methods and fluorescent probes. Mouse embryonic fibroblasts transfected with GFP-LC3 (green). After fixation cells were immune-labeled for mitochondrial DNA with a mouse monoclonal antibody (red) and counterstained using Alexa fluor 647 phalloidin (magenta) and Hoechst 33342 (blue). Images were acquired from z-planes and projected in maximum intensity. Scale bars: 20 µm.

ACKNOWLEDGEMENTS

This research work was supported by the National Research Foundation (NRF) and the Medical Research Council (MRC), South Africa. I wish to thank Dr Tavaré (University of Bristol, UK) for providing the CFP-DEVD-YFP construct, Prof Hiromi Imamura, Kyoto University, Japan for the provision of the ATP-FRET construct and Prof Noboru Mizushima, Tokyo University, for the provision of GFP-LC3 expressing mouse embryonic fibroblasts.

REFERENCES

1. Giepmans BN, Adams SR, Ellisman MH, Tsien RY (2006) The fluorescent toolbox for assessing protein location and function. Science 312(5771): 217-224.
2. Hood L, Heath JR, Phelps ME, Lin B (2004) Systems biology and new technologies enable predictive and preventative medicine. Science 306(5696): 640-643.
3. Loos B, Lockshin R, Klionsky DJ, Engelbrecht AM and Zakheri Z (2013) On the variability of autophagy and cell death susceptibility. Autophagy 9(9): 1270-1285.
4. Tyas L, Brophy VA, Pope A, Rivett AJ and Tavaré M (2000) Rapid caspase-3 activation during apoptosis revealed using fluorescenceresonance energy transfer, EMBO Rep 1: 266-270.
5. Loos B, Genade S, Ellis B, Lochner A, Engelbrecht AM (2011) At the core of survival: Autophagy delays the onset of both apoptotic and necrotic cell death in a model of ischemic cell injury. Exp Cell Res 317 (10): 1437-53.
6. Arrasate M, Mitra S, Schweitzer ES, Segal MR, Finkbeiner S (2004) Inclusion body formation reduces levels of mutant huntingtin and the risk of neuronal death. Nature 431(7010): 805-10.
7. Monastyrska I, Rieter E, Klionsky DJ (2009) Multiple roles of the cytoskeleton in autophagy. Biol Rev Camb Philos Soc 84: 431-448.
8. Rubinsztein DC, Ravikumar B, Acevedo-Arozena A (2005) Dyneins, autophagy, aggregation and neurodegeneration. Autophagy 1(3): 177-178.
9. Essmann F, Bantel H, Totzke G, Engels IH, Sinha B, Schulze-Osthoff K (2003) Staphylococcus aureus alpha-toxin-induced cell death: predominant necrosis despite apoptotic caspase activation, Cell Death Differ 10: 1260-1272.
10. Imamura H, Nhat KP, Togawa H, Saito K, Iino R, Kato-Yamada Y, Nagai T, Noji H (2009) Visualization of ATP levels inside single living cells with fluorescence resonance energy transfer-based genetically encoded indicators. Proc Natl Acad Sci U S A 106(37): 15651-15656.
11. Loos B, Du Toit A and Hofmeyr JH (2014) Defining and measuring autophagosome flux-concept and reality. Autophagy 10(11): 2087- 2096.
12. Mortiboys H, Kelly Jean T, Koopman WJH, Klaffke S, Abou-Sleiman P, et al. (2008) Mitochondrial Function and Morphology Are Impaired in parkin -Mutant Fibroblasts. Annals of Neurology 64(5): 555-65.
13. Karbowski M, Arnoult D, Chen H, Chan DC, Smith CL, et al. (2004) Quantitation of Mitochondrial Dynamics by Photolabeling of Individual Organelles Shows That Mitochondrial Fusion Is Blocked during the Bax Activation Phase of Apoptosis'. The Journal of Cell Biology 164(4): 493-99.
14. Jovanovic K, Loos B, Da Costa B, Penny C, Weiss S (2014) High resolution imaging study of interactions between the 37kDA/67kDa Laminin receptor and APP, beta-secretase and gamma-secretase in Alzheimer's disease. PLoS ONE 9(6): e100373.
15. Martone ME, Gupta A, Ellisman MH (2004) e-Neuroscience: challenges and triumphs in integrating distributed data from molecules to brains. Nature Neuroscience 7: 467-472.
16. Betzig E, Patterson GH, Sougrat R, Lindwasser OW, Olenych S, et al. (2006) Imaging intracellular fluorescent

proteins at nanometer resolution. Science 313(5793): 1642-5.

17. Pampaloni F, Ansari N and Stelzer EHK (2013) High-resolution deep imaging of live cellular spheroids with light-sheet-based fluorescence microscopy. Cell Tissue Res 352: 161-177

CHAPTER 17

MESENCHYMAL STEM CELL AND OSTEOARTHRITIS: A LITERATURE REVIEW

Zhaleh Shariati Sarabi (MD)[1], Jalil Tavakol Afshari (PhD)[2], Ali Ghassemi (MD)[3], Mehdi Yaghobi (MD)[4]

[1] Rheumatic Diseases Research Center, School of Medicine, Mashhad University of Medical Sciences, Mashhad, Iran

[2] Immunology Research Center, Avicenna Research Institute,Mashhad University of Medical Sciences, Mashhad, Iran

[3] Department of Bone marrow stem cell transplantation, School of Medicine, Mashhad University of Medical Sciences, Mashhad, Iran

[4]Department of Internal medicine, School of Medicine, Mashhad University of Medical Sciences, Mashhad, Iran

ABSTRACT

The most common disease in the aged population is osteoarthritis (OA) that is resulting in progressive dysfunction following isolated cartilage injuries, subchondral bone remodeling, tissue loss, marginal osteophytes, and loss of joint space. Mesenchymal stem cells (MSCs) are multipotent stem cells; they are able to produce many or all joint tissues. Bone marrow and adipose tissue are rich sources of mesenchymal cells that are useful for the reconstruction of injured tissues such as bone, cartilage, or cardiac muscle. Recently, some studies have been performed on the use of the direct intra-articular injection of mononuclear cells (MNCs) and MSCs as potential therapeutic targets in OA. In this review, the history of MSCs in the treatment of OA are explained. Injection of Bone Marrow Aspirates Concentrate (BMAC) has significantly improved both joint pain and function in radiologic findings; some studies suggested that the injection would be even more effective in early to moderate phases of OA. Injection of MSCs in combination with growth factors may be better solution for the treatment.

INTRODUCTION

Osteoarthritis (OA) is a chronic degenerative process, resulting in progressive dysfunction following isolated cartilage injuries, subchondral bone remodeling, tissue loss, marginal osteophytes, and loss of joint space (1). Cartilage damage is a main problem in OA. The treatments, used for cartilage repair in OA, include medication therapy, surgical joint replacement, microfracture, and osteochondral graft transplantation (mosaicplasty) (2). Elderly patients undergo a total knee replacement. In younger patients with less damaged joints, attempts have been made to introduce cartilage regeneration for the prevention of joint dysfunction and replacment. Currently, with tissue engineering methods, stem cells may be able to be used for autologous chondrocyte implantation (ACI) strategies that provide an abundant cell source to repair OA (3).

If the injection of mesenchymal stem cells (MSC) is effective for the treatment of OA, we could use this modality before other aggressive treatments such as joint surgery and replacement. MSCs are multipotent stem cells that differentiate into various functional cell types of mesodermal tissues and have shown the ability to engraft and migrate into multiple

Citation: Zhaleh Shariati Sarabi (MD), Jalil Tavakol Afshari (PhD), Ali Ghassemi (MD), Mehdi Yaghobi (MD), Mesenchymal stem cell and osteoarthritis: a literature review, http://rcm.mums.ac.ir/article_6830_0eabf0d102f318c3e0db1901e3aa4716.pdf.

musculoskeletal tissues (4,5). Bone marrow contains MSC progenitor cells of some mesenchymal tissues such as cartilage, bone, muscle, and fat (6). Tissue engineering combine MSCs with chondrogenic signals and various scaffolds to produce a functional tissue that could be used to repair focal cartilage defects (7).

LITERATURE REVIEW

Recent recommendations have focused on intra-articular injection of MSCs as a therapy for OA. Autologous bone marrow-derived MSCs have been injected in different animal models of OA including Murphy et al. (8) injected MSCs into the knee joints of the goats, Toghraie et al. (9) used a rabbit model and Black et al. (10) used dog models. Results showed that the direct intra-articular injection of MSCs reduced the development of advanced OA lesions. In vitro culture for osteogenic comparison, displayed properties of MSCs and detected sufficient bone matrix and bone-specific protein (osteoclacin) of bone marrow stem cells (BMSCs) under the osteogenic differentiation condition. In 2003, Ugarte et al. (11) reported no significant difference in osteogenic ability between human BMSCs and adipose tissue– derived MSCs (AMSCs). In 2005, Gun-II et al. (12) showed that human AMSCs have inferior osteogenic capabilities compared to BMSCs. In 2008, Hayeshi et al. (13) demonstrated that adult MSCs from bone marrow and periosteum could be considered ideal candidates for the regeneration of bone tissue. In 2006, Caplan et al. (14) and Agung et al. (15) revealed that intra-articular injection of MCSs into injured tissues could induce a host repair response to replace the injured tissue. The underlying mechanisms in the development of these novel therapeutic interventions modulate MSC to promote joint surface regeneration and influence the outcomes of joint disorders such as OA.

In clinical settings, MSCs could been also used for the regeneration and maintenance of articular cartilage in OA. In 2011, Pak reported stem cells injection into the right hip in two patients, obtained from adipose tissue; the following MRI images showed significant positive changes. They confirmed that stem cell therapy could improve OA and osteonecrosis of the femoral head (16). Wolfstadt et al. investigated 214 adults with OA in USA from 1990 to 2013 in seven different clinical studies. MSCs had the potential for improving function and decreasing inflammation in patients with OA (17).

Between 2005 to 2009, Centeno et al. studied 227 patients who were treated for various orthopedic conditions with culture-expanded, bone marrow-derived MSCs. The pre- and postprocedure MRI analysis revealed increased meniscus and cartilage volume, and the modified visual analogue scale (VAS) scores decreased by 95% at three-month follow-up. Two subsequent studies from the same group using the same treatment procedure were conducted on a large group of patients suffering from OA and other intra-articular pathology; thus, MSC-related complications were generally infrequent, transient, or remediated with simple therapeutic measures (18,19). In a study performed in Iran (2012) by Emadedin et al, intra-articular injection of culture-expanded MSCs was used for OA therapy in six females with radiological evidence of knee OA. Up to six months postinjection, all measured parameters improved (19). Wong et al. investigated two groups of patients with OA in 2013 at Singapore. In first group, as a controlled clinical study, microfracturing (MFX) was combined with intra-articular injection of MSCs that was used for the treatment of unicompartmental OA knees undergoing high tibial osteotomy (HTO), whereas the second group of 28 patients received an intra-articular injection with hyaluronic acid. One year after surgical intervention, MRI scans showed significantly better Magnetic Resonance Observation of Cartilage Repair Tissue (MOCART) scores for the cell-recipient group. This result demonstrated that intra-articular injection of cultured MSCs was effective in the improvement of both short-term clinical and MOCART outcomes in patients undergoing HTO and MFX for varus knees with cartilage defects (20). In a recent study in Germany, Kim et al. used combined intra-articular injection of autologous Bone Marrow Aspirates Concentrate (BMAC) with adipose tissue into the knees of 41 patients with OA. They showed significant improved pain and function of knee in all patients (21).

Davatchi et al. worked on four patients at Shariati Hospital, Tehran in 2011. The patients were aged 55, 57, 65, and 54 years, with moderate to severe knee OA and 30 mL of bone marrow was taken from each patient, and cultured for MSC growth. They showed pain improvement in three patients when walking and one patient remained unchanged (22). In 2002, Wakitani et al. studied twenty-four knees of 24 patients with OA in Japan. They cultured adherent cells in bone marrow aspirates and embedded in collagen gel, transplanted into the articular cartilage defect in the medial femoral condyle. After 6.3 weeks, defects were covered and were covered with harder hyaline cartilage-like tissue 42 weeks after transplantation. They recommended transplantation of autologous culture expanded MSCs for the repair of articular cartilage defects in human (6). In another study in 2013 at Poland, Skowroński et al. used Bone Marrow Concentrate (BMC) and collagen membrane in 54 patients with large cartilage lesions of the knee. All scales of 52 patients were significantly improved without infectious complication. They reported that bone marrow stem cell injection was an effective modality for the reconstruction of large cartilage lesions (23). Vangsness et al. performed treatment options for tissue restoration and the prevention of degenerative changes in the knee of 55 patients in 2014 at California, USA. Patients were randomized to one of the following treatment groups: in Group A, patients received 50 × 106 allogeneic MSCs; Group B, 150 × 106 MSCs; and the control group, a sodium hyaluronate. Twelve months after meniscectomy, there was a significant increase in meniscal volume in 24% patients of Group A and 6% patients of Group B. On the basis of VAS, MSC group experienced a significant reduction in pain compared with the control group (24).

Varma et al. selected 50 patients with mild to moderate knee OA in 2010 at India. They divided patients into two groups. Group A and B received arthroscopic debridement, but group B received mesenchymal stem cell concentrate injection as well. On the basis of VAS and OA outcome score, the technique used in this study considerably improved the overall OA outcome score, especially the quality of life (25).

CONCLUSION

Findings from the studies showed that BMC had important clinical implications which could be applied in OA treatment. The aforementioned methods have shown that the injection of MSCs alone or in combination with growth factors might be a better solution for the treatment of OA. MSCs injection was effective for the treatment of OA; thus, we could use this modality in clinical trials.

ACKNOWLEDGEMENT

We would like to thank Clinical Research Development Unit of Ghaem Hospital for their assistant in this manuscript.

REFERENCES

1. Wieland HA, Michaelis M, Kirschbaum BJ, et al. Osteoarthritis-an untreatable disease? Nat Rev Drug Discov. 2005;4:331-344.
2. Hunziker EB. Articular cartilage repair: basic science and clinical progress. A review of the current status and prospects. Osteoarthritis Cartilage. 2002;10:432-463.
3. Lawrence RC, Felson DT, Helmick CG, et al. Estimates of the prevalence of arthritis and other rheumatic conditions in the United States. Part II. Arthritis Rheum. 2008;58:26-35.
4. Chen FH, Tuan RS. Mesenchymal stem cells in arthritic diseases. Arthritis Res Ther. 2008;10:223.
5. Caplan AI, Bruder SP. Mesenchymal stem cells: building blocks for molecular medicine in the 21st century. Trends Mol Med. 2001;7:259-264.
6. Wakitani S, Imoto K, Yamamoto T, et al. Human autologous culture expanded bone marrow mesenchymal cell transplantation for repair of cartilage defects in osteoarthritic knees. Osteoarthritis Cartilage. 2002;10:199-206.
7. Tomita S, Li RK, Weisel RD, et al. Autologous transplantation of bone marrow cells improves damaged heart function. Circulation. 1999;100:II247-256.
8. Murphy JM, Fink DJ, Hunziker EB, et al. Stem cell therapy in a caprine model of osteoarthritis. Arthritis Rheum. 2003;48:3464-3474.
9. Frisbie DD, Kisiday JD, Kawcak CE, et al. Evaluation of adipose-derived stromal vascular fraction or bone marrow-derived mesenchymal stem cells for treatment of osteoarthritis. J Orthop Res. 2009;27:1675-1680.
10. Black LL, Gaynor J, Adams C, et al. Effect of intraarticular injection of autologous adipose-derived mesenchymal stem and regenerative cells on clinical signs of chronic osteoarthritis of the elbow joint in dogs. Vet Ther. 2008;9:192-200.
11. De Ugarte DA, Morizono K, Elbarbary A, et al. Comparison of multi-lineage cells from human adipose tissue and bone marrow.Cells Tissues Organs. 2003;174:101-109.
12. Im GI, Shin YW, Lee KB. Do adipose tissue-derived mesenchymal stem cells have the same osteogenic and chondrogenic potential as bone marrow-derived cells?Osteoarthritis Cartilage. 2005;13:845-853.
13. Hayashi O, Katsube Y, Hirose M, et al. Comparison of osteogenic ability of rat mesenchymal stem cells from bone marrow, periosteum, and adipose tissue. Calcif Tissue Int. 2008 ;82:238-247.
14. Caplan AI, Dennis JE. Mesenchymal stem cells as trophic mediators. J Cell Biochem. 2006;98:1076-1084.
15. Agung M, Ochi M, Yanada S, et al. Mobilization of bone marrow-derived mesenchymal stem cells into the injured tissues after intraarticular injection and their contribution to tissue regeneration. Knee Surg Sports Traumatol Arthrosc. 2006;14:1307-1314.
16. Pak J. Regeneration of human bones in hip osteonecrosis and human cartilage in knee osteoarthritis with autologous adipose-tissue-derived stem cells: a case series. J Med Case Rep. 2011;5:296.
17. Wolfstadt JI, Cole BJ, Ogilvie-Harris DJ, et al. Current concepts: the role of mesenchymal stem cells in the management of knee osteoarthritis. Sports Health. 2015;7:38-44.
18. Centeno CJ, Schultz JR, Cheever M, et al. Safety and complications reporting update on the re-implantation of culture-expanded mesenchymal stem cells using autologous platelet lysate technique. Curr Stem Cell Res Ther. 2011;6:368-378.
19. Emadedin M, Aghdami N, Taghiyar L, et al. Intra-articular injection of autologous mesenchymal stem cells in six patients with knee osteoarthritis. Arch Iran Med. 2012;15:422-428.
20. Wong KL, Lee KB, Tai BC, et al. Injectable cultured bone marrow-derived mesenchymal stem cells in varus knees with cartilage defects undergoing high tibial osteotomy: a prospective, randomized controlled clinical trial with 2 years' follow-up. Arthroscopy. 2013;29:2020-2028.
21. Kim JD, Lee GW, Jung GH, et al. Clinical outcome of autologous bone marrow aspirates concentrate (BMAC) injection in degenerative arthritis of the knee. Eur J Orthop Surg Traumatol. 2014;24:1505-1511.
22. Davatchi F, Abdollahi BS, Mohyeddin M, et al. Mesenchymal stem cell therapy for knee osteoarthritis. Preliminary report of four patients. Int J Rheum Dis. 2011;14:211-215.
23. Skowroński J, Skowroński R, Rutka M. Large cartilage lesions of the knee treated with bone marrow concentrate and collagen membrane--results. Ortop Traumatol Rehabil. 2013;15:69-76.
24. Vangsness CT Jr, Farr J 2nd, Boyd J, et al. Adult human mesenchymal stem cells delivered via intra-articular injection to the knee following partial medial meniscectomy: a randomized, double-blind, controlled study. J Bone Joint Surg Am. 2014;96:90-98.
25. Varma HS, Dadarya B, Vidyarthi A. The new avenues in the management of osteo-arthritis of knee--stem cells. J Indian Med Assoc. 2010;108:583-585.

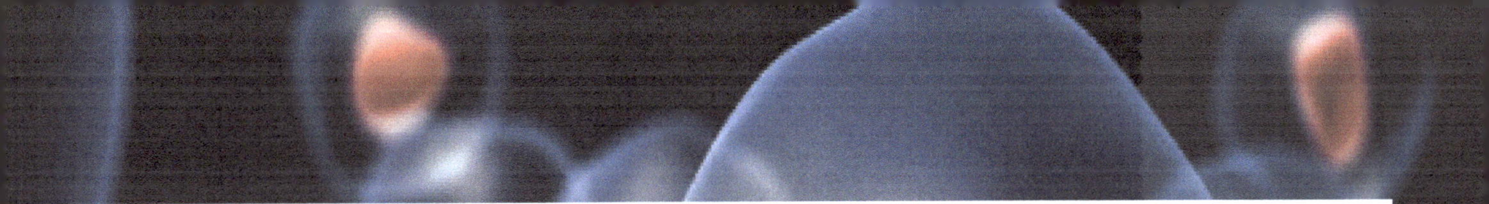

CHAPTER 18

RATIONAL DESIGN OF ORGANELLE COMPARTMENTS IN CELLS

Claudio Angione[1], Giovanni Carapezza[2], Jole Costanza[2], Pietro Lió[1], Giuseppe Nicosia[2]

[1] Computer Laboratory, University of Cambridge, Cambridge, United Kingdom

[2] Department of Mathematics and Computer Science, University of Catania, Catania, Ital

MOTIVATION AND OBJECTIVES

In recent years there is a growing interest in researching on mitochondria, chloroplasts and other mitochondrion-like organelles (e.g. hydrogenosomes, mitosomes and apicoplasts) because of the integrate bio-search for comorbidities- related genes, pathway dysfunctions, the energy balance in aging, inflammation and disease, and the discovery of novel factors involved in organelle division, movement, signaling and adaptation to varying environmental and pathogenic conditions. Furthermore, there is an impressive amount of mitochondria and chloroplasts sequence data (thousands of mitochondrial sequences from many species have been sequenced) that have been used in the last ten years to derive the history of species. Notably, there are no examples of examined eukaryotes without a mitochondrion- related organelle (Shiflett and Johnson, 2010). Despite these research efforts, there is a lack of knowledge about the relationships between the organelles in a cell and its metabolism. We aim at investigating and comparing the complexity of these organelles through a common framework that includes single- and multi-objective optimization, robustness analysis and sensitivity analysis. The possibility of multiobjective- optimization in organelles such as the mitochondrion may be related to the different tasks of maximizing the ATP or the heat, or intermediate compounds of the Krebs cycle in order to provide input for biosynthetic pathways (e.g. the amino acids synthesis). Furthermore, rather than focusing only on networks of molecules, we think of the cell as an integrated system (Yoneda et al, 2009). Indeed, the systems biology approach, i.e. taking into account only molecular networks, misses the analysis of the organization provided by organelles. An organelle can be viewed as a functional organization of macromolecules working to accomplish essential cellular functions. Many conditions depend on a variety of environmental and other factors, and therefore cannot be fully investigated by conventional molecular-level approaches.

Citation: Claudio Angione, Giovanni Carapezza, Jole Costanza, Pietro Lió, Giuseppe Nicosia, Rational design of organelle compartments in cells, DOI: http://dx.doi.org/10.14806/ej.18.B.537

METHODS

The cell contains many membrane-bound organelles, each specialized in one or more functions. External reactions can be thought of as links between organelles, since they involve metabolites found both in the cytoplasm and in the organelles. In our framework (Figure 1), we take into account complete models of the organelles in the cell, whose state space reflects its metabolism. The genetic algorithm underlying the optimization allows us to reach the optimal Pareto-front, i.e. to move the front towards the optimal point (e.g., maximum ATP and NADH), which is unfeasible if the two objective are negatively correlated with one another. The framework is implemented in MATLAB.

RESULTS AND DISCUSSION

The genetic and the energy-converting networks of mitochondria and chloroplasts are descended, with little modification, from those of their ancestor bacteria. In this regard, we explore how the optimization and the Pareto front analysis can provide interesting insights into the evolutionary dynamics leading to the formation of organelle compartmentalization in the single- and multi-celled life. Furthermore, the sensitivity and robustness analyses can detect clusters of parameters corresponding to clusters of chemical reactions, which are often found in the cell and reflect the presence of different pathways or membrane-bound organelles. The interplay between optimization, sensitivity and robustness is useful not merely to reach the optimal configuration for the organelles, but also to conduct tentative analyses on their parameters.

We have applied our framework to investigate models of organelles, e.g. mitochondria and chloroplasts. In the mitochondrial model (Bazil et al, 2010), we found that the most sensitive parameters are the Hexokinase max rate and the F1F0 ATP synthase activity. In the multi-objective optimization stage we analyzed the ATP-NADH Pareto front obtained with several calcium concentrations: if Ca^{2+} increases, we obtain an increase in NADH formation, while ATP remains constant; if Ca^{2+} drastically decreases, there is a lower ATP synthesis. Unexpectedly, with a slow decrease of Ca^{2+}, both objectives are maximized. Our results highlight also that the natural mitochondrion is more robust than the optimized one, as it features a global robustness value of 26.94% and a local value of 9.00%.

In the chloroplast model (Zhu et al, 2007), our framework detected RuBisCO and GAP dehydrogenase as the most sensitive enzymes of the C3 cycle. The Pareto fronts allowed us to find a trade-off between the maximization of the CO_2 uptake rate and the minimization of the nitrogen consumption, with the aim of absorbing more CO_2 while consuming less "leaf-fuel". RuBisCO and PGA kinase are the most robust enzymes.

The functional optimization is not partitioned or delegated to the organelles. The selection is on the phenotype and acts on the whole cell's compartmentalized genomes. Indeed, it is the whole protozoan cell that competes with other protozoa. The fact that an organelle is kept during the evolution means that its contribution to the overall protozoan's fitness is not marginal, i.e., the presence of the organelle ensures some advantages over losing it. The contribution of each organelle is both to maximize the energy production and to coevolve with the other cell structures, so as to ensure the maximum fitness of the cell. The organelles in a cell play also a key role in the neuronal degeneration. In this regard, neurotransmitters are found in vesicles, i.e. tiny organelles that allow to respond to packets ("units") of neuronal chemical signaling.

Following our framework and extending it, in the near future we plan to design, analyze and optimize the metabolism of systems composed of different species living and interacting in the same organism. In this project we have performed an in silico design that can explore the reaction network and seek in the search space the solutions that optimize two or more objectives. Therefore, our approach lies in the field of computational metabolic engineering. This kind of analysis could easily highlight the complementarity of different metabolic networks. For instance, mitochondria and chloroplasts are (usually) both found in plants, and are part of the same functional pipeline: starting from CO_2, the photosynthesis in the chloroplast creates glucose that enters the mitochondria to create ATP.

The model of the whole cell of a human pathogen (Karr et al, 2012) has opened new frontiers in this research field. The whole-cell model refers to the Mycoplasma genitalium and consists of 28 submodels accounting for all the biological functions of the cell. We have already applied our methodology to Flux Balance Analysis, Gene-Protein-Reaction associations, Ordinary Differential Equations (ODEs) and Differential Algebraic Equations (DAEs). Hence, our framework is suitable for general purpose or black-box analysis, enabling us to investigate not only the model of metabolism, but also the whole-cell model. Our final goal is to improve our methodology in order to tackle any BioCAD problem.

In each Pareto front we have considered the single organelle, while in the cell there are usually many organelles that could differ for activity depending on their location in the cell. In a network of organelles, most of the reactions involve more than one organelle; an appropriate approach would be to build a Pareto front where each metabolite belongs to a different organelle in the network, linking with a set of Delay Differential Equations (DDEs). DDEs differ from ODEs in that they allow rates of change to depend on the state of the system at an earlier time. In organelle systems, DDEs could account for diffusion processes and maturation events.

The integrated bio-search for a diseased, perturbed, misfunctional pathway often needs an accurate understanding of the relationships and interactions of that pathway with the organelle network system. The methodology proposed in our work could address most questions emerging in neurodegenerative and cancer disease investigation, which are focused on the interaction between organelle networks and cellular metabolism.

REFERENCES

1. Bazil J N, Buzzard G T, Rundell A E (2010) Modeling mitochondrial bioenergetics with integrated volume dynamics. PLoS computational biology, 6(1):e1000632
2. Karr J R, Sanghvi J C, et al (2012) A Whole-Cell Computational Model Predicts Phenotype from Genotype. Cell, 150(2):389-401
3. Shiflett A and Johnson P J (2010) Mitochondrion-related organelles in parasitic eukaryotes. Annual review of microbiology, 64:409
4. Stelling J, et al (2004) Robustness of Cellular Functions. Cell, 118(6):675-685
5. Stracquadanio G and Nicosia G (2011) Computational energy-based redesign of robust proteins. Computers & chemical engineering, 35(3):464–473

6. Yoneda Y, et al (2009) Frontier biomedical science underlying organelle network biology, http://www.fbs.osaka-u.ac.jp/organelle-network/eng/greeting/greeting/

7. Zhu X G, de Sturler E, Long S P (2007) Optimizing the distribution of resources between enzymes of carbon metabolism can dramatically increase photosynthetic rate. Plant Physiology, 145:513–526

CHAPTER 19

ORGAN-ON-A-CHIP: NEW PLATFORM FOR BIOLOGICAL ANALYSIS

Fan An,[1,2] Yueyang Qu,[1,2] Xianming Liu,[3] Runtao Zhong,[3] and Yong Luo[1,2]

[1]School of Pharmaceutical Science and Technology, Dalian University of Technology, Dalian, China.

[2]State Key Laboratory of Fine Chemicals, Department of Chemical Engineering, Dalian University of Technology, Dalian, China.

[3]Dalian Institute of Chemical Physics, Chinese Academy of Sciences, Dalian, China.

ABSTRACT

Direct detection and analysis of biomolecules and cells in physiological microenvironment is urgently needed for fast evaluation of biology and pharmacy. The past several years have witnessed remarkable development opportunities in vitro organs and tissues models with multiple functions based on microfluidic devices, termed as "organ-on-a-chip". Briefly speaking, it is a promising technology in rebuilding physiological functions of tissues and organs, featuring mammalian cell co-culture and artificial microenvironment created by microchannel networks. In this review, we summarized the advances in studies of heart-, vessel-, liver-, neuron-, kidney- and Multi-organs-on-a-chip, and discussed some noteworthy potential on-chip detection schemes.

INTRODUCTION

The capability of traditional platforms, such as well plate, does not match researchers' demands for precise and creative experiments any more. With various advantages, such as low cost, versatility, miniaturization, preciseness, and flexibility in handling biological samples, microfluidics has drawn tremendous attention in scientific and engineering academia and has become a great platform to replace the conventional one. Specially designed microfluidic devices, which are called organ-on-a-chip(s),[1] provide a controllable culture microenvironment for living cells in micrometersized chambers for modeling physiological functions of tissues and organs. Currently, a series of "suborgans" were created that imitated the construction of minimal functional units and kept partial function of organs or tissues in vivo. These emulative organ or tissue models provide a powerful tool for biological and pharmaceutical researches. In this mini review, we introduce some recent advances in the novel organ-on-a-chip researches that can be extended to pharmaceutical analysis and new drug development. Also, this review covers related techniques of organs and tissues that are of great interest to drug developers, such as heart, vessel, liver, neuron, and kidney.

Citation: Fan An, Yueyang Qu, Xianming Liu, Runtao Zhong, and Yong Luo Organ-on-a-Chip: New Platform for Biological Analysis. Analytical Chemistry Insights 2015:10 39–45 doi:10.4137/ACI.S28905.

HEART-ON-A-CHIP

Cardiovascular diseases (CVDs) are the first cause of death globally. It is estimated that 17.5 million people died from

CVDs in 2012, representing 31% of global death.[2] In a mammalian body, in the circulatory system, heart is responsible for pumping blood, and has a most important feature of autorhythmicity, modulated by the endocrine and nervous systems.[3] Autorhythmic heartbeat leads to cyclical fluctuant mechanical stress at the endothelial cell (EC) surface, which is important for vascular phenomena such as vessel remodeling, endothelial permeability, vasoregulation, blood formation, and vascular pathology, including platelet thrombus formation and atheroma. With the help of microfluidic chip technology, replicating functional cardiovascular organ model in vitro is becoming a reality.

In Xiao's report,[4] human embryonic stem cell (hESC) and primary neonatal rat cardiomyocytes were used to generate a microfabricated cardiac biowire bioreactor that was attractive for pharmaceutical testing. The spontaneous beating of the cardiac biowires could be retarded by the treatment of nitric oxide, which was carried by the medium into the chamber. Furthermore, the integrated carbon rod electrodes offered electrical stimulation for further improving the phenotype of cardiomyocytes (Fig. 1). Nguyen and colleagues reported a heart-on-chip methodology with which an accurate controllable physiologic mechanical stimulation was managed to rebuild a simulative physiological condition that holds significant promise for immature cardiomyocytes for generating functional cardiac patches in vitro for replacement of injured cardiac tissues.[5] In another example,[6] a microfluidic device that generates cardiac-like flow in a continuous closed culture system was reported with advantages of miniaturization, low circulatory volume (2–3 μL).

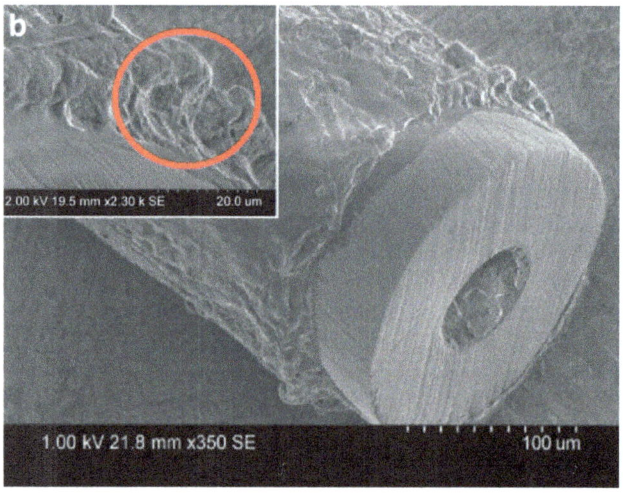

Figure 1: SEM image for the thick layer of cardiac tissue attached to the tubing surface after remodeling. Primary and hESC-derived cardiomyocytes were used to generate cardiac biowires and beat spontaneously. Reproduced from Xiao Y, Zhang B, Liu H, et al. Microfabricated ...

VESSEL-ON-A-CHIP

The network of blood vessels distributes throughout the whole body of vertebrates and connects all the organs and tissues. Even the structure and dimension of different vessels are distinct, and the foundation of the blood circulation system is still the vascular ECs that build selective permeability barrier among tissue and blood. ECs involve in multiple physiological functions (such as barrier, clotting, inflammation, angiogenesis, vasoconstriction–vasodilation, and tumor metastasis) and become a hotspot in organ-on-a-chip research.

A functional and perfusable 3D microvascular network on a chip was reported recently[7] (Fig. 2). In that report, ECs were spatially controlled cocultured with stromal fibroblasts, pericytes, or cancer cells to generate normal microvessel development and tumor vasculature angiogenesis. Further research in the regulating angiogenesis mechanism was achieved. It was realized with the powerful microfluidic system that incorporated the whole milieu of soluble factors, produced by cells in situ and allowed potential drugs to target angiogenesis in disease[8] (Fig. 3).

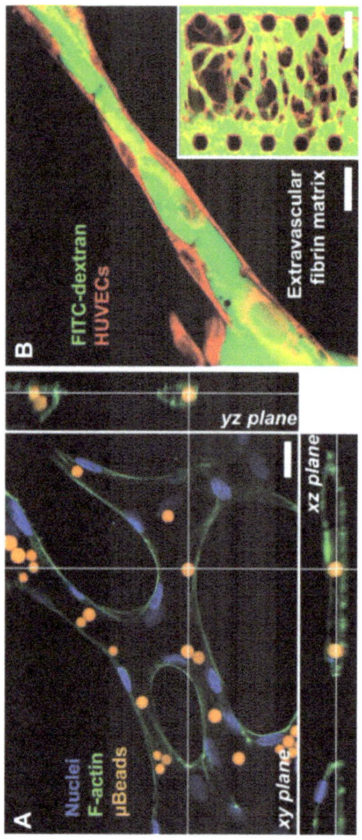

Figure 2: Introduction of solutions containing red fluorescent microbeads (A) or fluorescein isothiocyanate (FITC)-dextran (70 kDa, green) (B) into the perfusable microvessels. Human umbilical vein endothelial cells (HUVECs) were cocultured with human normal lung ...

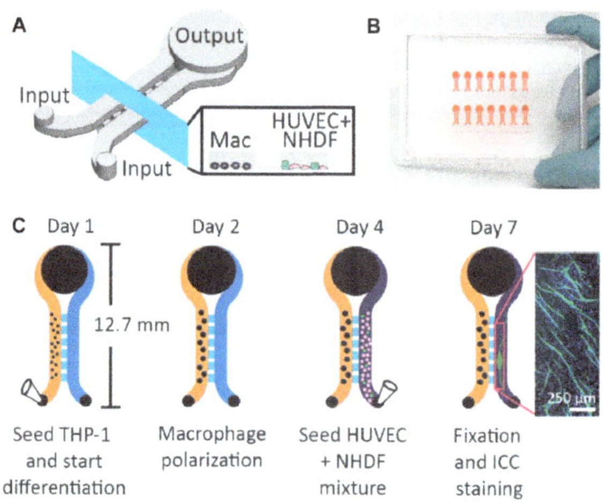

Figure 3: Microfluidic device design and workflow to analyze biomo-

lecular signaling in angiogenesis. (A) Macrophages and the HUVEC/Primary neonatal normal human dermal fibroblast (NHDF) mixture were seeded in left and right channels, respectively. (B) An array...

Vascular smooth muscle cell (VSMC) is another functional cell composing the majority of the wall of blood vessels, responsible for systolic–diastolic to regulate the blood volume and pressure. A vascular muscular thin film was generated, based on genipin and extracellular matrix (ECM) proteins modified PDMS, sufficiently mimicking the circumferential alignment of VSMCs in arteries and successfully maintaining tissue contractility and integrity throughout two weeks.[9] Three cell types forming different layers of the vessel wall were patterned on a membrane that was stress-induced, rolling into a tube, precisely mimicking blood vessels, and many other tubular structures.[10] This study provided a new strategy to rebuild tissues in vitro.

LIVER-ON-A-CHIP

In human body, liver is a major metabolism organ, and its functions include regulation of glycogen storage, decomposition of red blood cells, plasma protein synthesis, hormone production, and detoxification.[11] Ample vascular and bile network provides supplement of nutrient–oxygen and elimination of metabolites and waste products. The hepatocyte makes up to 70%–85% of the liver's mass and is the main functional parenchymal cell of the liver. Therefore, hepatocytes are seen as the representative model for liver in artificial liver in vitro. Unfortunately, even though it is well known that the liver keeps the capability to regenerate in vivo,[12,13] the isolated hepatocyte degeneration is unavoidable during the culture in vitro in weeks.

With the development of lab-on-a-chip technique, it has been demonstrated that a continuous medium flow allows a feasible exchange of oxygen and wastes, respectively, with the outside and creates a more physiological environment in stimulating the expression of detoxifying genes.[14] Moreover, with a liver-on-a-chip method, appropriate interaction between cell–cell, cell–ECM, and cell-soluble factors contribute to stabilize primary hepatocyte functions.[15,16] Also, micro-bioreactor based on hepatocytes is considered to be a promising in vitro model for artificial liver, in which hepatocytes were 3D cultured in different approaches, such as spheroids, sandwich gels, porous scaffolds, or encapsulation in natural or synthetic hydrogels.[17–26] In another report, rat hepatocytes were isolated by Li and coworkers, and these cells were then encapsulated in polyethylene glycol diacrylate, which maintained liver-specific function for over 50 days while responding to prototypic small molecules and drug–drug interactions. Even more, this microtissue could be manufactured using a microfluidic droplet generator[27] (Fig. 4).

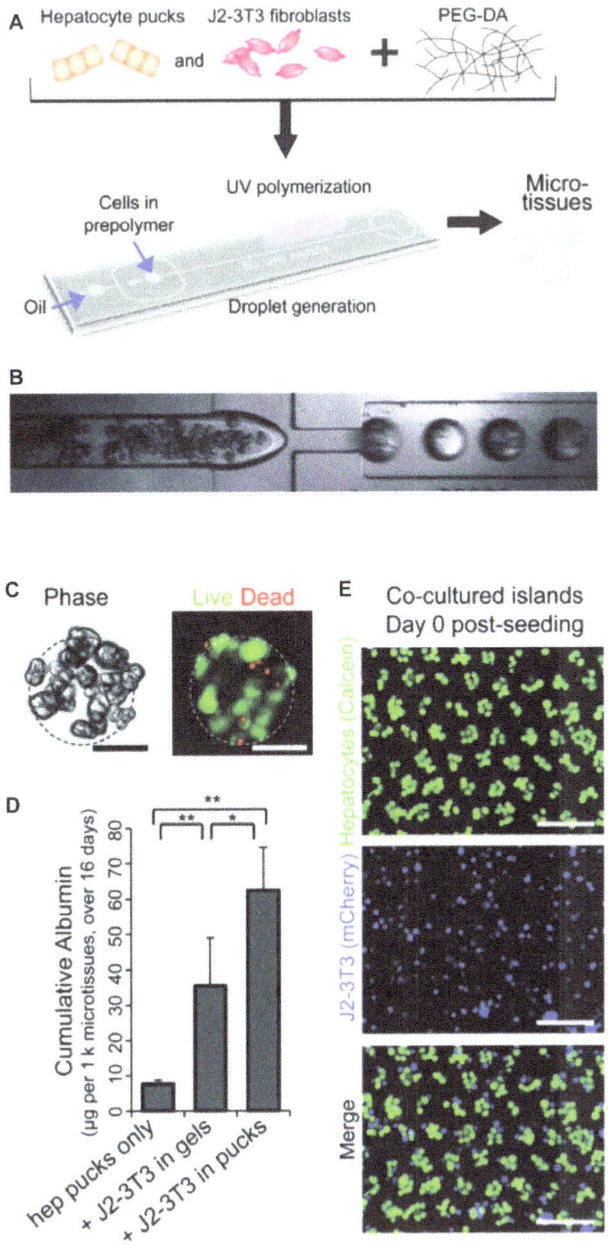

Figure 4: Microfluidic encapsulation of hepatocyte pucks. (A) A microtissue of hepatocyte pucks and fibroblast were generated using a (B) microfluidic droplet generating device. (C) Phase image and viability staining of an individual microtissue. (D) Albumin secretion ...

NEURON-ON-A-CHIP

Central nervous system (CNS) and peripheral nervous system (PNS) form the nervous system with the basic unit of neuron. Neuron, also called nerve cell, is electrically excitable. These highly differentiated cells do not undergo cell division, and therefore makes traumatic injuries and degenerative diseases in the CNS insufficiently or completely untreatable.[28] Neurons are generally generated by neural stem/progenitor cells (NSPCs),[29–31] and that makes researchers place great importance on NSPCs differentiation into the desired neural cell types when repairing the injury.[32–36]

On a microfluidic device fabricated by Shamloo and coworkers,[37] rat NSPCs were isolated and cultured within varying densities of collagen matrices, and then, the effect of varying concentrations of nerve growth factor on migration and differentiation was detected.[37] Moreover, a method for observing migration out of the embryonic medial ganglionic eminence (MGE) in real-time was described[38] (Fig. 5). Live neurons were costained with different intracellular probes to monitor the distribution of cellular organelles, including Golgi and nuclear. Then, an efficient method to transfect cells in MGE explants with adeno-associated virus expressing reporter proteins was established. In another work, NSPCs were cultured on an indium tin oxide surface, and cell differentiation, process development, and functionality of differentiated neuron were successfully controlled by the electrical stimulation.[39]

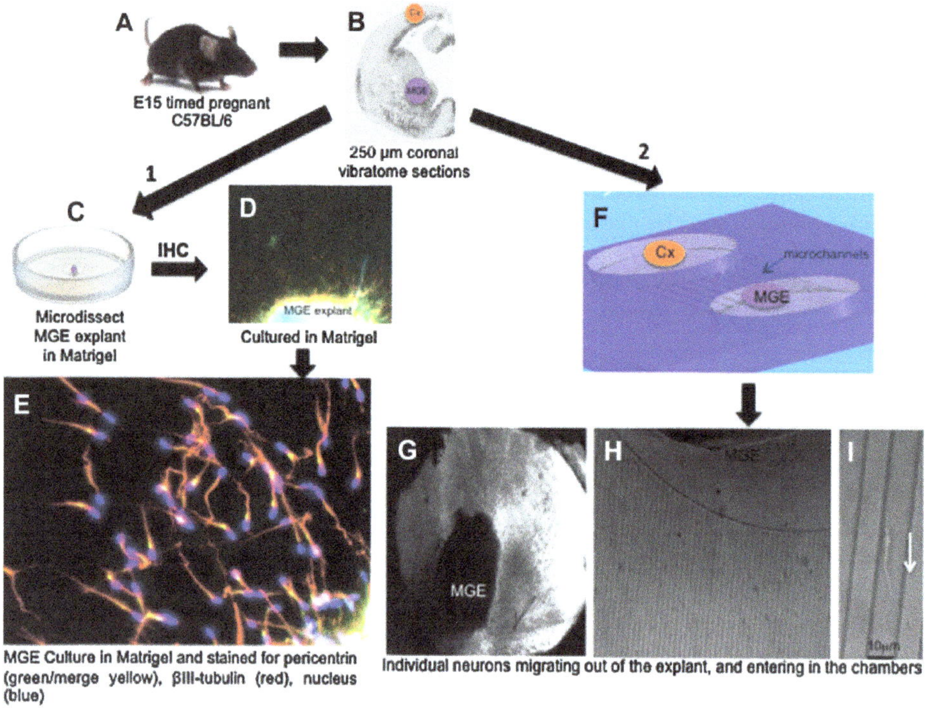

Figure 5: Experimental models to study MGE neurons migration from the explants of mouse embryos. (A) Brains of the E15 embryos isolated from female mice and sliced to dissect MGE explants (B). MGE explants were placed in a Matrigel coated glass dish (C) and neurons ...

KIDNEY-ON-A-CHIP

Kidney is the most important organ for drug metabolism and elimination; therefore, drug-induced nephrotoxicity is one of the most frequently addressed problems in drug safety evaluation.

There were some early works exploring on nephrotoxicity screening platform-based microfluidics. With the similar PDMS champing porous membrane structure, Jang and Suh used rats, collecting duct epithelial cells to build kidney tubules on chip and observed their transportation ability on the chip[40] (Fig. 6). In an independent report, Ferrell used opossum kidney epithelial cells to build the chip and analyzed the mechanism of albumin transportation.[41] Also, Jang KJ et al applied human primary renal tubules epithelial cells on their microfluidic device and measured a number of important parameters, such as albumin transport, glucose reabsorption, brush border alkaline phosphatase activity, cisplatin toxicity, and P-gp efflux transporter activity[42] on chip. All of these works used only one kind of cells in forming partial physiological environment, and then characterized their systems with a series of biomarkers.

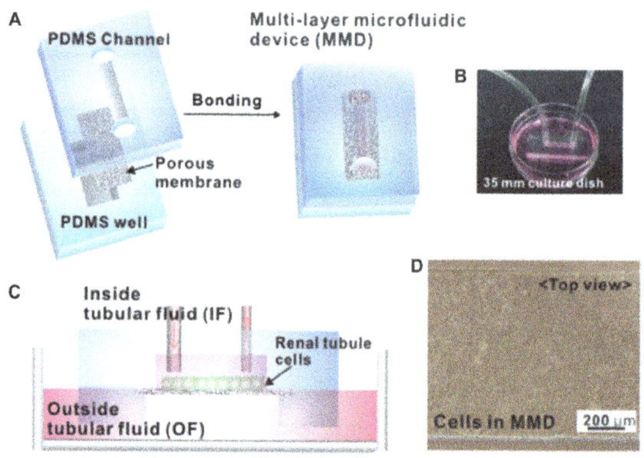

Figure 6: Fabrication and operation of a multilayer microfluidic device (MMD) for efficient culture and analysis of renal tubular cells. (A) Sandwich structure of MMD. Photograph (B) and schematic of the device (C) on a culture dish containing outside tubular fluid. ...

MULTI-ORGANS-ON-A-CHIP

Not only efficacy but also pharmacokinetics, toxicity, and security characteristics decide the fates of candidate drugs. Unfortunately, the high throughput, traditional drug screening platforms, such as 96-well plate, are excessively simple, which is greatly different from the complex conditions in vivo and incapable of evaluating drugs accurately and comprehensively. Numerous candidates are eliminated during animal and clinical test, which is one of the reasons why drug development process is so costly, time-consuming, and inefficient.

Multi-organ-on-a-chip is probably a novel strategy to increase efficiency and precision of drug screening before animal test. Being different from single organ chips, these elaborate microfluidic devices integrate several cells from different organs and tissues according to the way in vivo, attempting to simulate human body. For example,[43] in Abacia and Shuler's report, a multi-organ-on-a-chip that was designed according to actual blood velocity and organ proportions had been proved to be a platform suitable for studying physiologically based pharmacokinetics/pharmacodynamics. In another example,[44]intestine, liver, skin, and kidney tissues were cocultured and sustainable over at least 28 days for absorption, distribution, metabolism, and excretion profiling. In this way, multi-organs-on-a-chip devices are capable of providing comprehensive information of candidate drugs and may bring revolutionary changes in the pharmaceutical industry.

CONCLUSION AND PERSPECTIVES

At present, more and more research groups regard microfluidic device as an important platform for their potential in-life sciences. In cell- and tissue-level researches in vitro, the primary cells, immortalization cell lines, and stem cells are the major experiment models. Among them, immortalization cell lines are most easily cultured but anamorphic frequently happens on well plate and resulted in deviation from reality. Primary cells are more effective models but hard to maintain in traditional culture method and also induction of stem cells is under scrutiny. The microfluidic device has shown its ability on controlling microenvironment, which is urgently needed for primary and stem cells research. On the other hand, considered to be the substitute of experiment animal in physiological and pharmaceutical research, "human-on-chip" is the ultimate goal with the combination of different functional organ chips. Multiple elements can be integrated in one chip, such as different cells, ECM, fluid, droplet, valve, even sensors, and more useful features. Benefited by "design as you wish," prospect of microfluidic device is limitless.

Organ-on-a-chip holds many advantages over traditional in vitro cytology experiment tools (such as 96-well plate and transwell chamber) and general cell culture systems based on microchip. For example, the cell microenvironment is highly maneuverable and closer to it is in vivo; organ-on-a-chip keeps capacity to reproduce the function and structure of the organ; multi-organs-on-a-chip are able to simultaneously evaluate efficacy, toxicity, and pharmacokinetics, which are otherwise impossible to evaluate with traditional drug screening tools. Although powerful, at the current stage, organ-on-a-chip cannot completely replace the traditional drug screening tools. One important reason is lack of on-chip detection schemes. Off-chip HPLC and mass spectrometer, which are of low efficiency, were widely used for determination of biomolecules on the organ-on-a-chip. Separation of molecules of interest cannot be conducted on the organ-on-a-chip yet. Direct observation of cells on a chip with an optical microscope is somehow difficult because of low transparency of the membrane and thickness of organ-on-a-chip beyond the working distance of the objective. And it is analysts' responsibility to solve these problems on the way, by cooperating with the biologists. And here we would expect some solutions, for example, electrophoresis of different modes can be integrated with an organ-on-a-chip to address separation issue; the design of an organ-on-a-chip can be updated to fit the short working distance of high magnification objective; coupling of organ-on-a-chip with mass spectrometer can be carried out; electrode can be integrated in the organ-on-a-chip for in situ monitoring of biological process, etc. In summary, development of an on-chip detection approach will accelerate the commercialization of organ-on-a-chip, and it will eventually speed up the progress in pharmacological studies and new drug development.

FUNDING

This study was funded by the key deployment project of the Chinese Academy of Sciences (KSZD-EW-Z-021) and the National Key Scientific Instrument and Equipment Development Project, China (No. 2011YQ03012404). The authors confirm that the funder had no influence over the study design, content of the article, or selection of this journal.

COMPETING INTERESTS

\Paper subject to independent expert blind peer review. All editorial decisions made by independent academic editor. Upon submission manuscript was subject to anti-plagiarism scanning. Prior to publication all authors have given signed confirmation of agreement to article publication and compliance with all applicable ethical and legal requirements, including the accuracy of author and contributor information, discloure of competing interests and funding sources, compliance with ethical requirements relating to human and animal study participants, and compliance with any copyright requirements of third parties. This journal is a member of the Committee on Publication Ethics (COPE).

REFERENCES

1. Bhatia SN, Ingber DE. Microfluidic organs-on-chips. Nat Biotechnol. 2014;32(8):760–772.
2. World Health Organization Cardiovascular Diseases (CVDs)Available from:http://www.who.int/mediacentre/factsheets/fs317/en/
3. Reinecke M. Neurotensin. Immunohistochemical localization in central and peripheral nervous system and in endocrine cells and its functional role as neurotransmitter and endocrine hormone. Prog Histochem Cytochem. 1985;16(1):1–172.
4. Xiao Y, Zhang B, Liu H, et al. Microfabricated perfusable cardiac biowire: a platform that mimics native cardiac bundle. Lab Chip. 2014;14(5):869–882.
5. Nguyen MD, Tinney JP, Ye F, et al. Effects of physiologic mechanical stimulation on embryonic chick cardiomyocytes using a microfluidic cardiac cell culture model. Anal Chem.2015;87(4):2107–2113.
6. Chen H, Cornwell J, Zhang H, et al. Cardiac-like flow generator for long-term imaging of endothelial cell responses to circulatory pulsatile flow at microscale. Lab Chip. 2013;13:2999–3007.
7. Kim S, Lee H, Chung M, Jeon NL. Engineering of functional, perfusable 3D microvascular networks on a

chip. Lab Chip.2013;13(8):1489–1500.

8. Theberge AB, Yu J, Young EW, Ricke WA, Bushman W, Beebe DJ. Microfluidic multiculture assay to analyze biomolecular signaling in angiogenesis. Anal Chem. 2015;87(6):3239–3246.
9. Hald ES, Steucke KE, Reeves JA, Win Z, Alford PW. Long-term vascular contractility assay using genipin-modified muscular thin films. Biofabrication. 2014;6(4):045005.
10. Yuan B, Jin Y, Sun Y, et al. A strategy for depositing different types of cells in three dimensions to mimic tubular structures in tissues. Adv Mater. 2012;24(7):890–896.
11. Maton A, Jean H, Charles W, et al. Human Biology and Health.Englewood Cliffs, NJ, USA: Prentice Hall; 1993.
12. Michalopoulos GK, DeFrances MC. Liver regeneration. Science. 1997;276:60–66.
13. Michalopoulos GK. Liver regeneration after partial hepatectomy: critical analysis of mechanistic dilemmas. Am J Pathol. 2010;176:2–13.
14. Vinci B, Duret C, Klieber S, et al. Modular bioreactor for primary human hepatocyte culture: medium flow stimulates expression and activity of detoxification genes. Biotechnol J.2011;6:554–564.
15. Fraczek J, Bolleyn J, Vanhaecke T, Rogiers V, Vinken M. Primary hepatocyte cultures for pharmaco-toxicological studies: at the busy crossroad of various antidedifferentiation strategies. Arch Toxicol. 2013;87(4):577–610.
16. LeCluyse EL. Human hepatocyte culture systems for the in vitro evaluation of cytochrome P450 expression and regulation. Eur J Pharm Sci. 2001;13(4):343–368.
17. Tong JZ, Bernard O, Alvarez F. Long-term culture of rat-liver cell spheroids in hormonally defined media. Exp Cell Res.1990;189(1):87–92.
18. Hamilton G, Westmoreland C, George E. Effects of medium composition on the morphology and function of rat hepatocytes cultured as spheroids and monolayers. In Vitro Cell Dev Biol Anim. 2001;37(10):656–667.
19. Dvir-Ginzberg M, Elkayam T, Aflalo ED, Agbaria R, Cohen S. Ultrastructural and functional investigations of adult hepatocyte spheroids during *in vitro* cultivation. Tissue Eng. 2004;10(11–12):1806–1817.
20. Dunn JC, Tompkins RG, Yarmush ML. Long-term *in vitro*function of adult hepatocytes in a collagen sandwich configuration.Biotechnol Prog. 1991;7(3):237–245.
21. Richert L, Binda D, Hamilton G, et al. Evaluation of the effect of culture configuration on morphology, survival time, antioxidant status and metabolic capacities of cultured rat hepatocytes. Toxicol In Vitro. 2002;16(1):89–99.
22. Glicklis R, Shapiro L, Agbaria R, Merchuk JC, Cohen S. Hepatocyte behavior within three-dimensional porous alginate scaffolds. Biotechnol Bioeng. 2000;67(3):344–353.
23. Kaufmann PM, Heimrath S, Kim BS, Mooney DJ. Highly porous polymer matrices as a three-dimensional culture system for hepatocytes. Cell Transplant. 1997;6(5):463–468.
24. Underhill GH, Chen AA, Albrecht DR, Bhatia SN. Assessment of hepatocellular function within PEG hydrogels. Biomaterials.2007;28(2):256–270.
25. Chen AA, Thomas DK, Ong LL, Schwartz RE, Golub TR, Bhatia SN. Humanized mice with ectopic artificial liver tissues.Proc Natl Acad Sci U S A. 2011;108(29):11842–11847.
26. Nishikawa Y, Tokusashi Y, Kadohama T, Nishimori H, Ogawa K. Hepatocytic cells form bile duct-like structures within a three-dimensional collagen gel matrix. Exp Cell Res. 1996;223(2):357–371.
27. Li CY, Stevens KR, Schwartz RE, Alejandro BS, Huang JH, Bhatia SN. Micropatterned cell-cell interactions enable functional encapsulation of primary hepatocytes in hydrogel microtissues.Tissue Eng Part A. 2014;20(15–16):2200–2212.
28. Spinal cord injury facts and figures at a glance. J Spinal Cord Med. 2014;37(2):243–244.
29. Jagasia R, Song H, Gage FH, Lie DC. New regulators in adult neurogenesis and their potential role for repair. Trends Mol Med.2006;12(9):400–405.
30. Riess P, Zhang C, Saatman KE, et al. Transplanted neural stem cells survive, differentiate, and improve neurological motor function after experimental traumatic brain injury. Neurosurgery.2002;51(4):1043–1052.
31. Yamashita T, Ninomiya M, Acosta PH, et al. Subventricular zone-derived neuroblasts migrate and differentiate into mature neurons in the post-stroke adult striatum. J Neurosci.2006;26:6627–6636.
32. Nait-Oumesmar B, Picard-Riera N, Kerninon C, Baron-Van Evercooren A. The role of SVZ-derived neural precursors in demyelinating diseases: from animal models to multiple sclerosis. J Neurol Sci. 2008;265(1–2):26–31.
33. Rice A. Proliferation and neuronal differentiation of mitotically active cells following traumatic brain injury. Exp Neurol.2003;183(2):406–417.
34. Sherafat MA, Heibatollahi M, Mongabadi S, Moradi F, Javan M, Ahmadiani A. Electromagnetic field stimulation potentiates endogenous myelin repair by recruiting subventricular neural stem cells in an experimental model of white matter demyelination. J Mol Neurosci. 2012;48(1):144–153.
35. Thau-Zuchman O, Shohami E, Alexandrovich AG, Leker RR, Cereb J. Vascular endothelial growth factor increases neurogenesis after traumatic brain injury. J Cereb Blood Flow Metab.2010;30(5):1008–1016.
36. Thored P, Arvidsson A, Cacci E, et al. Persistent production of neurons from adult brain stem cells during recovery after stroke.Stem Cells. 2006;24(3):739–747.
37. Shamloo A, Heibatollahi M, Mofrad MR. Directional migration and differentiation of neural stem cells within three-dimensional microenvironments. Integr Biol (Camb) 2015;7(3):335–344.
38. Nery FC, da Hora CC, Yaqub U, et al. New methods for investigation of neuronal migration in embryonic brain explants. J Neurosci Methods. 2015;239:80–84.

39. Lei KF, Lee IC, Liu YC, Wu YC. Successful differentiation of neural stem/progenitor cells cultured on electrically adjustable indium tin oxide (ITO) surface. Langmuir. 2014;30(47):14241–14249.

40. Jang KJ, Suh KY. A multi-layer microfluidic device for efficient culture and analysis of renal tubular cells. Lab Chip.2010;10(1):36–42.

41. Ferrell N, Ricci KB, Groszek J, Marmerstein JT, Fissell WH. Albumin handling by renal tubular epithelial cells in a microfluidic bioreactor. Biotechnol Bioeng. 2012;109(3):797–803.

42. Jang KJ, Mehr AP, Hamilton GA, et al. Human kidney proximal tubule-on-a-chip for drug transport and nephrotoxicity assessment. Integr Biol (Camb) 2013;5(9):1119–1129.

43. Abacia HE, Shuler ML. Human-on-a-chip design strategies and principles for physiologically based pharmacokinetics/pharmacodynamics modeling. Integr Biol (Camb) 2015;7(4):383–391.

44. Maschmeyer I, Lorenz AK, Schimek K, et al. A four-organ-chip for interconnected long-term co-culture of human intestine, liver, skin and kidney equivalents. Lab Chip. 2015;15(12):2688–2699.

CHAPTER 20

ADVANCES IN IMPROVING MAMMALIAN CELLS METABOLISM FOR RECOMBINANT PROTEIN PRODUCTION

Claudia Altamirano[1,2], Julio Berrios[1], Mauricio Vergara[1], Silvana Becerra[2]

[1]Pontificia Universidad Católica de Valparaíso, Escuela de Ingeniería Bioquímica, Valparaíso, Chile

[2]Creas Conicyt Regional GORE Valparaíso, Chile

ABSTRACT

Background: The production of recombinant proteins for therapeutic use represents a great impact on the biotechnology industry. In this context, established mammalian cell lines, especially CHO cells, have become a standard system for the production of such proteins. Their ability to properly configure and excrete proteins in functional form is an enormous advantage which should be contrasted with their inherent technological limitations. These cell systems exhibit a metabolic behaviour associated with elevated cell proliferation which involves a high consumption of glucose and glutamine, resulting in the rapid depletion of these nutrients in the medium and the accumulation of ammonium and lactate. Both phenomena contribute to the limitation of cell growth, the triggering of apoptotic processes and the loss of quality of the recombinant protein.

Results: In this review, the use of alternative substrates and genetic modifications (host cell engineering) are analyzed as tools to overcome those limitations. In general, the results obtained are promising. However, metabolic and physiological phenomena involved in CHO cells are still barely understood. Thus, most of publications are focused on specific modifications rather than giving a systemic perspective.

Conclusions: A deeper insight in the integrated understanding of metabolism and cell mechanisms is required in order to define complementary strategies at these two levels, so providing effective means to control nutrients consumption, reduce by-products and increase process productivity.

INTRODUCTION

The market of biopharmaceutical products or drugs based on recombinant proteins (r-proteins) is growing rapidly, with total sales reaching more than 138 billion US dollars in 2010 (Visiongain, 2012). Since the first therapeutic application of a r-protein produced in mammalian cells was approved in 1986 (human tissue plasminogen activator, tPA, Genentech), biopharmaceuticals have changed the landscape for the treatment of many diseases, including many types of cancer,

Citation: Claudia Altamirano, Julio Berrios, Mauricio Vergara, Silvana Becerra, Advances in improving mammalian cells metabolism for recombinant protein production, DOI.: 10.2225/vol16-issue3-fulltext-2.

rheumatic conditions and orphan diseases. Moreover, the biopharmaceutical market has grown at an annual average rate of 35% since 2001, indicating a financially sound future for the biopharmaceutical industry (Aggarwal, 2007).

Although there are a variety of alternative expression systems available, including those using bacteria and transgenic animals, mammalian cells have remained during the last 25 years as the main expression system for the commercial production of therapeutic proteins, as this system creates efficient post-translational modifications on the r-proteins (Durocher and Butler, 2009). The Chinese hamster ovary (CHO) cell line is the standard expression system for the production of recombinant products for therapeutic applications. However, other cell lines such as NS0, baby hamster kidney (BHK) and human embryo kidney-293 (HEK-293) cells have also received regulatory approval for r-protein production on a commercial scale (Jayapal et al. 2007; Mohan et al. 2008; Kim et al. 2012). The use of CHO cells has been favoured over other cell lines due to their ability to produce r-proteins with complex post-translational modifications (e.g., glycosylation), which are essential to biologic functionality and security (Wurm, 2004). However, despite the enormous potential of established cell lines, these production systems have inherent technological limitations: low specific growth rate, low cell concentration and high consumption of the main sources of carbon and energy (usually glucose and glutamine). The consequence of the latter aspect is the high generation of undesirable metabolites (mainly lactate and ammonium), limiting the life-span of the culture and the quality of the protein of interest. These limits result in low productivity for the production process (Schröder, 2008; Lim et al. 2010; Omasa et al. 2010; Sunley and Butler, 2010; Dietmair et al. 2011).

Most reports usually include information about glucose and glutamine consumption and lactate and ammonium production, intending to characterise the response of the system in a general way. However, there are rather few studies that provide more insight from a systemic point of view with respect to the metabolic behaviour of mammalian cells cultivated in vitro. In these cases, the most commonly used tool is metabolic flux analysis, which allows distinguishing changes in the distribution of intracellular flux as a result of specific perturbations or alternative metabolic routes (Boghigian et al. 2010; Quek et al 2010; Ahn and Antoniewicz, 2012). However, this tool does not provide information concerning changes in genic expression or regulation, existing only few publications regarding this point (Korke et al. 2004; Baik et al. 2006; Gupta and Lee, 2007; Jacob et al. 2010; Becker et al. 2011; Kim, 2012).

Strategies aimed specifically at reducing the production of lactate and ammonium ion have been mainly approached in two ways: the use of slowly metabolised substrates and cell engineering (directed to attenuate or overexpress transport proteins and/or enzymes).

This review presents a description of the main characteristics of the metabolic behaviour of mammalian cells cultivated in vitro, mainly referred to CHO cells, and the analyses of medium and host cell engineering strategies to improve the efficiency in the use of glucose and glutamine.

CHARACTERISTICS OF THE ENERGETIC METABOLISM OF MAMMALIAN CELL LINES

Alterations of the energetic metabolism have been well described and largely documented for the majority of mammalian cells presenting rapid cell growth, such as carcinogenic cell lines (e.g. HeLa, MCF-7, hepatomas) and transformed mammalian lines (e.g. CHO, BHK, HEK-293, hybridoma). These alterations involve an increase in the glycolytic capacity, even in the presence of high oxygen concentrations (aerobic glycolysis), and glutamine catabolism (glutaminolysis) that exceeds purely biosynthetic requirements. Evidence shows that there is a significant metabolic flexibility between glycolysis, glutaminolysis and oxidative phosphorylation for adapting the energy production mechanisms to microenvironmental changes and/or biosynthetic needs.

Aerobic glycolysis and glutaminolysis

The main mechanisms that have been suggested for explaining aerobic glycolysis and glutaminolysis in highly proliferative cells (Figure 1) are related to the overexpression of normal proteins or unusual isoforms (although characteristic of highly proliferative cells), the down-regulation of key proteins for the transport of NADH from the cytosol to mitochondria and pyruvate oxidation through the tricarboxylic acid cycle (TCAc).

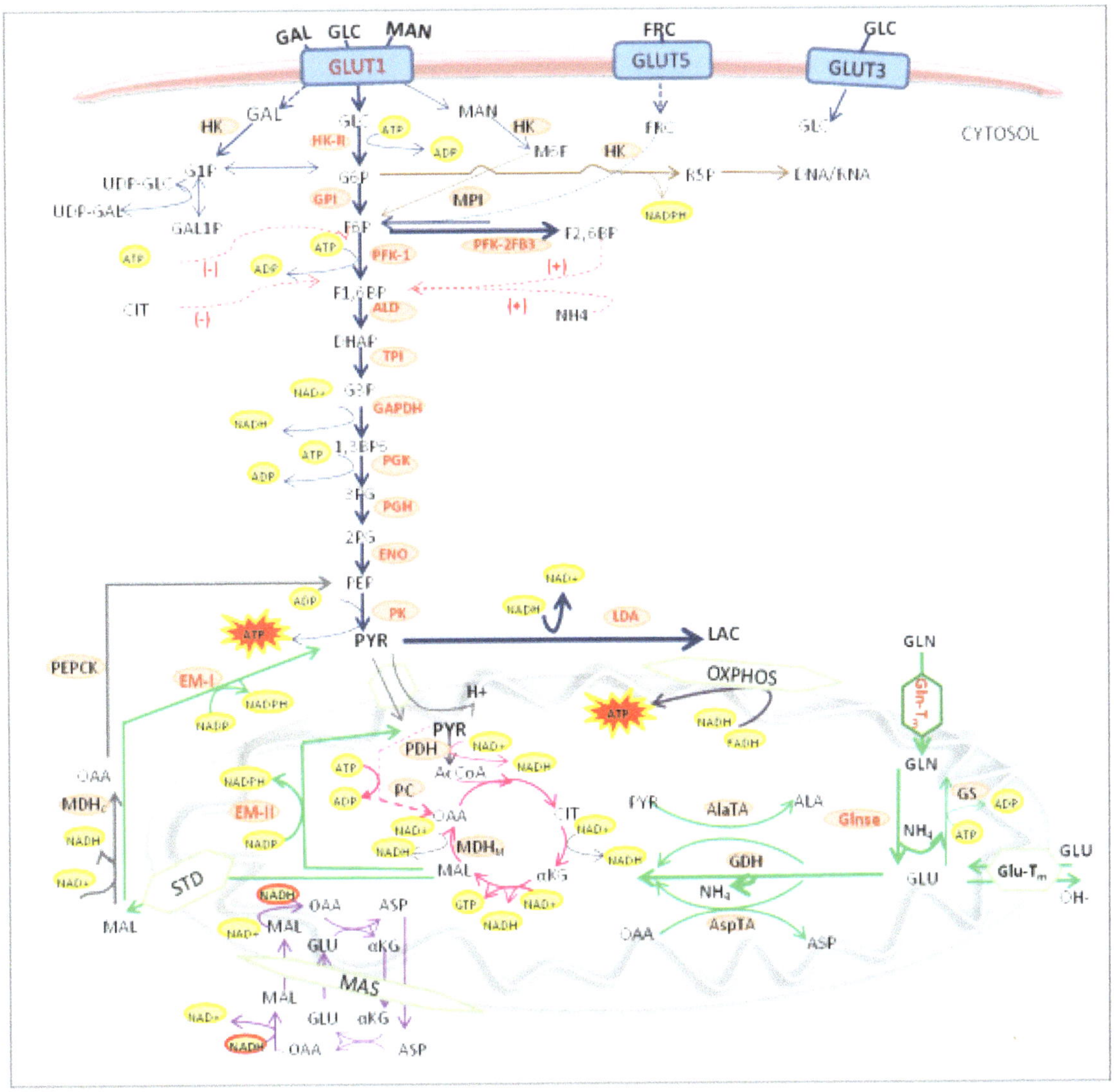

Figure 1: Scheme of glycolytic and glutaminolysis pathways, tricarboxylic acid cycle and main mitochondrial transport systems. Figure legend: blue lines: reaction steps of glycolysis pathway; green lines: reaction steps of glutaminolysis pathway; pink lines: pyruvate dehydrogenase reaction plus reaction steps of tricarboxylic acid cycle (TCAc); purple lines: reaction steps of the malate aspartate shuttle; red words: overexpressed enzymes in highly proliferative cells; dotted red line: positive (+) or negative (-) regulation.

Metabolites:	Enzymes:	Transporters:
AcCoA = Acetyl-CoA	AlaTA = Alanine transaminase	Gln-T = Glutamine transporter
ALA = Alanine	ALD = Aldolase	Glu-T = Glutamate transporter
ASP = Aspartate	AspTA = Aspartate transaminase	GLUT1 = Glucose transporter (Isoform 1)
CIT = Citrate	CPS-I = carbamoyl phosphate synthetase I	GLUT1 = Glucose transporter (Isoform 1)
DHAP = Dihydroxyacetone phosphate	EM-I = Malic enzyme type I	GLUT3 = Glucose transporter (Isoform 3)
F1,6BP = Frucote-1,6-biphosphate	EM-II = Malic enzyme type II	GLUT5 = Glucose transporter (Isoform 5)
F2,6BP = Frucote-2,6-biphosphate	ENO = Enolase	MAS = Malate aspartate shuttle
F6P = Fructose-6-phosphate	F6P = Fructose-6-phosphate isomerase	OXPHOS = Oxidative phosphorilation
FRC = Fructose	GAPDH = Glyceraldehyde-3-phosphate dehydrogenase	STD = Standart transport of dicarboxylate
G1P = glucose-1-phosphate	GDH = Glutamate dehydrogenase	
G3P = Glyceraldehyde-3-phosphate	Glnse = Glutaminase	
G6P = Glucose-6-phosphate	GPI = Glucose-6-phosphate isomerase	
GAL = Galactose	GS = Glutamine syntetase	
GAL1P = Galactose-1-phosphate	HK = Hexokinase	
GDP-GAL = UDP-galactose	HK-II = Hexokinase (Isoform II)	
GLC = Glucose	LDA = Lactate dehydrogenase	
GLN = Glutamine	MDHc = Malate dehydrogenase (citoplasmatic)	
GLU = Glutamate	MDHm = Malate dehydroganse (mitochondrial)	
LAC = Lactate	MPI = Mannose-6-phosphate isomerase	
M6F = Mannose-6-phosphate	PC = Pyruvate caboxylase	
MAL = Malate	PDH = Pyruvate dehydrogenase	
MAN = Mannose	PDHK = Pyruvate dehydrogenase kinase	
OAA = Oxaloacetate	PEPCK = Phosphoenolpyruvate carboxykinase	
PEP = Phosphoenolpyruvate	PFK-1 = Phosphofructokinase-1	
PYR = Pytuvate	PFK-2FB3 = Phosphofructokinase-2 isoform FB3	
R5P = Ribose-5-phosphate	PGH = Phosphoglyceromutase	
UDP-GLC = UDP-glucose	PGK = Phosphoglycerate kinase	
1,3BPG = 1,3-biphosphoglicerate	PK = Pyruvate kinase	
2PG = 2-phosphoglycerate	PYC = Pyruvate carboxylase from yeast	
3PG = 3-phosphoglycerate	TPI = Triosephosphate isomarase	
αKG = alfa-ketoglutarate		

In highly proliferative cells, the high glycolytic flux can be explained by an overexpression of the glucose transporters GLUT1 and GLUT3 (Macheda et al. 2005) and all the enzymes associated with the glycolitic pathway (Moreno-Sánchez et al. 2007). In the latter case, the overexpression has been as high as 300-fold for hexokinase II (HK-II) and phosphofructokinase type 1 (PFK-1). The overexpression of HK-II may favour its association with mitochondria and may increase its access to synthesized ATP through oxidative phosphorylation (OXPHOS). The overexpression of PFK-1, together with the PFK-2FB3 version, favours the synthesis of F1. This is explained by the fact that PFK-1 is activated by fructose-2,6-bisphosphate, and PFK-2FB3 has a low affinity for F2, 6BP (unlike the normal isoform PFK-2), thereby favouring the accumulation of F2, 6BP (Yalcin et el. 2009). The activation of PFK-1 by F2, 6BP can be antagonized bycitrate and ATP. Moreover, it is important to note that PFK-1 is positively modulated by ammonium ion in normal mammalian cells (Sanders and Ezzat, 1971) and CHO cells (Urbano et al. 2000).

A high glycolytic flux may depend on the rapid regeneration of NAD^+ from $NADH^+$ through the conversion of pyruvate to lactate and by NADH shuttle systems. Usually, the regeneration of NAD^+ occurs in the mitochondria. However, because the mitochondrial membrane is impermeable to NADH, the cell uses shuttle systems to indirectly incorporate NADH equivalents in the mitochondria, followed by oxidation through the respiratory chain. The malate-aspartate shuttle (MAS) has been identified in many cell lines (Greenhouse and Lehninger, 1976) (Figure 1). The flux through the MAS under determined conditions has been shown to limit the velocity of NAD^+ regeneration in the cytosol (Eigenbrodt et al. 1985). Although the mechanism of these shuttle systems has not been studied in detail, some reports have shown that in some cases, its efficiency and transport rate could be lower than the NADH production rate in glycolysis (Rubi et al. 2004).

Regarding the key enzymes connecting glycolysis to the TCAc - pyruvate dehydrogenase and carboxylase pyruvate - in two different lines of animal cells (BHK and CHO), neither of them have significant activity (Neermann and Wagner, 1996). Based on this evidence, the authors explain that the flux in the formation of pyruvate is similar to the flux in the formation of lactate. Using spectrophotometric assays, (Fitzpatrick et al. 1993) determined that the activity of PDH was undetectable in batch cultures of B-lymphocyte hybridoma cells. At the mid-exponential phase, they showed that 96% of the glucose was catabolized via glycolysis and 36% of the glutamine was oxidized to CO2. Interestingly, the energy contribution from the catabolism of these two substrates was similar: 55% by glutamine and 45% by glucose. Such behaviour has been verified already in many studies (Quek et al. 2010), showing a low flux through PDH and insignificant changes, even under very distinct culture conditions. Ahn and Antoniewicz (2011) and Sengupta et al. (2011) have recently reported that the metabolic pattern characteristic of the CHO growth phase in batch culture suffer substantial changes in the stationary phase. In the latter phase, the metabolic pattern is characterized by a low glycolytic flux and the activation of the oxidative pentose phosphate pathway flux. However, the fluxes through PDH were similar in the exponential and stationary phases in both case (Ahn and Antoniewicz, 2011; Sengupta et al. 2011). The activities of TCAc and OXPHOS are maintained due to the expression of unusual mitochondrial proteins, such as NADP+-dependent malic enzyme, glutaminase and glutamine transporter, that have high activity and affinity for their substrates (Figure 1).

Three isoforms of malic enzymes are known, the most common being the type II (EM-II) found in carcinogenic cells (Moreadith and Lehninger, 1984; Sauer et al. 1980). EM-II is a mitochondrial enzyme (NADP+ / NADPH dependent) that irreversibly catalyzes the formation of pyruvate from malate, and its activity is 10- to 20-fold higher in these cells than in normal cells. Sauer et al. (1980) suggested that this enzyme is important for the formation of mitochondrial pyruvate from glutamine by determining that its activity increases simultaneously with glutaminase activity. In carcinogenic cells, glutamine is transported into the mitochondria at a rate 4- to 10-fold higher than in normal cells and is then oxidized to glutamate by a Pi-dependent glutaminase, with a 10- to 20-fold higher activity than the normal enzyme. This Pi-dependent glutaminase may be located in the intermembrane space of the mitochondria and/or in the mitochondrial matrix (Molina et al. 1995).

Glutamate can follow two alternative pathways, either via glutamate deshydrogenase or via transaminases (TA). In the GDH pathway, glutamate is deaminated into a-ketoglutarate, releasing an additional ammonium ion, resulting in 2 moles of ammonium ion being generated from 1 molecule of glutamine. Through this route, the carbon skeleton of glutamine can be completely oxidised to CO2 due to the participation of EM-II or part of it can be released from the mitochondria as malate to be later converted into pyruvate (Bonarius et al. 2001; Chong et al. 2010). Usually, the conversion from malate to pyruvate in the cytosol can be catalysed by phosphoenolpyruvate carboxykinase (PEPCK) or by cytosolic malic enzyme type I (EM-I). Many studies have shown a deficiency of PEPCK in hybridomas and in the CHO and BHK cell lines (Fitzpatrick et al. 1993; Neermann and Wagner, 1996). However, metabolic flux analysis using isotopic labelling with 13C revealing a significant activity of EM-I (Bonarius et al. 2001; Quek et al. 2010).

Through the TA pathway, glutamate is converted into α-ketoglutarate via aspartate transaminase or via alanine transaminase. Because oxaloacetate or pyruvate, which are transaminated, are obtained from the glutamate channelled through the TCAc, TA pathways have a closed loop structure. They can be used by cells as an internal mechanism to balance the overproduction of ammonium ion due to the rapid consumption of glutamine.

As previously described, the metabolism of glutamine can be channelled through different pathways that share many metabolites and transport systems in the mitochondria. In terms of energy, the GDH pathway is more efficient than the TA pathway (27 ATP moles vs. 9). This pathway is usually preferred when the energy demand is not fulfilled because of glucose exhaustion (Häggström, 1991; Martinelle et al. 1998) or because of the use of alternative, slowly metabolisable substrates (e.g., glutamate) (Altamirano et al. 2001; Bonarius et al. 2001).

The metabolic flexibility observed in various cell lines shows that the cells have the ability to adjust the metabolic flux depending on the specific culture conditions. This adjustment is a result of both regulatory systems that act on the allosteric modulation of enzyme activity and the transcription of one or more enzymes. The latter is probably the most robust regulatory system in terms of control.

One of the most studied regulatory mechanisms regarding transcription and translation, as far as aerobic glycolysis and glutaminolysis are concerned, is the hypoxia inducible factor 1 (HIF-1) pathway. HIF-1 is a transcription dimeric factor, which is highly stable under aerobiosis. Under such condition, one of its units is degraded by the process of ubiquitin-mediated proteolysis (Wood et al. 1998), causing HIF-1 inactivation. HIF-1 has been shown to promote the expression of many glycolytic enzymes, leading to a stimulation of glycolytic flux (Jose et al. 2011). HIF-1a also favours glycolytic flux by increasing the expression of pyruvate dehydrogenase kinase 1 (PDHK-1), which inhibits the complex activity of PDH by phosphorylation (Kim et al. 2006; Hitosugi et al. 2011). Unfortunately, there are very few studies connecting HIF-1 to the regulation of aerobic glycolysis and glutaminolysis in CHO cells (Wood et al. 1998).

The c-Myc transcription factor exhibits similar characteristics to those of HIF-1 because it activates glucose transporter genes (GLUT-1) and glycolytic genes in several carcinogenic cells (Moreno-Sánchez et al. 2007). Moreover, c-Myc has been determined to potentially activate the transport of mitochondrial glutamine and glutaminase, its overexpression being correlated with an increase in the lactate production generated by glutaminolysis (Wise et al. 2008; Gao et al. 2009). c-Myc is also a factor participating in the regulation of cell proliferation and apoptosis (Mulukutla et al. 2010; Jose et al. 2011), which has been described to play that role in CHO cells (Kuystermans et al. 2010).

Lactate and ammonium ion: Undesirable metabolites

One of the main consequences of the metabolic behaviour of animal cells cultivated in vitro is the exacerbated formation of lactate and ammonium ion (Lao and Toth, 1997; Sanfeliu et al. 1997; Wagner, 1997). When these products exceed certain critical concentrations, they inhibit cell growth and decrease r-protein production (Lao and Toth, 1997; Wagner, 1997; Gambhir et al. 2003).

Lactate can cause a significant reduction of the pH of the medium, which inhibit the cell growth, but this effect can be avoided with a controlling system that maintains the pH within an optimal range. However, the principal mechanism by which lactate appears to negatively affect cultures is by an increase in the osmolarity of the culture media (Cruz et al. 2000). In general, concentrations lower than 20 mM lactate do not affect cell growth or productivity. A lactate concentration of 20-40 mM impairs the productivity and more than 40 mM inhibits cell growth (Wagner, 1997).

Ammonium ion has a stronger impact on cell culture than lactate because concentrations between 2 and 5 mM ammonium ion can inhibit growth and impair productivity. The precise mechanism of ammonium ion toxicity in cells is not clear, even though this effect has been shown to depend on pH (Schneider et al. 1996). There is evidence that the increase in the ammonium ion concentration in the culture medium produces an increase in the activated N-acetylhexosamine pool (UDP-N-acetylglucosamine and UDP-N-acetylgalactosamine; UDP-GNAc). This increase involves a decrease in the UTP concentration and subsequent growth inhibition (Ryll et al. 1994). Ammonium ion also appears to compete with cations, such as potassium, for transport inside the cell through the Na+/K+-ATPase and Na+K+2Cl-cotransporter (Martinelle et al. 1998). Moreover, ammonium ion may adversely affect glycosylation in mammalian cell cultures since it causes a decrease of sialic acid (NeuAc) and galactose (Gal) (Gawlitzek et al. 2000; Yang and Butler, 2002), variations of site occupancy (Borys et al. 1993) and increase of branching (Gawlitzek et al. 1998). However, it is not clear if these effects are due to alterations in glucosamine and galactosamine levels (Koch et al. 1979), the UDP-GNAc pool (Ryll et al. 1994) or changes in the intracellular pH (Grammatikos et al. 1998). An increase in glycosilation branching is observed when the UDP-GNAc extracellular level increases, without affecting the NeuAc content of the glycoprotein (Grammatikos et al. 1998; Baker et al. 2001), whereas high ammonium ion concentrations lead not only to an increase in glycosylation branching but also to a loss of NeuAc in the glycoproteins (Grammatikos et al 1998; Gawlitzek et al. 2000; Yang and Butler, 2002). In contrast, an increase in ammonium ion generates an increase in glucosamine and galactosamine levels. However, the impact of these aminated sugars on glycosylation is quite different. Glucosamine feeding causes a reduction in glycosylation (Koch et al. 1979), whereas galactosamine feeding increases both the mannosylation and sialylation of proteins (Koch et al. 1979; Pels-Rijcken et al. 1995) suggested that the negative impact of glucosamine on glycosylation might be due to glucosamine itself rather than a nucleotide sugar, as the reversibility of the glucosamine effect was correlated with glucosamine concentration during the period when UDP-Glc, UDP-GlcNAc and GDP-Man did not change significantly. This observation might lead to the hypothesis that the use of other carbon sources (galactose or mannose) instead of glucose can have different and potentially favourable consequences on the protein glycosylation pattern. In this sense, supplementation of the culture media with 10 mM galactose has been shown to produce a 5-fold increase in the UDP-Gal levels and a slight increase in the galactosylation level (Hills et al. 2001). However, it is interesting to note that in this study the carbon source was glucose, with galactose added as a supplement. No reports have been found on the effect on glycosylation when galactose is the only carbon source and on the effect of mannose under the same conditions.

In the majority of studies that have assessed the effect of ammonium ion on the culture of mammalian cells, particularly on CHO cells, the experimental model studied has used exogenous ammonium ion sources (e.g. NH4Cl) (Ryll et al. 1994; Lao and Thoh, 1997; Cayli et al. 1999; Yang and Butler, 2002). However, the effect of endogenous ammonium ion could produce a different response in cell metabolism (Schneider et al. 1996). Such effect has not been analysed yet.

Due to the impact that lactate and ammonium ion have on the production of r-proteins in mammalian cells, multiple strategies have been designed to reduce their negative effects. These strategies are discussed below.

IMPROVING METABOLISM: ALTERNATIVE METABOLISABLE SUBSTRATES AND CELL ENGINEERING

There are two main approaches that have been used to limit glucose and glutamine consumption (Figure 2). One is based on substrates that, because of their nature, are less available or slowly metabolised. The other one is based on using genetic modifications (cell engineering) for changing the metabolic behaviour by attenuating or stimulating the overexpression of key metabolic proteins. Strategies that involve the controlled feeding of glucose and glutamine such as fed-batch cultures are also possible, but are out of the scope of this review.

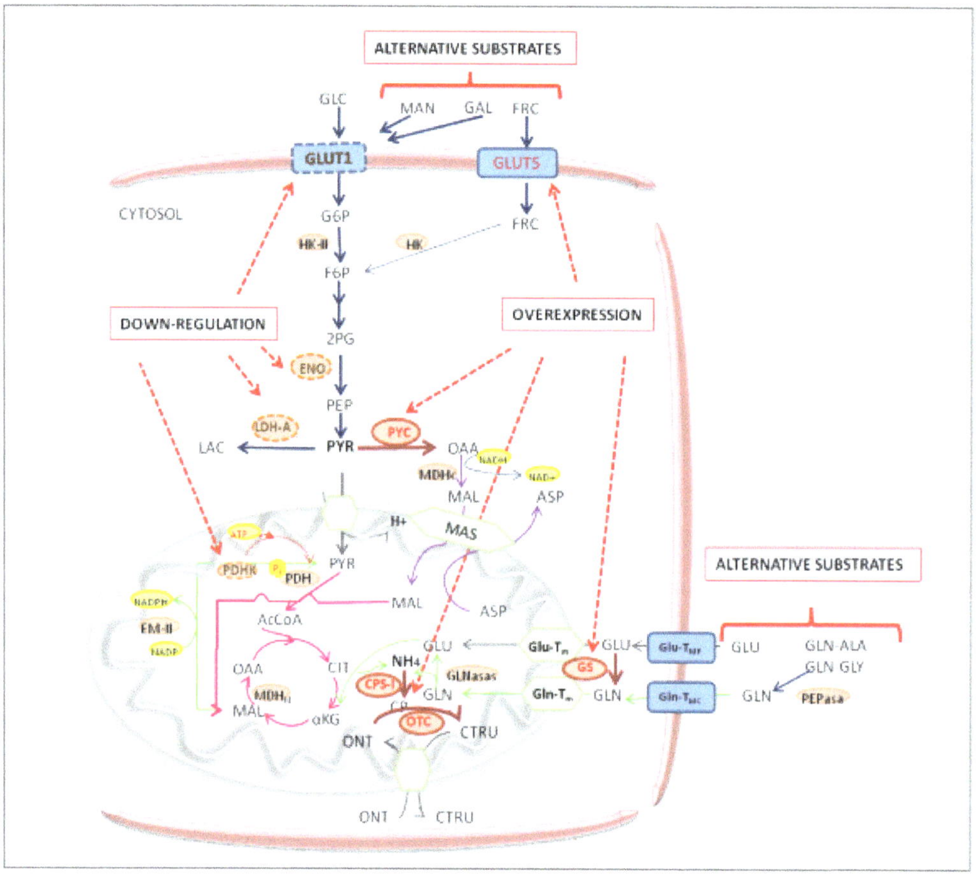

Figure 2: Schematic representation of the main strategies used to improve cell metabolism. Figure legend: blue lines: reaction steps of glycolysis pathway; green lines: reaction steps of glutaminolysis pathway; pink lines: pyruvate dehydrogenase reaction plus reaction steps of tricarboxylic acid cycle (TCAc); purple lines: reaction steps of the malate aspartate shuttle; red words: overexpressed enzymes/transporter; brown letters: down-expressed enzymes/transporter.

Metabolites	Enzymes
AcCoA = Acetyl-CoA	CPS-I = carbamoyl phosphate synthetase I
ASP = Aspartate	EM-II = Malic enzyme type II
CIT = Citrate	ENO = Enolase
CP = Carbamoyl phosphate	F6P = Fructose-6-phosphate isomerase
CTRU = Citrulline	GS = Glutamine syntetase
F6P = Fructose-6-phosphate	HK = Hexokinase
FRC = Fructose	HK-II = Hexokinase (Isoform II)
G1P = glucose-1-phosphate	LDH-A = Lactate dehydrigenase isoform A
G3P = Glyceraldehyde-3-phosphate	MDHc = Malate dehydrogenase (citoplasmatic)
G6P = Glucose-6-phosphate	MDHm = Malate dehydroganse (mitochondrial)
GAL = Galactose	OTC = ornithine transcarbamoylase
GLC = Glucose	PDH = Pyruvate dehydrogenase
GLN = Glutamine	PDHK = Pyruvate dehydrogenase kinase
GLN-ALA = Alanyl-Glutamine	PYC = Pyruvate carboxylase from yeast
GLN-GLY = Glycyl-Glutamine	**Transporters**
GLU = Glutamate	Gln-T = Glutamine transporter
LAC = Lactate	Glu-T = Glutamate transporter
MAL = Malate	GLUT1 = Glucose transporter (Isoform 1)
MAN = Mannose	GLUT1 = Glucose transporter (Isoform 1)
OAA = Oxaloacetate	GLUT5 = Glucose transporter (Isoform 5)
ONT = Ornithine	MAS = Malate aspartate shuttle
PEP = Phosphoenolpyruvate	
PYR = Pytuvate	
2PG = 2-phosphoglycerate	
αKG = alfa-ketoglutarate	

Reducing lactate production

The use of galactose or fructose instead of glucose has been shown to be an alternative to prevent lactate overproduction because the affinity of hexokinase for these sugars is approximately 20-fold lower than for glucose (Wagner, 1997). In particular, the use of galactose has been studied as an alternative for reducing lactate production levels (Wagner et al. 1991; Altamirano et al. 2000; Altamirano et al. 2004; Wilkens et al. 2011). Galactose is metabolised by a series of reactions that constitute the Leloir pathway, which consists of three enzymatic activities. UDP-glucose pyrophosphorylase (Lai and Elsas, 2000) also participates in the metabolism of galactose, which is part of other metabolic pathways. Galactose metabolism is slower than glucose, allowing a reduction in lactate generation. Galactose metabolism has also been shown to increase cell viability and culture duration (Altamirano et al. 2000). However, a significant reduction of cell growth occurs. Noticeably, lactate produced when growing on galactose is re-metabolized by the cells. These results have been used as a basis for the study of fed-batch (Clark et al. 2005; Altamirano et al. 2006) and perfusion cultures (Vergara et al. 2012).

On the other side, various cell engineering strategies have been tested to reduce both glucose consumption and lactate generation. Paredes et al. (1999) reported that knocking down the glucose transporter GLUT1 in hybridoma cells reduced glucose uptake by almost 50% and lactate/glucose yield by 12%. Recently, tests have been conducted regarding the transfection of CHO cells with the fructose transporter GLUT5 to enable growth on fructose (Wlaschin and Hu, 2007). The clone with the lowest GLUT5 expression had the lowest lactate/hexose yield (ΔL/ΔHexose), whereas the clone with the highest GLUT5 expression had the highest ΔL/ΔHexose. This result suggests that the abundance of fructose transporter is important and that limiting sugar uptake may increase metabolic efficiency. The clone with the lowest lactate production achieved 3-fold higher cell densities in fructose fed-batch cultures compared with glucose. Using a similar approach, the recombinant IgG production of human myeloma cells cultured in a fructose-based medium exhibited an approximately 2-fold increase compared with that in a glucose-based medium (Inoue et al. 2010).

The down regulation of α-enolase in hybridoma cells, which converts 2-phosphoglycerate to phosphoenolpyruvate (Figure 1) through iRNA has been assessed. This intervention allowed reducing glucose uptake by 22% but did not significantly reduce the lactate/glucose yield (Paredes et al. 1999). Works using that approach have not been reported since.

Sanfeliu et al. (1997) used oxamate, a specific lactate-dehydrogenase (LDH) inhibitor, to modulate the activity of this enzyme in hybridoma cells. They concluded that this activity is fundamental for the maintenance of cell viability because it helps recycling the NADH produced and accumulated during glycolysis. However, it should be considered that oxamate inhibits any LDH isoform. Most recently, hybridoma cells were genetically manipulated to obtain mutants having a reduction of 50% or more in LDH activity (Chen et al. 2001). It has been shown that the LDH isotype of CHO cell exists as A3B and A2B2 tetramers (Jeong et al. 2001). Viability and growth was improved when the cells were genetically manipulated, and the production of the r-protein increased by a factor of 3. Similarly, Kim and Lee (2007) found that LDH knockdown in CHO cells reduced specific lactate production by as much as 79% and improved product titer up to 2.2-fold.

The genetic manipulation of cells has also improved the connection between glycolysis and TCAc by the inclusion of pyruvate carboxylase (PC), which has reduced lactate levels considerably (Irani et al. 1999; Elias et al. 2003). The expression of PC in BHK cells reduced glucose and glutamine consumption and decreased lactate production up to 2.6-fold. In perfusion cultures, a 2-fold increase in the erythropoietin titer was obtained (Irani et al. 2002). Similarly, the overexpression of PYC in HEK293 cells decreased glucose and glutamine consumption, with a reduction in lactate and ammonia production up to 4- and 3-fold, respectively (Elias et al. 2003). Although CHO cells overexpressing PYC did not exhibit significant decrease in glucose uptake rate, lactate production decreased by 1.2- to 1.5-fold, and the specific rate of oxygen consumption increased by up to 1.9-fold (Fogolin et al. 2004). Additionally, the culture duration was extended up to 5 days, resulting in a 2.5-fold improvement in the yield of rhGM-CSF (recombinant human granulocyte macrophage colony stimulating factor).

Wilkens and Gerdtzen (2011) transfected CHO cells with GLUT5 and PYC genes (CHO-PYC-Slc2a5). These cells had the highest specific growth rate among all the tested clones (CHO-PYC, CHO-Slc2a5 and CHO wild type). The CHO-PYC-Slc2a5 cells had a ΔL/ΔHexose 50% lower than wild type cells and 35% more than CHO-Slc2a5 cells (cells only transfected with GLUT5). Using a similar approach, Zhou et al. (2011) attenuated the expression of LDH-A and PDHK in CHO cells through iRNA. In fed-batch cultivation of the genetically modified cells, a strong reduction in lactate production (90%), and increase in specific and volumetric productivity of r-protein (approximately 75% and 68%, respectively) was obtained, without a significant impact on cell growth. These results indicate that the cellular system is highly flexible and additional improvements in metabolism are feasible.

Reducing ammonium ion production

The efforts to avoid the effects of an increase in ammonium ion concentration have basically focused on the substitution of glutamine by another compound able to enter the TCAc. The replacement of glutamine by glutamate is an example of this approach, having the following advantages (Altamirano et al. 2000): glutamate has a higher chemical stability than glutamine, glutamine can be directly synthesized from glutamate in CHO cells because they express glutamine synthetase (GS), and glutamate is less ammoniagenic because it has only one amine group. Indeed, the replacement of glutamine by glutamate in the culture media has proved to be a good strategy for CHO cells, resulting in a cell growth improvement of 17% and reduction in the accumulated ammonium ion of 70% (Altamirano et al. 2000). Hong et al. (2010) recently showed that this substitution caused the concentration of an IgG produced in CHO cells to increase by 1.7-fold. In contrast, the level of r-protein galactosylation increased depending on the proportion of galactosylated glycans.

The substitution of glutamine by dipeptides such as alanyl-glutamine and glycyl-glutamine has also been used to reduce ammonium ion production. These dipeptides, that are more stable than the amino acids in the culture medium, are intra or extracellularly cleaved by peptidases, thereby releasing glutamine and either alanine or glycine. The availability of glutamine is therefore dependent on the peptidase activity. In hybridoma cell cultures, high cell yields were obtained in the presence of 6 mM ala-gln or 20 mM gly-gln (Christie and Butler, 1994). The final cell yield in gly-gln was 14% higher than in gln. However, the productivity of monoclonal antibody was comparable in gln, ala-gln or gly-gln. The specific consumption rates of glucose and six amino acids were reduced, and additionally the accumulation of ammonia and lactate was significantly lower (Christie and Butler, 1994). In all these cases, a significant reduction in specific glucose consumption and specific lactate production was observed. The reason for slower glucose consumption is still unclear. One explanation could be related to the participation of PFK-1, which has been determined to be activated by ammonium ion in CHO K1 cells (Urbano et al. 2000). Therefore, in agreement with Schneider et al. (1996), a decrease in the intracellular ammonium ion concentration could reduce the glycolytic flux and thus the lactate production.

The use of mannose alone has also been used as a strategy to reduce ammonium ion effects (Ryll et al. 1994; Cayli et al. 1999; Berríos et al. 2009; Berríos et al. 2011). Mannose enters the cells using a hexose transporter in mammalian cells (Rodriguez et al. 2005), merging with the glycolytic pathway at the F6P level. In CHO cells, mannose consumption occurs at the same rate as glucose, generating similar levels of lactate (Altamirano et al. 2000; Berríos et al. 2009; Berríos et al. 2011). However, intracellular metabolism has been shown to respond in a very different manner when mannose is in the culture medium (Ryll et al. 1994, Cayli et al. 1999, Berríos et al. 2011), especially with respect to the UDP-GNAc pool levels (Ryll et al. 1994; Cayli et al. 1999), as mentioned above. This phenomenon would be mediated by M6F, which acts as an inhibitor of glucosamine-6-phosphate isomerase (Figure 3). This enzyme catalyses the reaction between F6P and ammonium ion to produce GlcN-6P, a precursor of UDP-GNAc. The use of galactose or fructose instead of glucose has proved to reduce the UDP-GNAc pool, even though this change has been attributed to a low rate of glucose synthesis (Ryll et al. 1994).

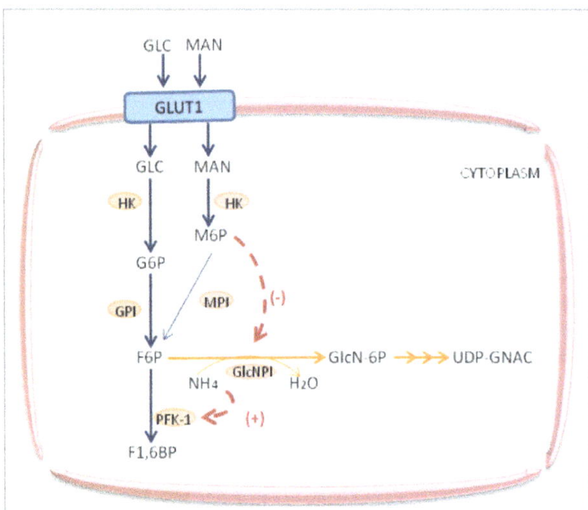

Figure 3: Possible role of the mannose-6-phosphate (M6P) in the regulation of glucosamine phosphate isomerase (GlcNPI) and generation of activated sugars. Figure legend: blue lines: reaction steps of glycolysis pathway; orange lines: nucleotide-activated sugars pathways (precursors of oligosaccharides); dotted red line: positive (+) or negative (-) regulation.

Metabolites	Enzymes	Transporters
F1,6BP = Frucote-1,6-biphosphate	F6P = Fructose-6-phosphate isomerase	Gln-T = Glutamine transporter
F6P = Fructose-6-phosphate	GlcNPI = Glucosamine phosphoisomerase	Glu-T = Glutamate transporter
GLC = Glucose	HK = Hexokinase	GLUT1 = Glucose transporter (Isoform 1)
GlcN-6P = Glucosamine-6-phosphate	MPI = Mannose-6-phosphate isomerase	GLUT1 = Glucose transporter (Isoform 1)
MAN = Mannose	PFK-1 = Phosphofructokinase-1	
OAA = Oxaloacetate		
UDP-GNAC = UDP-N-acetylhexosamine		

Engineering hybridoma cells with GS, which converts glutamate to glutamine, allowed cell growth in glutamine free medium (Paredes et al. 1999), with results similar to those described previously (that is, when glutamine is replaced by glutamate).

The urea cycle is likely to be inactive in CHO cells, since only trace amounts of urea have been detected (Zamorano et al. 2010). In this regard, the expression of the carbamoyl phosphate synthetase I (CPS-I) and ornithine transcarbamoylase (OTC) (that are mitochondrial enzymes that catalyse the first and second steps of the urea cycle) in CHO cells reduced ammonium ion production by 33%, cell growth rate increasing in a similar percentage (Park et al. 2000). In a later study, it was determined that tPA produced in CHO cells overexpressed CPS-I and OTC was highly active which is quite valuable for the biopharmaceutical industry (Kim and Kim, 2006).

CONCLUDING REMARKS

Culture longevity, productivity and quality of the r-protein have been improved by strategies based on modifications of the metabolic behaviour of established mammalian cell lines by using either alternative substrates or host cell engineering. However, success largely depends on the host cell, the produced r-protein and the conditions of the production process (e.g. fed-batch or perfusion culture). A deeper insight in the integrated understanding of metabolism and cell mechanisms is required in order to define complementary strategies at these two levels, so providing effective means to control nutrients consumption, reduce by-products and increase process productivity.

REFERENCES

1. AGGARWAL, S. (2007). What's fueling the biotech engine? *Nature Biotechnology*, vol. 25, no. 10, p. 1097-1104.
2. AHN, W.S. and ANTONIEWICZ, M.R. (2011). Metabolic flux analysis of CHO cells at growth and non-growth phases using isotopic tracers and mass spectrometry. *Metabolic Engineering*, vol. 13, no. 5, p. 598-609.
3. AHN, W.S. and ANTONIEWICZ, M.R. (2012). Towards dynamic metabolic flux analysis in CHO cell cultures. *Biotechnology Journal*, vol. 7, no. 1, p. 61-74.
4. ALTAMIRANO, C.; PAREDES, C.; CAIRÓ, J.J. and GÒDIA, C. (2000). Improvement of CHO cell culture medium formulation: Simultaneous substitution of glucose and glutamine. *Biotechnology Progress*, vol. 16, no. 1, p. 69-75.
5. ALTAMIRANO, C.; ILLANES, A.; CASABLANCAS, A.; GÁMEZ, X.; CAIRÓ, J.J. and GÒDIA, F. (2001). Analysis of CHO cells metabolic redistribution in a glutamate-based defined medium in continuous culture. *Biotechnology Progress*, vol. 17, no. 6, p. 1032-1041.
6. ALTAMIRANO, C.; PAREDES, C.; ILLANES, A.; CAIRÓ, J.J. and GÒDIA, F. (2004). Strategies for fed-batch cultivation of t-PA producing CHO cells: Substitution of glucose and glutamine and rational design of culture medium. *Journal of Biotechnology*, vol. 110, no. 2, p. 171-179.
7. ALTAMIRANO, C.; ILLANES, A.; BECERRA, S.; CAIRÓ, J.J. and GÒDIA, F. (2006). Considerations on the lactate consumption by CHO cells in the presence of galactose. *Journal of Biotechnology*, vol. 125, no. 4, p. 547-556.
8. BAIK, J.Y.; LEE, M.S.; AN, S.R.; YOON, S.K.; JOO, E.J.; KIM, Y.H.; PARK, H.W. and LEE, G.M. (2006). Initial transcriptome and proteome analyses of low culture temperature-induced expression in CHO cells producing erythropoietin. *Biotechnology and Bioengineering*, vol. 93, no. 2, p. 361-371.
9. BAKER, K.N.; RENDALL, M.H.; HILLS, A.E.; HOARE, M.; FREEDMAN, R.B. and JAMES, D.C. (2001). Metabolic control of recombinant protein n-glycan processing in NS0 and CHO cells. *Biotechnology and Bioengineering*, vol. 73, no. 3, p. 188-202.
10. BECKER, J.; TIMMERMANN, C.; JAKOBI, T.; RUPP, O.; SZCZEPANOWSKI, R.; HACKL, M.; GOESMANN, A.; TAUCH, A.; BORTH, N.; GRILLARI, J.; PÜHLER, A.; NOLL, T. and BRINKROLF, K. (2011). Next-generation sequencing of the CHO cell transcriptome. *BMC Proceedings*, vol. 5, no. 8, p. 6.
11. BERRIOS, J.; DÍAZ-BARRERA, A.; BAZÁN, C. and ALTAMIRANO, C. (2009). Relationship between tissue plasminogen activator production and specific growth rate in Chinese Hamster Ovary cells cultured in mannose at low temperature. *Biotechnology Letters*, vol. 31, no. 10, p. 1493-1497.
12. BERRIOS, J.; ALTAMIRANO, C.; OSSES, N. and GONZÁLEZ, R. (2011). Continuous CHO cell cultures with improved recombinant protein productivity by using mannose as carbon source: Metabolic analysis and scale-up simulation. *Chemical Engineering Science*, vol. 66, no. 11, p. 2431-2439.
13. BOGHIGIAN, B.A.; SETH, G.; KISS, R. and PFEIFER, B.A. (2010). Metabolic flux analysis and pharmaceutical production. *Metabolic Engineering*, vol. 12, no. 2, p. 81-95.
14. BONARIUS, H.P.J.; ÖZEMRE, A.; TIMMERARENDS, B.; SKRABAL, P.; TRAMPER, J.; SCHMID, G. and HEINZLE, E. (2001). Metabolic-flux analysis of continuously cultured hybridoma cells using $^{13}CO_2$mass spectrometry in combination with ^{13}C-lactate nuclear magnetic resonance spectroscopy and metabolite balancing. *Biotechnology and Bioengineering*, vol. 74, no. 6, p. 528-538.
15. BORYS, M.C.; LINZER, D.I.H. and PAPOUTSAKIS, E.T. (1993). Culture pH affects expression rates and glycosylation of recombinant mouse placental lactogen proteins by Chinese hamster ovary (CHO) cells. *Nature Biotechnology*, vol. 11, no. 6, p. 720-724.
16. CAYLI, A.; HIRSCHMANN, F.; WIRTH, M.; HAUSER, H. and WAGNER, R. (1999). Cell lines with reduced UDP-N-Acetylhexosamine pool in the presence of ammonium. *Biotechnology and Bioengineering*, vol. 65, no. 2, p. 192-200.
17. CHEN, K.; LIU, Q.; XIE, L.; SHARP, P.A. and WANG, D.I.C. (2001). Engineering of a mammalian cell line for reduction of lactate formation and high monoclonal

antibody production. *Biotechnology and Bioengineering*, vol. 72, no. 1, p. 55-61.

18. CHONG. W.P.K.; REDDY, S.G.; YUSUFI, F.N.K.; LEE, D.Y.; WONG, N.S.C.; HENG, C.K.; YAP, M.G.S. and HO, Y.S. (2010). Metabolomics-driven approach for the improvement of Chinese hamster ovary cell growth: Overexpression of malate dehydrogenase II. *Journal of Biotechnology*, vol. 147, no. 2, p. 116-121.
19. CHRISTIE, A. and BUTLER, M. (1994). Glutamine-based dipeptides are utilized in mammalian cell culture by extracellular hydrolysis catalysed by a specific peptidase. *Journal of Biotechnology*, vol. 37, no. 3, p. 277-290.
20. CLARK, K.J.R.; GRIFFITHS, J.; BAILEY, K.M. and HARCUM, S.W. (2005). Gene-expression profiles for five key glycosylation genes for galactose-fed CHO cells expressing recombinant IL-4/13 cytokine trap. *Biotechnology and Bioengineering*, vol. 90, no. 5, p. 568-577.
21. CRUZ, H.J.; FREITAS, C.M.; ALVES, P.M.; MOREIRA, J.L. and CARRONDO, M.J.T. (2000). Effects of ammonia and lactate on growth, metabolism, and productivity of BHK cells. *Enzyme and Microbial Technology*, vol. 27, no. 1-2, p. 43-52.
22. DIETMAIR, S.; NIELSEN, L.K. and TIMMINS, N.E. (2011). Engineering a mammalian super producer. *Journal of Chemical Technology and Biotechnology*, vol. 86, no. 7, p. 905-914.
23. DUROCHER, Y. and BUTLER, M. (2009). Expression systems for therapeutic glycoprotein production. *Current Opinion in Biotechnology*, vol. 20, no. 6, p. 700-707.
24. EIGENBRODT, E.; FISTER, P. and REINACHER, M. (1985). New perspectives on carbohydrate metabolism in tumor cells, In: BEITNER, R. ed. *Regulation of carbohydrate metabolism*. CRC Press, Boca Raton, vol. 2, p. 141-179.
25. ELIAS, C.B.; CARPENTIER, E.; DUROCHER, Y.; BISSON, L.; WAGNER, R. and KAMEN, A. (2003). Improving glucose and glutamine metabolism of human HEK 293 and *trichoplusiani* insect cells engineered to express a cytosolic pyruvate carboxylase enzyme. *Biotechnology Progress*, vol. 19, no. 1, p. 90-97.
26. FITZPATRICK, L.; JENKINS, H.A. and BUTLER, M. (1993). Glucose and glutamine metabolism of a murine B-lymphocyte hybridoma grown in batch culture. *Applied Biochemistry and Biotechnology*, vol. 43, no. 2, p. 93-116.
27. FOGOLIN, M.B.; WAGNER, R.; ETCHEVERRIGARAY, M. and KRATJE, R. (2004). Impact of temperature reduction and expression of yeast pyruvate carboxylase on hGM-CSF-producing CHO cells. *Journal of Biotechnology*, vol. 109, no. 1-2, p. 179-191.
28. GAMBHIR, A.; KORKE, R.; LEE, J.; FU, P.C.; EUROPA, A. and HU, W.S. (2003). Analysis of cellular metabolism of hybridoma cells at distinct physiological states. *Journal of Bioscience and Bioengineering*, vol. 95, no. 4, p. 317-327.
29. GAO, P.; TCHERNYSHYOV, I.; CHANG, T.C.; LEE, Y.S.; KITA, K.; OCHI, T.; ZELLER, K.I.; DE MARZO, A.M.; VAN EYK, J.E.; MENDELL, J.T. and DANG, C.V. (2009). c-Myc suppression of miR-23a/b enhances mitochondrial glutaminase expression and glutamine metabolism. *Nature*, vol. 458, no. 7239, p. 762-765.
30. GAWLITZEK, M.; RYLL, T.; LOFGREN, J. and SLIWKOWSKI, M.B. (2000). Ammonium alters N-glycan structures of recombinant TNFR-IgG: Degradative versus biosynthetic mechanisms. *Biotechnology and Bioengineering*, vol. 68, no. 6, p. 637-643.
31. GAWLITZEK, M.; VALLEY, U. and WAGNER, R. (1998). Ammonium Ion and glucosamine dependent increases of oligosaccharide complexity in recombinant glycoproteins secreted from cultivated BHK-21 cells. *Biotechnology and Bioengineering*, vol. 57, no. 5, p. 518-528.
32. GRAMMATIKOS, S.I.; VALLEY, U.; NIMTZ, M.; CONRADT, H.S. and WAGNER, R. (1998). Intracellular UDP-N-Acetylhexosamine pool affects N-glycan complexity: A mechanism of ammonium action on protein glycosylation. *Biotechnology Progress*, vol. 14, no. 3, p. 410-419.
33. GREENHOUSE, W. and LEHNINGER, A. (1976). Occurrence of the malate-aspartate shuttle in various tumor types. *Cancer Research*, vol. 36, no. 4, p. 1392-1396.
34. GUPTA, P. and LEE, K.H. (2007). Genomics and proteomics in process development: Opportunities and challenges. *Trends in Biotechnology*, vol. 25, no. 7, p. 324-330.
35. HÄGGSTRÖM, L. (1991). Energetic of glutaminolysis a theorical evaluation, In: SPIER, R.E.; GRIFFITHS, J.B. and MEIGNIER, B. eds. *Production of biological from animal cells in culture*. Butterworth-Heinemann, Ltd., Oxford. p. 79-81.
36. HILLS, A.E.; PATEL, A.; BOYD, P. and JAMES, D.C. (2001). Metabolic control of recombinant monoclonal antibody N-glycosylation in GS-NS0 cells. *Biotechnology and Bioengineering*, vol. 75, no. 2, p. 239-251.
37. HITOSUGI, T.; FAN, J.; CHUNG, T.W.; LYTHGOE, K.; WANG, X.; XIE, J.; GE, Q.; GU, T.L.; POLAKIEWICZ, R.D.; ROESEL, J.L.; CHEN, G.Z.; BOGGON, T.J.; LONIAL, S.; FU, H.; KHURI, F.R.; KANG, S. and CHEN, J. (2011). Tyrosine phosphorylation of mitochondrial pyruvate dehydrogenase kinase 1 is important for cancer metabolism. *Molecular Cell*, vol. 44, no. 6, p. 864-877.
38. HONG, J.K.; CHO, S.M. and YOON, S.K. (2010). Substitution of glutamine by glutamate enhances production and galactosylation of recombinant IgG in Chinese hamster ovary cells. *Applied Microbiology and Biotechnology*, vol. 88, no. 4, p. 869-876
39. INOUE, Y.; TSUKAMOTO, Y.; YAMANAKA, M.; NAKAMURA, S.; INOUE, A.; NISHINO, N. and KAWAHARA, H. (2010). Efficient production of recombinant IgG by metabolic control and

co-expression with GLUT5 in a fructose-based médium. *Cytotechnology*, vol. 62, no. 4, p. 301-306.

40. IRANI, N.; WIRTH, M.; VAN DEN HEUVEL, J. and WAGNER, R. (1999). Improvement of the primary metabolism of cell cultures by introducing a new cytoplasmic pyruvate carboxylase reaction. *Biotechnology and Bioengineering*, vol. 66, no. 4, p. 238-246.

41. IRANI, N.; BECCARIA, A.J. and WAGNER, R. (2002). Expression of recombinant cytoplasmic yeast pyruvate carboxylase for the improvement of the production of human erythropoietin by recombinant BHK-21 cells. *Journal of Biotechnology*, vol. 93, no. 3, p. 269-282.

42. JACOB, N.M.; KANTARDJIEFF, A.; YUSUFI, F.N.K.; RETZEL, E.F.; MULUKUTLA, B.C.; CHUAH, S.H.; YAP, M. and HU, W.S. (2010). Reaching the depth of the Chinese hamster ovary cell transcriptome.*Biotechnology and Bioengineering*, vol. 105, no. 5, p. 1002-1009.

43. JAYAPAL, K.; WLASHIN, K.; HU, W. and YAP, M. (2007). Recombinant protein therapeutics from CHO cells 20 years and counting. *Chemical Engineering Progress*, vol. 103, no. 10, p. 40-47.

44. JEONG, D.W.; KIM, T.S.; LEE, J.W.; KIM, T.K.; KIM, H.J.; KIM, I.H. and KIM, I.Y. (2001). Blocking of acidosis-mediated apoptosis by a reduction of lactate dehydrogenase activity through antisense mRNA expression. *Biochemical and Biophysical Research Communications*, vol. 289, no. 5, p. 1141-1149.

45. JOSE, C.; BELLANCE, N. and ROSSIGNOL, R. (2011). Choosing between glycolysis and oxidative phosphorylation: A tumor›s dilemma? *Biochimica et Biophysica Acta (BBA)-Bioenergetics*, vol. 1807, no. 6, p. 552-561.

46. KIM, H.J. and KIM, H.J. (2006). Glycosylation variant analysis of recombinant human tissue plasminogen activator produced in urea-cycle-enzyme-expressing Chinese hamster ovary (CHO) cell line. *Journal of Bioscience and Bioengineering*, vol. 102, no. 5, p. 447-451.

47. KIM, J.Y.; KIM, Y.G. and LEE, G.M. (2012). CHO cells in biotechnology for production of recombinant proteins: Current state and further potential. *Applied Microbiology and Biotechnology*, vol. 93, no. 3, p. 917-930.

48. KIM, J.W.; TCHERNYSHYOV, I.; SEMENZA, G.L. and DANG, C.V. (2006). HIF-1-mediated expression of pyruvate dehydrogenase kinase: A metabolic switch required for cellular adaptation to hypoxia.*Cell Metabolism*, vol. 3, no. 3, p. 177-185.

49. KIM, S.H. and LEE, G.M. (2007). Down-regulation of lactate dehydrogenase-A by siRNAs for reduced lactic acid formation of Chinese hamster ovary cells producing thrombopoietin. *Applied Microbiology and Biotechnology*, vol. 74, no. 1, p. 152-159.

50. KIM, Y.G. (2012). Omics-Based CHO cell engineering - entrance into post-genomic era. *Advances in Genetic Engineering & Biotechnology*, vol. 1, no. 1.

51. KOCH, H.U.; SCHWARZ, R.T. and SCHOLTISSEK, C. (1979). Glucosamine itself mediates reversible inhibition of protein glycosylation. *European Journal of Biochemistry*, vol. 94, no. 2, p. 515-522.

52. KORKE, R.; DE LEÓN GATTI, M.; LAU, A.L.Y.; LIM, J.W.E.; SEOW, T.K.; CHUNG, M.C.M. and HU, W.S. (2004). Large scale gene expression profiling of metabolic shift of mammalian cells in culture.*Journal of Biotechnology*, vol. 107, no. 1, p. 1-17.

53. KUYSTERMANS, D.; DUNN, M.J. and AL-RUBEAI, M. (2010). A proteomic study of cMyc improvement of CHO culture. *BMC Biotechnology*, vol. 10, no. 25 p. 1-13.

54. LAI, K. and ELSAS, L.J. (2000). Overexpression of human UDP-glucose pyrophosphorylase rescues galactose-1-phosphate uridyltransferase-deficient yeast. *Biochemical and Biophysical Research Communications*, vol. 271, no. 2, p. 392-400.

55. LAO, M.S. and TOTH, D. (1997). Effects of ammonium and lactate on growth and metabolism of a recombinant Chinese hamster ovary cell culture. *Biotechnology Progress*, vol. 13, no. 5, p. 688-691.

56. LIM, Y.; WONG, N.S.C.; LEE, Y.L.; KU, S.C.Y.; WONG, D.C.F. and YAP, M.G.S. (2010). Engineering mammalian cells in bioprocessing - current achievements and future perspectives. *Biotechnology Applied Biochemistry*, vol. 55, no. 4, p. 175-189.

57. MACHEDA, M.L.; ROGERS, S. and BEST, J.D. (2005). Molecular and cellular regulation of glucose transporter (GLUT) proteins in cancer. *Journal of Cellular Physiology*, vol. 202, no. 3, p. 654-662.

58. MARTINELLE, K.; DOVERSKOG, M.; JACOBSSON, U.; CHAPMAN, B.E.; KUCHEL, P.W. and HÄGGSTRÖM, L. (1998). Elevated glutamate dehydrogenase flux in glucose-deprived hybridoma and myeloma cells: Evidence from 1H/15N NMR. *Biotechnology and Bioengineering*, vol. 60, no. 4, p. 508-517.

59. MOHAN, C.; KIM, Y.G.; KOO, J. and LEE, G.M. (2008). Assessment of cell engineering strategies for improved therapeutic protein production in CHO cells. *Biotechnology Journal*, vol. 3, no. 5, p. 624-630.

60. MOLINA, M.A.; SEGURA, J.C.; ALEDO, J.; MEDINA, M.A.; NUNEZ DE CASTRO, I. and MARQUEZ, J. (1995). Glutamine transport by vesicles isolated from tumour-cell mitochondrial inner membrane.*Biochemical Journal*, vol. 308, no. 2, p. 629-633.

61. MOREADITH, R.W. and LEHNINGER, A.L. (1984). Purification, kinetic behavior, and regulation of NAD(P)+-malic enzyme of tumor mitochondria. *The Journal of Biological Chemistry,* vol. 259, no. 10, p. 6222-6227.

62. MORENO-SÁNCHEZ, R.; RODRÍGUEZ-ENRÍQUEZ, S.; MARÍN-HERNÁNDEZ, A. and SAAVEDRA, E. (2007). Energy metabolism in tumor cells. *FEBS Journal*, vol. 274, no. 6, p. 1393-1418.

63. MULUKUTLA, B.C.; KHAN, S.; LANGE, A. and HU, W.S. (2010). Glucose metabolism in mammalian cell culture: New insights for tweaking vintage pathways. *Trends in Biotechnology*, vol. 28, no. 9, p. 476-484.

64. NEERMANN, J. and WAGNER, R. (1996). Comparative analysis of glucose and glutamine metabolism in transformed mammalian cell lines, insect and primary liver cells. *Journal of Cellular Physiology*, vol. 166, no. 1, p. 152-169.

65. OMASA, T.; ONITSUKA, M. and KIM, W.D. (2010). Cell engineering and cultivation of Chinese hamster ovary (CHO) cells. *Current Pharmaceutical Biotechnology*, vol. 11, no. 3, p. 233-240.

66. PAREDES, C.; PRATS, E.; CAIRO, J.J.; AZORÍN, F.; CORNUDELLA, L.I. and GÓDIA, F. (1999). Modification of glucose and glutamine metabolism in hybridoma cells through metabolic engineering. *Cytotechnology*, vol. 30, no. 1-3, p. 85-93.

67. PARK, H.S.; KIM, I.H.; KIM, I.Y.; KIM, K.H. and KIM, H.J. (2000). Expression of carbamoyl phosphate synthetase I and ornithine transcarbamoylase genes in Chinese hamster ovary *dhfr*-cells decreases accumulation of ammonium ion in culture media. *Journal of Biotechnology*, vol. 81, no. 2-3, p. 129-140. .

68. PELS-RIJCKEN, W.R.; FERWERDA, W.; VAN DEN EIJNDEN, D.H. and OVERDIJK, B. (1995). Influence of D-galactosamine on the synthesis of sugar nucleotides and glycoconjugates in rat hepatocytes. *Glycobiology*, vol. 5, no. 5, p. 495-502.

69. QUEK, L.E.; DIETMAIR, S.; KRÖMER, J.O. and NIELSEN, L.K. (2010). Metabolic flux analysis in mammalian cell culture. *Metabolic Engineering*, vol. 12, no. 2, p. 161-171.

70. RODRIGUEZ, P.; RIVAS, C.I.; GODOY, A.; VILLANUEVA, M.; FISCHBARG, J.; VERA, J.C. and REYES, A.M. (2005). Redefining the facilitated transport of mannose in human cells: Absence of a glucose-insensitive, high-affinity facilitated mannose transport system. *Biochemistry*, vol. 44, no. 1, p. 313-320.

71. RUBI, B.; DEL ARCO, A.; BARTLEY, C.; SATRUSTEGUI, J. and MAECHLER, P. (2004). The malate-aspartate NADH shuttle member Aralar1 determines glucose metabolic fate, mitochondrial activity, and insulin secretion in beta cells. *The Journal of Biological Chemistry*, vol. 279, no. 53, p. 55659-55666.

72. RYLL, T.; VALLEY, U. and WAGNER, R. (1994). Biochemistry of growth inhibition by ammonium ions in mammalian cells. *Biotechnology and Bioengineering*, vol. 44, no. 2, p. 184-193.

73. SANDERS, A.L. and EZZAT, Y.S. (1971). Modulation of the kinetic properties of phosphofructokinase by ammonium ions. *The Journal of Biological Chemistry*, vol. 246, no. 8, p. 2464-2467.

74. SANFELIU, A.; PAREDES, C.; CAIRÓ, J.J. and GÒDIA, F. (1997). Identification of key patterns in the metabolism of hybridoma cells in culture. *Enzyme and Microbial Technology*, vol. 21, no. 6, p. 421-428.

75. SAUER, A.; DAUCHY, T.; NAGEL, O. and MORRIS, P. (1980). Mitochondrial malic enzyme: Mitochondrial NAD(P)+ dependent malic enzyme activity and malate-dependent pyruvate formation are progression-linked in morris hepatomas. *The Journal of Biological Chemistry*, vol. 255, no. 9, p. 3844-3848.

76. SCHNEIDER, M.; MARISON, I.W. and VON STOCKAR, U. (1996). The importance of ammonia in mammalian cell culture. *Journal of Biotechnology*, vol. 46, no. 3, p. 161-185.

77. SCHRÖDER, M. (2008). Engineering eukaryotic protein factories. *Biotechnology Letters*, vol. 30, no. 2, p. 187-196.

78. SENGUPTA, N.; ROSE, S.T. and MORGAN, J.A. (2011). Metabolic flux analysis of CHO cell metabolism in the late non-growth phase. *Biotechnology and Bioengineering,* vol. 108, no. 1, p. 82-92.

79. SUNLEY, K. and BUTLER, M. (2010). Strategies for the enhancement of recombinant protein production from mammalian cells by growth arrest. *Biotechnology Advances*, vol. 28, no. 3, p. 385-394.

80. URBANO, A.M.; GILLHAM, H.; GRONER, Y. and BRINDLE, K.M. (2000). Effects of overexpression of the liver subunit of 6-phosphofructo-1-kinase on the metabolism of a cultured mammalian cell line. *Biochemical Journal*, vol. 352, no. 3, p. 921-927.

81. VERGARA, M.; BECERRA, S.; DÍAZ-BARRERA, A.; BERRIOS, J. and ALTAMIRANO, C. (2012). Simultaneous environmental manipulations in semi-perfusion cultures of CHO cells producing rh-tPA. *Electronic Journal Biotechnology*, vol. 15, no. 6.

82. VISIONGAIN. (2012). Biosimilars and Follow-On Biologics: Global Market Outlook 2010-2022, 213 p.

83. WAGNER, A.; MARC, A.; ENGASSER, J.M. and EINSELE, A. (1991). Growth and metabolism of human tumor kidney cells on galactose and glucose. *Cytotechnology*, vol. 7, no. 1, p. 7-13.

84. WAGNER, R. (1997). Metabolic Control of Animal Cell Culture Processes. In: HAUSER, H. and WAGNER, R. eds. *Mammalian cell biotechnology in protein production*, Walter de Gruyter, Berlin, p. 233-276.

85. WILKENS, C.A. and GERDTZEN, Z.P. (2011). Engineering CHO cells for improved central carbon and energy metabolism. *BMC Proceedings*, vol. 5, no. 8, p. 120.

86. WILKENS, C.A.; ALTAMIRANO, C. and GERDTZEN, Z.P. (2011). Comparative metabolic analysis of lactate for CHO cells in glucose and galactose. *Biotechnology and Bioprocess Engineering*, vol. 16, no. 4, p. 714-724.

87. WISE, D.R.; DEBERARDINIS, R.J.; MANCUSO, A.; SAYED, N.; ZHANG, X.Y.; PFEIFFER, H.K.; NISSIM, I.; DAIKHIN, E.; YUDKOFF, M.; MCMAHON, S.B. and THOMPSON, C.B. (2008). Myc regulates a transcriptional program that stimulates mitochondrial glutaminolysis and leads to glutamine addiction. *Proceedings of the National Academy of Sciences of the United States of America*, vol. 105, no. 48, p. 18782-18787.

88. WLASCHIN, K.F. and HU, W.S. (2007). Engineering cell metabolism for high-density cell culture via manipulation of sugar transport. *Journal of Biotechnology*, vol. 131, no. 2, p. 168-176.

89. WOOD, S.M.; WIESENER, M.S.; YEATES, K.M.; OKADA, N.; PUGH, C.W.; MAXWELL, P.H. and RATCLIFFE, P.J. (1998). Selection and analysis of a mutant cell line defective in the hypoxia-inducible factor-1 alpha-subunit (HIF-1alpha). Characterization of hif-1alpha-dependent and -independent hypoxia-inducible gene expression. *The Journal of Biology Chemistry*, vol. 273, no. 14, p. 8360-8368.

90. WURM, F.M. (2004). Production of recombinant protein therapeutics in cultivated mammalian cells.*Nature Biotechnology*, vol. 22, no. 11, p. 1393-1398.

91. YALCIN, A.; TELANG, S.; CLEM, B. and CHESNEY, J. (2009). Regulation of glucose metabolism by 6-phosphofructo-2-kinase/fructose-2,6-bisphosphatases in cancer. *Experimental and Molecular Pathology*, vol. 86, no. 3, p. 174-179.

92. YANG, M. and BUTLER, M. (2002). Effects of ammonia and glucosamine on the heterogeneity of erythropoietin glycoforms. *Biotechnology Progress*, vol. 18, no. 1, p. 129-138.

93. ZAMORANO, F.; VANDE-WOUWER, A. and BASTIN, G. (2010). A detailed metabolic flux analysis of an underdetermined network of CHO cells. *Journal of Biotechnology*, vol. 150, no. 4, p. 497-508.

94. ZHOU, M.; CRAWFORD, Y.; NG, D.; TUNG, J.; PYNN, A.F.J.; MEIER, A.; YUK, I.H.; VIJAYASANKARAN, N.; LEACH, K.; JOLY, J.; SNEDECOR, B. and SHEN, A. (2011). Decreasing lactate level and increasing antibody production in Chinese hamster ovary cells (CHO) by reducing the expression of lactate dehydrogenase and pyruvate dehydrogenase kinases. *Journal of Biotechnology*, vol. 153, no. 1-2, p. 27-34.

INDEX